New Frontiers in Hydroinformatics

New Frontiers in Hydroinformatics

Edited by Roman Morris

SYRAWOOD
PUBLISHING HOUSE

New York

Published by Syrawood Publishing House,
750 Third Avenue, 9th Floor,
New York, NY 10017, USA
www.syrawoodpublishinghouse.com

New Frontiers in Hydroinformatics
Edited by Roman Morris

International Standard Book Number: 978-1-64740-146-7 (Hardback)

Cataloging-in-Publication Data

New frontiers in hydroinformatics / edited by Roman Morris.
    p. cm.
Includes bibliographical references and index.
ISBN 978-1-64740-146-7
1. Hydrology--Data processing. 2. Hydrology--Computer simulation.
3. Geographic information systems. I. Morris, Roman.
GB656.2.E43 N49 2022

551.480 285--dc23

TABLE OF CONTENTS

Preface..VII

Chapter 1 **Intensity–duration–frequency curves from remote sensing rainfall**
estimates: comparing satellite and weather radar over the eastern Mediterranean....................1
Francesco Marra, Efrat Morin, Nadav Peleg, Yiwen Mei and Emmanouil N. Anagnostou

Chapter 2 **Seasonal streamflow forecasts in the Ahlergaarde catchment, Denmark: the effect**
of preprocessing and post-processing on skill and statistical consistency............................17
Diana Lucatero, Henrik Madsen, Jens C. Refsgaard, Jacob Kidmose and
Karsten H. Jensen

Chapter 3 **Combining satellite data and appropriate objective functions for improved spatial**
pattern performance of a distributed hydrologic model 34
Mehmet C. Demirel, Juliane Mai, Gorka Mendiguren, Julian Koch, Luis Samaniego
and Simon Stisen

Chapter 4 **Impact of remotely sensed soil moisture and precipitation on soil moisture**
prediction in a data assimilation system with the JULES land surface model 51
Ewan Pinnington, Tristan Quaife and Emily Black

Chapter 5 **ERA-5 and ERA-Interim driven ISBA land surface model simulations: which one**
performs better?...65
Clement Albergel, Emanuel Dutra, Simon Munier, Jean-Christophe Calvet,
Joaquin Munoz-Sabater, Patricia de Rosnay and Gianpaolo Balsamo

Chapter 6 **The challenge of forecasting impacts of flash floods: test of a simplified hydraulic**
approach and validation based on insurance claim data...83
Guillaume Le Bihan, Olivier Payrastre, Eric Gaume, David Moncoulon and
Frédéric Pons

Chapter 7 **A geostatistical data-assimilation technique for enhancing macro-scale**
rainfall–runoff simulations 101
Alessio Pugliese, Simone Persiano, Stefano Bagli, Paolo Mazzoli, Juraj Parajka,
Berit Arheimer, René Capell, Alberto Montanari, Günter Blöschl and Attilio Castellarin

Chapter 8 **The benefit of seamless forecasts for hydrological predictions over Europe** 117
Fredrik Wetterhall and Francesca Di Giuseppe

Chapter 9 **Hybridizing Bayesian and variational data assimilation for high-resolution**
hydrologic forecasting..129
Felipe Hernández and Xu Liang

Chapter 10 **Informing a hydrological model of the Ogooué with multi-mission remote**
sensing data 150
Cecile M. M. Kittel, Karina Nielsen, Christian Tøttrup and Peter Bauer-Gottwein

Chapter 11   **A hydrological routing scheme for the Ecosystem Demography model (ED2+R) tested in the Tapajós River basin in the Brazilian Amazon**..170
Fabio F. Pereira, Fabio Farinosi, Mauricio E. Arias, Eunjee Lee, John Briscoe and Paul R. Moorcroft

Chapter 12   **Rainfall–runoff modelling using Long Short-Term Memory (LSTM) networks**...................190
Frederik Kratzert, Daniel Klotz, Claire Brenner, Karsten Schulz and Mathew Herrnegger

Chapter 13   **Assessment of actual evapotranspiration over a semiarid heterogeneous land surface by means of coupled low-resolution remote sensing data with an energy balance model: comparison to extra-large aperture scintillometer measurements**..............................208
Sameh Saadi, Gilles Boulet, Malik Bahir, Aurore Brut, Émilie Delogu, Pascal Fanise, Bernard Mougenot, Vincent Simonneaux and Zohra Lili Chabaane

Permissions

List of Contributors

Index

PREFACE

This book aims to highlight the current researches and provides a platform to further the scope of innovations in this area. This book is a product of the combined efforts of many researchers and scientists, after going through thorough studies and analysis from different parts of the world. The objective of this book is to provide the readers with the latest information of the field.

The application of information and communications technologies in order to address the crucial problems of the efficient use of water falls under the domain of hydroinformatics. It uses techniques such as artificial neural networks, support vector machines and genetic programming. It primarily integrates the concepts of hydraulics, hydrology and environmental engineering. Hydroinformatics lays emphasis on the social nature of the problems related to water management and associated decision-making processes. The social processes by which technologies are brought into use are also analyzed in this domain. This book covers in detail some existent theories and innovative concepts revolving around hydroinformatics. Different approaches, evaluations, methodologies and advanced studies have been included herein. The extensive content of this book provides the readers with a thorough understanding of the subject.

I would like to express my sincere thanks to the authors for their dedicated efforts in the completion of this book. I acknowledge the efforts of the publisher for providing constant support. Lastly, I would like to thank my family for their support in all academic endeavors.

Editor

Intensity–duration–frequency curves from remote sensing rainfall estimates: comparing satellite and weather radar over the eastern Mediterranean

Francesco Marra[1], **Efrat Morin**[1], **Nadav Peleg**[2], **Yiwen Mei**[3], and **Emmanouil N. Anagnostou**[3]

[1]Institute of Earth Sciences, Hebrew University of Jerusalem, 91904, Jerusalem, Israel

[2]Institute of Environmental Engineering, Hydrology and Water Resources Management, ETH Zurich, Zurich, Switzerland

[3]Department of Civil and Environmental Engineering, University of Connecticut, Storrs, CT, USA

Correspondence to: Francesco Marra (marra.francesco@mail.huji.ac.il)

Abstract. Intensity–duration–frequency (IDF) curves are widely used to quantify the probability of occurrence of rainfall extremes. The usual rain gauge-based approach provides accurate curves for a specific location, but uncertainties arise when ungauged regions are examined or catchment-scale information is required. Remote sensing rainfall records, e.g. from weather radars and satellites, are recently becoming available, providing high-resolution estimates at regional or even global scales; their uncertainty and implications on water resources applications urge to be investigated. This study compares IDF curves from radar and satellite (CMORPH) estimates over the eastern Mediterranean (covering Mediterranean, semiarid, and arid climates) and quantifies the uncertainty related to their limited record on varying climates. We show that radar identifies thicker-tailed distributions than satellite, in particular for short durations, and that the tail of the distributions depends on the spatial and temporal aggregation scales. The spatial correlation between radar IDF and satellite IDF is as high as 0.7 for 2–5-year return period and decreases with longer return periods, especially for short durations. The uncertainty related to the use of short records is important when the record length is comparable to the return period (~ 50, ~ 100, and $\sim 150\%$ for Mediterranean, semiarid, and arid climates, respectively). The agreement between IDF curves derived from different sensors on Mediterranean and, to a good extent, semiarid climates, demonstrates the potential of remote sensing datasets and instils confidence on their quantitative use for ungauged areas of the Earth.

1 Introduction

Intensity–duration–frequency (IDF) curves are widely used in hydrological design and as decision support information in flood risk and water management (Watt and Marsalek, 2013). They are distribution functions of rain intensity maxima conditioned on duration and allow for linking the characteristics of a rainfall event to the probability of its occurrence (Chow et al., 1988; Eagleson, 1970). Derivation of IDF curves generally relies on historical precipitation data and consists in fitting an extreme values distribution to extreme rainfall values. Rain gauges, providing long records of precipitation data, are traditionally used to estimate IDF curves at the gauge locations. Nevertheless, owing to the sparseness of gauge networks worldwide (Kidd et al., 2017), this approach raises important issues when design applications require information at the catchment scale. Moreover, the representativeness of gauge-derived IDFs decreases moving away from the gauge location and no or very sparse information is available for the many ungauged locations of the Earth.

In the last decades a body of research has been devoted to these issues. Areal reduction factors and design storms assume homogeneity of rainfall extreme climatology to adapt the point-IDF values estimated by rain gauges to wider areas, such as catchments, based either on the climatology of the region (Sivapalan and Blöschl, 1998), on spatial precipitation observations (Bacchi and Ranzi, 1996; Durrans et al., 2002; Allen and DeGaetano, 2005; Lombardo et al.,

2006; Overeem et al., 2010; Wright et al., 2014), or stochastic model simulations (Peleg et al., 2017b, a). In principle, areal reduction factors may depend on a number of factors, such as geographic location, characteristics of the examined catchment, analysed duration and period, season, meteorological conditions, and others (Svensson and Jones, 2010). Their derivation is thus hampered by many sources of uncertainty. Regionalization and interpolation techniques continue assuming spatial homogeneity to derive IDF curves for ungauged locations (Ceresetti et al., 2012). However, homogeneity of rainfall extremes is a weak assumption when wide areas are considered or when, even at the small scales, the data record length is limited, especially in the case of short durations (Peleg et al., 2017b).

Remote sensing instruments, such as weather radars or satellites, provide high spatio-temporal resolution (i.e. 1–10 km and 5–60 min), distributed, regional, or even global rainfall estimates. These datasets allow for capturing dynamics and variability of rainfall extremes at scales that cannot be represented by rain gauges (Amitai et al., 2011; Chen et al., 2013; Marra et al., 2016; Panziera et al., 2016; Tapiador et al., 2012) and permit one to overcome the issues related to the conversion of point to areal information. Given the quantitative uncertainty related to remote sensing precipitation products (Berne and Krajewski, 2013; Mei et al., 2016; Stampoulis et al., 2013) and their limited records, unable to provide adequate samples of the climatology, the possibility of using such datasets for rainfall frequency analysis has only recently started to be explored. The main focus of these studies was the assessment of their potential (Eldardiry et al., 2015), for poorly instrumented regions (Awadallah et al., 2011; Nastos et al., 2013) and areas characterized by strong climatic gradients (Marra and Morin, 2015). Few authors, so far, provided quantitative IDF curves from remote sensing instruments, using some kind of regionalization approach (Endreny and Imbeah, 2009; Overeem et al., 2009; Wright et al., 2013; Panziera et al., 2016).

Since quantitatively accurate information is essential for design applications and the derivation of IDF curves based on historical records does not require short latency in the data, these studies made use of gauge-adjusted products and assessed the accuracy of the IDF curves derived from remote sensing datasets, using rain gauge curves as a reference. However, this approach neglected two important aspects. First, early warning systems, e.g. for flash floods (Borga et al., 2011, 2014; Villarini et al., 2010), urban floods (Yang et al., 2016), landslides/debris flows (Tiranti et al., 2014; Borga et al., 2014; Segoni et al., 2015), or heavy rain (Panziera et al., 2016), need to operate in real time and rely on short-latency remote sensing measurements. In these situations, calculating the frequency of near-real-time estimates using IDF curves derived from gauge-adjusted data could provide misleading results. It is therefore useful to analyse the characteristics of IDF curves derived from non-adjusted rainfall data, which are expected to represent the frequencies of near-real-time estimates. Second, areal IDFs provided by remote sensing instruments are expected to differ from point IDFs (Peleg et al., 2017b), and the use of different records (i.e. different samples of the climate) introduces further differences. No exact match between remote sensing and rain gauge IDFs should be expected a priori. Moreover, evaluating unadjusted data is important for demonstrating the value of satellite products in ungauged areas.

The aim of this study is to advance knowledge on the use of remote sensing precipitation estimates for rainfall frequency analysis. IDF curves computed from different datasets, namely the National Oceanic and Atmospheric Administration (NOAA) Climate Prediction Center morphing (CMORPH) technique (Joyce et al., 2004), the gauge-adjusted CMORPH (Xie et al., 2011), and the ground-based C-Band Shacham weather radar archive (Marra and Morin, 2015), are compared over the eastern Mediterranean using common spatio-temporal scales and records. The effect of spatial and temporal aggregation of rainfall is analysed. The uncertainty related to the use of short records in the climatic contexts characterizing the study region is assessed using long records of rain gauge observations. To the authors' knowledge this is the first study in which at-site IDF curves derived from different gridded remote sensing datasets are compared.

The study area and the rainfall datasets are presented in Sect. 2. Section 3 describes the methods used for computing and comparing satellite and radar IDF curves and for estimating the uncertainty related to the record length. In Sect. 4 the results of the study are presented and discussed. Section 5 provides the conclusions and suggestions for the future use of remote sensing datasets for rainfall frequency analysis.

2 Study area and data

2.1 Study area

The present study is focused on the eastern Mediterranean, in particular on the area covered by the Shacham weather radar, shown in Fig. 1. The orography presents a longitudinal organization that, going east from the Mediterranean Sea, encounters a coastal plain, a hilly region (up to ~ 800 m a.s.l.), the Jordan rift valley (~ 400 m below the sea level), and the Jordan Plateau (~ 1000 m a.s.l.). Mountains in the north of the study area rise up to 2800 m a.s.l.

Three climatic regions, the Mediterranean, semiarid, and arid, can be identified in the area, corresponding to the Csa, BSh, and BWh Köppen–Geiger definitions, respectively (Peel et al., 2007). The criteria used to define these classes are reported in Table 1. In this study, we follow the classification by Srebro and Soffer (2011). The Mediterranean climate characterizes the north-western coastal region and the northern part of the Jordan Plateau. The arid climate characterizes the south-western and southern portions of the area, and the Jordan rift valley. These two regions are separated by a strip characterized by semiarid climate that can sometimes be very narrow (Fig. 1). Climatic gradients are important with mean

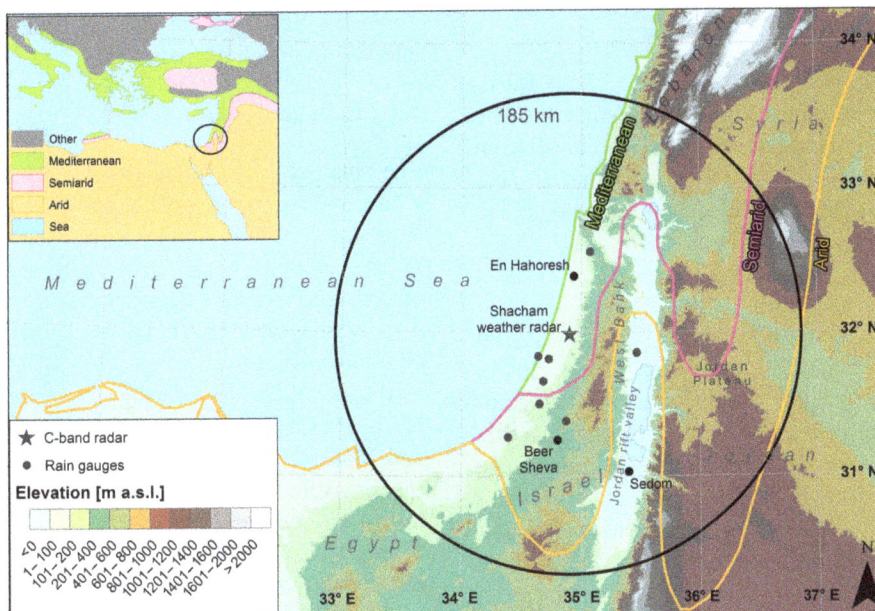

Figure 1. Map of the study area showing terrain elevation, climatic classification from Srebro and Soffer (2011), location of the radar, and rain gauges used in the study. The small map shows the location of the study area within the eastern Mediterranean.

Table 1. Köppen–Geiger classification (Peel et al., 2007; Srebro and Soffer, 2011) and number of pixels analysed for each climatic region according to the Köppen–Geiger classification. MAT: mean annual temperature; MAP: mean annual precipitation; T_{hot}: temperature of the hottest month; T_{cold}: temperature of the coldest month; P_{ds}: precipitation of the driest month in summer; P_{ww}: precipitation of the wettest month in winter.

Climatic region	Köppen–Geiger definition	Köppen–Geiger criteria	Number of pixels (fraction of the total)
Mediterranean	Temperate, dry, hot summer (Csa)	$T_{hot} \geq 22\,°C$ $0\,°C < T_{cold} < 18\,°C;$ $P_{ds} < 40\,mm$ $P_{ds} < P_{ww}/3$	318 (20.1 %)
Semiarid	Arid, steppe, hot (BSh)	$10 \cdot MAT \leq MAP \leq 20 \cdot MAT;$ $MAT \geq 18\,°C$	404 (25.5 %)
Arid	Arid, desert, hot (BWh)	$MAP < 10 \cdot MAT;$ $MAT \leq 18\,°C$	225 (14.2 %)
Sea	–	–	635 (40.1 %)

annual precipitation exceeding $1000\,mm\,yr^{-1}$ in the northern part of the area and dramatically dropping from 600 to $100\,mm\,yr^{-1}$ in just 25 km distance along the narrow strip west of the hilly region (Alpert and Shafir, 1989; Goldreich, 1994). The summer is dry, with almost no rain from June to August. Rainfall over the Mediterranean areas is generally brought by cold fronts and post-frontal systems, associated with mid-latitude cyclones (Goldreich, 2003; Peleg and Morin, 2012), and by Syrian low and active Red Sea Trough, occasionally correlated with flash floods, in the semiarid and arid climates (Dayan et al., 2001; Dayan and Morin, 2006; Kahana et al., 2002; de Vries et al., 2013). On rare occasions

the subtropical jet, or tropical plume, brings widespread rainfall over the whole region (Dayan and Morin, 2006; Kahana et al., 2002; Rubin et al., 2007; Tubi and Dayan, 2014; Dayan et al., 2015). Important gradients have been reported also for the climatology of extreme rainfall. Low return-period intensities were found to be scaled with the mean annual precipitation. Conversely, the more arid the climate is, the more skewed the extreme values distribution is, with long return-period intensities for arid areas being higher than the corresponding values for semiarid and Mediterranean areas, especially for short durations (Ben-Zvi, 2009; Marra and Morin, 2015).

2.1.1 Radar data

The Shacham weather radar is a C-Band (5.35 cm wavelength), non-Doppler instrument, operational since the late 1980s. Observations from this radar have been archived from October 1990 to March 2014, providing a unique 23-year record and have been extensively used for climatologic and hydrological studies (e.g. Morin et al., 1995, 2001, 2009; Morin and Gabella, 2007; Peleg and Morin, 2012; Peleg et al., 2013, 2017b; Yakir and Morin, 2011).

The Shacham radar record has recently been reanalysed for rainfall frequency analysis (Marra and Morin, 2015). Radar quantitative precipitation estimation is obtained combining physically based correction algorithms and quantitative adjustments based on the comparison with rain gauge measurements. The procedure is discussed in details in Marra and Morin (2015) and included (i) checking of antenna pointing, (ii) ground clutter filtering, corrections for the effects of (iii) wet radome attenuation, (iv) attenuation due to the propagation of the radar beam, (v) beam blockage, and (vi) vertical variations of reflectivity, together with a (vii) hail filter. A two-step bias adjustment was then applied combining a yearly range-dependent and an event-based mean field bias adjustments based on comparison with quality-checked rain gauge data. Radar rain rates exceeding 150 mm h^{-1} have been set to this cap value in order to avoid contaminations due to hail. The readers are referred to Marra et al. (2014) and Marra and Morin (2015) for an extensive description of the correction procedures and the quantitative assessment of the archive.

This study is based on the hourly radar archive. Each hourly radar product is created when at least 60 % of radar scans are available during the 1 h time interval. This quality check on the data availability ensures a good coverage of the examined time interval and allows for exploring durations longer than the ones analysed by Marra and Morin (2015).

2.2 Satellite data

The satellite precipitation products selected for this study are the high-resolution CMORPH (HRC) and its gauge-adjusted version (CHRC) available from NOAA CPC. The two products, with $0.073° \times 0.073°$ half-hourly estimates, offer high-resolution, quasi-global, long-term records of satellite rainfall estimates (available, as in November 2015, for 16 years: from January 1998 to December 2013). CMORPH integrates multiple satellite-based microwave rain estimates (Ferraro, 1997; Ferraro et al., 2000; Kummerow et al., 2001) in space and time using motion vectors derived from infrared images (Joyce et al., 2004). The newly available gauge-adjusted CMORPH applies daily gauge adjustment using estimates from $\sim 30\,000$ gauges worldwide (Xie et al., 2011). The gauge-adjusted CMORPH product is using gauge data from ~ 12 gauges in the region (Chen et al., 2008), which may also be used in the adjustment of the radar-rainfall dataset.

The half-hourly original data has been aggregated to hourly intervals to match the radar archive resolution.

2.3 Rain gauge data

The small-scale representativeness of rain gauge measurements makes them not suitable for a large-scale quantitative assessment of remote sensing products (Gires et al., 2014; Peleg et al., 2017b). Here, we take advantage of their long records to empirically quantify the uncertainty in rainfall frequency analysis related to the use of short records in different climatic conditions.

We identified 11 rain gauges, operated by the Israeli Meteorological Service, among the ones already used by Marra and Morin (2015). Rain gauge data are available as storm maxima values extracted from digitized charts for durations up to 4 h (before 2000) or automatic measurements with 10 min (from 2000 up to 2005) and 1 min (from 2006 on) resolution. This limits our analysis of rain gauge data to durations shorter than 4 h. When possible, rain gauge data have been aggregated into hourly blocks to be as consistent as possible with the remote sensing datasets. Stationarity of the records has been tested at 0.1 significance level (Phillips and Perron, 1988). The selected rain gauges belong to different climatic regions with five, four, and two rain gauges available for Mediterranean, semiarid, and arid climates, respectively (Fig. 1). The length of the analysed records ranges between 30 and 67 years and is at least double that of the satellite datasets record (16 years) with just one exception (30 years) belonging to the semiarid climatic region.

3 Methods

3.1 Derivation of intensity–duration–frequency curves

In this study, the generalized extreme value (GEV) distribution (Appendix A) is used to fit the annual maxima series (AMS) of average rain intensities observed over 1, 3, 6, 12, and 24 h durations. The use of AMS ensures independency of the elements of the series and, rather than peak over threshold series, is suitable for this study because it does not require the definition of thresholds, problematic operation on highly variable climatic conditions, and potentially undermining the interpretation of the comparison between different datasets. The GEV distribution is a three-parameter extreme values distribution used worldwide to model rainfall extremes (Fowler and Kilsby, 2003; Gellens, 2002; Koutsoyiannis, 2004; Overeem et al., 2008). It is described by the location, scale, and shape parameters, representing mean, dispersion and skewness of the distribution, respectively. The GEV distribution has been shown to fit the AMS better than other distributions, such as the Pearson type III or the generalized logistic (Alila, 1999; Kysely and Picek, 2007; Perica et al., 2013; Schaefer, 1990). It is being widely used for rainfall frequency analyses based on remote sensing data (Eldardiry et al., 2015; Marra and Morin, 2015; Overeem et al., 2009; Paixao et al., 2015; Panziera et al., 2016; Peleg et al., 2017b), owing to its ability to include all the three asymptotic extreme value types (Gumbel, Fréchet and Weibull) into one (Katz et al., 2002).

In order to have radar and satellite data on a common grid suitable for the comparison, the full archive of hourly radar data was remapped by spatially averaging the 1 km × 1 km radar pixels to the corresponding ~8 km × 8 km CMORPH pixels. We identified the AMS of calendar years for the examined durations using a moving window with 1 h jumps. Even if using hydrologic years is more natural, we chose to use calendar years to exploit the full CMORPH record, available from January 1998 to December 2013. A minimum time lag of 48 h between annual maxima observed in different years was set to fulfil the independency requirements of the GEV theory. All the available rainfall estimates have been included in the IDF estimation, even if data from the other sources was missing during a given storm. No co-occurrence of the annual maxima is thus imposed. There are, in fact, potential situations in which radar or rain gauges missed storms due to technical problems or power issues that cannot be directly identified from the data itself (Morin et al., 1998; Ben-Zvi, 2009; Marra and Morin, 2015). At-site GEV parameters (i.e. pixel by pixel) were identified using the maximum likelihood method (MATLAB statistics toolbox). Situations for which the fitting was not successful due to convergence problems or excessive number of iterations were discarded. Less than 0.7 % of the cases were discarded for this reason. The IDF values were calculated for a set of five return periods (2, 5, 10, 15, and 25 years).

The effect of spatial and temporal aggregation of rainfall estimates, key issue when dealing with remote sensing instruments, is analysed spatially aggregating the original radar record (23 years, 1 km × 1 km) on grid sizes gradually increasing from 2 × 2 to 64 km × 64 km and for durations from 1 to 24 h using moving windows with 1 km and 1 h jumps. In this analysis, radar is preferred over CMORPH due to its ability to provide more direct rainfall estimates at the small spatial and temporal scales.

3.2 Comparison between intensity–duration–frequency maps

IDF maps obtained from the satellite precipitation datasets (HRC and CHRC) are compared to the ones obtained from the radar archive during corresponding years (16 years: 1998–2013). The comparison is extended over an analysis domain defined excluding the pixels that are known to be not reliable. In particular, pixels located closer than 27 km or farther than 185 km from the radar or behind the hilly region are excluded due to insufficient reliability of the radar data, and pixels located in proximity of major lakes are excluded due to false rainfall signals in the CMORPH estimates (e.g. Guo et al., 2015). The number of pixels analysed for each climatic region is reported in Table 1. A limited number of cases with problematic data, providing unrealistic outcomes, could still be found in all the products. In order to avoid single problematic situations dominating the results, we analysed the distribution of the GEV parameters over the whole study area and excluded from the comparison the pixels for which any of them was outside of the 1–99th quantiles interval of the corresponding distribution.

Three widely used non-dimensional, normalized metrics are selected to compare radar-IDF and satellite-IDF maps: correlation coefficient (CC), measuring the spatial correlation of the maps; multiplicative bias (bias), measuring the mean quantitative agreement of the maps; and normalized standard difference (NSD), measuring the variability of the residuals of the normalized maps. Additional information on the metrics is provided in Appendix B.

3.3 Uncertainty related to the record length

We assumed the records of rain gauge data to be a complete sample of the climatology of extremes for return periods comparable to the remote sensing data record length. Synthetic records of rain gauge data were created by randomly sampling years, without replacement, from the full rain gauge record, and the corresponding IDF values were calculated. We focused on synthetic records of 10, 15, 20, and 25 years and bootstrapped the operation 999 times for each rain gauge and for each synthetic record length. The 5–95th quantile interval of the obtained distributions was used to measure the uncertainty related to the record length (Overeem et al., 2008).

4 Results and discussion

4.1 Comparison between satellite- and radar-derived GEV parameters

The distribution of the GEV parameters (25–75th quantile intervals and median over the whole study area) derived from radar and satellite datasets in the three climatic regions and over the Mediterranean Sea (sea, from here on) are presented in Fig. 2. We recall here that the scale, location, and shape parameters provide a measure of the mean, dispersion, and skewness of the underlying distribution, respectively. Location parameters from HRC (CHRC) estimates are smaller (larger) than the ones from radar over Mediterranean climate and over the sea, meaning that extreme values from HRC (CHRC) are in general lower (higher) than radar extreme values while in semiarid and arid climates HRC and CHRC generally identify higher parameters than the radar (i.e. higher extreme values). Differences in the location parameters can be associated to the bias between extreme values in the datasets. The scale parameters are normalized over the corresponding location parameters to appreciate the relative differences. Normalized scale parameters from HRC

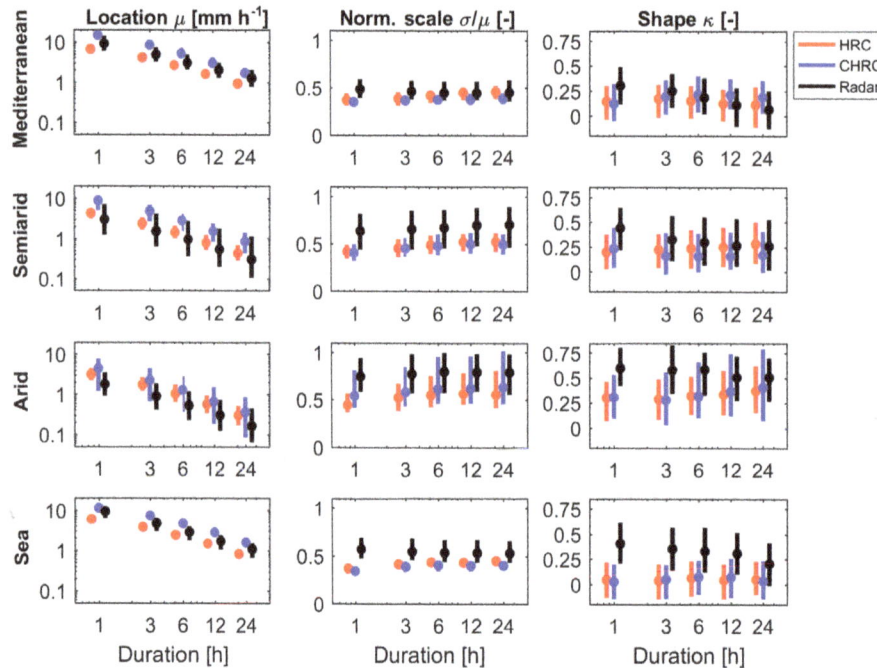

Figure 2. Distribution (median and 25–75th quantiles) of the GEV parameters derived from satellite (HRC and CHRC) and radar datasets. Note that scale parameters are normalized over the corresponding location parameters. The parameters for different products are represented around the corresponding duration; therefore, the logarithmic scale in the *x* axes should be interpreted accordingly.

and CHRC are similar and lower than the ones derived from radar. Normalized scale parameters, together with their variability, tend to increase when moving from sea to Mediterranean, semiarid, and arid climates. The drier climate, the larger the dispersion of the GEV distribution. A slight increase of the normalized scale parameters with duration can be noticed in the HRC/CHRC data.

The shape parameters are mostly greater than zero, suggesting, in line with previous studies (Katz et al., 2002; Papalexiou and Koutsoyiannis, 2013), that type II extreme values distribution should be considered for the area. Both radar and satellite products derive high shape parameters for arid climate that decrease when moving to semiarid and to Mediterranean climates, thus suggesting that drier regions are characterized by thicker-tailed distributions. Over the sea, satellite shape parameters are close to zero, while radar shape parameters tend to be higher. Increasing duration, radar-derived parameters are decreasing, while satellite-derived parameters tend to increase, becoming comparable for durations longer than ~ 6 h. In general, the shape parameters derived from radar are higher than the ones derived from CMORPH so that, in arid climate, the radar dataset includes few cases with a shape parameter greater than 0.5 and, in single occasions, even greater than 1. Among the many possible causes, this can be associated to the measurement uncertainty of radar estimates and to the record intermittency due to radar shutdowns (Marra and Morin, 2015). In fact, the rainfall regime of arid climate is characterized by short and

intense events (Karklinsky and Morin, 2006), so that gaps in the record can cause the missing of important storms leading to particularly low estimates of some annual maxima and, consequently, to thicker-tailed distributions. It should also be noted that missing of short-duration extremes by CMORPH due to the overpasses frequency of microwave satellites could contribute to the differences observed between CMORPH and radar.

4.1.1 Effect of spatial and temporal aggregation

The location and scale parameters consistently decreased as the spatial and temporal aggregation scales increased. This is an expected effect, caused by the smoothing of rainfall fields operated by the spatial averaging; therefore, results are not reported in this paper. Conversely, it is interesting to analyse the shape parameter. The distributions of the shape parameters derived from the full radar record aggregated on grid sizes increasing from 2 km × 2 km to 64 km × 64 km and on durations increasing from 1 to 24 h are presented in Fig. 3. The shape parameters consistently decrease both with spatial and temporal aggregation. This means that the smoothing effect due to the spatial and temporal aggregation of rainfall measurement depends on the return period, and is more pronounced for longer return periods. These results relate to the spatial–temporal scales of extreme precipitation, usually analysed using the areal reduction factors, and suggest a non-homogeneity of the scales of rainfall extremes with return

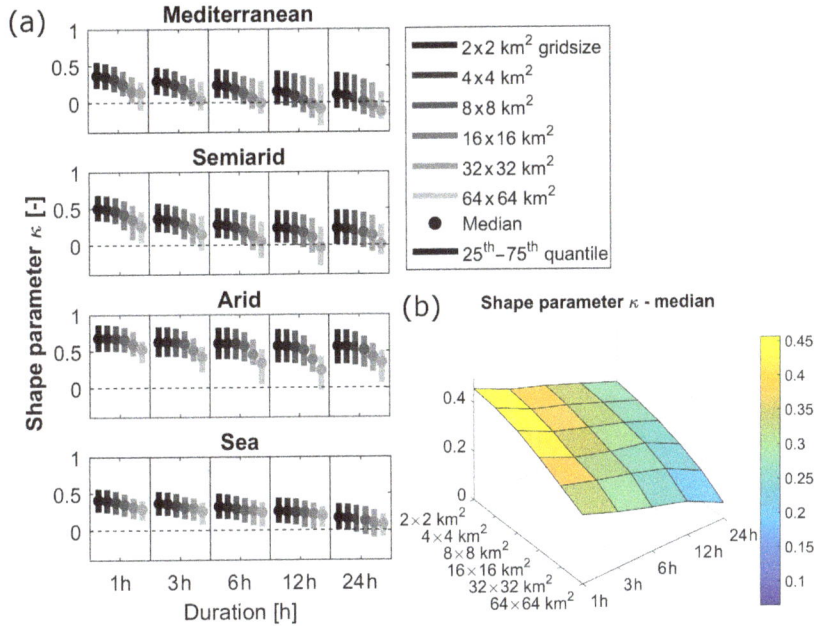

Figure 3. (a) Distribution (median and 25–75th quantiles) of the shape parameter derived aggregating radar estimates on grid sizes increasing from 2 km× 2 km to 64 km× 64 km. **(b)** Median of the shape parameters (all climates) derived aggregating radar estimates as a function of the spatial and of the temporal aggregation scales

period. When using higher spatial and temporal resolutions, it is more probable to observe, in the relatively short archive of radar data, higher extreme events, since they are likely to be more localized in both space and time. This supports previous findings on areal reduction factors derived using radar data (Bacchi and Ranzi, 1996; Durrans et al., 2002; Allen and DeGaetano, 2005; Lombardo et al., 2006; Overeem et al., 2010). However, it should be noted that other studies reported no clear dependency of the areal reduction factors on return period (Wright et al., 2014). Magnitude and combination of spatial and temporal effects are shown to depend on the climate, confirming that the spatio-temporal scales of rainfall extremes are highly dependent on the climatic conditions.

4.2   Satellite IDF and radar IDF

A visual comparison between IDF curves derived from rain gauges and from the co-located radar and satellite pixels is presented in Fig. 4. Corresponding data periods (16 years, 1 h block) are used over the three climatic regions (Mediterranean – En Hahoresh, semiarid – Beer Sheva and arid – Sedom; Fig. 1) for 1, 6, and 24 h durations. The reported cases, discussed in Marra and Morin (2015), are known to have good radar visibility, with the exception of Sedom that, being farther from the radar and behind the hilly region, can be subject to overshooting of the radar beam. Radar reproduces better the skewness of the IDF curves and radar IDFs are, with the exception of the arid case, within the rain gauge confidence interval. Note that these results represent the local

scale; while interpreting them, one should take into account the different scales of rain gauges (point scale) and remote sensing datasets ($\sim 8 \times 8\,\text{km}^2$, in this case) and the natural variability that extreme rainfall presents when relatively short records are used (Gires et al., 2014; Peleg et al., 2017b).

Examples of IDF maps for 3 h duration from HRC, CHRC, and radar products are presented in Fig. 5. It is shown that the spatial variability of IDF values increases with return period, owing to the larger uncertainty associated to longer return periods. Noteworthy, HRC IDFs are lower than the corresponding CHRC and radar IDFs, while the CHRC IDFs seem to be larger than radar IDFs for 2-year return period and comparable for 25-year return period. A quantitative analysis of the differences between HRC/CHRC-IDF maps and radar-IDF maps is provided in the next section.

4.3   Comparison between satellite IDF and radar IDF

A comprehensive quantitative comparison between radar-IDF and CMORPH-IDF maps is presented in Fig. 6. In general, very similar patterns of CC and NSD are observed for HRC and CHRC while substantially different bias patterns are observed. This confirms the point made by Mei et al. (2016) that daily gauge adjustment influences the magnitude rather than the space–time organization of annual extremes.

The CC, measuring the spatial correlation of the IDF maps, is as high as ~ 0.70–0.76 for 2- and 5-year return periods and decreases when longer return periods are examined, especially for shorter durations (~ 0.27–0.29 for 1 h, ~ 0.40–0.55 for 24 h). CHRC has higher CC with radar than HRC.

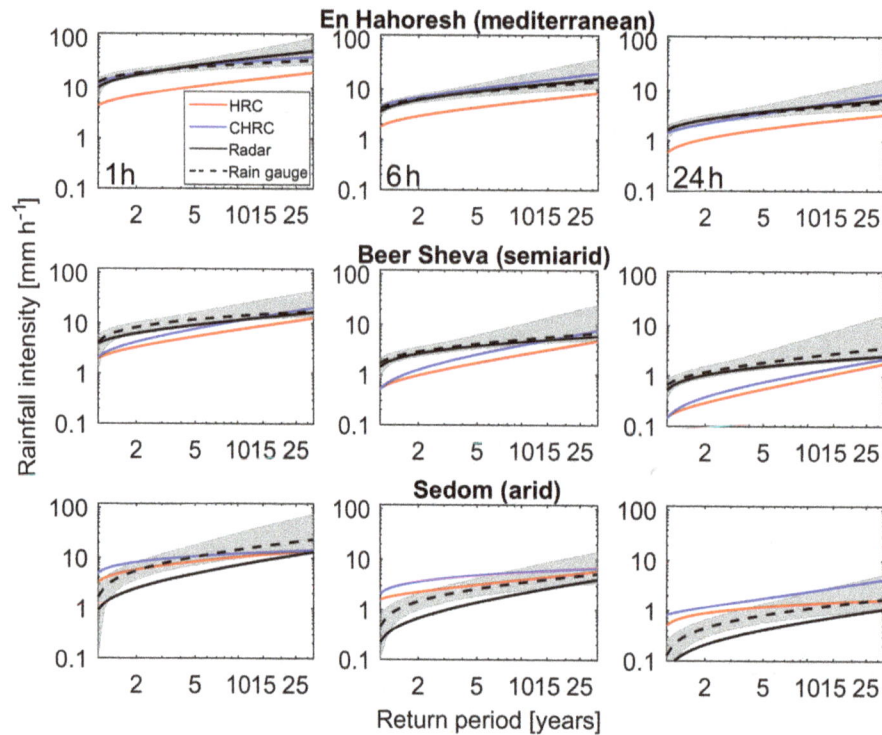

Figure 4. Visual comparison of the annual maxima and of the IDF curves derived for the examined 16 years (1998–2013) from HRC (red), CHRC (blue), radar (solid black), and rain gauge (dashed black). The shaded area represents the 95 % confidence interval of the rain gauge IDF. Three example locations in the Mediterranean (En Hahoresh), semiarid (Beer Sheva) and arid (Sedom) climates and for 1, 6, and 24 h durations are shown (see gauge location in Fig. 1).

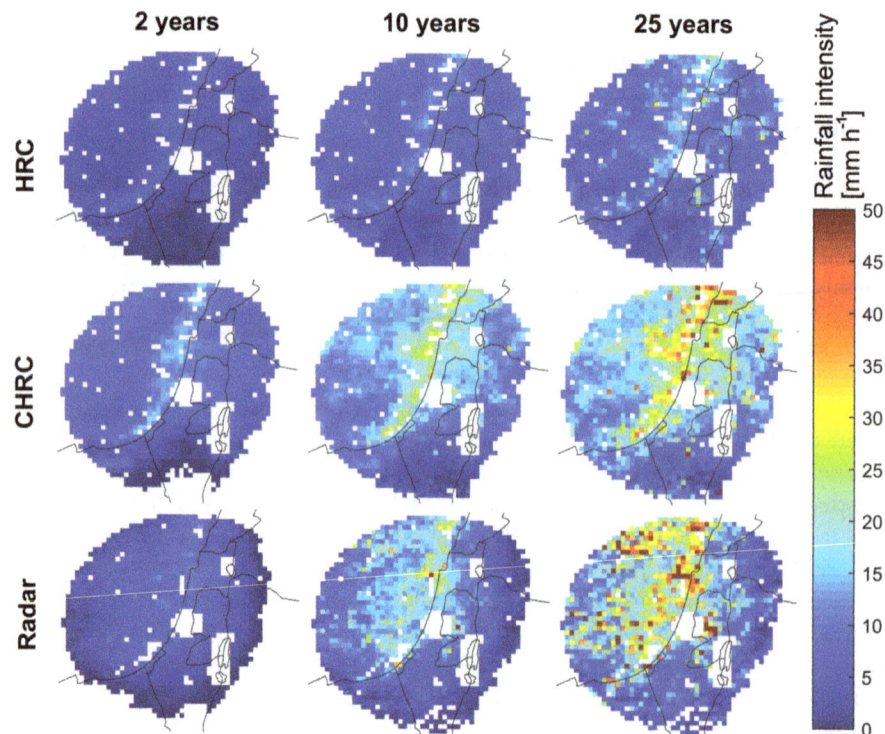

Figure 5. Example of IDF maps for 3 h duration from HRC, CHRC, and radar for 2-, 10-, and 25-year return periods. Only the pixels included in the analyses are shown.

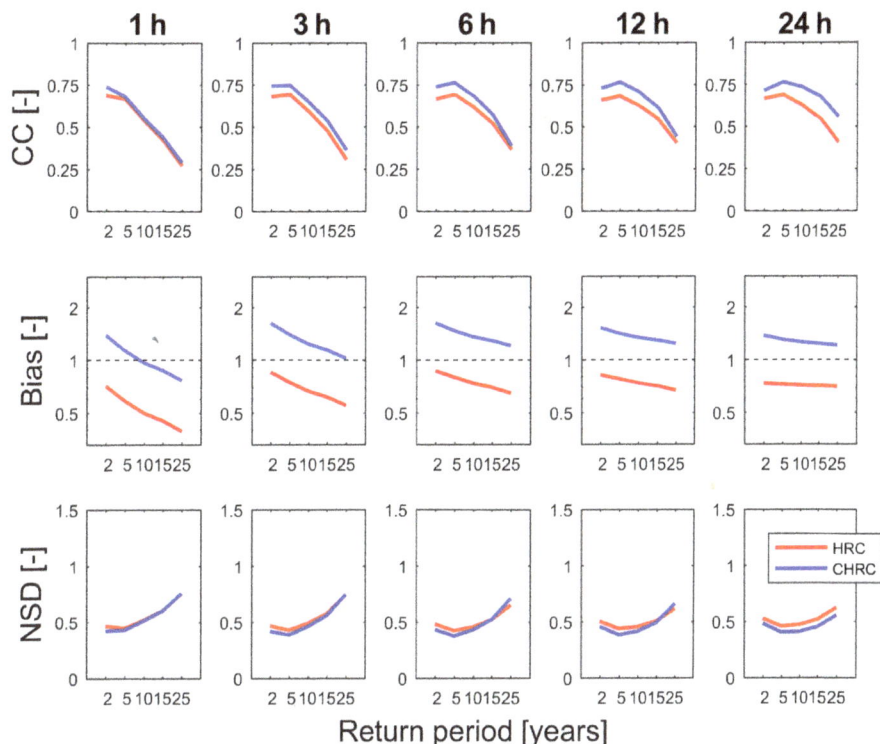

Figure 6. Comparison of IDF values between HRC and radar (red) and CHRC and radar (blue). The upper row of panels shows the CC for different durations, the middle row the bias and the lower row the NSD.

Bias results show that CHRC tends to overestimate while HRC tends to underestimate relative to radar IDFs. For short durations, the bias for both satellite datasets (gauge adjusted and unadjusted) strongly decreases with increasing return period. As the duration increases, this trend diminishes so that, for longer durations, the bias is almost consistent for all return periods. This reflects what observed for the shape parameters: the higher shape parameters estimated from radar for short durations lead the radar-IDF curves to be thicker tailed, i.e. to predict larger intensities for longer return periods with respect to longer durations. Moreover, this suggests that, for sufficiently long durations, it is possible to identify a factor linking radar IDFs to satellite IDFs (~ 0.7 for HRC, ~ 1.2 for CHRC at 24 h). The NSD patterns are very similar and tend to increase with return period (up to ~ 0.75 for 25 years, 1 h) and to decrease with duration (~ 0.55–0.62 for 25 years, 24 h). Figures 7 and 8 show the metrics calculated between HRC/CHRC-IDF and radar-IDF maps for the three climatic regions and over the sea. The spatial correlation between radar- and CHRC-IDF maps is larger, with respect to HRC-IDF maps, for all the climatic regions and is very high over arid areas/short durations and semiarid areas/long durations. On the contrary, CC between HRC-IDF and radar-IDF maps is very low for long durations (~ 0.05 for 25 years, 24 h), probably due to the very low values measured by CMORPH when no gauge adjustment is applied.

Note that the CC between both HRC/CHRC IDFs and radar IDFs is low over the sea, especially for shorter durations, and that the difference becomes less important for longer durations. This is not coming as a surprise since no gauge data are available for the adjustment of satellite or radar data over the sea. As pointed out above, gauge adjustment is only weakly impacting the space–time organization of CMORPH extreme estimates, while it is a crucial step in radar quantitative precipitation estimation. This observation, together with the increased reliability of satellite-based estimations over the sea (Kidd and Levizzani, 2011), suggests that spatial distribution of IDF values indicated by satellite products should be considered more accurate.

Overestimation of CHRC-IDF maps with respect to radar-IDF maps is more marked for Mediterranean and semiarid climates, with a factor ~ 1–1.6. For durations longer than 3 h, the bias for Mediterranean and semiarid climates for both HRC and CHRC shows almost no trend with return period, meaning that the trend observed in the general case is due to arid climate and sea area, where the differences between radar- and HRC/CHRC-shape parameters is larger (Fig. 2). This confirms that radar and HRC/CHRC almost agree in the identification of the skewness of IDF curves over Mediterranean and semiarid climates for durations larger than 3 h. NSD over arid and semiarid areas is larger than for Mediterranean climates and sea.

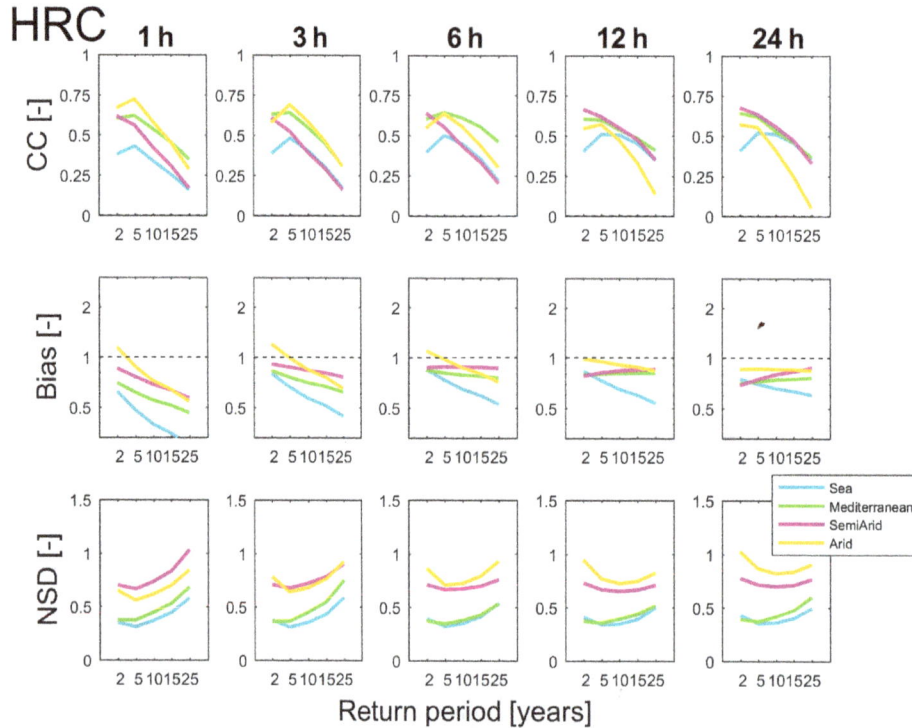

Figure 7. Comparison of IDF values between HRC and radar for different climatic regions. Blue lines represent sea areas while green, pink, and orange lines represent the Mediterranean, semiarid, and arid climates respectively. The upper row of panels shows the CC for different durations, the middle row the bias, and the lower row the NSD.

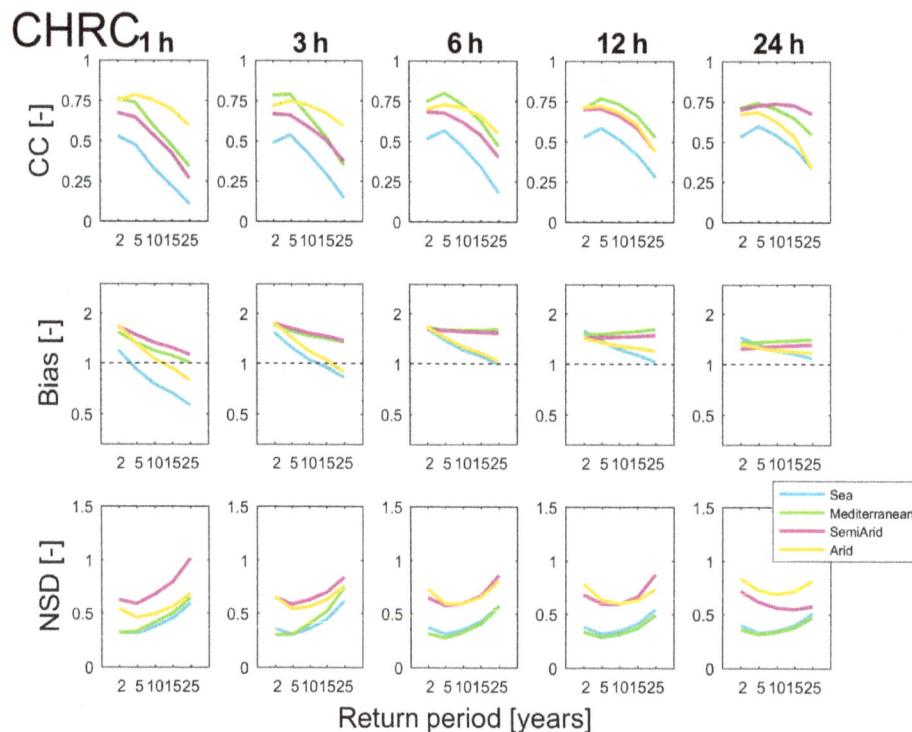

Figure 8. Comparison of IDF values between CHRC and radar for different climatic regions. Blue lines represent sea areas while green, pink, and orange lines represent the Mediterranean, semiarid, and arid climates respectively. The upper row of panels shows the CC for different durations, the middle row the bias, and the lower row the NSD.

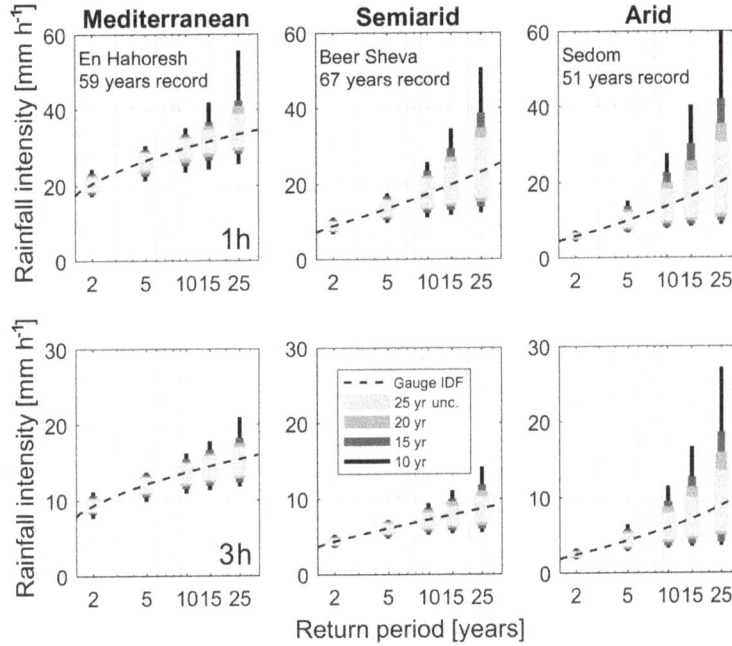

Figure 9. Uncertainty related to the record length for 1 and 3 h durations for three example cases in the Mediterranean (En Hahoresh), semiarid (Beer Sheva) and arid (Sedom) climates. The dashed lines show the IDF curves from the full records (59, 67, and 51 years, respectively) and the vertical bars show the width of the 5–95th quantile interval of the 999 bootstrap sampling repetitions for record lengths of 10, 15, 20, and 25 years. Width and light of the colour of the bars increase with the record length.

4.4 Uncertainty related to the record length

In this section, we present the results of the bootstrap sampling of long rain gauge records used to quantify the uncertainty related to the record length of remote sensing datasets. The uncertainty presented here is the component related to the under-sampling of rainfall climatology due to the use of short data records and is quantified as the 5–95th quantile interval of the bootstrap sampling. The uncertainty for two example cases is presented in Fig. 9. The figure reports the 5–95th quantile interval of the bootstrap sampling of 10, 15, 20, and 25 years of data out of the whole record of the three cases shown in Fig. 4. As expected, uncertainties become important when the record length is similar, or smaller, than the estimated return period. Arid climates are characterized by larger uncertainties, especially when short records and short durations are examined; this is probably due to the low number of rain events per year, and to the thicker-tailed characteristic of arid IDF curves. Short records are shown to be more likely overestimating, rather than underestimating, the IDF values.

Figure 10 shows the relative uncertainty (width of the 5–95th quantile interval of the 999 bootstrap sampling repetitions normalized over the long record rain gauge-IDF value) as a function of the ratio between return period and record length. Uncertainties for 1 and 3 h durations are comparable, with the uncertainty for 3 h duration being smaller. This suggests that time aggregation potentially decreases part of the

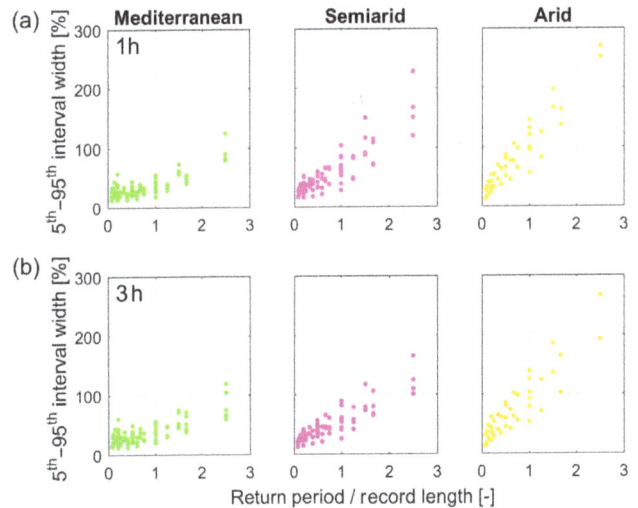

Figure 10. Relative uncertainty (width of the 5–95th quantile interval of the 999 bootstrap sampling repetitions normalized over the rain gauge-IDF value) plotted against the ratio between the estimated return period and the record length. The first row of panels shows the results for 1 h duration, the second row for 3 h. Columns of panels show results for each climatic region.

issues related to the use of short records. Uncertainty is larger for return periods longer than the record length, especially for short durations and drier climates. For the Mediterranean climate, the uncertainty is generally lower than 50 % when the return period is shorter than the record length and reaches up to ~ 125 % for return periods more than twice the data record length. In the semiarid climate, the uncertainty is shown to be larger, exceeding 100 % (30–100 %) even for return periods comparable to the data record length and exceeding 200 % for return periods more than twice the data record length. In the arid climate, the uncertainty is even larger, reaching 150 % for return periods equal to the data record length and exceeding 250 % for return periods more than twice the data record length.

5   Conclusions

This study compared the use of rainfall estimates from a ground-based C-band weather radar and from a high-resolution satellite precipitation product, CMORPH (HRC), and its gauge-adjusted version (CHRC), for the identification of intensity–duration–frequency (IDF) curves. IDF curves were computed using the above products over the eastern Mediterranean (Mediterranean, semiarid, and arid climates and over the sea) and the uncertainties due to the limited record length of the remote sensing datasets were quantified basing on long records of rain gauge measurements. Our findings can be summarized as follows:

1. The shape parameters of the generalized extreme values distribution, as derived from radar, HRC and CHRC, are mostly greater than zero; drier climates are characterized by higher shape parameters, suggesting that thicker-tailed distributions better describe rainfall extremes of drier areas. In general, the shape parameters derived from radar are higher than the ones from CMORPH, especially for arid climate and over the sea.

2. The shape parameter tends to decrease when rainfall estimates are aggregated in space and/or time. The effect is related to a non-homogeneity of spatial and temporal scales of rainfall extremes with return period. This non-homogeneity depends on the climatic conditions.

3. The spatial correlation coefficient between corresponding radar-IDF and HRC/CHRC-IDF maps is between 0.70 and 0.76 for short return periods, but decreases with increase in return period, especially for short durations. In general, for both HRC and CHRC the correlation is higher in arid climate for durations up to 3 h and in Mediterranean and semiarid climates for longer durations (6–24 h). Low correlations are observed over the sea.

4. HRC IDFs and CHRC IDFs are, respectively, lower and higher than radar IDFs. In both cases the observed bias decreases with return period, especially for short durations and arid climate. For longer durations and Mediterranean/semiarid climates, the decreasing trend

almost disappears so that the bias can be considered independent from the return period (~ 0.7 for HRC, ~ 1.2 for CHRC for 24 h).

5. Comparison of HRC IDF and CHRC IDF against radar IDF shows consistent patterns of correlation and dispersion, and different biases. This means that gauge adjustment influences the magnitude rather than the space–time organization of annual extremes and suggests that HRC IDF can potentially be used to estimate the frequencies of CMORPH estimates in near-real-time early warning systems.

6. The uncertainty related to the use of short records becomes important when the record length is shorter or comparable to the examined return period. This is particularly true for drier climates and shorter durations, with potential uncertainty of ~ 50, ~ 100, and ~ 150 % for return periods comparable to the record length in Mediterranean, semiarid, and arid climates, respectively.

Rainfall frequency analysis by means of remote sensing rainfall estimates remains a challenging task, especially when dry climates are explored. Nevertheless, the agreement between IDF curves derived from different sensors on Mediterranean and, to a good extent, semiarid climates, demonstrates their potential for the description of small-scale spatial patterns of IDF curves and instils confidence on their quantitative use for ungauged areas of the Earth. Spatial and temporal aggregation of rainfall information represent viable ways to take advantage of remote sensing datasets and decrease the uncertainties related to the derived IDF curves. In particular, remote sensing rainfall archives can provide important information when 2–10-year return periods and 12–24 h durations are requested, scales that are relevant for both flood risk management (e.g. issuing of warning) and hydrological design (e.g. sewer systems design, large-scale drainage planning).

Appendix A: GEV distribution

The GEV cumulative distribution function can be written as (Coles, 2001):

$$
\begin{cases}
F\left(I; \mu, \sigma, \kappa\right) = \exp\left\{-\left[1 + \frac{\kappa}{\sigma}(I - \mu)\right]^{-\frac{1}{\kappa}}\right\} \\
\quad \text{for } \kappa \neq 0 \\
F\left(I; \mu, \sigma, \kappa\right) = \exp\left\{-\exp\left[-\frac{1}{\sigma}(I - \mu)\right]\right\} \\
\quad \text{for } \kappa = 0
\end{cases}
\tag{A1}
$$

where I is the average intensity over a given duration and μ, σ and κ are respectively the location, scale and shape parameters of the distribution. The shape parameter is directly related to the tail of the extreme values distribution: low (high) shape parameters are related to lower (higher) probability of having large extremes. When the shape parameter is lower than 0 (type III distribution), an upper limit to the extreme values is expected. When the shape parameter is greater (type II) or equal (type I distribution) to 0, no upper limit to the values is expected. In particular, when the shape parameter is greater than 0.5, the standard deviation of the distribution is infinite and when it is greater than 1, the mean of the distribution is infinite. Under hypotheses on the regularity of the tail of the distribution, the Fisher–Tippet theorem demonstrates that GEV distribution is the only possible limit distribution for the extreme values of independent and identically distributed random variables.

Appendix B: Comparison metrics

Correlation coefficient (CC) measures the spatial correlation of the derived maps. It is calculated as

$$
\mathrm{CC} = \frac{\mathrm{Cov}\left(\mathbf{r}, \mathbf{s}\right)}{\mathrm{SD}(\mathbf{r}) \cdot \mathrm{SD}(\mathbf{s})},
\tag{B1}
$$

where $\mathbf{r}$ and $\mathbf{s}$ refer to the radar-IDF maps and satellite-IDF maps (HRC or CHRC), respectively. It ranges between -1 and 1 with positive (negative) values indicating positive (negative) correlation.

Multiplicative bias measures the mean quantitative agreement of two maps. It is calculated as

$$
\mathrm{Bias} = \frac{\sum_i^N s_i}{\sum_i^N r_i},
\tag{B2}
$$

where r_i and s_i are the radar-IDF and satellite-IDF estimates, respectively, and i is the index of a pixel within the N pixels composing the maps. It may assume any positive value with 1 indicating perfect mean quantitative agreement, value higher (lower) than 1 indicating that the mean satellite estimate is higher (lower) than the mean radar estimate.

Normalized standard difference (NSD) measures the standard deviation of the residuals of the normalized maps and is calculated as

$$
\mathrm{NSD} = \sqrt{\frac{1}{N} \sum_i^N \left(\frac{s_i}{\hat{s}} - \frac{r_i}{\hat{r}}\right)^2},
\tag{B3}
$$

where $\hat{s}$ and $\hat{r}$ are the mean values of satellite and radar maps. NSD may assume any value ≥ 0.

Competing interests. The authors declare that they have no conflict of interest.

Acknowledgements. Radar data were provided by E.M.S. (Mekorot Company) and rain gauge data were provided by the Israel Meteorological Service. The study was partially funded by the Lady Davis Fellowship Trust [project: RainFreq], by the Israel Science Foundation [grant no. 1007/15], by the PALEX DFG project, and by NSF-BSF grant [BSF 2016953]. This work is a contribution to the HyMeX program. We thank Luca Panziera and two anonymous reviewers for contributions that improved the quality of this paper.

Edited by: B. Su

References

Alila, Y.: A hierarchical approach for the regionalization of precipitation annual maxima in Canada, J. Geophys. Res., 104, 31645–31655, doi:10.1029/1999JD900764, 1999.

Allen, R. J. and DeGaetano, A. T.: Considerations for the use of radar-derived precipitation estimates in determining return intervals for extreme areal precipitation amounts, J. Hydrol., 315, 203–219, doi:10.1016/j.jhydrol.2005.03.028, 2005.

Alpert, P. and Shafir, H.: Meso c-scale distribution of orographic precipitation: numerical study and comparison with precipitation derived from radar measurements, J. Appl. Meteorol., 28, 1105–1117, doi:10.1175/1520-0450(1989)028<1105:MSDOOP>2.0.CO;2, 1989.

Amitai, E., Petersen, W. A., Llort, X., and Vasilof, S.: Multi-platform comparisons of rain intensity for extreme precipitation events, IEEE T. Geosci. Remote, 99, 1–12, doi:10.1109/TGRS.2011.2162737, 2011.

Awadallah, A., ElGamal, M., ElMostafa, A., and ElBadry, H.: Developing intensity–duration–frequency curves in scarce data region: an approach using regional analysis and satellite data, Engineering, 3, 215–226, doi:10.4236/eng.2011.33025, 2011.

Bacchi, B. and Ranzi, R.: On the Derivation of the Areal Reduction Factor of Storms, Atmos. Res., 42, 123–35, 1996.

Ben-Zvi, A.: Rainfall Intensity–Duration–Frequency Relationships Derived from Large Partial Duration Series, J. Hydrol., 367, 104–114, doi:10.1016/j.jhydrol.2009.01.007, 2009.

Berne, A. and Krajewski, W. F.: Radar for hydrology: unfulfilled promise or unrecognized potential?, Adv. Water Resour., 51, 357–366, doi:10.1016/j.advwatres.2012.05.005, 2013.

Borga, M., Anagnostou, E. N., Blöschl, G., and Creutin, J. D.: Flash Flood Forecasting, Warning and Risk Management: The HYDRATE Project, Environ. Sci. Policy, 14, 834–44, 2011.

Borga M., Stoffel, M., Marchi, L., Marra, F., and Jacob, M.: Hydrogeomorphic response to extreme rainfall in headwater systems: flash floods and debris flows, J. Hydrol., 518, 194–205, doi:10.1016/j.jhydrol.2014.05.022, 2014.

Ceresetti, D., Ursu, E., Carreau, J., Anquetin, S., Creutin, J. D., Gardes, L., Girard, S., and Molinié, G.: Evaluation of classical spatial-analysis schemes of extreme rainfall, Nat. Hazards Earth Syst. Sci., 12, 3229–3240, doi:10.5194/nhess-12-3229-2012, 2012.

Chen, M., Shi, W., Xie, P., Silva, V. B. S., Kousky, V. E., Wayne Higgins, R., and Janowiak, J. E.: Assessing objective techniques for gauge-based analyses of global daily precipitation, J. Geophys. Res., 113, D04110, doi:10.1029/2007JD009132, 2008.

Chen, S., Hong, Y., Cao, Q., Kirstetter, P. E., Gourley, J. J., Qi, Y., Zhang, J., Howard, K., Hu, J., and Wang, J.: Performance evaluation of radar and satellite rainfalls for Typhoon Morakot over Taiwan: are remote-sensing products ready for gauge denial scenario of extreme events, J. Hydrol., 506, 4–13, doi:10.1016/j.jhydrol.2012.12.026, 2013.

Chow, V. T., Maidment, D. R., and Mays, L. W.: Applied Hydrology, McGraw-Hill Inc., New York, 572 pp., 1988

Coles, S.: An Introduction to Statistical Modeling of Extreme Values, Springer-Verlag, London, 2001.

Dayan, U. and Morin, E.: Flash flood-producing rainstorms over the Dead Sea: a review, Geol. S. Am. S., 401, 53–62, doi:10.1130/2006.2401(04), 2006.

Dayan, U., Ziv, B., Margalit, A., Morin, E., and Sharon, D.: A severe autumn storm over the middle-east: synoptic and mesoscale convection analysis, Theor. Appl. Climatol., 69, 103–122, doi:10.1007/s007040170038, 2001.

Dayan, U., Nissen, K., and Ulbrich, U.: Review Article: Atmospheric conditions inducing extreme precipitation over the eastern and western Mediterranean, Nat. Hazards Earth Syst. Sci., 15, 2525–2544, doi:10.5194/nhess-15-2525-2015, 2015.

de Vries, A. J., Tyrlis, E., Edry, D., Krichak, S. O., Steil, B., and Lelieveld, J.: Extreme precipitation events in the Middle East: dynamics of the Active Red Sea Trough, J. Geophys. Res.-Atmos., 118, 7087–7108, doi:10.1002/jgrd.50569, 2013.

Durrans, S. R., Julian, L. T., and Yekta, M.: Estimation of Depth-Area Relationships Using Radar-Rainfall Data, J. Hydrol. Eng., 7, 356–367, 2002.

Eagleson, P. S.: Dynamic Hydrology, McGraw-Hill, New York, 462 pp., 1970.

Eldardiry, H., Habib, E., and Zhang, Y.: On the use of radar-based quantitative precipitation estimates for precipitation frequency analysis, J. Hydrol., 531, 441–453, doi:10.1016/j.jhydrol.2015.05.016, 2015.

Endreny, T. A. and Imbeah, N.: Generating robust rainfall intensity–duration–frequency estimates with short-record satellite data, J. Hydrol., 371, 182–191, doi:10.1016/j.jhydrol.2009.03.027, 2009.

Ferraro, R. R.: SSM/I derived global rainfall estimates for climatological applications, J. Geophys. Res., 102, 16715–16735, 1997.

Ferraro, R. R., Weng, F., Grody, N. C., and Zhao, L.: Precipitation characteristics over land from the NOAA-15 AMSU sensor, Geophys. Res. Lett., 27, 2669–2672, 2000.

Fowler, H. J. and Kilsby, C. G.: A regional frequency analysis of United Kingdom extreme rainfall from 1961 to 2000, Int. J. Climatol., 23, 1313–1334, doi:10.1002/joc.943, 2003.

Gellens, D.: Combining regional approach and data extension procedure for assessing GEV distribution of extreme precipitation in Belgium, J. Hydrol., 268, 113–126, doi:10.1016/S0022-1694(02)00160-9, 2002.

Gires, A., Tchiguirinskaia, I., Schertzer, D., Schellart, A., Berne, A., and Lovejoy, S.: Influence of small scale rainfall variability on standard comparison tools between radar and rain gauge data, Atmos. Res., 138, 125–138, 2014.

Goldreich, Y.: The spatial distribution of annual rainfall in Israel – a review, Theor. Appl. Climatol., 50, 45–59, doi:10.1007/BF00864902, 1994.

Goldreich, Y.: The Climate of Israel, Observations, Research and Applications, Springer, New York, 2003.

Guo, H., Chen, S., Bao, A., Hu, J., Gebregiorgis, A. S., Xue, X., and Zhang, X.: Inter-Comparison of High-Resolution Satellite Precipitation Products over Central Asia, Remote Sens., 7, 7181–7211, doi:10.3390/rs70607181, 2015.

Joyce, R. J., Janowiak, J. E., Arkin, P. A., and Xie, P.: CMORPH: A Method that Produces Global Precipitation Estimates from Passive Microwave and Infrared Data at High Spatial and Temporal Resolution, J. Hydrometeorol., 5, 487–503, doi:10.1175/1525-7541(2004)005<0487:CAMTPG>2.0.CO;2, 2004.

Kahana, R., Ziv, B., Enzel, Y., and Dayan, U.: Synoptic climatology of major floods in the Negev Desert, Israel, Int. J. Climatol., 22, 867–882, doi:10.1002/joc.766, 2002.

Karklinsky, M. and Morin, E.: Spatial characteristics of radar-derived convective rain cells over southern Israel, Meteorol. Z., 15, 513–520, doi:10.1127/0941-2948/2006/0153, 2006.

Katz, R. W., Parlange, M. B., and Naveau, P.: Statistics of extremes in hydrology, Adv. Water Resour., 25, 1287–1304, 2002.

Kidd, C. and Levizzani, V.: Status of satellite precipitation retrievals, Hydrol. Earth Syst. Sci., 15, 1109–1116, doi:10.5194/hess-15-1109-2011, 2011.

Kidd, C., Becker, A., Huffman, G., Mullaer, C., Joe, P., Skofronick-Jackson, G., and Kirshbaum, D.: So, How Much of the Earth's Surface Is Covered by Rain Gauges?, B. Am. Meteorol. Soc., 98, 69–78, doi:10.1175/BAMS-D-14-00283.1, 2017.

Koutsoyiannis, D.: Statistics of extremes and estimation of extreme rainfall: II. Empirical investigation of long rainfall records, Hydrolog. Sci. J., 49, 591–610, doi:10.1623/hysj.49.4.575.54430, 2004.

Kummerow, C., Hong, Y., Olson, W. S., Yang, S., Adler, R. F., McCollum, J., Ferraro, R., Petty, G., Shin, D. B., and Wilheit, T. T.: Evolution of the Goddard profiling algorithm (GPROF) for rainfall estimation from passive microwave sensors, J. Appl. Meteorol., 40, 1801–1820, 2001.

Kysely, J. and Picek, J.: Regional growth curves and improved design value estimates of extreme precipitation events in the Czech Republic, Clim. Res., 33, 243–255, 2007.

Lombardo, F., Napolitano, F., and Russo, F.: On the use of radar reflectivity for estimation of the areal reduction factor, Nat. Hazards Earth Syst. Sci., 6, 377–386, doi:10.5194/nhess-6-377-2006, 2006.

Marra, F. and Morin, E.: Use of radar QPE for the derivation of Intensity–Duration–Frequency curves in a range of climatic regimes, J. Hydrol., 531, 427–440, doi:10.1016/j.jhydrol.2015.08.064, 2015.

Marra, F., Nikolopoulos, E. I., Creutin, J. D., and Borga, M.: Radar rainfall estimation for the identification of debris-flow occurrence thresholds, J. Hydrol. 519, 1607–1619, doi:10.1016/j.jhydrol.2014.09.039, 2014.

Marra, F., Nikolopoulos, E. I., Creutin, J. D., and Borga, M.: Space-time organization of debris flows-triggering rainfall and its effect on the identification of the rainfall threshold relationship, J. Hydrol., 541, 246–255, doi:10.1016/j.jhydrol.2015.10.010, 2016.

Mei, Y., Nikolopoulos, E. I., Anagnostou, E. N., Zoccatelli, D., and Borga, M.: Error analysis of satellite precipitation-driven modeling of flood events in complex alpine terrain, Remote Sens., 8, 293, doi:10.3390/rs8040293, 2016.

Morin, E. and Gabella, M.: Radar-based quantitative precipitation estimation over Mediterranean and dry climate regimes, J. Geophys. Res., 112, D20108, doi:10.1029/2006JD008206, 2007.

Morin, E., Enzel, Y., Shamir, U., and Garti, R.: The characteristic time scale for basin hydrological response using radar data, J. Hydrol., 252, 85–99, doi:10.1016/S0022-1694(01)00451-6, 2001.

Morin, E., Jacoby, Y., Navon, S., and Ben-Halachmi, E.: Towards flash-flood prediction in the dry Dead Sea region utilizing radar rainfall information, Adv. Water Resour., 32, 1066–1076, doi:10.1016/j.advwatres.2008.11.011, 2009.

Morin, J., Rosenfeld, D., and Amitai, E.: Radar rain field evaluation and possible use of its high temporal and spatial resolution for hydrological purposes, J. Hydrol., 172, 275–292, doi:10.1016/0022-1694(95)02700-Y, 1995.

Morin, J., Sharon, D., Rubin, S.: Rainfall Intensity in Israel, Selected Stations, Report 1/94, IMS, Bet Dagan, Israel, 1998.

Nastos, P. T., Kapsomenakis, J., Douvis, K. C.: Analysis of precipitation extremes based on satellite and high-resolution gridded data set over Mediterranean basin, Atmos. Res., 131, 46–59, 2013.

Overeem, A., Buishand, T. A., and Holleman, I.: Rainfall Depth–Duration–Frequency Curves and Their Uncertainties, J. Hydrol., 348, 124–134, doi:10.1016/j.jhydrol.2007.09.044, 2008.

Overeem, A., Buishand, T. A., and Holleman, I.: Extreme rainfall analysis and estimation of depth–duration–frequency curves using weather radar, Water Resour. Res., 45, W10424, doi:10.1029/2009WR007869, 2009.

Overeem, A., Buishand, T. A., Holleman, I., and Uijlenhoet, R.: Extreme Value Modeling of Areal Rainfall from Weather Radar, Water Resour. Res., 46, W09514, doi:10.1029/2009WR008517, 2010.

Paixao, E., Mirza, M. M. Q., Shephard, M. W., Auld, H., Klaassen, J., and Smith, G.: An integrated approach for identifying homogeneous regions of extreme rainfall events and estimating IDF curves in Southern Ontario, Canada: incorporating radar observations, J. Hydrol., 528, 734–750, doi:10.1016/j.jhydrol.2015.06.015, 2015.

Panziera, L., Gabella, M., Zanini, S., Hering, A., Germann, U., and Berne, A.: A radar-based regional extreme rainfall analysis to derive the thresholds for a novel automatic alert system in Switzerland, Hydrol. Earth Syst. Sci., 20, 2317–2332, doi:10.5194/hess-20-2317-2016, 2016.

Papalexiou, S. M. and Koutsoyiannis, D.: Battle of Extreme Value Distributions: A Global Survey on Extreme Daily Rainfall, Water Resour. Res., 49, 187–201, doi:10.1029/2012WR012557, 2013.

Peel, M. C., Finlayson, B. L., and McMahon, T. A.: Updated world map of the Köppen–Geiger climate classification, Hydrol. Earth Syst. Sci., 11, 1633–1644, doi:10.5194/hess-11-1633-2007, 2007.

Peleg, N. and Morin, E.: Convective rain cells: radar-derived spatiotemporal characteristics and synoptic patterns over the eastern Mediterranean, J. Geophys. Res., 117, D15116, doi:10.1029/2011JD017353, 2012.

Peleg, N., Ben-Asher, M., and Morin, E.: Radar subpixel-scale rainfall variability and uncertainty: lessons learned from observations of a dense rain-gauge network, Hydrol. Earth Syst. Sci., 17, 2195–2208, doi:10.5194/hess-17-2195-2013, 2013.

Peleg, N., Blumensaat, F., Molnar, P., Fatichi, S., and Burlando, P.: Partitioning the impacts of spatial and climatological rainfall variability in urban drainage modeling, Hydrol. Earth Syst. Sci., 21, 1559–1572, doi:10.5194/hess-21-1559-2017, 2017a.

Peleg, N., Marra, F., Fatichi, S., Paschalis, A., Molnar, P., and Burlando, P.: Spatial variability of rainfall at radar subpixel scale, J. Hydrol., in press, doi:10.1016/j.jhydrol.2016.05.033, 2017b.

Perica, S., Martin, D., Pavlovic, S., Roy, I., St. Laurent, M., Trypaluk, C., Unruh, D., Yekta, M., and Bonnin, G.: Precipitation-Frequency Atlas of the United States, NOAA Atlas 14, Volume 9, Version 2.0: Southeastern States (Alabama, Arkansas, Florida, Georgia, Louisiana, Mississippi), U.S. Department of Commerce, National Oceanic and Atmospheric Adminstration, National Weather Service, Silver Spring, Maryland, 2013.

Phillips, P. and Perron, P.: Testing for a Unit Root in Time Series Regression, Biometrika, 75, 335–346, doi:10.1175/2007MWR1919.1, 1988.

Rubin, S., Ziv, B., and Paldor, N.: Tropical plumes over eastern North Africa as a source of rain in the Middle East, Mon. Weather Rev., 135, 4135–4148, doi:10.1175/2007MWR1919.1, 2007.

Schaefer, M. G.: Regional analyses of precipitation annual maxima in Washington State, Water Resour. Res., 26, 119–131, doi:10.1029/WR026i001p00119, 1990.

Segoni, S., Battistini, A., Rossi, G., Rosi, A., Lagomarsino, D., Catani, F., Moretti, S., and Casagli, N.: Technical Note: An operational landslide early warning system at regional scale based on space-time-variable rainfall thresholds, Nat. Hazards Earth Syst. Sci., 15, 853–861, doi:10.5194/nhess-15-853-2015, 2015.

Sivapalan, M. and Blöschl, G.: Transformation of point rainfall to areal rainfall: intensity–duration–frequency curves, J. Hydrol., 204, 150–167, doi:10.1016/S0022-1694(97)00117-0, 1998.

Srebro, H. and Soffer, T.: The New Atlas of Israel: The National Atlas, Survey of Israel and The Hebrew University of Jerusalem, Jerusalem, Israel, 2011.

Stampoulis, D., Anagnostou, E. N., and Nikolopoulos, E. I.: Assessment of high-resolution satellite-based rainfall estimates over the Mediterranean during heavy precipitation events, J. Hydrometeorol., 14, 1500–1514, doi:10.1175/JHM-D-12-0167.1, 2013.

Svensson, C. and Jones, D. A.: Review of Methods for Deriving Areal Reduction Factors, Journal of Flood Risk Management 3, 232–245, doi:10.1111/j.1753-318X.2010.01075.x, 2010.

Tapiador, F. J., Turk, F. J., Petersen, W., Hou, A. Y., García-Ortega, E., Machado, L. A. T., Angelis, C. F., Salio, P., Kidd, C., Huffman, G. J., and de Castro, M.: Global precipitation measurement: methods, datasets and applications, Atmos. Res., 104–105, 70–97, doi:10.1016/j.atmosres.2011.10.021, 2012.

Tiranti, D., Cremonini, R., Marco, F., Gaeta, A. R., and Barbero, S.: The DEFENSE (Debris Flows triggEred by Storms – Nowcasting System): An Early Warning System for Torrential Processes by Radar Storm Tracking Using a Geographic Information System (GIS), Comput. Geosci., 70, 96–109, doi:10.1016/j.cageo.2014.05.004, 2014.

Tubi, A. and Dayan, U.: Tropical Plumes over the Middle East: Climatology and synoptic conditions, Atmos. Res., 145–146, 168–181, doi:10.1016/j.atmosres.2014.03.028, 2014.

Villarini, G., Krajewski, W., Ntelekos, A. A., Georgakakos, K. P., and Smith, J. A.: Towards Probabilistic Forecasting of Flash Floods: The Combined Effects of Uncertainty in Radar-Rainfall and Flash Flood Guidance, J. Hydrol., 394, 275–84, doi:10.1016/j.jhydrol.2010.02.014, 2010.

Watt, E. and Marsalek, J.: Critical review of the evolution of the design storm event concept, Can. J. Civil. Eng., 40, 105–113, doi:10.1139/cjce-2011-0594, 2013:

Wright, D. B., Smith, J. A., Villarini, G., and Baeck, M. L.: Estimating the frequency of extreme rainfall using weather radar and stochastic storm transposition, J. Hydrol., 488, 150–165. doi:10.1016/j.jhydrol.2013.03.003, 2013.

Wright, D. B., Smith, J. A., and Baeck, M. L.: Flood Frequency Analysis Using Radar Rainfall Fields and Stochastic Storm Transposition, Water Resour. Res., 50, 1592–1615, doi:10.1002/2013WR014224, 2014.

Xie, P., Yoo, S.-H., Joyce, R. J., and Yarosh, Y.: Bias-corrected CMORPH: A 13-year analysis of high-resolution global precipitation, EGU General Assembly, Vienna, Austria, 4–8 April 2011, EGU2011-1809, 2011.

Yakir, H. and Morin, E.: Hydrologic response of a semi-arid watershed to spatial and temporal characteristics of convective rain cells, Hydrol. Earth Syst. Sci., 15, 393–404, doi:10.5194/hess-15-393-2011, 2011.

Yang, T.-H., Hwang, G.-D., Tsai, C.-C., and Ho, J.-Y.: Using rainfall thresholds and ensemble precipitation forecasts to issue and improve urban inundation alerts, Hydrol. Earth Syst. Sci., 20, 4731–4745, doi:10.5194/hess-20-4731-2016, 2016.

Seasonal streamflow forecasts in the Ahlergaarde catchment, Denmark: the effect of preprocessing and post-processing on skill and statistical consistency

Diana Lucatero[1], Henrik Madsen[2], Jens C. Refsgaard[3], Jacob Kidmose[3], and Karsten H. Jensen[1]

[1]Department of Geosciences and Natural Resource Management, University of Copenhagen, Copenhagen, Denmark
[2]DHI, Hørsholm, Denmark
[3]Geological Survey of Denmark and Greenland (GEUS), Copenhagen, Denmark

Correspondence: Diana Lucatero (diana.lucatero@ign.ku.dk)

Abstract. In the present study we analyze the effect of bias adjustments in both meteorological and streamflow forecasts on the skill and statistical consistency of monthly streamflow and yearly minimum daily flow forecasts. Both raw and preprocessed meteorological seasonal forecasts from the European Centre for Medium-Range Weather Forecasts (ECMWF) are used as inputs to a spatially distributed, coupled surface–subsurface hydrological model based on the MIKE SHE code. Streamflow predictions are then generated up to 7 months in advance. In addition to this, we post-process streamflow predictions using an empirical quantile mapping technique. Bias, skill and statistical consistency are the qualities evaluated throughout the forecast-generating strategies and we analyze where the different strategies fall short to improve them. ECMWF System 4-based streamflow forecasts tend to show a lower accuracy level than those generated with an ensemble of historical observations, a method commonly known as ensemble streamflow prediction (ESP). This is particularly true at longer lead times, for the dry season and for streamflow stations that exhibit low hydrological model errors. Biases in the mean are better removed by post-processing that in turn is reflected in the higher level of statistical consistency. However, in general, the reduction of these biases is not sufficient to ensure a higher level of accuracy than the ESP forecasts. This is true for both monthly mean and minimum yearly streamflow forecasts. We discuss the importance of including a better estimation of the initial state of the catchment, which may increase the capability of the system to forecast streamflow at longer leads.

1 Introduction

Seasonal streamflow forecasting encompasses a variety of methods that range from purely data-based to entirely model-based or hybrid methods that exploit the benefits of each (Mendoza et al., 2017). Data-driven methods find empirical relationships between streamflow and a variety of predictors. These relationships are then used to derive forecasts for the upcoming seasons. Different predictors can be used depending on the relative importance they have for the regional hydroclimatic conditions. Predictors that have been used include large-scale climate indicators such as El Niño or the North Atlantic Oscillation, (Schepen et al., 2016; Shamir, 2017; Wang et al., 2009; Olsson et al., 2016), precipitation and land temperature (Córdoba-Machado et al., 2016), the state of the catchment in the form of streamflow, soil moisture, groundwater storages or snow storages that can be derived either by the use of a hydrological model, hence the term "hybrid" (Robertson et al., 2013; Rosenberg et al., 2011), or by means of observed antecedent conditions (Robertson and Wang, 2012).

Model-based systems include a hydrological model in the forecasting chain. Differences between forecasting frameworks may arise in the forcings, the initialization framework and/or the hydrological model structure and parameters. Focusing on the forcing, one can either use observed meteorology from previous years, a method that is commonly known as ensemble streamflow prediction (ESP) (Day, 1985), or outputs from general circulation models (GCMs) (Crochemore et al., 2016; Wood et al., 2002, 2005;

Wood and Lettenmaier, 2006; Yuan et al., 2011, 2013, 2015, 2016). In principle, the latter should be more suitable in providing skillful forecasts as they are able to capture the evolving chaotic behavior of the atmosphere, whereas the ESP approach assumes that what has been observed in the past can be used as a proxy for what will happen in the future, an assumption that requires stationary climate conditions. On the other hand, the lack of reliability of GCMs in forecasting atmospheric patterns at long lead times precludes their use in weather-impacted sectors (Bruno Soares and Dessai, 2016; Weisheimer and Palmer, 2014). For example, a previous study on the skill of the European Centre for Medium-Range Weather Forecasts (ECMWF) System 4 in Denmark concluded that, in general, the precipitation forecast bias in the catchment area was in general around -25% (Lucatero et al., 2017). This bias, together with the sharpness of forecasts, led to a mild positive skill limited to the first month lead time (Lucatero et al., 2017). These results are in accordance with skill studies with focus on a similar area (Crochemore, et al., 2017). This is the reason why preprocessing and post-processing should be performed when using GCM forecasts to force a hydrological model to eliminate biases intrinsic to climate and hydrological models. In the context of this study, preprocessing refers to any method that improves the forcings, i.e., precipitation and temperature, used in the hydrological forecasting system. Post-processing refers to the improvements achieved in the outputs of the hydrological model, e.g., streamflow. In this respect, post-processing also corrects errors in hydrological models that cannot be eliminated through calibration (Shi et al., 2008; Yuan et al., 2015; Yuan and Wood, 2012).

A couple of studies have quantified the effects on streamflow skill by preprocessing either seasonal (Crochemore et al., 2016) or medium-range (Verkade et al., 2013) forecasts. Other studies have assessed the efficiency of post-processing streamflow forecasts only (Bogner et al., 2016; Madadgar et al., 2014; Ye et al., 2015; Zhao et al., 2011; Wood and Schaake, 2008). To the best of our knowledge, only Roulin and Vannitsem (2015), Yuan and Wood (2012) and Zalachori et al. (2012) have compared the additional gain in skill of doing both preprocessing and post-processing. The previous studies have shown that improvements made by preprocessing the forcings do not necessarily translate into improvements in streamflow forecasts (Verkade et al., 2013; Zalachori et al., 2012). Improvements are larger when post-processing is done, and a combination of preprocessing and post-processing provides the best results (Yuan and Wood, 2012; Zalachori et al., 2012). To the best of our knowledge, only Yuan and Wood (2012) have made this evaluation in the context of seasonal forecasting.

The present study focuses on the following aspects: (i) the evaluation of the use of a GCM to generate seasonal streamflow forecasts, (ii) the study of the effect that preprocessing and post-processing have on streamflow forecasts 1–7 months ahead, and (iii) the effect of hydrological model

biases in forecast skill evaluations. This is done by a combination of the following methodological choices. First, we make use of seasonal meteorological forecasts of ECMWF System 4 (Molteni, et al., 2011). Secondly, the hydrological simulations use an integrated physically based and spatially distributed model based on the MIKE SHE code (Graham and Butts, 2005). Thirdly, our evaluation focuses on three forecast qualities: bias, skill and statistical consistency. Skill is measured using ESP as a reference and focusing on both accuracy and sharpness. Finally, the focus here is to evaluate forecasts of monthly average streamflow throughout the year and minimum daily flows during the summer. The catchment serving as a basis of our study is groundwater-dominated and is located in a region where seasonal forecasting is a challenging endeavor (Lucatero et al., 2017). The following questions are then addressed.

1. How do GCM-generated forecasts compare to those of the ESP approach?

2. What is the effect of preprocessing and post-processing on streamflow forecasts in terms of bias, skill and statistical consistency? And more specifically, is there one single approach, or a combination of several, that reduces the bias and augments skill and statistical consistency?

3. What is the effect that hydrological model bias has on the evaluation of preprocessed and post-processed streamflow forecasts?

2 Data and methods

The following sections give a description of the methodology followed in this study. A graphical depiction of the steps carried out can be seen in Fig. 1.

2.1 Area of study, observational data and hydrological model

The present study is carried out for the Ahlergaarde catchment located in West Jutland, Denmark (Fig. 2), which has a size of $1044\,km^2$. It is located in one of the most irrigated zones in Denmark, with 55 % of the area covered with agricultural crops such as barley, grass, wheat, maize and potatoes. The remaining area is distributed in categories as follows: grass (30 %), forest (7 %), heath (5 %), urban (2 %) and other (1 %) (Jensen and Illangasekare, 2011).

The climatology of the area is shown in Fig. 3. Climate in the Ahlergaarde region is mainly influenced by its proximity to the sea towards the west. The mean annual precipitation, reference evapotranspiration and discharge for the period 1990–2013 is 983, 540 and 500 mm, respectively. The hydrology of the catchment is groundwater-dominated due to the high permeability of the top geological layer, which consists mainly of sand and gravel. Another consequence of the

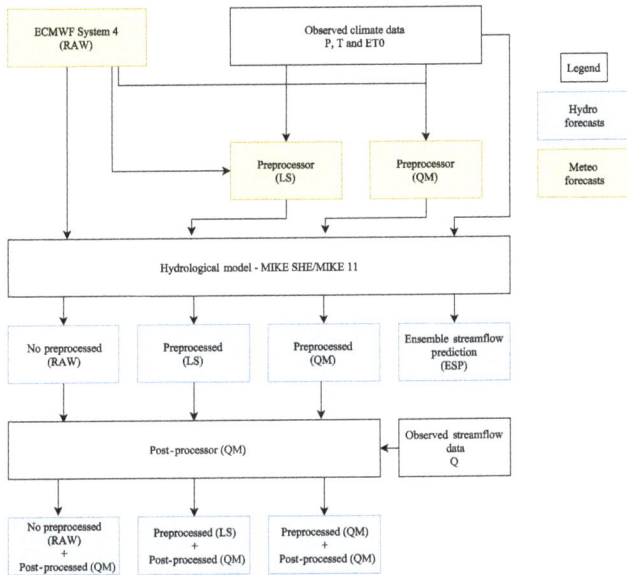

Figure 1. Diagram of generation of forecasts and verification procedures. RAW refers to the uncorrected ECMWF System 4 forecasts, while LS and QM refer to forecasts (either meteorological or hydrological) that are corrected using the linear scaling/delta change or quantile mapping method, respectively, for precipitation (P), temperature (T) and reference evapotranspiration (E_{T0}). Preprocessed refers to streamflow forecasts generated using corrected meteorological forecasts while post-processed refers to corrected streamflow forecasts.

geological composition of the surface layer is that overland flow rarely happens (Jacob Kidmose, personal communication, 2014). Daily precipitation (P), temperature (T) and reference evapotranspiration (E_{T0}) data are retrieved from the Danish Meteorological Institute (DMI; Scharling and Kern-Hansen, 2012). The dataset spatial domain covers Denmark with a 10 km grid resolution for P and a 20 km resolution for T and E_{T0}. P is corrected for systematic under-catch due to wind effects (Stisen et al., 2011, 2012), and E_{T0} is derived using the Makkink formulation (Hendriks, 2010). Finally, daily streamflow observations are retrieved from the Danish Hydrological Observatory (HOBE) (Jensen and Illangasekare, 2011) datasets.

The hydrological simulations for this study are grounded on a physically based, spatially distributed, coupled surface–subsurface model that simulates the main hydrological processes such as evapotranspiration, overland flow, unsaturated, saturated and streamflows and their interactions. The model is based on the MIKE SHE code (Graham and Butts, 2005). Groundwater flow is described by the governing equation for three-dimensional groundwater flow based on Darcy's law. Drain flow is considered when the groundwater table exceeds a drain level. Surface water flow in streams is simulated by a one-dimensional channel flow model based on kinematic routing, while a two-dimensional diffusive wave approximation of the St. Venant equations is used for over-

land flow routing. Finally, a two-layer approach is used for the simulations of unsaturated flow and evapotranspiration (Graham and Butts, 2005). Snow is not an important process in the study area; therefore, the model takes snowmelt into account by using a simple degree-day model formulation. The horizontal numerical discretization is 200 m, whereas the vertical discretization is based on six numerical layers whose dimension depends on the geological stratigraphy. Model parameters were calibrated against groundwater head and discharge using an automated optimizer, PEST (parameter estimation) version 11.8 (Doherty, 2016) for the 2006–2009 period. Parameters to be calibrated were selected based on a sensitivity analysis study. These are hydraulic conductivities for 10 geological units, specific yield, specific storage, drain time constant, detention storage, river-groundwater conductance and root depth of 10 vegetation types. The reader is referred to Zhang et al. (2016) for further details on the calibration procedure.

2.2 Forecast generation: GCM-based and ESP

As seen in Fig. 1, P, T and E_{T0} forecasts are taken from the ECMWF System 4 (RAW), preprocessed ECMWF System 4 (linear scaling, LS, and quantile mapping, QM), and historical observations (ESP). The European Centre for Medium-Range Weather Forecasts (ECMWF) offers a seasonal forecasting product that currently is in its version number 4 (Molteni et al., 2011). An attempt to reduce the biases intrinsic in ECMWF System 4 led to what we refer to as preprocessed forecasts. The reader is referred to Lucatero et al. (2017) for details of the evaluation of both ECMWF System 4 and preprocessed forecasts for Denmark. The spatial resolution of the raw forecasts is 0.7° in latitude and longitude. Forecasts were interpolated to a 10 km grid to match the resolution of the observed grid. For the Ahlergaarde catchment, forecast–observation data for the 1990–2013 period are extracted for 24 grid points covering the study area, leading to a sample size of 24 years. Finally, E_{T0} is computed using the Makkink formulation (Hendriks, 2010) that takes T and incoming shortwave solar radiation from the ECMWF System 4 forecasts as inputs.

Daily raw and preprocessed forecasts are initialized on the first day of each calendar month with a 7-month lead time. The number of ensemble members varies by month, 15 for January, March, April, June, July, September, October and December, and 51 for the remaining months. The number of ensembles is higher for February, May, August and November to aid in improving forecasts for the most predictable seasons. ESP forcings are taken from the observation record, with each year acting as an ensemble member. Values are taken from the start of each calendar month, with a 7-month lead time in order to match the lead time of the ECMWF System 4 forecasts. Since the year to be forecasted is left out of the ensemble, the number of ensemble members for the ESP is 23. Both the ECMWF System 4-generated forecasts and

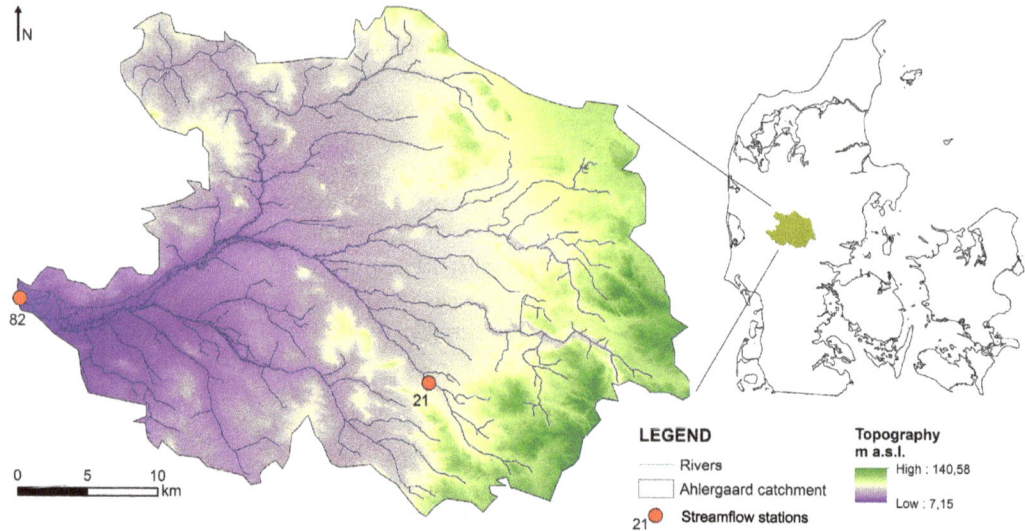

Figure 2. Location and topography of the Ahlergaarde catchment. The outlet station (82) and the upstream sub-catchment (21) are used in the study.

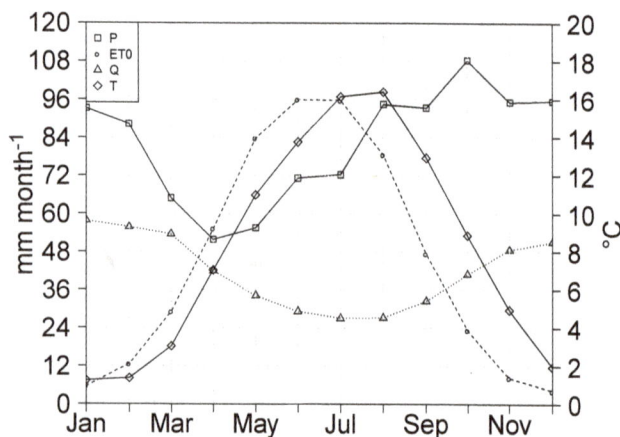

Figure 3. Climatology of the Ahlergaarde catchment. Values for precipitation (P), reference evapotranspiration (E_{T0}), streamflow (Q) and temperature (T) are monthly average values over the period 1990–2013.

ESP share the same hydrological initial conditions for forecasts initiated in the same month. These are computed from a spin-up run starting in January 1990 and up until 2013. Initial states are saved on the first day of each calendar month. Forecasts are then run on a daily basis up to 7 months.

2.3   Preprocessor and post-processor

Preprocessed forcings for the hydrological model were retrieved from data of the companion paper, Lucatero et al. (2017). The authors used two well-known bias correction techniques, LS and QM. In LS the ensemble is adjusted with a scaling factor, either by multiplication (for P and E_{T0}) or addition (T). The scaling factor is computed as the ratio or

difference between the averages of the ensemble mean and the observed mean for a specific month, lead time and location, with the sole purpose of adjusting the mean.

QM (Zhao et al., 2017) matches the quantiles of the ensemble distribution with the quantiles of the observed distribution in the following way:

$$f_{k,i}^* = G^{-1}\left(F\left(f_{k,i}\right)\right), \tag{1}$$

where G and F represent the observed and the ensemble distribution functions, respectively, for forecast–observation pair i, for $i = 1, \ldots, M$, with M being the number of forecast–observation pairs. $f_{k,i}$ represents ensemble member k, $k = 1, \ldots, N$, where N is the ensemble size and $f_{k,i}^*$ represents the corrected ensemble member k. F is an empirical distribution function trained with all ensemble members in a given month for a given lead time and location. G and F are fitted on a leave-one-out cross-validation mode, i.e., the forecast–observation pair i is left out of the sample. For example, for a forecast of the target month April, initialized in February, F is computed using all ensemble members, comprising 30 (days) times 23 (number of years in the training sample minus the year to be corrected), times the ensemble size of that particular month (15 or 51). The same is done for G. Linear extrapolation is applied to approximate the values between the bins of F and G and to map ensemble values and quantiles that are outside the training sample.

QM is the only method used for post-processing in the present study as no striking differences in either bias or skill were found between LS and QM in Lucatero et al. (2017). Moreover, QM shows more satisfactory results for the correction of forecasts in the lower tail of the distribution and for the correction of forecasts that also exhibit underdispersion (Lucatero et al., 2017).

2.4 Performance metrics

The performance of raw, preprocessed and post-processed forecasts is evaluated. Our main focus is the following four qualities: bias, skill in regards to accuracy and sharpness and statistical consistency. Bias is the measure of under- or over-estimation of the mean of the ensemble in comparison with the observed values (Yapo et al., 1996):

$$\text{PBias} = \left(\frac{\sum_{i=1}^{M} \overline{f}_i}{\sum_{i=1}^{M} y_i} - 1 \right) \cdot 100, \tag{2}$$

where $\overline{f}_i$ and y_i represent, respectively, the ensemble mean and the observed values for forecast–observation pair i of a particular month, lead time and location. If the value in Eq. (2) is negative, we have an underprediction, and conversely an overprediction, if the value is positive.

Secondly, we compute the continuous rank probability score CRPS (Hersbach, 2000) as a general measure of the accuracy of the forecasts. The computation of the score is as follows:

$$\text{CRPS} = \frac{1}{M} \sum_{i=1}^{M} \int_{-\infty}^{\infty} \left[P_i(x) - H(x - y_i) \right]^2 dx, \tag{3}$$

where $P_i(x)$ represents the cumulative distribution function (CDF) of the ensemble for forecast–observation pair i, $H(x - y_i)$ is the Heaviside function that takes the value 0 when $x < y_i$ or 1 otherwise. y_i is the verifying observation of forecast–observation pair i. Sharpness for forecast–observation pair i is measured as the difference between the 25 and the 75 % percentiles. The average of these differences along the forecast–observation record is then used as a measure of sharpness. Both the CRPS and sharpness scores are then given in the units of the variable of interest, i.e., $m^{-3}\,s^{-1}$ for streamflow. Both scores are positive oriented; i.e., the lower the value, the more accurate or sharper a forecast. A skill score can then be computed in the following manner:

$$\text{Skill} = 1 - \frac{\text{Score}_{\text{sys}}}{\text{Score}_{\text{ref}}}, \tag{4}$$

where, for the present study, $\text{Score}_{\text{sys}}$ is the score of streamflow forecasts generated either using raw, preprocessed ECMWF System 4 or post-processed forecasts. $\text{Score}_{\text{ref}}$ is the score value of our reference system, the ESP. The range of the skill score in Eq. (4) is from $-\infty$ to 1, and values closer to 1 are preferred. Negative values indicate that, on average, the system being evaluated does not perform better than the ESP used as reference. Hereafter, we denote the skill with respect to accuracy as CRPSS and the skill in terms of sharpness as SS. In order to evaluate the statistical significance of the differences of skill between GCM-generated forecasts

and ESP, we use a two-sided Wilcoxon–Mann–Whitney test (WMW test) at the 5 % significance level (see Hollander et al., 2014).

Since the number of ensemble members varies from month to month, the value of the skill scores for months with a larger ensemble size will be more favorable. Although the purpose of the present study is not to make an in-depth analysis of the effect of changing ensemble size, we utilized a bootstrapping technique to make the reader aware of the possible gains in skill due to increased ensemble size. This is accomplished by computing the skill scores of a random selection of 15 of the 51 ensemble members for February, May, August and November as in Jaun et al. (2008). This step is performed 1000 times. The final value of the skill score of interest is then the average of these. Note that the bootstrapping is not applied to the ESP forecasts with an ensemble size of 23 members.

Finally, in order to evaluate the statistical consistency between predictive and observed distribution functions, we use the probability integral transform (PIT) diagram. The PIT diagram is the CDF of $z_i = P(X \le y_i)$, where z_i is the value of the cumulative distribution function that the observed value attains within the ensemble distribution for each forecast–observation pair i. Note that the PIT diagram is the continuous equivalent of the rank histograms (Friederichs and Thorarinsdottir, 2012) and it is mainly used to evaluate statistical consistency of a continuous predictive CDF. However, in this study, the z_i's are based on the empirical CDF of the ensemble members at a given lead time. Note that the evaluation of the appropriateness of the choice of PIT diagrams over rank histograms for ensemble forecasts is beyond the scope of the present study. For a forecasting system to be statistically consistent, meaning that the observations can be seen as a draw of the predictive CDF, the CDF of the z_i's should be close to the CDF of a uniform distribution in the [0, 1] range. Deviations from the uniform distribution signify bias in the ensemble mean and spread (see Laio and Tamea, 2007). Finally, in order to make the test for uniformity formal, we make use of the Kolmogorov confidence bands. The bands are two straight lines, parallel to the 1 : 1 diagonal and at a distance $q(\alpha)/\sqrt{M}$, where $q(\alpha)$ is a coefficient that depends on the significance level of the test, i.e., $q(\alpha = 0.05) = 1.358$ (see Laio and Tamea, 2007; D'Agostino and Stephens, 1986), and M is again the number of forecast–observation pairs. The test for uniformity is not rejected if the CDF of the z_i's lies within these bands.

2.5 Forecasts of minimum daily flow within a year

Annual minimum daily flow forecasts can be used for optimizing groundwater extractions for irrigation. The years for which the predicted minimum daily flows are above the prescribed minimum can be exploited and utilized for crops with a higher irrigation demand that may increase economic returns. Here we focus on forecasts initiated in April. For the purposes of this study, minimum daily flows are defined as the flow of the day with the minimum yearly discharge ($m^{-3} s^{-1}$) that usually happens during July to September (Fig. 3). Note that timing errors are not an issue here due to the computation choice of minimum daily flows. Observed minimum daily flow is computed as the flow of the day with the minimum discharge over the 7-month forecasting period (April–October). Forecasted minimum daily flow (for each ensemble) is computed in the same manner. Timing errors will only be visible if forecasted minimum daily flow was chosen to be the discharge values of the day where minimum daily flow was observed, which is not the case here. Studies that have focused their attention on situations of low flow or hydrological drought in the context of seasonal forecasting exist (Fundel et al., 2013; Demirel et al., 2015; Trambauer et al., 2015), documenting the possibility of extracting skillful forecasts months ahead for low flow/drought scenarios. Finally, minimum daily flow forecasts are evaluated using the same skill scores as for monthly flow forecasts, i.e., using ESP as a reference forecast.

3 Results

3.1 Hydrological model evaluation

Figure 4 shows the results for simulated streamflow at the upstream station 21 and the downstream station 82. The focus of the evaluation is done for daily values during the period from 2000 to 2003. As a preliminary evaluation, we computed the percent bias (PBias) and the Nash–Sutcliffe model efficiency coefficient (NSE) for the complete observed–simulated record (1990–2013). There is, in general, a good agreement in timing between observed and simulated values. The visual inspection of the hydrographs reveal, however, an amplitude error that is more pronounced at the upstream station 21, especially during the winter season. Evidence for this is also reflected by the high values of bias and the negative NSE for this station (NSE $= -0.85$). Furthermore, a scatterplot of simulated and observed minimum daily flows for the 24 years shows an overestimation of the minimum daily flows that is more pronounced at the upstream station (Fig. 4). At the outlet station 82 there is a better behavior in terms of bias and NSE, with an overestimation of only 1.7 % and a NSE of 0.73. Moreover, for this station there is a better agreement in both the high and low flows through the year. The latter can be verified by looking at the scatterplot of the minimum daily flows (Fig. 4), with the majority of points lying close to the 1 : 1 diagonal.

Due to the poor performance at the upstream station 21, in the following Sect. 3.2–3.4 we will discuss the skill and consistency of the different approaches for forecast improvement, with a focus on the outlet station only. The large biases in the upstream station, combined with the structural biases of the meteorological forecasts, seem to inflate the skill of the streamflow forecasts. This will be further discussed in Sect. 3.5.

3.2 Streamflow forecasts forced with raw meteorological forecasts

The bias and skill of the monthly streamflow forecasts forced with raw ECMWF forecasts are shown in the first row of Fig. 5. The x axis represents the different lead times in months, while the y axis represents the target month. For example, the bias of November with a lead time of 2 represents the value of bias for a forecast initiated on 1 October for the target month of November. This bias is in the $[-30, -20\%]$ range. In general, the absolute bias increases with lead time, and usually moves from an overprediction (or mild underprediction) to a large negative bias at longer lead times.

Figure 5 also shows the skill of accuracy and sharpness. The months with statistically significant differences in skill between the ESP and ECMWF System 4 forecasts are represented with a black circle. There is a connection between bias and skill of accuracy in the sense that months with a higher bias tend to be the ones with lower or nonexistent skill (e.g., September, October, November). The opposite also holds; i.e., months with milder bias tend to be the months when the forecast is improving over the reference forecast to a higher degree (e.g., December, January, February). This is by no means surprising, as the CRPS penalizes forecasts that have biases.

The CRPSS is negative, except for some months during winter and at short lead times for which a forecast generated using raw ECMWF System 4 forcings improves accuracy up to 40 % compared to ESP. As for the case of bias, skill depends on lead time, reaching its most negative values for forecasts generated 7 months in advance. One important feature is the high skill that a forecast generated using ECMWF raw forcings has in terms of sharpness (SS). Figure 5 shows that this quality is present in the majority of target months and lead times. Note, however, that sharpness is only a desirable property when biases are low. In our case study, the width of the raw forecasts is smaller than that of the ESP, indicating overconfidence when biases are high.

The results of the bootstrapping procedure for the computation of the skill score due to accuracy (CRPSS) indicate that, by reducing the ensemble size to 15, there is a reduction of skill as expected (not shown). However, this reduction does not change the main conclusion. Months with positive skill due to accuracy remain, in general, positive. For example, the CRPSS of February streamflow forecasts at a lead time of 1 is 0.31 for 51 ensemble members. After the bootstrapping experiment with the reduction to 15 ensemble members, the skill score is mildly reduced to 0.29. In order to make the reader aware of the possible increase of skill due to increased ensemble size, green crosses in Fig. 5 (and subsequent figures dealing with skill due to accuracy) represent the target months and lead times with 51 ensemble members.

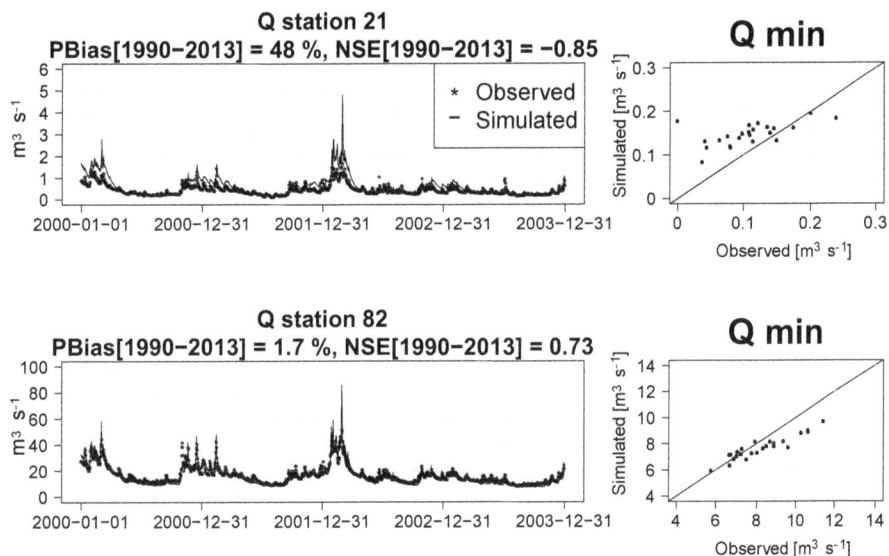

Figure 4. Hydrographs for the 2000–2003 period. Percentage bias (PBias) and Nash–Sutcliffe efficiency score (NSE) are computed using the daily observed–simulated values for the complete 1990–2013 period. The scatterplots represent the observed–simulated annual minimum daily flow values.

Statistical consistency of the raw forecasts is visualized on the first column of the PIT diagrams in Fig. 6 for winter (December, January and February) and summer (June, July and August) (first and second row, respectively) at a lead time of 1. Kolmogorov confidence bands are also plotted for a graphical test of uniformity at the $\alpha = 0.05$ level. For the sake of brevity, the remaining seasons and lead times are not shown. For the particular seasons and lead time shown, statistical consistency seems to be achieved only for the wettest months (December–February). The explanation for this particular behavior will be given in Sect. 3.5. Early spring and November forecasts are also able to pass the uniformity test at a lead time of 1 month (not shown). Summer forecasts together with late spring and autumn months (May, September and October, not shown) show a significant underprediction, which prevents them from passing the uniformity test. Statistical consistency worsens as the lead time increases, in accordance with the deterioration of the bias in Fig. 5.

3.3 Streamflow forecasts forced with preprocessed meteorological forecasts

The second and third rows of Fig. 5 show the bias and skill of streamflow forecasts generated using preprocessed forcings from ECMWF System 4 using the LS and the QM method, respectively.

Several conclusions can be drawn when comparing forecasts using the preprocessed and raw forcings. First, biases are clearly improved, especially for longer lead times. For example, for October forecasts from a lead time of 3 to 7 months, biases are reduced from the $[-40, -30\,\%]$ to the $[-20, 10\,\%]$ range for LS and to the $[-15, 20\,\%]$ range for

QM. There are, however, no obvious differences between the two preprocessing methods, which seem to perform equally well in reducing biases. Secondly, three features of accuracy are seen. The first one is that, also for accuracy, there are no obvious differences in skill between the two preprocessing methods. Furthermore, there seems to be a reduction of skill for the winter months and March in the first month lead time. These months are the only ones with a statistically significant skill using the raw forecasts. This feature is a consequence of the reduction of the forcing biases, a situation that will be further discussed in Sect. 3.5. The last feature is that the improvement of the forcings can help to reduce the negative skill in streamflow forecasts. For example, April to November forecasts at longer lead times, generated using raw ECMWF System 4 forcings, exhibit a highly negative and statistically significant skill, sometimes lower than -1.0. Streamflow forecasts generated using preprocessed forcings for those months tend to have a neutral skill. This in turn implies that their accuracy is not different from the accuracy of ESP forecasts. The final conclusion is related to sharpness. As we can see in Fig. 5, streamflow forecasts generated using preprocessed forcings have an ensemble range that is wider than the reference ESP forecasts. This indicates that preprocessing the forcings also leads to a reduction of sharpness in comparison to forecasts generated using raw forcings (Sect. 3.2.).

The second and third columns in Fig. 6 show the PIT diagrams of streamflow forecasts generated using preprocessed forcings for the winter and summer forecasts in the first month lead time. The statistical consistency for the winter months is worse than the consistency of forecasts generated using raw forcings. The same degree of deteriora-

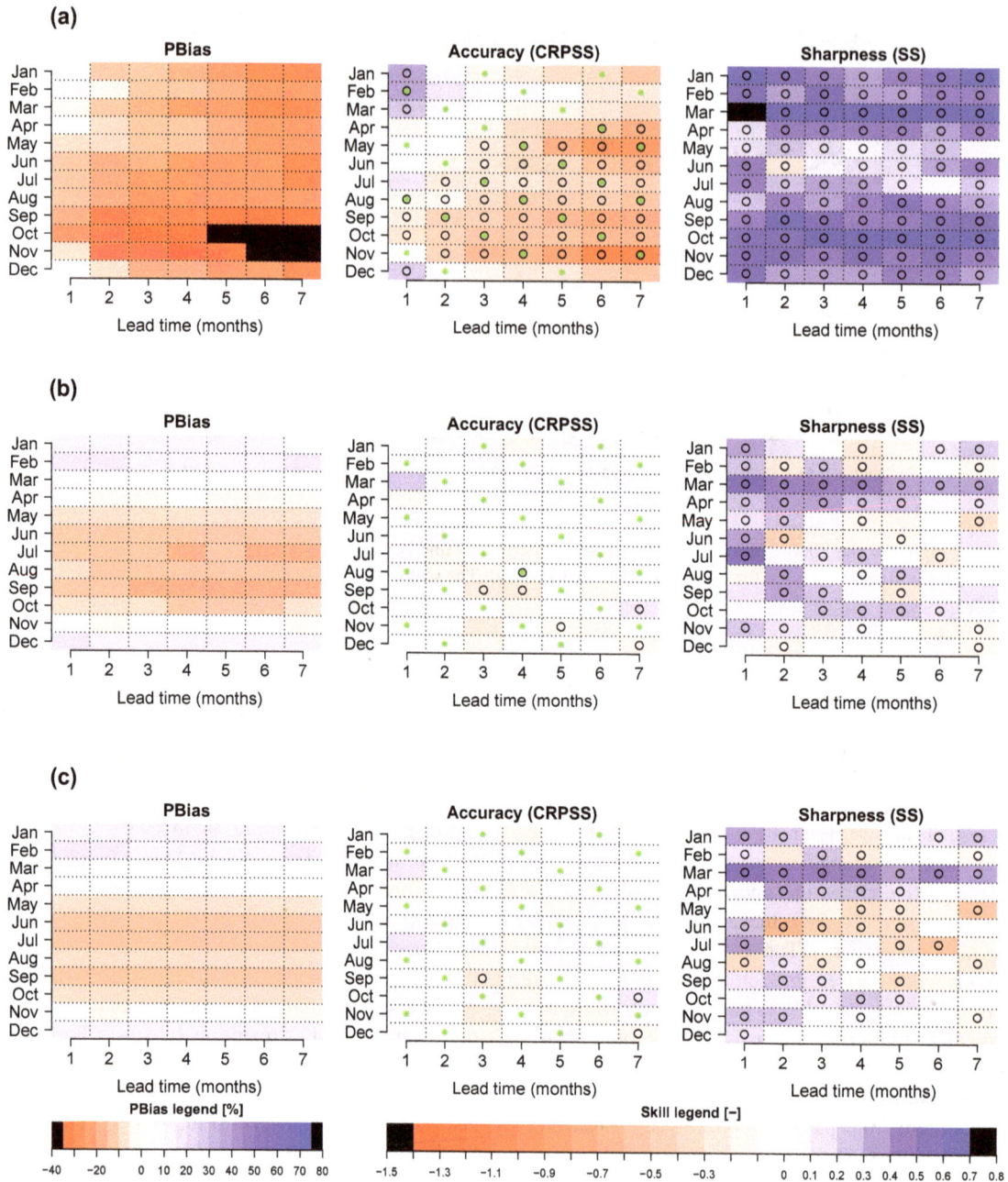

Figure 5. PBias and skill in terms of accuracy and sharpness of monthly means of daily streamflow of raw and preprocessed forecasts at station 82. Streamflow forecasts are generated using (**a**) raw meteorological forecasts and preprocessed meteorological forecasts with the (**b**) linear scaling/delta change (LS) and (**c**) quantile mapping (QM) methods. The *y* axis represents the target month, and the *x* axis represents the different lead times at which target months are forecasted. Values in the blue range show a positive bias/skill and values in red a negative bias/skill. Circles represent the cases where the distribution of the accuracy and/or sharpness for ESP differs from that of the ECMWF System 4-generated forecasts at a 5 % significance level using the WMW test. Green crosses represent the months/lead times for which the ensemble size is 51.

tion is seen for both preprocessing methods. This is caused by compensational errors that will be further discussed in Sect. 3.5. Besides that particular season, improvements in consistency after preprocessing can be seen during the autumn (not shown) and August, although to a lesser degree.

For spring (not shown) and early summer forecasts, the same level of consistency is observed for both the raw and preprocessed forecasts. At longer lead times, the benefit of preprocessing for statistical consistency is clearer, with most of the months passing the uniformity test (not shown).

Seasonal streamflow forecasts in the Ahlergaarde catchment, Denmark: the effect of preprocessing...

25

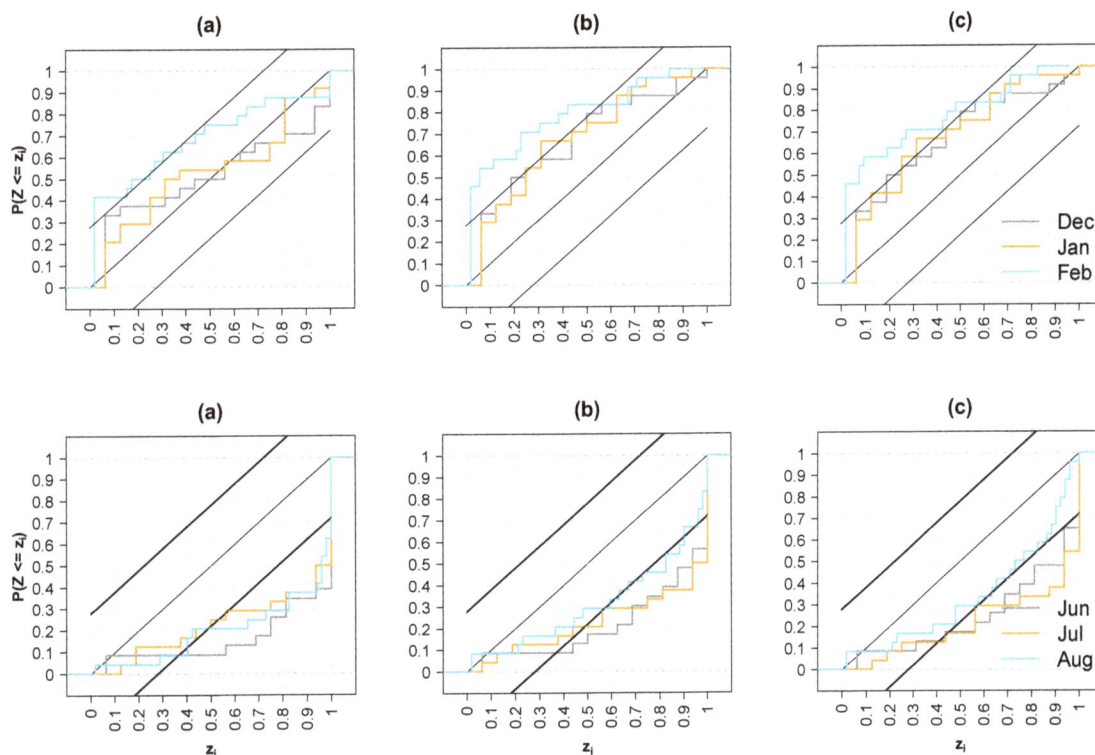

Figure 6. PIT diagrams of monthly means of daily streamflow forecasts for winter (upper row) and summer (bottom row) of station 82 for **(a)** raw meteorological forecasts and preprocessed meteorological forecasts with **(b)** linear scaling/delta change (LS) and **(c)** quantile mapping (QM). The lead time is 1 month. Different colors represent different months in the season. The black lines parallel to the 1 : 1 diagonal are the Kolmogorov bands at the 5 % significance level.

3.4 Post-processed streamflow forecasts

The final step in the analysis is the post-processing of streamflow forecasts generated using raw and preprocessed ECMWF System 4 forcings. Figure 7 shows the verification results that can be directly compared to the results in Fig. 5.

The first column in Fig. 7 shows a clear reduction of the absolute bias compared to the raw and preprocessed generated forecasts. Bias lies within the range [−10, 10 %], for all months and lead times. Furthermore, the majority of the CRPSS values for all months and lead times are positive, while a small negative skill is seen during the autumn. Note, however, that the differences in accuracy between ESP and the post-processed forecasts are only significant at the 5 % level for few target months and lead times. In general, there seems to be a worsening of the sharpness after post-processing (Fig. 5). However, this deterioration is lower when comparing preprocessed versus post-processed forecasts. Furthermore, the degree of the deterioration varies according to the target month. For example, summer months (June and July) exhibit a larger deterioration of sharpness; i.e., the forecast spread is larger than that of the ESP. On the other hand, forecasts for late autumn and early December appear to be narrower than ESP forecasts after post-processing.

Figure 8 shows the PIT diagrams for the months of the summer and winter seasons in the first month lead time of post-processed streamflow forecasts. The plot can be directly compared to Fig. 6. As seen from the PIT diagram, all months in those seasons pass the uniformity test, indicating that after post-processing, the observations can be considered as random samples of the predictive distribution. The remaining PIT diagrams for spring and autumn and lead times of 2–7 months (not shown in Fig. 8) show that statistical consistency is present for all months and lead times. At longer lead times, the CDFs of the $z_i's$ are closer to the 1 : 1 diagonal. This is achieved due to two factors: (i) the additional reduction of bias after post-processing and (ii) the worsening of sharpness for long lead times, when the larger ensemble spread encloses a larger portion of observed values.

3.5 Effect of hydrological model bias in skill evaluations

As mentioned in Sect. 3.1., hydrological model biases, which are larger for the upstream station 21 (Fig. 4), combined with structural biases in GCMs, can lead to a situation with a high skill resulting from compensational errors providing "the right forecast for the wrong reasons". In order to illustrate this point, Fig. 9a and b show the CRPSS for, respectively,

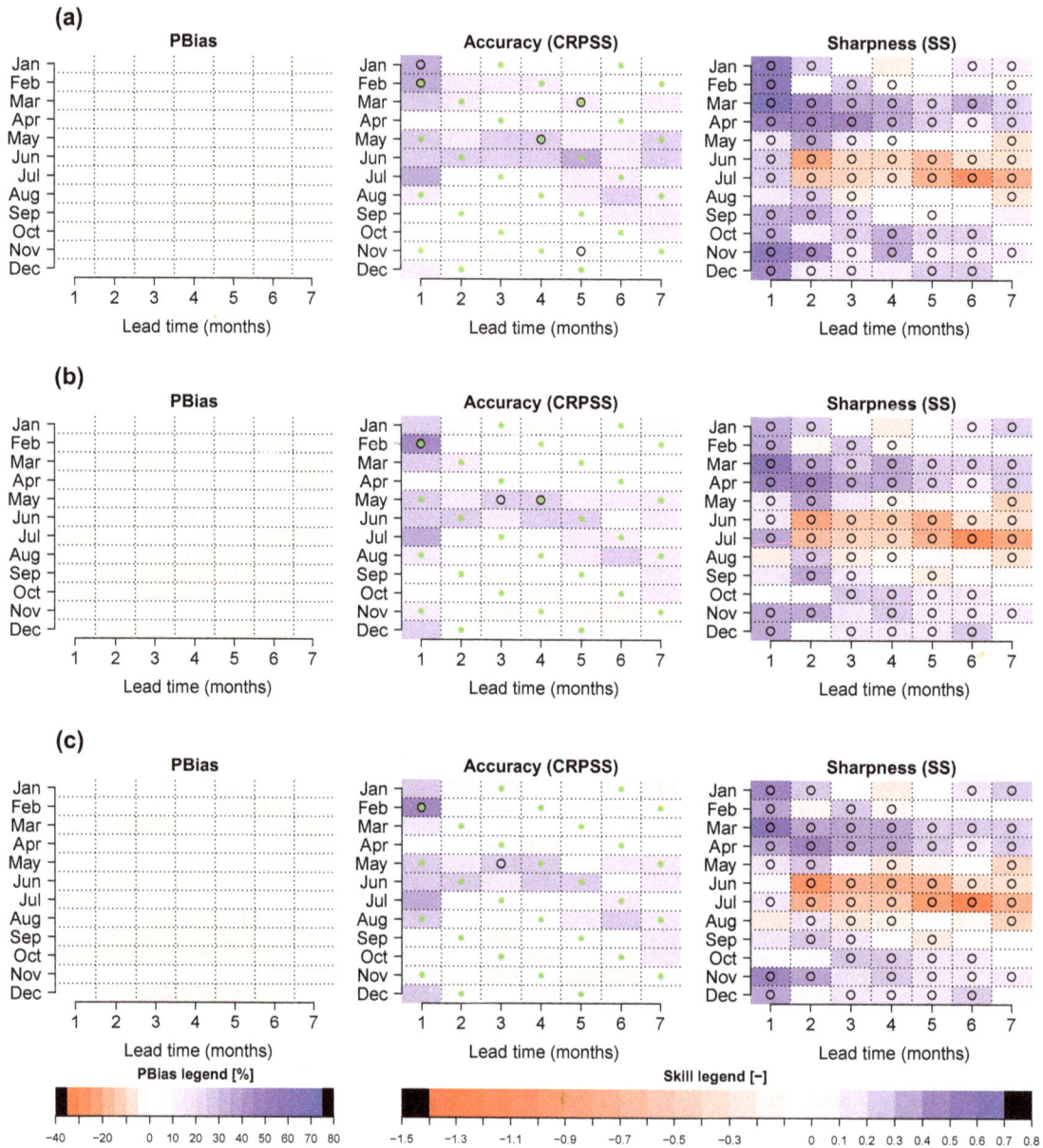

Figure 7. PBias and skill (sharpness and accuracy) of daily monthly mean streamflow forecasts for post-processed forecasts using the quantile mapping (QM) method for predictions generated using raw (**a**) and preprocessed meteorological forcings with the linear scaling/delta change (**b**) and quantile mapping (**c**) methods. Legend is the same as Fig. 5. Green crosses represent the months/lead times were the ensemble size is 51.

station 21 with a large bias (PBias = 48 %, Fig. 4) and station 82 with a small bias (PBias = 1.7 %, Fig. 4). The figure shows CRPSS for forecasts generated using raw ECMWF forcings and preprocessed forcings with the LS method for the target months January–December at a lead time of 4 (e.g., January forecasts initiated in October). In addition to the computation of bias and accuracy of ECMWF-based streamflow forecasts and ESP forecasts using observed streamflow, we also include a computation of bias and accuracy against simulated streamflows (continuous run of the Ahlergaarde

model with observed meteorological forcings, Fig. 4). This is done in order to remove the effect of hydrological model bias and hence focus the analyses on the biases coming from forcings alone.

The high skill against observed streamflows is more visible during the wettest months (November–April) for station 21 where hydrological model biases are highest (Fig. 4). Once the comparison is made against simulated streamflows, the high positive skill becomes highly negative (Fig. 9a). The deterioration of skill when compared against simulated

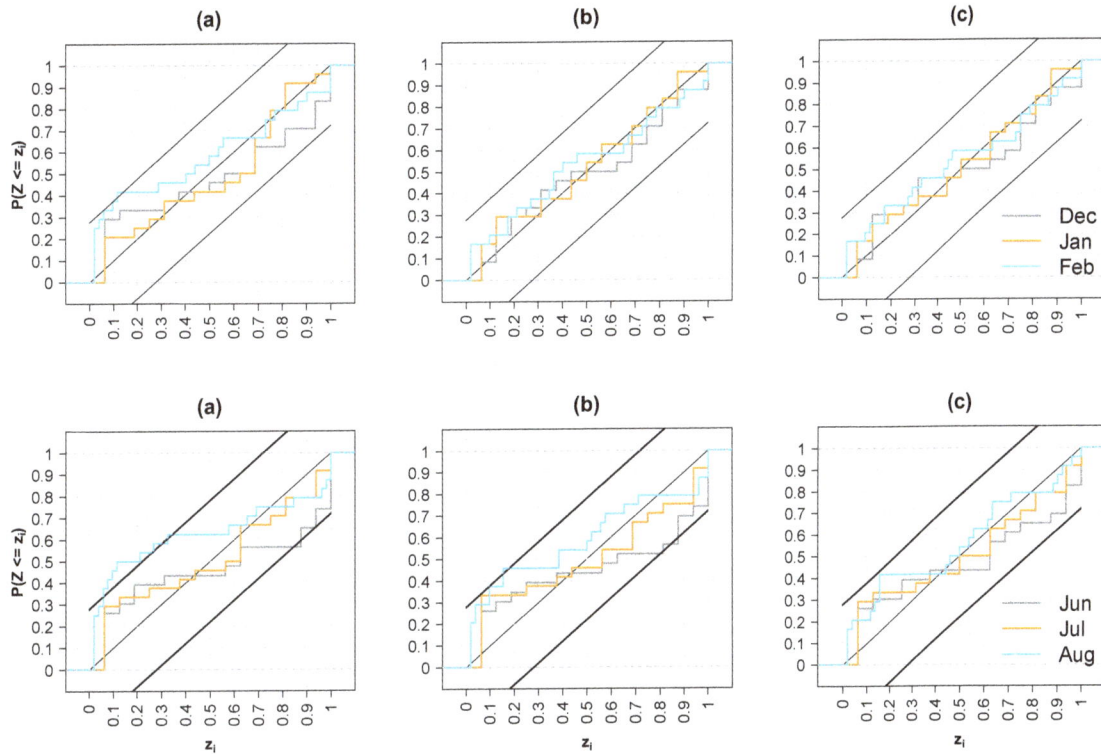

Figure 8. PIT diagrams of daily monthly mean streamflow post-processed forecasts for summer (upper row) and winter (bottom row) of station 82. Streamflow forecasts are post-processed using the quantile mapping (QM) method for predictions generated using raw (**a**) and preprocessed forcings with the linear scaling/delta change (**b**) and quantile mapping (**c**) methods. Lead time 1 month. The black lines parallel to the 1 : 1 diagonal are the Kolmogorov bands at the 5 % significance level.

streamflows is also seen at station 82 for December–March, although to a lesser extent (Fig. 9b). To illustrate why this happens, Fig. 9c and d show the monthly streamflow forecasts for all 24 years for the target month of December of forecasts initialized in September (lead time 4). Both ESP and raw (Fig. 9c) and preprocessed (Fig. 9d) forecasts are shown, along with their respective skill scores of accuracy, when the comparison is made against observed (CRPSS) and simulated (CRPSS.s) values.

Figure 9c shows two issues. First, the large hydrological model bias causes ESP to have a deviation from the observations, leading to a high CRPS for the reference forecast in Eq. (4). Secondly, for the winter months, precipitation from the raw ECMWF System 4 forecasts exhibits a negative bias of around -25 % (Lucatero et al., 2017). This compensates the biased streamflow forecasts and results in a low $CRPS_{Sys}$ value in Eq. (4). The CRPSS then becomes positive and large (0.54). However, when the comparison is done against simulated values, the skill score becomes highly negative (CRPSS.s $= -0.41$). Once the biases in the forcings are removed (Fig. 9d), then the hydrological model bias takes over, leading these forecasts to the same level as the ESP, increasing its CRPS, which in turn reduces the skill score (CRPSS $= -0.04$).

Note that the opposite situation arises, i.e., "the wrong forecast for the wrong reason", when the hydrological model error is small and precipitation forecast bias is large. Biases in precipitation forecasts will propagate through streamflow forecasts, leading to a streamflow bias of equal sign and of similar magnitude as the precipitation bias. The streamflow bias is then reduced when the meteorological forecast bias is removed (Fig. 5, second and third row). This situation appears during summer or autumn (Fig. 5, first row), when hydrological model errors are smaller than in winter.

The apparent skill trend along target months of raw GCM-based streamflow forecasts (Fig. 5, first row) is a product of the above explained error interactions, rather than the existence (or lack) of predictability during the given months. Further analysis linking concurrent and/or previous hydrometeorological processes (i.e., accumulation of snowpack) to streamflow forecast skill would require additional research as discussed later. Moreover, the preprocessed meteorological forecasts' bias is invariant along lead time, and in terms of forecast accuracy, only mild improvements are found over ensemble climatology during the first month lead time (Lucatero et al., 2017). This situation, together with the reduction of error interactions negatively affecting streamflow forecasts at longer leads, produces the flattening of the trend in skill along lead time (Fig. 5, second and third row).

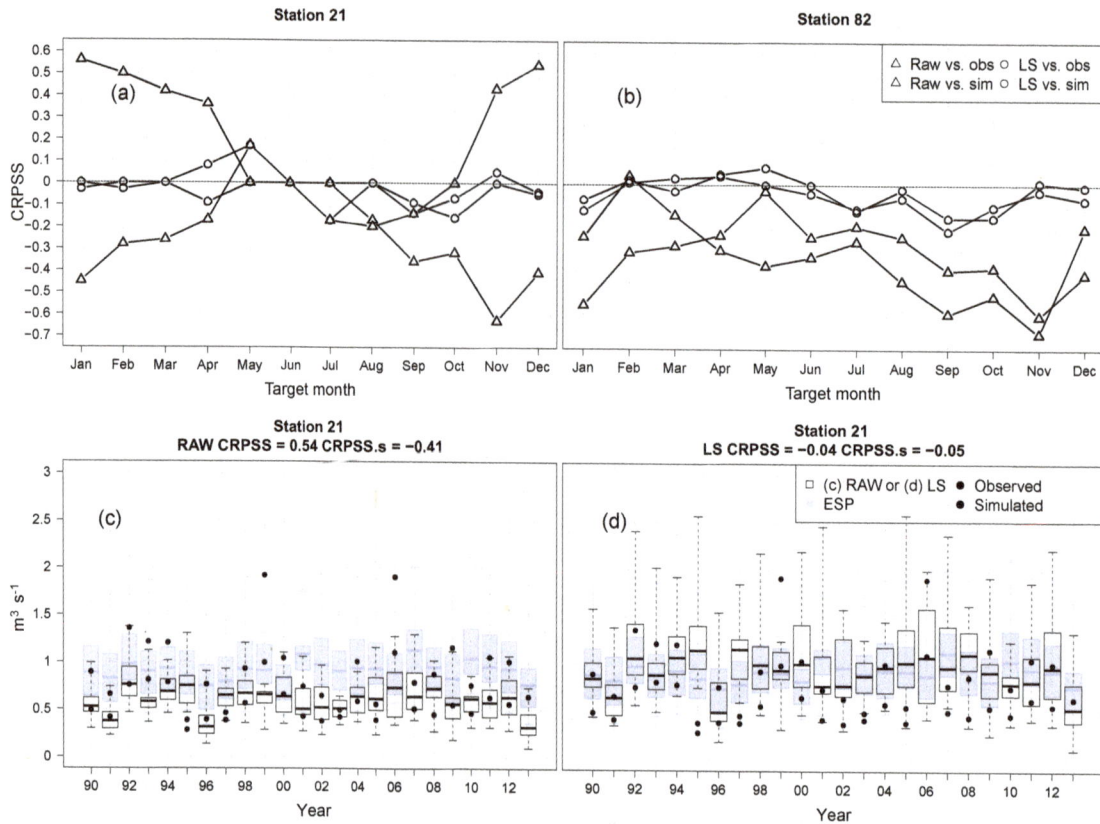

Figure 9. (a, b) Skill of accuracy (CRPSS) for upstream station 21 and outlet station 82 for target months January–December at a lead time of 4. Triangles and circles represent the forecasts generated using raw ECMWF System 4 forcings and preprocessed with LS, respectively, whereas black and blue lines represent the comparison against observed and simulated streamflow, respectively. The second row shows the monthly forecasts of December streamflow initialized in September (4-month lead time) for predictions using raw **(c)** and preprocessed **(d)** forcings for all years in the record (1990–2013) for station 21.

Stations like 21 could benefit the most from post-processing, removing hydrological model biases that calibration alone could not remove. This is illustrated with the visualization of the CRPSS of the different forecasts in Fig. 10a–d. The comparison is made against observations. Figure 10b shows a reduction of skill after raw forcings have been preprocessed, as a result of the compensation errors discussed above. However, once the hydrological biases are removed with post-processing (Fig. 10c and d), the skill is positive and significant throughout November to April. Note, however, that the high skill at this particular station is mainly driven by the poor performance of the reference ESP, due to the large bias of the hydrological model (Fig. 4). It is also worth noting the lack of differences in skill between Fig. 10c and d, showing that, for this particular location, a combination of preprocessing plus post-processing is just as good as post-processing of the forecasts generated using raw forcings alone.

3.6 Forecasts of the minimum daily flow within a year

In addition to the evaluation of the monthly streamflow forecasts, we have assessed whether the use of GCM forecasts can add value to the forecasting of annual minimum daily flows compared to ESP. Figure 11 shows forecasts of the minimum daily flow in each year of the study period, considering forecasts issued on 1 April for the next 7 months. Forecasts are for the outlet station 82 for both raw forecasts and the different preprocessing and post-processing strategies. Black box plots represent the forecast generated using the raw outputs of the ECMWF System 4 (Fig. 11a), the preprocessed forecasts (Fig. 11c and e) and the post-processed forecasts (Fig. 11b, d, f). The box plots in the background (blue) represent the ESP forecasts and the red dots represent the yearly observed minimum discharges. When we look at Fig. 10a, several features can be highlighted. First, despite the underprediction of the raw generated forecasts and, to a lesser extent, the ESP forecasts of the highest minimum daily flows in the 2000s, the year-to-year variability is replicated well. Secondly, even though the raw generated forecasts are sharper than the ESP by about 10 % (SS = 0.11), they do not

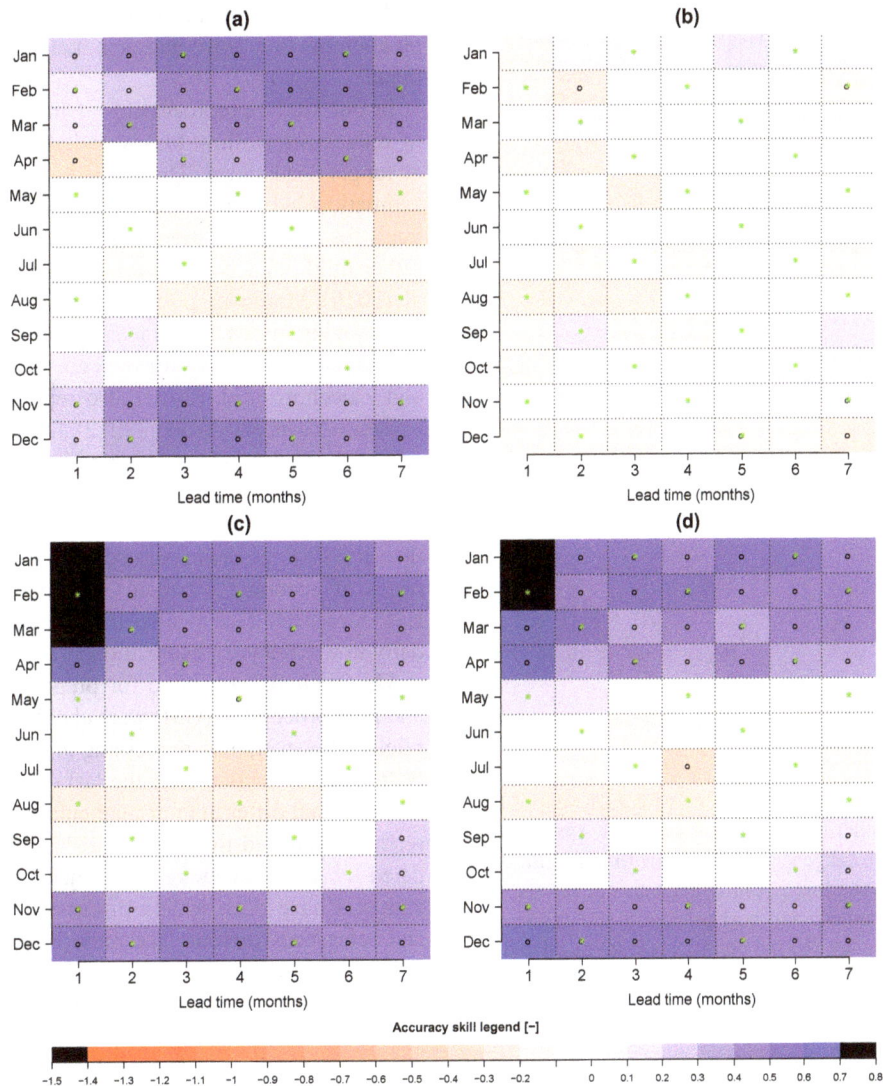

Figure 10. CRPSS of station 21 for forecasts generated using raw (**a**) and preprocessed (**b**) forcings, in addition to the post-processed streamflow forecasts using the quantile mapping method (QM) for raw meteorological (**c**) and preprocessed meteorological forcings using the linear scaling/delta change method (**d**).

manage to perform better than ESP in terms of skill of accuracy (CRPSS $= -0.14$); i.e., they are overconfident.

Preprocessing meteorological forecasts seems to have a positive effect on minimum daily flow forecasting, reducing the CRPSS from -0.14 to -0.01 when using the LS preprocessor. This happens because of the loss of sharpness (from 0.11 to -0.11), which allows the forecasts to better capture the higher minimum daily flows during the 00s. However, it is still difficult to outperform the ESP. Post-processing seems to have a similar effect: a loss of sharpness and decrease in bias that allow the forecasts to capture the high minimum daily flows in the 2000s and 2010s. This situation, however, leads to a loss in skill in forecasting minimum daily flows in the 1990s, leveling out the skill to a similar score (CRPSS $= -0.12$) as the forecasts generated using

raw ECMWF forcings (CRPSS $= -0.14$). Thus, it seems that an attempt to reduce meteorological and hydrological biases through processing the forcings and/or the streamflow will result in only a modest increase in skill of minimum daily flow predictions on average. ESP remains a reference forecast system difficult to outperform.

4　Discussion

Monthly streamflow forecasts derived for raw, preprocessed meteorology and post-processed streamflow in general show limited skill beyond a 1-month lead time for the Ahlergaarde catchment in Denmark. This is not a surprising result, given the limited skill of meteorological forecasts in the region (Lucatero et al., 2017). Similar results have been documented

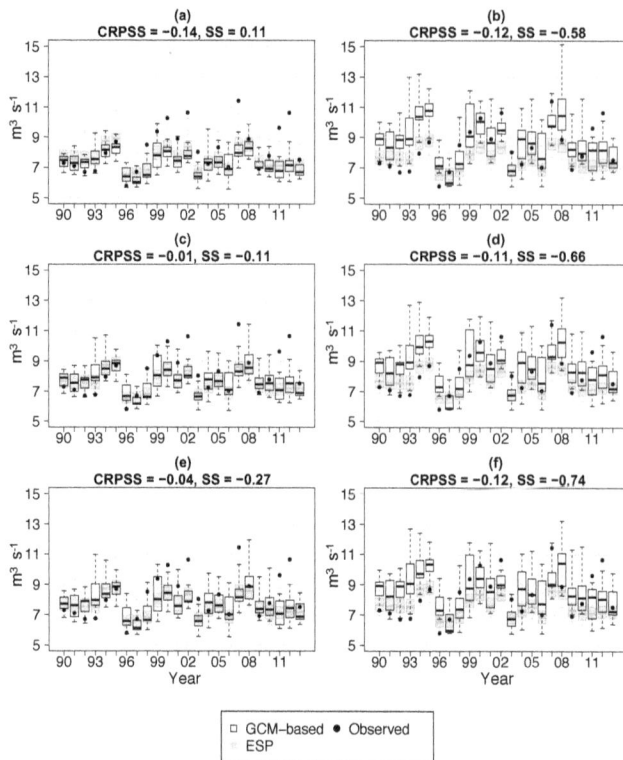

Figure 11. Forecasts of minimum daily flows for each year of the period 1990–2013, considering a forecast issued on 1 April for the next 7 months. Forecasts are generated using raw forcings (**a**), pre-processed forcings with the linear scaling/delta change (**c**) and the quantile mapping (**e**) methods and post-processed streamflow for forecasts generated using raw (**b**) and preprocessed inputs (**d** and **f**). Blue shaded box plots are ESP forecasts. CRPSS and SS are computed using Eq. (4) with ESP as reference.

in Wood et al. (2005), Yuan et al. (2013) and, more recently, Crochemore et al. (2016) for France. GCM-based streamflow forecasts could then be of potential use if the end user is interested in gaining accuracy of forecasts for the next month only. Moreover, we were able to demonstrate that, at least for a groundwater-dominated catchment located in a region with temperate climate, the GCM ability to improve forecasts of minimum daily flows within a year is also limited, regardless of any attempts to correct forcings and/or streamflow forecasts. Further research could focus on the usefulness of GCM forecasts for drought forecasting, i.e., magnitude, duration and severity (Fundel, 2013) in comparison to forecasts generated using the ESP method.

Furthermore, caution must be taken when hydrological model errors are large, as it may lead to erroneous evaluations of skill when hydrological model biases are neutralized by opposite GCM errors, e.g., forecasts of monthly streamflow during the winter in the study region. This is an issue somewhat underexplored in studies of forecast skill and should be evaluated especially when calibration objective functions focus on attributes that differ from the ones

looked for in the final forecast quantity of interest, and when no attempts to remove biases in meteorological forecasts are made.

In our study, preprocessing of the forcings alone helped to reduce streamflow biases and reduce the negative skill at longer lead times. The reduction of the under- or overestimation led to forecasts with a higher statistical consistency for most of the months and lead times considered. This rather mild enhancement was also found by Crochemore et al. (2016). Moreover, post-processing alone does a better job in removing biases in the mean, which, in turn, helps to ameliorate issues with the statistical consistency. Ye et al. (2015) and Zalachori et al. (2012) also report the above behavior, whereas Yuan and Wood (2012) found a better correction of statistical consistency after both preprocessing and postprocessing. The removal of biases of both forcings and hydrological model did not ensure a higher level of accuracy than the ESP, as demonstrated by the nonsignificant differences of accuracy between GCM-based forecasts and the ESP forecasts. This is also true for forecasts of minimum daily flows in a year, as mentioned above.

The methods used here for preprocessing (LS and QM) and post-processing (QM) were chosen because of their simplicity. However, post-processing in general is a field that has been gaining traction over the last decade, with a variety of methods that differ in their mathematical sophistication. The reader is referred to Li et al. (2017) for a detailed and updated literature review on the subject. Moreover, QM disadvantages have been widely discussed in Zhao et al. (2017) and references therein. The main issue discussed concerns the fact that when the forecast–observation linear relationship is weak, or nonexistent, QM has difficulties creating forecasts that are consistent (i.e., that have skill at least as good as the reference forecast). Other methods could have been used that allow for correction of both statistical consistency together with consistency. However, the benefits of the more sophisticated methods might be dampened due to the limited sample size, which is often the case in hydrometeorological forecasting. Nevertheless, our present study could be extended by analyzing the added skill gained by the increased complexity of processing methods, using the same reforecast dataset, such as the case of Mendoza et al. (2017), although with its application focused on statistical forecasting.

Another obvious omission of the study is the exploitation of storages in the form of snow, soil moisture or/and groundwater and taking advantage of the hydrological memory that may increase skill at longer leads. This has been the routine for snow-dominated catchments in the western United States by means of ESP (Wood and Lettenmaier, 2006). However, a preliminary evaluation of the relationship between groundwater levels in winter and minimum daily flows during the summer in the Ahlergaarde catchment studied here showed that relatively high correlations exist in large parts of the catchment (Jacob Kidmose, personal communication, 2014).

This correlation can be further explored in the forecasting mode to extend the positive skill lead time by means of data assimilation (Zhang et al., 2016) or by statistical post-processing of streamflow forecasts (Mendoza et al., 2017). Moreover, predictability attribution studies exist that quantify the sensitivity of the skill of a forecasting system relative to different degrees of uncertainty, either in the forcing or the initial conditions. Wood et al. (2016) developed a framework to detect where to concentrate on improvements, e.g., either the initial conditions, usually by means of data assimilation (Zhang et al., 2016), or the seasonal climate forecasts. This might shed light on, and possibly reinforce, the hypothesis that for groundwater-dominated catchments and forecasting of low flows, initial conditions will have a higher influence on forecast skill at longer lead times (Paiva et al., 2012; Fundel et al., 2013).

5   Conclusions

Seasonal forecasts of streamflows initiated in each calendar month for the 1990–2013 period were generated for a groundwater-dominated catchment located in a region where seasonal atmospheric forecasting is a challenge. We analyzed the bias and statistical consistency of monthly streamflow forecasts forced with ECMWF System 4 seasonal forecasts along all calendar months throughout the year. In addition to this, we evaluated their accuracy and sharpness relative to that of the forecasts generated using an ensemble of historical meteorological observations, the ESP. Monthly streamflow forecasts generated using raw ECMWF System 4 forcings show skill only during the winter months in the first month lead time. Nevertheless, it was shown that the apparent large skill can be an effect of compensational errors between meteorological forecasts and the hydrological model. Due to biases of GCM-based meteorological seasonal forecasts and errors in the hydrological model that calibration alone cannot defuse, both preprocessing and post-processing using two popular and simple correction techniques were used to remove them: LS and QM. Finally, we also estimated the skill that the different forecast generation approaches have on forecasting the minimum yearly daily discharge. Our results show that post-processing streamflow allows for the most gain in skill and statistical consistency. However, monthly streamflow and annual minimum daily discharge forecasts generated using forcings from GCM still show difficulties in outperforming ESP forecasts, especially at lead times longer than 1 month.

Data availability. ECMWF seasonal reforecasts are available under a range of licences; for more information visit http://www.ecmwf.int (last access: 27 June 2018). The hydrological model forcing data (temperature, precipitation and reference evapotranspiration) are from the Danish Meteorological Institute (https://www.dmi.dk/vejr/arkiver/vejrarkiv/, last access: 27 June 2018). The streamflow data are available on the HOBE data platform (http://www.hobe.dk/index.php/data/live-data, last access: 27 June 2018). A more detailed description of the data usage can be found on the Hydrocast project website (http://hydrocast.dhigroup.com/, last access: 27 June 2018) and the HOBE project website (http://hobe.dk/, last access: 27 June 2018).

Competing interests. The authors declare that they have no conflict of interest.

Special issue statement. This article is part of the special issue "Sub-seasonal to seasonal hydrological forecasting". It is not associated with a conference.

Acknowledgements. This study was supported by the project "HydroCast – Hydrological Forecasting and Data Assimilation", contract no. 0603-00466B (http://hydrocast.dhigroup.com/, 27 June 2018), funded by the Innovation Fund Denmark. Special thanks to Florian Pappenberger for providing the ECMWF System 4 reforecast and Andy Wood and Pablo Mendoza for hosting the first author at NCAR. We also express our gratitude to Massimiliano Zappa and one anonymous reviewer for their comments that improved the quality of this paper.

Edited by: Maria-Helena Ramos

References

Bogner, K., Liechti, K., and Zappa, M.: Post-processing of stream flows in Switzerland with an emphasis on low flows and floods, Water (Switzerland), 8, 115, https://doi.org/10.3390/w8040115, 2016.

Bruno Soares, M. and Dessai, S.: Barriers and enablers to the use of seasonal climate forecasts amongst organisations in Europe, Clim. Change, 137, 89–103, https://doi.org/10.1007/s10584-016-1671-8, 2016.

Córdoba-Machado, S., Palomino-Lemus, R., Gámiz-Fortis, S. R., Castro-Díez, Y., and Esteban-Parra, M. J.: Seasonal streamflow prediction in Colombia using atmospheric and oceanic patterns, J. Hydrol., 538, 1–12, https://doi.org/10.1016/j.jhydrol.2016.04.003, 2016.

Crochemore, L., Ramos, M.-H., and Pappenberger, F.: Bias correcting precipitation forecasts to improve the skill of seasonal streamflow forecasts, Hydrol. Earth Syst. Sci., 20, 3601–3618, https://doi.org/10.5194/hess-20-3601-2016, 2016.

D'Agostino, R. B. and Stephens, A. M.: Goodness-of-fit techniques, Dekker, New York, 1986.

Day, G. N.: Extended stream flow forecasting Using NWSRFS, J. Water Res. Pl., 111, 157–170, 1985.

Demirel, M. C., Booij, M. J., and Hoekstra, A. Y.: The skill of seasonal ensemble low-flow forecasts in the Moselle River for three different hydrological models, Hydrol. Earth Syst. Sci., 19, 275–291, https://doi.org/10.5194/hess-19-275-2015, 2015.

Doherty, J.: PEST, Model-independent parameter estimation, User manual: 5th Edn., Watermark Numerical Computing, 2010.

Friederichs, P. and Thorarinsdottir, T. L.: Forecast verification for extreme value distributions with an application to probabilistic peak wind prediction, Environmetrics, 23, 579–594, https://doi.org/10.1002/env.2176, 2012.

Fundel, F., Jörg-Hess, S., and Zappa, M.: Monthly hydrometeorological ensemble prediction of streamflow droughts and corresponding drought indices, Hydrol. Earth Syst. Sci., 17, 395–407, https://doi.org/10.5194/hess-17-395-2013, 2013.

Graham, D. N. and Butts, M. B.: Flexible, integrated watershed modelling with MIKE SHE, in: Watershed Models, edited by: Singh, V. P. and Frevert D. K., 245–272, CRC Press, Florida, 2005.

Hendriks, M.: Introduction to Physical Hydrology, Oxford University Press, Oxford, 2010.

Hersbach, H.: Decomposition of the Continuous Ranked Probability Score for Ensemble Prediction Systems, Weather Forecast., 15, 559–570, https://doi.org/10.1175/1520-0434(2000)015<0559:DOTCRP>2.0.CO;2, 2000.

Hollander, M., Wolfe, D. A., and Chicken, E.: Nonparametric statistical methods, 3 Edn., Wiley Series in Probability and Statistics, Hoboken, New Jersey, 2014.

Jaun, S., Ahrens, B., Walser, A., Ewen, T., and Schär, C.: A probabilistic view on the August 2005 floods in the upper Rhine catchment, Nat. Hazards Earth Syst. Sci., 8, 281–291, https://doi.org/10.5194/nhess-8-281-2008, 2008.

Jensen, K. H. and Illangasekare, T. H.: HOBE: A Hydrological Observatory, Vadose Zone J., 10, 1–7, https://doi.org/10.2136/vzj2011.0006, 2011.

Laio, F. and Tamea, S.: Verification tools for probabilistic forecasts of continuous hydrological variables, Hydrol. Earth Syst. Sci., 11, 1267–1277, https://doi.org/10.5194/hess-11-1267-2007, 2007.

Li, W., Duan, Q., Miao, C., Ye, A., Gong, W., and Di, Z.: A review on statistical postprocessing methods for hydrometeorological ensemble forecasting, WIREs Water, 4, e1246, https://doi.org/10.1002/wat2.1246, 2017.

Lucatero, D., Madsen, H., Refsgaard, J. C., Kidmose, J., and Jensen, K. H.: On the skill of raw and postprocessed ensemble seasonal meteorological forecasts in Denmark, Hydrol. Earth Syst. Sci. Discuss., https://doi.org/10.5194/hess-2017-366, in review, 2017.

Madadgar, S., Moradkhani, H., and Garen, D.: Towards improved post-processing of hydrologic forecast ensembles, Hydrol. Process., 28, 104–122, https://doi.org/10.1002/hyp.9562, 2014.

Mendoza, P. A., Wood, A. W., Clark, E., Rothwell, E., Clark, M. P., Nijssen, B., Brekke, L. D., and Arnold, J. R.: An intercomparison of approaches for improving operational seasonal streamflow forecasts, Hydrol. Earth Syst. Sci., 21, 3915–3935, https://doi.org/10.5194/hess-21-3915-2017, 2017.

Molteni, F., Stockdale, T., Balmaseda, M., Balsamo, G., Buizza, R., Ferranti, L., Magnusson, L., Mogensen, K., Palmer, T., and Vitart, F.: The new ECMWF seasonal forecast system (System 4), ECMWF Technical Memorandum 656, November, 49, 2011.

Olsson, J., Uvo, C. B., Foster, K., and Yang, W.: Technical Note: Initial assessment of a multi-method approach to spring-flood forecasting in Sweden, Hydrol. Earth Syst. Sci., 20, 659–667, https://doi.org/10.5194/hess-20-659-2016, 2016.

Paiva, R. C. D., Collischonn, W., Bonnet, M. P., and de Gonçalves, L. G. G.: On the sources of hydrological prediction uncertainty in the Amazon, Hydrol. Earth Syst. Sci., 16, 3127–3137, https://doi.org/10.5194/hess-16-3127-2012, 2012.

Robertson, D. E. and Wang, Q. J.: A Bayesian Approach to Predictor Selection for Seasonal Streamflow Forecasting, J. Hydrometeorol., 13, 155–171, https://doi.org/10.1175/JHM-D-10-05009.1, 2012.

Robertson, D. E., Pokhrel, P., and Wang, Q. J.: Improving statistical forecasts of seasonal streamflows using hydrological model output, Hydrol. Earth Syst. Sci., 17, 579–593, https://doi.org/10.5194/hess-17-579-2013, 2013.

Rosenberg, E. A., Wood, A. W., and Steinemann, A. C.: Statistical applications of physically based hydrologic models to seasonal streamflow forecasts, Water Resour. Res., 47, 3, https://doi.org/10.1029/2010WR010101, 2011.

Roulin, E. and Vannitsem, S.: Post-processing of medium-range probabilistic hydrological forecasting: Impact of forcing, initial conditions and model errors, Hydrol. Process., 29, 1434–1449, https://doi.org/10.1002/hyp.10259, 2015.

Scharling, M. and Kern-Hansen, C.: Climate Grid Denmark – Dataset for use in research and education, DMI Tech. Rep., 1–12, available at: https://www.dmi.dk/vejr/arkiver/vejrarkiv/ (last access: 27 June 2018), 2012.

Schepen, A., Zhao, T., Wang, Q. J., Zhou, S., and Feikema, P.: Optimising seasonal streamflow forecast lead time for operational decision making in Australia, Hydrol. Earth Syst. Sci., 20, 4117–4128, https://doi.org/10.5194/hess-20-4117-2016, 2016.

Shamir, E.: The value and skill of seasonal forecasts for water resources management in the Upper Santa Cruz River basin, southern Arizona, J. Arid Environ., 137, 35–45, https://doi.org/10.1016/j.jaridenv.2016.10.011, 2017.

Shi, X., Wood, A. W., and Lettenmaier, D. P.: How Essential is Hydrologic Model Calibration to Seasonal Streamflow Forecasting?, J. Hydrometeorol., 9, 1350–1363, https://doi.org/10.1175/2008JHM1001.1, 2008.

Stisen, S., Sonnenborg, T. O., Højberg, A. L., Troldborg, L., and Refsgaard, J. C.: Evaluation of Climate Input Biases and Water Balance Issues Using a Coupled Surface-Subsurface Model, Vadose Zone J., 10, 37–53, https://doi.org/10.2136/vzj2010.0001, 2011.

Stisen, S., Højberg, A. L., Troldborg, L., Refsgaard, J. C., Christensen, B. S. B., Olsen, M., and Henriksen, H. J.: On the importance of appropriate precipitation gauge catch correction for hydrological modelling at mid to high latitudes, Hydrol. Earth Syst. Sci., 16, 4157–4176, https://doi.org/10.5194/hess-16-4157-2012, 2012.

Trambauer, P., Werner, M., Winsemius, H. C., Maskey, S., Dutra, E., and Uhlenbrook, S.: Hydrological drought forecasting and skill assessment for the Limpopo River basin, southern Africa, Hydrol. Earth Syst. Sci., 19, 1695–1711, https://doi.org/10.5194/hess-19-1695-2015, 2015.

Verkade, J. S., Brown, J. D., Reggiani, P., and Weerts, A. H.: Post-processing ECMWF precipitation and temperature ensemble reforecasts for operational hydrologic forecasting at various spatial scales, J. Hydrol., 501, 73–91, https://doi.org/10.1016/j.jhydrol.2013.07.039, 2013.

Wang, Q. J., Robertson, D. E., and Chiew, F. H. S.: A Bayesian joint probability modeling approach for seasonal forecasting of streamflows at multiple sites, Water Resour. Res., 45, 1–18, https://doi.org/10.1029/2008WR007355, 2009.

Weisheimer, A. and Palmer, T. N.: On the reliability of seasonal climate forecasts, J. R. Soc. Interface, 11, 20131162, https://doi.org/10.1098/rsif.2013.1162, 2014.

Wood, A. W. and Lettenmaier, D. P.: A test bed for new seasonal hydrologic forecasting approaches in the western United States, B. Am. Meteorol. Soc., 87, 1699–1712, https://doi.org/10.1175/BAMS-87-12-1699, 2006.

Wood, A. W. and Schaake, J. C.: Correcting Errors in Streamflow Forecast Ensemble Mean and Spread, J. Hydrometeorol., 9, 132–148, https://doi.org/10.1175/2007JHM862.1, 2008.

Wood, A. W., Maurer, E. P., Kumar, A., and Lettenmaier, D. P.: Long-range experimental hydrologic forecasting for the eastern United States, J. Geophys. Res.-Atmos., 107, 1–15, https://doi.org/10.1029/2001JD000659, 2002.

Wood, A. W., Kumar, A., and Lettenmaier, D. P.: A retrospective assessment of National Centers for Environmental prediction climate model-based ensemble hydrologic forecasting in the western United States, J. Geophys. Res.-Atmos., 110, 1–16, https://doi.org/10.1029/2004JD004508, 2005.

Wood, A. W., Hopson, T., Newman, A., Brekke, L., Arnold, J., and Clark, M.: Quantifying Streamflow Forecast Skill Elasticity to Initial Condition and Climate Prediction Skill, J. Hydrometeorol., 17, 651–668, https://doi.org/10.1175/JHM-D-14-0213.1, 2016.

Yapo, P. O., Gupta, H. V., and Sorooshian, S.: Automatic calibration of conceptual rainfall-runoff models: sensitivity to calibration data, J. Hydrol., 181, 23–48, https://doi.org/10.1016/0022-1694(95)02918-4, 1996.

Ye, A., Duan, Q., Schaake, J., Xu, J., Deng, X., Di, Z., Miao, C., and Gong, W.: Post-processing of ensemble forecasts in low-flow period, Hydrol. Process., 29, 2438–2453, https://doi.org/10.1002/hyp.10374, 2015.

Yuan, X.: An experimental seasonal hydrological forecasting system over the Yellow River basin – Part 2: The added value from climate forecast models, Hydrol. Earth Syst. Sci., 20, 2453–2466, https://doi.org/10.5194/hess-20-2453-2016, 2016.

Yuan, X. and Wood, E. F.: Downscaling precipitation or bias-correcting streamflow? Some implications for coupled general circulation model (CGCM)-based ensemble seasonal hydrologic forecast, Water Resour. Res., 48, 1–7, https://doi.org/10.1029/2012WR012256, 2012.

Yuan, X., Wood, E. F., Luo, L., and Pan, M.: A first look at Climate Forecast System version 2 (CFSv2) for hydrological seasonal prediction, Geophys. Res. Lett., 38, 1–7, https://doi.org/10.1029/2011GL047792, 2011.

Yuan, X., Wood, E. F., Roundy, J. K., and Pan, M.: CFSv2-Based seasonal hydroclimatic forecasts over the conterminous United States, J. Clim., 26, 4828–4847, https://doi.org/10.1175/JCLI-D-12-00683.1, 2013.

Yuan, X., Roundy, J. K., Wood, E. F., and Sheffield, J.: Seasonal forecasting of global hydrologic extremes: System development and evaluation over GEWEX basins, B. Am. Meteorol. Soc., 96, 1895–1912, https://doi.org/10.1175/BAMS-D-14-00003.1, 2015.

Zalachori, I., Ramos, M.-H., Garçon, R., Mathevet, T., and Gailhard, J.: Statistical processing of forecasts for hydrological ensemble prediction: a comparative study of different bias correction strategies, Adv. Sci. Res., 8, 135–141, https://doi.org/10.5194/asr-8-135-2012, 2012.

Zhang, D., Madsen, H., Ridler, M. E., Kidmose, J., Jensen, K. H., and Refsgaard, J. C.: Multivariate hydrological data assimilation of soil moisture and groundwater head, Hydrol. Earth Syst. Sci., 20, 4341–4357, https://doi.org/10.5194/hess-20-4341-2016, 2016.

Zhao, L., Duan, Q., Schaake, J., Ye, A., and Xia, J.: A hydrologic post-processor for ensemble streamflow predictions, Adv. Geosci., 29, 51–59, https://doi.org/10.5194/adgeo-29-51-2011, 2011.

Zhao, T., Bennett, J., Wang, Q. J., Schepen, A., Wood, A., Robertson D., and Ramos, M.-H.: How suitable is quantile mapping for postprocessing GCM precipitation forecasts? J. Clim., 30, 3185–3196. https://doi.org/10.1175/JCLI-D-16-0652.1, 2017.

Combining satellite data and appropriate objective functions for improved spatial pattern performance of a distributed hydrologic model

Mehmet C. Demirel[1,6], **Juliane Mai**[2,4], **Gorka Mendiguren**[1,5], **Julian Koch**[1,3], **Luis Samaniego**[2], and **Simon Stisen**[1]

[1]Geological Survey of Denmark and Greenland, Øster Voldgade 10, 1350 Copenhagen, Denmark

[2]Department Computational Hydrosystems, UFZ – Helmholtz Centre for Environmental Research, Leipzig, Germany

[3]Department of Geosciences and Natural Resource Management, University of Copenhagen, Copenhagen, Denmark

[4]Department of Civil and Environmental Engineering, University of Waterloo, Waterloo, Canada

[5]Department of Environmental Engineering, Technical University of Denmark, 2800 Kgs. Lyngby, Denmark

[6]Department of Civil Engineering, Istanbul Technical University, 34469 Maslak, Istanbul, Turkey

Correspondence: Mehmet C. Demirel (mecudem@yahoo.com) and Simon Stisen (sst@geus.dk)

Abstract. Satellite-based earth observations offer great opportunities to improve spatial model predictions by means of spatial-pattern-oriented model evaluations. In this study, observed spatial patterns of actual evapotranspiration (AET) are utilised for spatial model calibration tailored to target the pattern performance of the model. The proposed calibration framework combines temporally aggregated observed spatial patterns with a new spatial performance metric and a flexible spatial parameterisation scheme. The mesoscale hydrologic model (mHM) is used to simulate streamflow and AET and has been selected due to its soil parameter distribution approach based on pedo-transfer functions and the build in multi-scale parameter regionalisation. In addition two new spatial parameter distribution options have been incorporated in the model in order to increase the flexibility of root fraction coefficient and potential evapotranspiration correction parameterisations, based on soil type and vegetation density. These parameterisations are utilised as they are most relevant for simulated AET patterns from the hydrologic model. Due to the fundamental challenges encountered when evaluating spatial pattern performance using standard metrics, we developed a simple but highly discriminative spatial metric, i.e. one comprised of three easily interpretable components measuring co-location, variation and distribution of the spatial data.

The study shows that with flexible spatial model parameterisation used in combination with the appropriate objective functions, the simulated spatial patterns of actual evapotranspiration become substantially more similar to the satellite-based estimates. Overall 26 parameters are identified for calibration through a sequential screening approach based on a combination of streamflow and spatial pattern metrics. The robustness of the calibrations is tested using an ensemble of nine calibrations based on different seed numbers using the shuffled complex evolution optimiser. The calibration results reveal a limited trade-off between streamflow dynamics and spatial patterns illustrating the benefit of combining separate observation types and objective functions. At the same time, the simulated spatial patterns of AET significantly improved when an objective function based on observed AET patterns and a novel spatial performance metric compared to traditional streamflow-only calibration were included. Since the overall water balance is usually a crucial goal in hydrologic modelling, spatial-pattern-oriented optimisation should always be accompanied by traditional discharge measurements. In such a multi-objective framework, the current study promotes the use of a novel bias-insensitive spatial pattern metric, which exploits the key information contained in the observed patterns while allowing the water balance to be informed by discharge observations.

1 Introduction

Reliable estimations of spatially distributed actual evapotranspiration (AET) are useful for various sustainable water resources management practices such as irrigation planning, agricultural drought monitoring and water demand forecasting in large cultivated areas (Wei et al., 2017). Distributed hydrologic models can potentially provide this insight since evapotranspiration (ET) is a major part of the water cycle. In spite of their ability to simulate detailed spatial patterns of a range of hydrological state variables and fluxes, distributed model evaluation remains focused on temporal aspects of the aggregated streamflow variable (Demirel et al., 2013; Schumann et al., 2013). We are interested in including spatial AET patterns in the model calibration using spatial parameterisations and complementary objective functions. Different methods exist that utilise satellite-based land surface temperature data to derive spatially detailed estimates of latent heat fluxes from the land surface and canopy on a scale relevant for catchment modelling (Kalma et al., 2008). Since AET cannot be measured directly by satellite, surface energy balance models are developed to estimate AET based on data from a range of spectral and thermal bands (Guzinski et al., 2013; Norman et al., 1995; Su, 2002). While these satellite-based estimates are usually employed as a tool to understand and improve the model parameterisations (Conradt et al., 2013; Hunink et al., 2017; Schuurmans et al., 2011), they can also be used to calibrate models (Crow et al., 2003; Immerzeel and Droogers, 2008; Zhang et al., 2009). Therefore, adding satellite-based observations to model calibration is not novel; however, specifically evaluating spatial patterns in the calibration has rarely been done (Stisen et al., 2011b). Interesting examples exist where model calibration could benefit from the spatial pattern information of actual evapotranspiration (Githui et al., 2016; Li et al., 2009; Zhang et al., 2009) and satellite-based recharge patterns (Hendricks Franssen et al., 2008). This paper utilises monthly patterns of AET first to understand and organize ET-related model spatial parameterisations and then to pursue a calibration. This is because adding only temporal aspects of the spatial observations to the objective function is not sufficient for achieving significant improvements in simulated spatial patterns if model parameterisation is not flexible enough to physically adjust to the observed pattern. Besides, the model structure, parameterisations and calibration schemes have usually been designed for streamflow optimisations (Vazquez et al., 2011; Velázquez et al., 2010). In order to ensure compatibility between the spatial pattern calibration target and model parameterisation, the flexibility of the spatial model parameterisation needs to be reconsidered. Recently, inadequate representation of spatial variability and hydrologic connectivity of a well-known distributed model (VIC) has been reported by Melsen et al. (2016). The mesoscale hydrologic model (mHM) has the flexibility to alter the spatial patterns via pedo-transfer function (PTF) parameters and by including a multi-scale parameter regionalisation (MPR) scheme (Kumar et al., 2013; Samaniego et al., 2010). Mizukami et al. (2017) incorporated this MPR approach with VIC to estimate parameters for large domains based on geophysical data for 531 basins. The multi-basin calibration results using MPR revealed physically meaningful parameter fields without patchiness (discontinuities). The study by Loosvelt et al. (2013) is one of few other examples that incorporate PTFs for soil texture and moisture components of a hydrologic model.

All calibration strategies rely on the selection of performance metrics indicating the goodness of fit of the model to be optimized. Choosing an appropriate set of objective functions is crucial to build a robust calibration strategy, since there will be trade-offs between different objective functions or redundant information. In the hydrology literature, there are a range of different temporal metrics for hydrograph matching while metrics designed for spatial pattern matching are less common (Koch et al., 2017; Rees, 2008). For distributed models, spatial metrics usually evaluate cell-to-cell correlation and deviations (e.g. Pearson's R and bias). The use of multi-component metrics as described for discharge by Gupta et al. (2009) is, however, rare for spatial pattern evaluation. An essential feature of our study is introducing a new spatial efficiency (SPAEF) metric that contains three components, i.e. correlation, variance and histogram intersection, providing reliable bias-insensitive pattern information unlike other traditional metrics focusing on only one aspect like correlation, mean squared error or bias.

Prior to model calibration, sensitivity analysis is usually conducted to attribute response of the model outputs to the changes in model parameters (Shin et al., 2013), which can enhance our understanding of both temporal and spatial model behaviour (Berezowski et al., 2015). In the context of spatial model calibration, the sensitivity analysis should not only identify the parameters that affect the water balance and hydrograph dynamics but also the parameters that shape the spatial patterns of the simulated states and fluxes. To achieve this, we have to design objective functions that reflect the spatial pattern of the models and utilise these in model parameter sensitivity analysis.

In light of the well-known equifinality problems in model calibration (Beven and Freer, 2001) spatial pattern evaluation can be useful for selecting the most appropriate parameter set from a group of sets leading to both reasonable streamflow performance and physically meaningful AET pattern. Immerzeel and Droogers (2008) showed how a semi-distributed model of a basin in southern India could be constrained by using spatially distributed observations with a monthly temporal resolution. Cornelissen et al. (2016) highlighted the need to identify which model parameters influence the simulated spatial pattern and showed that spatial patterns of simulated evapotranspiration were most sensitive to the land-use parameterisation, whereas precipitation was the most sensitive input data with respect to temporal dynamics of the model. Rakovec et al. (2016) used a total water storage (TWS) anomaly from the Gravity Recovery and Climate Experiment (GRACE) satellites and evapotranspiration estimates from FLUXNET data (https://fluxnet.ornl.gov/) to improve model parameterisations for discharge simulations. They showed that adding TWS anomalies to the calibration led to a reasonably good performance for continental 83 European basins with different climatology.

The main objectives of this study are to incorporate spatial patterns of satellite-based actual evapotranspiration data in the model calibration and validation. In order to improve AET simulations, we use transfer functions in the spatial model parameterisation that combine a priori maps of soil and vegetation properties with few global calibration parameters in order to enhance the spatial parameterisation flexibility and allow the parameter field to adjust to an observed spatial patterns of AET from the catchment. We also design a new multi-component metric specifically suited for comparing spatial patterns of two continuous variables. Here, we prioritise three main data properties, which are co-location, variation and distribution. The calibration is conducted using three strategies for objective function selection. First, streamflow metrics and spatial pattern metrics are used in isolation during calibration and subsequently they are combined in a more balanced model optimisation. In this way we can investigate the trade-offs and robustness of the different approaches by evaluating the performances regarding both streamflow and spatial patterns during calibration and validation.

2　Study area and data

2.1　Study area

The Skjern river basin is one of the most popular research basins in Denmark as it is highly instrumented for hydrological monitoring, including eddy-flux towers, a dense soil moisture network and other state-of-the-art monitoring of hydrological variables (Jensen and Illangasekare, 2011). The basin area is approximately $2500\,km^2$, containing mostly sandy soils (Fig. 1). The river is the largest in Denmark by flow volume and located in the western part of the Jutland peninsula, a region dominated by agriculture and forests together covering $\sim 80\,\%$ of the domain (Larsen et al., 2016). The basin is mostly flat with a maximum altitude of 130 m and it receives a mean annual precipitation of around 1000 mm (Stisen et al., 2011a). The mean annual streamflow is around 475 mm and monthly mean temperatures vary from 2 up to $17\,°C$ (Jensen and Illangasekare, 2011).

2.2　Satellite-based data

The Moderate Resolution Imaging Spectroradiometer (MODIS) polar orbiting platforms, Terra and Aqua, observe mid-latitude regions 4 times per day at a spatial

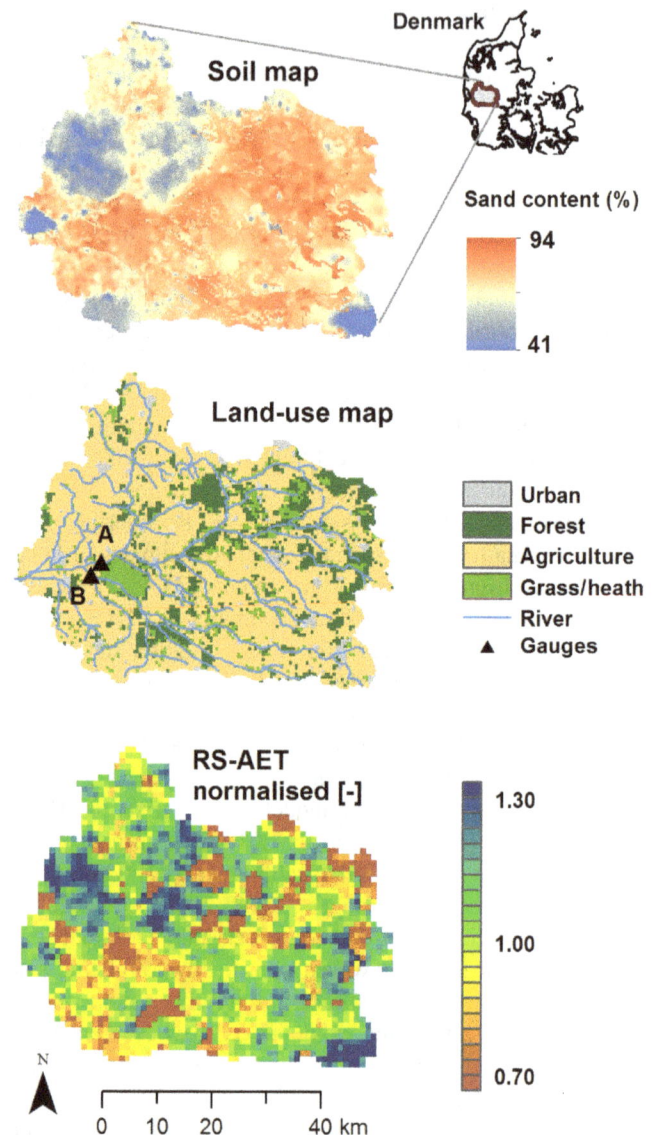

Figure 1. Skjern river basin location, soil type and land-use characteristics. An average pattern of satellite-based actual evapotranspiration for June (average of all years from 2001 until 2008) is presented to illustrate the interaction between soil type and land use that generate the land surface flux patterns.

resolution of approximately $1\,km \times 1\,km$. The two-source energy balance (TSEB) model proposed by Norman et al. (1995) based on the Priestley–Taylor approximation (Priestley and Taylor, 1972) is used in this study to calculate AET based on MODIS data under cloud-free conditions. The model inputs are land surface temperature (LST), solar zenith angle (SZA), and albedo and height of canopy, all derived from MODIS observations (Mendiguren et al., 2017). Additional inputs such as climate variables of air temperature and incoming radiation are obtained from ERA-Interim reanalysis data (Dee et al., 2011). The main

Table 1. Overview of morphological and meteorological data used as input for mHM. Acronyms: BIOS – BioScience Aarhus University, DMI – Danish Meteorological Institute, GEUS – Geological Survey of Denmark and Greenland, MODIS – Moderate Resolution Imaging Spectroradiometer, DGA – Danish Geodata Agency.

Variable	Description	Spatial resolution	Source
Q (daily)	Streamflow	Point	BIOS
P (daily)	Precipitation	10 km	DMI
ET_{ref} (daily)	Reference evapotranspiration	20 km	GEUS and DMI
T_{avg} (daily)	Average air temperature	20 km	GEUS and DMI
LAI	Fully distributed 12-monthly values based on 8-day time-varying leaf area index (LAI) dataset	1 km	MODIS and Mendiguren et al. (2017)
Land cover	Forest, agriculture and urban	250 m	GEUS
DEM-related data	Slope, aspect, flow accumulation and direction	250 m	DGA
Geology class	Two main geological formations	250 m	GEUS
Soil class	Fully distributed soil texture data	250 m	Greve et al. (2007)

motivation of preparing a new AET dataset based on land surface temperature is that most other available products are based mainly on vegetation index data which may not be sufficient to assess the complicated interplay among climate, soil and vegetation dynamics on the AET patterns, especially during the growing season. For more details on our newly produced AET data for Denmark, including equations, parameterisation, calibration and validation, please refer to the recent study by Mendiguren et al. (2017).

In this study, all remote-sensing-based AET data were averaged for each month during the growing season across all years for the model calibration period (2001–2008), resulting in six monthly mean maps from April to September representing AET under cloud-free conditions. This ensures that in spite of uncertainty in the individual instantaneous midday estimates of AET, the monthly maps represent the general spatial pattern for each month under cloud-free conditions. The individual daily AET patterns are evaluated for temporal consistency by calculating the Pearson correlation between each daily pattern and the monthly mean pattern for the given month. This analysis showed that the overall average correlation between an individual day and the monthly mean was 0.82. The satellite-based monthly AET maps are validated against eddy-covariance measurements for three different land cover types (forest, cropland and wetland) within the Skjern catchment and display good agreement on the monthly timescale (Mendiguren et al., 2017). Despite not being pure observations but rather estimates from an energy balance model based on satellite observations, we will refer to these AET maps as reference observations. Based on the sensitivity analysis in Mendiguren et al. (2017), which showed that the TSEB is largely controlled by the satellite input of LST, which can be considered an observation, it is assumed that the TSEB AET estimates represent spatial patterns of AET that are suitable for pattern evaluation of the hydrological model.

3   Hydrologic model

The mesoscale hydrologic model is a distributed model providing various simulated spatial outputs, fluxes and states at different spatio-temporal model resolutions (Samaniego et al., 2010, 2017). The model includes pedo-transfer functions for soil parameterisation and originally contains 53 global parameters that can be adjusted during calibration. In this study, some parameters are fixed at a default value and others have been added from the new spatial model parameterisations resulting in a total of 48 global parameters for further analysis. The model simulates major components of the hydrologic cycle, i.e. interception, infiltration, snow accumulation and melting, evapotranspiration, groundwater storage, seepage, and runoff generation. The readers are referred to the study by Samaniego et al. (2010) for full model description, assumptions, limitations and process formulations.

Table 1 provides a summary of the modelling data used in this study. As shown in the table, meteorological data can be on a different spatial scale than both morphological data and the model scale. This flexibility arises from the fact that mHM incorporates a multi-parameter regionalisation technique to swap between different scales while calculating all fluxes and routing streamflow on a preferred model scale. We run the model on 1 km × 1 km spatial scale and at daily time step. Some processes like ET are calculated at an hourly time step then the final results are aggregated to daily values. All morphological data are prepared on 250 m × 250 m scale. All three meteorological datasets, i.e. P, ET_{ref} and T_{avg}, were originally at 10–20 km resolution. We re-sampled them to 1 km × 1 km using cubic interpolation. This interpolation method is used to avoid patchiness in model simulations due to coarse grids on the native scale of the metrological data. We use 12 monthly leaf area index (LAI) maps to represent the climatology for both interception and PET correction for the entire period (2001–2014) and the model warm-up period (1997–2000).

3.1 Spatial model parameterisation

In order to facilitate a meaningful spatial-pattern-oriented calibration of a distributed model, we need to compromise between comprehensive (each cell in the basin) and lumped (one cell – one basin) parameterisations, as the first approach may require immense computer resources during calibration and the latter approach usually results in a uniform pattern. For instance, in a detailed calibration study by Corbari et al. (2013), each pixel in the catchment is represented by a parameter whereas, in a coarse parameterisation, a uniform parameter represents the entire catchment (Stisen et al., 2017). In this study, we follow an intermediate level of parameterisation comprised of several flexible spatial parameters and nonlinear equations, allowing us to stretch the spatial contrast of simulated actual evapotranspiration based on soil and vegetation properties. This level of parameterisation is still physically meaningful as the parameters are tied to the land surface characteristics of the basin via transfer functions.

3.1.1 Distributed root fraction coefficient

Root distribution with depth is generally perceived as being a function of vegetation type (Jackson et al., 1996), and our spatial parameterisation of root fraction distribution is initially separated based on land covers of forest and agricultural crops. However, following the site-specific soil and plant physical literature (Jensen et al., 2001; Madsen and Platou, 1983), we subdivide the root fraction coefficient for agricultural crops as a function of field capacity (FC). Here, spatial model parameterisation is implemented to the root fraction calculation in the original mHM structure which follows the asymptotic equation for vertical root distribution (Eq. 1) proposed by Jackson et al. (1996).

$$Y = 1 - (\beta_c)^d, \tag{1}$$

where Y is the cumulative root fraction from soil surface to depth d (cm), and β_c is the root fraction coefficient. We substituted the root fraction coefficient for agricultural crops (non-forest) with two new root fraction parameters, i.e. one root fraction for maximum FC (clay) and one for minimum FC (sand), which allow for full spatial distribution of root fraction with varying FC. This relation between soil characteristics and effective rooting depth is based on a site-specific database with more than 100 soil and root profiles collected in Denmark (Table 19.4 in Jensen et al., 2001) and the literature focusing on soil texture and effective rooting depths in Denmark (Madsen, 1985, 1986; Madsen and Platou, 1983). The approach is not necessarily globally valid, but designed for the specific region of western Denmark where very sandy soils (Fig. 1) are cultivated for agricultural purposes even though the soil properties influence root development. These parameters are used to form the root fraction coefficient for soil with agriculture ($\beta_{agriculture}$) based on field-capacity-dependent root fraction in Eqs. (2) and (3).

$$FC_{norm} = \frac{FC_i - FC_{min}}{FC_{max} - FC_{min}}, \tag{2}$$

where FC_{norm} is the normalised field capacity ranging from 0 to 1.

$$\beta_{agriculture} = (FC_{norm} \cdot \beta_{max}) + (1 - FC_{norm}) \cdot \beta_{min}, \tag{3}$$

where $\beta_{agriculture}$ is the new root fraction for soil with agriculture comprised of root fraction for clay (β_{max}) and root fraction for sand (β_{min}).

3.1.2 Dynamic ET_{ref} scaling function

As a second spatial parameterisation step, we incorporated remotely sensed vegetation information, to downscale coarse climatological reference evapotranspiration (ET_{ref}) to the model scale. This was done to emphasise the effect of vegetation on the simulated spatial patterns of AET. The original scaling factor in mHM is based on a lumped minimum correction and an aspect-driven additional term. Using aspect ratio for ET_{ref} correction makes sense in mountainous areas; however, this is found to be irrelevant for the Skjern basin which is characterized by a low topographical variation. The dynamic scaling function introduced here allows the modeller to superimpose the imprint of LAI on the simulated AET patterns via a downscaling of the ET_{ref}. The concept of a dynamic scaling function (DSF) is similar to the concept of a crop coefficient used to convert ET_{ref} to a potential evapotranspiration (ET_{pot}) for a given vegetation that differs from the reference crop. Our implementation follows the equation for estimating the crop coefficient for natural vegetation originally proposed by Allen et al. (1998). Similarly, Hunink et al. (2017) compared different applications of crop coefficients based on remotely sensed vegetation indices in hydrologic modelling. They found that the effect of crop coefficient parameterisations on the water balance is trivial and constant throughout the year; however, it has a major effect on seasonal evapotranspiration and soil moisture fluxes, showing the role of crop coefficients in spatial calibration. The DSF, shown in Eq. (65), is simply a time–space variable implementation of the crop coefficient for natural vegetation, parameterized through a spatio-temporal LAI (no unit) component accounting for the effects of characteristics that separate the actual vegetation from a reference grass (well-watered 10 cm height and albedo of 0.23). These characteristics include specific land cover, albedo and aerodynamic resistance (Allen et al., 1998; Liu et al., 2017). This ensures a physically meaningful downscaling from a coarse (here 20 km) ET_{ref} grid to the model resolution (here 1 km).

$$ET_{pot} = DFS \cdot ET_{ref}, \tag{4}$$

$$DSF = a + b \left(1 - e^{(-c \cdot LAI)} \right), \tag{5}$$

where a in the model (ET_{ref}-a) is the intercept term representing uniform scaling, b (ET_{ref}-b) represents the vegetation dependent component, and c (ET_{ref}-c) describes the degree of nonlinearity in the LAI dependency.

4 Methods

In this study, we applied a recently developed sequential screening method (Cuntz et al., 2015) to select important parameters for calibration. Since different parameters can be sensitive to different hydrologic processes, we tested three different performance metrics to evaluate process–parameter relationships. Two of these metrics are derived from the hydrograph, i.e. Kling–Gupta efficiency (KGE, Gupta et al., 2009) and KGE of only below-average streamflow (KGE_{low}), whereas the spatial efficiency metric focuses on the spatial pattern of actual evapotranspiration.

4.1 Objective functions

As an objective function for streamflow performance, we chose the Kling–Gupta efficiency, shown in Eq. (6) (Kling and Gupta, 2009), and applied it to both the entire time series and to the low-flow part of the hydrograph (below mean discharge).

$$KGE = 1 - \sqrt{\left(\alpha_Q - 1\right)^2 + \left(\beta_Q - 1\right)^2 + \left(\gamma_Q - 1\right)^2},$$
$$\alpha_Q = \rho(S, O) \text{ and } \beta_Q = \frac{\sigma_S}{\sigma_O}, \text{ and } \gamma_Q = \frac{\mu_S}{\mu_O}, \quad (6)$$

where α_Q is the Pearson correlation coefficient between observed and simulated discharge time series, β_Q is the relative variability based on the fraction of standard deviation in simulated and in observed values, and γ_Q is the bias term normalised by the standard deviation in the observed data.

Since comparison of two spatial pattern maps is of obvious importance, a bias-insensitive spatial performance metric is developed and used in this study. In this context, we adopted the structure of the Kling–Gupta efficiency while substituting the standard deviation term by a term based on the coefficient of variation σ_O/σ_S and replacing the bias term with a histogram comparison index to compare the intersection percentage of two histograms of observed and simulated spatial maps. The histogram intersect is performed after normalisation of the observed and simulated maps to a mean of 0 and standard deviation of 1 (z score). This ensures that the histogram comparison is unaffected by any bias or variance differences and solely reflects the agreement in distribution of the variable in space. The main utility of the histogram comparison is that it distinguishes between different soil and vegetation groups reflected in the spatial pattern results. This unique feature of being sensitive to clusters in the data compliments the other two components in the equation, in particular the correlation coefficient (α in Eq. 7) since α is highly vulnerable to very distinct clusters of points aligned on a diagonal axis. This can result in high correlation coefficient values in spite of low correlation inside the individual clusters inevitably misleading the model calibration. The separated clusters often occur in environmental models where different land-use classes and soil classes etc. can produce patchy spatial patterns. The new spatial efficiency metric (optimal value equals to 1) is defined as follows:

$$SPAEF = 1 - \sqrt{(\alpha - 1)^2 + (\beta - 1)^2 + (\gamma - 1)^2}$$
$$\alpha = \rho(A, B) \text{ and } \beta = \left(\frac{\sigma_A}{\mu_A}\right) / \left(\frac{\sigma_B}{\mu_B}\right) \text{ and }$$
$$\gamma = \frac{\sum_{j=1}^{n} \min\left(K_j, L_j\right)}{\sum_{j=1}^{n} K_j}, \quad (7)$$

where α is the Pearson correlation coefficient between the observed AET map (A) and simulated AET map (B) for a particular month, β is the fraction of coefficient of variations representing spatial variability, and γ is the percentage of histogram intersection (Swain and Ballard, 1991). The gamma (γ) is calculated for a given histogram K of the observed AET map (A) and the histogram L of the model simulated AET map (B), each one containing n bins, i.e. herein 100 bins. The maps are standardised to a mean of 0 and a standard deviation equal to 1 (z score) to avoid the effect of different units. In this study, we compare AET from TSEB (in W m^{-2}) based on instantaneous satellite data with daily averaged AET (mm day^{-1}) simulated by the model and regard the satellite-based AET maps as the "observation" even though they are more accurately AET "estimates" based on satellite observations. Attempts to use numerous other spatial metrics including Mapcurves, fractions skill score (FSS), Goodman and Kruskal's lambda, Theil's Uncertainty, empirical orthogonal functions (EOFs) and Cramér's V (Cramér, 1946; Koch et al., 2015; Rees, 2008) did not distinguish the general AET patterns or the spatial efficiency metric. The strength of the spatial efficiency metric is that each component contains different and non-overlapping information. Moreover, the components are straightforward compared to the aforementioned metrics. While the correlation term (α) expresses only the spatial correlation of AET values, the coefficient of variation term (β) expresses only the range or contrast in the image while the histogram term (γ) only expresses the agreement on histogram shape without considering either variation or correlation. Since all three terms are bias-insensitive, the spatial efficiency only constrains the model simulations with the pattern information in the satellite data while leaving the water balance (bias) to be constrained by streamflow metrics.

4.2 Sequential screening of the model parameters

We applied the variance-based sequential screening (SS) method introduced by Cuntz et al. (2015) to identify, with a low computational budget, the parameters which are most informative regarding a certain model output M.

For this approach the parameters are sampled in trajectories as initially described by Morris (1991) and improved by Campolongo et al. (2007). Each trajectory consists of $(N + 1)$ parameter sets, assuming that N is the total number of model parameters. The first parameter set in each trajectory is sampled randomly while all the subsequent sets i ($i > 1$) differ to the prior set ($i - 1$) in exactly one parameter value. Therefore, the whole trajectory is a path through the parameter space. Trajectories allow us to sample the whole parameter space efficiently and consider parameter interactions to certain extents. In the approach of Cuntz et al. (2015), only a small number (M_1) of such trajectories are sampled to lower the computational burden. The resulting $(M_1 \times (N + 1))$ model outputs are derived and the elementary effects (EEs) are computed for each parameter. The EEs are then used to identify the most informative parameters by deriving a threshold splitting the parameters into a set N_u of uninformative and a set N_i of informative ones. In the following, the first parameter set is again sampled randomly but then only the uninformative parameters are perturbed meaning that the new trajectory only consists of ($N_u + 1$) parameter sets. The derivation of model output and calculation of EEs is repeated. The major step is to determine whether one of the previously uninformative parameters is now above the threshold and if so it is added to the set of informative parameters N_i. These steps are repeated until no further parameter is added to the set N_i. At the end M_2 trajectories are sampled to confirm that the set of uninformative parameters N_u is stable and no further parameter would be found to be informative.

4.3 Model calibration and validation

We calibrated the 1 km daily mHM for the Skjern basin in Denmark using the well-known global search algorithm Shuffled Complex Evolution method from the University of Arizona (SCE-UA) (Duan et al., 1992). The SCE-UA algorithm is configured with two complexes running in parallel with 53 ($2n + 1$) parameter sets in each complex and 27 ($n + 1$) parameter sets per sub-complex. Moreover, the maximum relative objective function change is set to 1 % over five iterations as the model convergence criterion. This criterion was usually reached after 3500 runs; in rare cases up to 8000 runs were necessary. We evaluated the differences between monthly AET estimates from the TSEB reference data and simulated AET from the hydrologic model for the calibration period (2001–2008) and validation period (2009–2014).

The two streamflow stations are defined separately to follow the improvements in each metric throughout the calibrations. After testing different combinations of streamflow and spatial metrics, we chose two streamflow metrics (KGE and KGE_{low}) and one spatial efficiency metric, given by Eqs. (6)

and (7), respectively. These objective functions are used individually or combined in three model calibration cases based on (i) only streamflow using equally weighted KGE and KGE_{low}, (ii) only spatial patterns of AET using spatial efficiency, (iii) both equally weighted streamflow and spatial pattern matching using all three metrics. It should be noted that the case 2 calibration is designed as a benchmark to explore how good the pattern match can get when not considering streamflow performance, even though the solution might not be interesting from a hydrological perspective, since the bias insensitive spatial pattern metric does not secure a reasonable water balance. To test the overall robustness of the calibration framework we use an ensemble of nine calibrations for case 1 and nine calibrations for case 3, each started from a different seed number. In order to fairly weigh the objective functions, we retrieve the residuals (ε) from the three objective functions based on a random initial model run (Eqs. 8–10). We calculate the new weights which will result in equal contribution to the total error (Φ_{total}), i.e. 50 % from spatial metric and 50 % from the two streamflow metrics. Ideally, if it exists the optimiser searches a parameter set resulting in zero Φ_{total} otherwise the closest point to zero will be considered as the optimum solution.

$$\Phi_Q = \sum_{i=1}^{2} \left(\varepsilon_{KGE_i} \cdot \omega_{KGE_i} \right)^2 + \sum_{i=1}^{2} \left(\varepsilon_{KGE_{low,i}} \cdot \omega_{KGE_{low,i}} \right)^2, \quad (8)$$

$$\Phi_{Spatial} = \sum_{m=1}^{6} \left(\varepsilon_{SPAEF_m} \cdot \omega_{SPAEF_m} \right)^2, \quad (9)$$

$$\Phi_{total} = \Phi_Q + \Phi_{Spatial}, \quad (10)$$

where Φ_Q is the total Φ for streamflow of the two streamflow gauges and $\Phi_{Spatial}$ is the total Φ for spatial performances of six summer months. For Q-only calibration, the weight for SPAEF (ω_{SPAEF}) becomes zero whereas for spatial-only calibration the weights for KGE and KGE_{low} become zero.

5 Results

5.1 Sequential screening of the model parameters

Table 2 shows the sequential screening results based on KGE, KGE_{low} and SPAEF. Each objective function reflects on different spatio-temporal dynamics of the catchment. While KGE and KGE_{low} evaluate high and low streamflow dynamics and biases, the bias-insensitive SPAEF focuses on only spatial patterns of AET. From the results it is clear that some of the highly sensitive parameters for streamflow dynamics, especially interflow-related parameters, groundwater-related geology parameters and single routing parameters, have minor to zero influence on the spatial patterns of AET. The new ET parameters, ET_{ref}-a (non-forest), -af (forest), -b and -c are identified to be informative based on all objective functions. The root fraction coefficient for forest (rotfrcoffore) appeared to be not very important for streamflow metrics

Table 2. Selected 26 parameters for calibration and their normalised sensitivity indices sorted based on SPAEF column. Zero values are highlighted in italic. The three bold values are the highest values of the three sensitivity indices.

Parameter	Description	Normalised sensitivity		
		KGE	KGElow	SPAEF
ET_{ref}-af	Intercept – forest	0.022	0.117	**0.646**
ET_{ref}-c	Exponent coefficient	0.031	0.732	0.490
ET_{ref}-b	Base coefficient	0.439	3.013	0.317
rotfrcoffore	Root fraction coefficient for forest areas	0.011	0.013	0.162
$ETref$ – a	Intercept – non-forest	0.308	3.235	0.157
ptfhigconst	Constant in pedo-transfer function for soils with sand content higher than 66.5 %	0.063	0.223	0.096
rotfrcofclay	Root fraction coefficient for clay in agricultural cropland	0.101	0.274	0.094
ptfhigdb	Coefficient for bulk density in pedo-transfer function for soils with sand content higher than 66.5 %	0.036	0.257	0.070
rotfrcofsand	Root fraction coefficient for sand in agricultural cropland	0.120	0.439	0.061
canintfact	Canopy interception factor	0.004	0.029	0.018
orgmatforest	Organic matter content for forest	0.136	0.893	0.014
ptfhigclay	Coefficient for clay content in pedo-transfer function	0.008	0.033	0.011
infshapef	Infiltration shape factor	0.103	0.099	0.006
ptfkssand	Coefficient for sand content in pedo-transfer function for hydraulic conductivity	0.415	2.780	0.002
ptfksconst	Constant in pedo-transfer function for hydraulic conductivity of soils with sand content higher than 66.5 %	0.236	0.842	0.001
snotrestemp	Snow temperature threshold for rain and snow separation	0.034	0.206	*0.000*
ptfksclay	Coefficient for clay content in pedo-transfer function for hydraulic conductivity	0.040	0.313	*0.000*
orgmatimper	Organic matter content for impervious zone	0.009	0.020	*0.000*
expslwintflw	Exponent slow interflow	0.412	**3.490**	*0.000*
slwintreceks	Slow interception	0.872	1.296	*0.000*
intrecesslp	Interflow recession slope	0.602	1.105	*0.000*
rechargcoef	Recharge coefficient	**0.935**	0.666	*0.000*
geoparam1	Parameter for geological formation 1	0.328	0.138	*0.000*
geoparam2	Parameter for geological formation 2	0.558	0.207	*0.000*
strcelerity	Streamflow celerity for routing	0.364	0.062	*0.000*
intstorcapf	Interflow storage capacity factor	0.198	0.010	*0.000*

whereas it is crucial for SPAEF. Similarly, the two newly introduced parameters, i.e. root fraction coefficient for sand and clay (i.e. rotfrcofs and rotfrcofclay) soil, are informative based on all three objective functions. Organic matter for forest (orgmatforest) is especially important for low flows whereas organic matter for impervious areas (orgmatimper) has zero influence on spatial patterns of AET. The exponent slow interflow (expslwintflw) parameter is found to be most informative for low flows while recharge coefficient (rechargcoef) is most informative for streamflow and ET_{ref}-af is most informative for calibrating spatial patterns of AET.

On average 475 model evaluations are required to split the total number of 48 parameters into informative and uninformative ones. However, the number of iterations is dependent on objective function; therefore, 449 model runs were required for KGE, 431 model runs for KGE$_{low}$ and 544 model runs for SPAEF. This is in close agreement with the computational budget of $10N$ model evaluations already reported by Cuntz et al. (2015). This also makes the sequential screening method computationally very attractive compared to other global search methods. However, the computational advantage is at the cost of exploring a larger part of the parameter space, and hence the sequential screening is mostly valuable for identifying informative and non-informative parameters prior to calibration or further assessment of the parameter be-

haviour. Overall, these results show that there are 26 parameters above the threshold of 1 % of at least one case (Table 2). In principal the parameters with zero sensitivity (SPAEF column) can be fixed at some value during calibration, which may lead to faster convergence, with a lower degree of freedom. However, we include the same set of 26 parameters in all three calibration cases for consistency.

5.2 Model calibration and validation

The mHM model is calibrated using streamflow records (gauges A and B in Fig. 1) from an 8-year period (2001–2008) and validated for a recent period (2009–2014). A preceding 4-year period (1997–2000) is used for model warm-up. We prepared remotely sensed monthly averaged AET pattern maps calculated for these years considering only cloud-free days from summer months. AET patterns of winter months are not considered since it is mostly cloudy and ET is very low and uniform (energy limited) in winter.

The 26 selected parameters from SS are used in the following three calibration strategies: (1) only streamflow-oriented (Q-only) calibration using equally weighted KGE and KGE$_{low}$, (2) only spatial-pattern-oriented calibration using SPAEF, and (3) streamflow and spatial patterns of AET together using all three objective functions with equal weights of 50 % on spatial metric and 50 % for the two

Table 3. Summary of the calibration results for three cases. Median and standard deviation (SD) refer to the calibration ensemble ranked based on their total Φ.

Metrics	Gauge	Q-only		Spatial-only	Q and spatial	
		Median (SD)	Best	Single cal.	Median (SD)	Best
KGE $(-)$	(A)	0.83 (0.03)	0.97	-1.47	0.88 (0.01)	0.89
KGE $(-)$	(B)	0.94 (0.02)	0.97	-1.03	0.92 (0.01)	0.93
KGE_{low} $(-)$	(A)	0.81 (0.02)	0.86	-3.12	0.81 (0.02)	0.82
KGE_{low} $(-)$	(B)	0.81 (0.02)	0.85	-2.66	0.79 (0.02)	0.8
BIAS (%)	(A)	-6.24 (2.23)	-0.98	38.56	-2.44 (0.62)	-1.25
BIAS (%)	(B)	1.51 (0.83)	1.83	46.87	1.42 (0.93)	3.86
April – SPAEF		-0.88 (0.33)	-0.17	0.5	0.57 (0.02)	0.51
May – SPAEF		-0.62 (0.26)	-0.12	0.47	0.35 (0.07)	0.35
June – SPAEF		-0.59 (0.23)	-0.09	0.36	0.27 (0.11)	0.27
July – SPAEF		-0.39 (0.10)	-0.36	0.51	0.38 (0.04)	0.43
August – SPAEF		-0.29 (0.10)	-0.36	0.53	0.48 (0.05)	0.49
September – SPAEF		0.02 (0.19)	0.30	0.40	0.33 (0.02)	0.32

streamflow metrics (25 % each). Table 3 provides the overall picture of the three different calibration strategies where two of these strategies are based on an ensemble of nine calibrations. Therefore, the basic descriptive statistics are also given as robustness indicators. The results show that the combined calibration (Q and Spatial) produces similar results to both Q only and Spatial-only calibrations focusing on streamflow and spatial patterns of AET respectively. whereas the single-metric calibrations gave very different results for the opposite objective functions, e.g. SPAEF versus streamflow metrics. It is interesting that when comparing the calibration ensemble with the median performance there is very limited trade-off between the Q-only and the combined Q and Spatial calibrations, which have very similar average KGE values. When looking specifically at the best-performing ensemble member with lowest total Φ, there is a more pronounced trade-off between the Q-only and Q and Spatial together calibrations, as the streamflow performance is poorer when SPAEF is included in the group of objective functions. The differences in the streamflow metrics indicate that each objective function carries relevant but slightly conflicting information. Moreover, the results show that the hydrologic model simulates the best AET patterns in different months for different ensemble calibrations. In other words, while one ensemble member has the best performance for April, other calibrations may have the best performance for May and June. This is a secondary trade-off which illustrates that the calibration might benefit from temporal variability in the parameters controlling the spatial parameterisation scheme. It should be noted that ranking of the calibrations within the two ensembles is based on the overall Φ that is comprised of all objective functions for the corresponding calibration. For that reason, the best member of Q and Spatial calibration holds the lowest total Φ comprised of the highest possible KGE, KGE_{low} and SPAEF at the same time but not necessarily the highest SPAEF alone.

This resulted in a slightly lower SPAEF mean of 0.395 for the best member compared to the median member with a SPAEF mean of 0.396 (Table 3).

The results of the Q-only model calibration using only KGE and KGE_{low} reveal very poor simulated patterns of AET, with negative SPAEF for all months. This is not surprising since this calibration is not constrained regarding the spatial patterns, but also illustrates that discharge observations alone contain no spatial pattern information of AET. In contrast, the spatial-only calibration using only SPAEF shows a very poor water balance, with negative KGE and a large bias. We are aware that spatial-only calibration is not applicable or meaningful for hydrologic studies.

The model performance development through the calibrations ($9 + 9$) and optimum points are shown using scatter plots in Fig. 2, which displays all model runs with Φ values inside the specified plot ranges. The scatter plots illustrate trade-offs between objective functions and consistency among calibration ensemble members. The performance regarding spatial patterns ($\Phi_{Spatial}$) displays a high degree of trade-off with all combined calibrations achieving $\Phi_{Spatial}$ values around 0.8 whereas the Q-only calibrations achieve $\Phi_{Spatial}$ values ranging from 2.8 to 4.4. There are two main clusters in the Q-only calibrations: one around 0.11 Φ_Q and the other around 0.25 Φ_Q, whereas all nine Q and spatial calibrations follow a similar level on the y axis ($\Phi_{Spatial}$). It is surprising to see that SCE-UA did not always find the same optimum solution with varying seed number, which is the case mainly for the Q-only calibration. Perhaps more consistent optimum solutions for the Q-only calibrations could have been achieved with tighter stopping rules and the same initial parameter sets.

Similarly, the grey shades in Fig. 3 show the ensemble range of simulated hydrographs for the Q-only and Q and Spatial calibrations. From the hydrographs it is clear that the

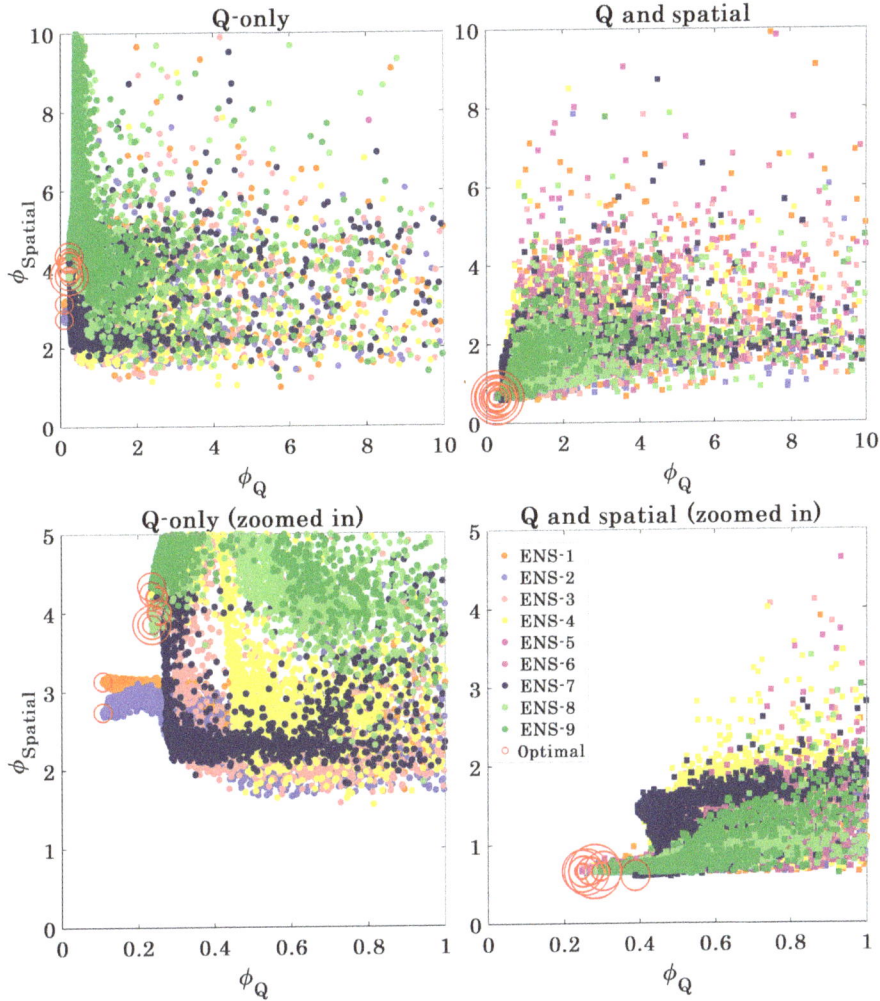

Figure 2. Scatter plots of total $\Phi_{Spatial}$ versus total Φ_Q for all nine calibration ensemble members. First and second row sub-plots are the same figures except for different extent, i.e. [10 10] and [1 5] to zoom into the edge of the search space. Different radius of red circles is used to show the optimum points for all nine ensemble members clearly.

ensemble range for station A is generally larger than that for station B, indicating larger uncertainty for sub-basin A. Interestingly, Fig. 3 also illustrates that the Q and Spatial calibration constrains the solution better, not only in AET simulations, but also in streamflow simulations, as indicated by the slightly narrower range in simulated streamflow for the Q and Spatial calibrations. However, even though the range of hydrographs is slightly narrower the simulations are also further from the observed measurements during summer months.

The corresponding simulated AET maps for the results presented in Table 3 are shown in Fig. 4. This figure illustrates the monthly mean maps across all years of actual evapotranspiration for the cloud-free days available for the remote sensing estimates. Only the best-performing members from the two ensembles are presented in this figure. The maps are normalised with their mean value to use one represen-

tative colour bar in the legend. As indicated in Table 3, the resultant maps from Spatial-only (third row in Fig. 4) and Q and Spatial calibrations (fourth row in Fig. 4) are obviously more similar to the reference monthly maps (first row in Fig. 4) than the maps of Q-only calibration (second row in Fig. 4). The results clearly show that the model can simulate month-to-month variations in AET patterns reasonably well. The poor AET performance in the Q-only maps is obvious in the second row of Fig. 4, where we see only a uniform simulated AET pattern except for the forest areas revealing very little information about variability in AET and the influence of soil and vegetation. This is due to the fact that the KGE and KGE_{low} objective functions contain no information on the patterns of AET resulting in an unconstrained optimisation regarding spatial pattern and variability. Therefore, the optimiser randomly moves in the SPAEF solution space and picks the best streamflow performance with no regard to

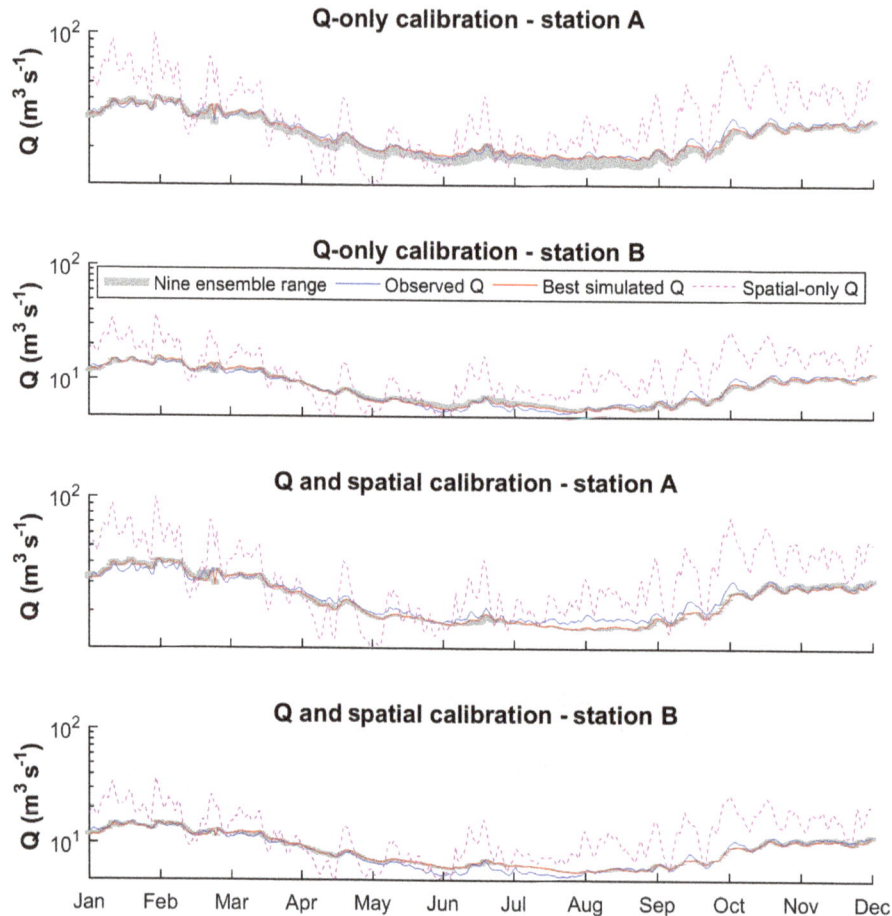

Figure 3. Average hydrograph of all years in the calibration period (2001–2008) to illustrate the ensemble of nine model calibrations with different seed numbers.

AET patterns. Although not perfect (average SPAEF = 0.46 and 0.40), the simulated pattern match in the last two rows of Fig. 4 is quite good compared to the remote-sensing-based estimate since the simulation is able to represent the general pattern influenced by soil, vegetation and land cover while maintaining a similar variance and smoothness.

Table 4 shows the same results as Table 3 but for the validation period spanning from 2009 until 2014. Obviously, the results are somewhat poorer than those for the calibration period. A drop in performance for spatial-only and combined metrics is mainly seen for KGE_{low} and the total bias, whereas the SPAEF for Spatial-only and Q and Spatial remains similar to the calibration periods with average SPAEF around 0.4. Interestingly, there is no real trade-off for streamflow metrics between Q-only and Q and Spatial calibrations for the validation period, even for the best-performing ensemble member. Although a better streamflow performance could be achieved by Q-only calibration during calibration, this cannot be sustained during validation, indicating some overfitting when using streamflow metric only. In contrast, the SPAEF performance does not drop during validation for

the combined Q and Spatial optimisation, indicating less overfitting and a more robust model parameterisation.

6 Discussion

In the initial phase of the study numerous flawed calibrations were carried out in an attempt to produce simulated spatial patterns of AET similar to the satellite-based reference patterns. However, the inability to produce similar patterns was found to be caused by limitations in spatial model parameterisation and spatial performance metric choice. Regarding the spatial parameterisation, the initial model was based on a spatially uniform parameterisation of root fraction coefficient and PET correction factor, two parameters with major control on the simulated AET. Therefore, more flexible yet physically meaningful parameterisations were implemented where full spatial variability was enabled by combining 2–3 calibration parameters to initial spatial distributions of soil type and LAI. Regarding the use of appropriate spatial performance metrics, the initial attempts using standard metrics of correlation coefficient, Mapcurves (Hargrove et al., 2006), coef-

Table 4. Summary of the validation results for three cases. Median and standard deviation (SD) refer to the validation ensemble ranked based on their total Φ.

Metrics		Q-only		Spatial-only	Q and spatial	
	Gauge (SD)	Median	Best cal.	Single (SD)	Median	Best
KGE (−)	(A)	0.83 (0.01)	0.86	−1.65	0.86 (0.01)	0.88
KGE (−)	(B)	0.89 (0.02)	0.93	−1.40	0.87 (0.01)	0.88
KGE_{low} (−)	(A)	0.70 (0.04)	0.79	−3.84	0.72 (0.02)	0.76
KGE_{low} (−)	(B)	0.65 (0.02)	0.66	−3.72	0.64 (0.03)	0.65
BIAS (%)	(A)	−15.76 (2.26)	−10.38	29.64	−11.89 (1.15)	−8.85
BIAS (%)	(B)	5.23 (1.04)	5.39	55.86	5.00 (1.45)	9.13
April – SPAEF		−0.76 (0.30)	−0.11	0.47	0.51 (0.02)	0.53
May – SPAEF		−0.65 (0.28)	−0.09	0.56	0.51 (0.03)	0.51
June – SPAEF		−0.50 (0.19)	−0.13	0.38	0.27 (0.04)	0.27
July – SPAEF		−0.56 (0.17)	−0.30	0.59	0.48 (0.09)	0.50
August – SPAEF		−0.15 (0.10)	−0.33	0.18	0.19 (0.07)	0.21
September – SPAEF		−0.12 (0.13)	−0.31	0.44	0.35 (0.02)	0.37

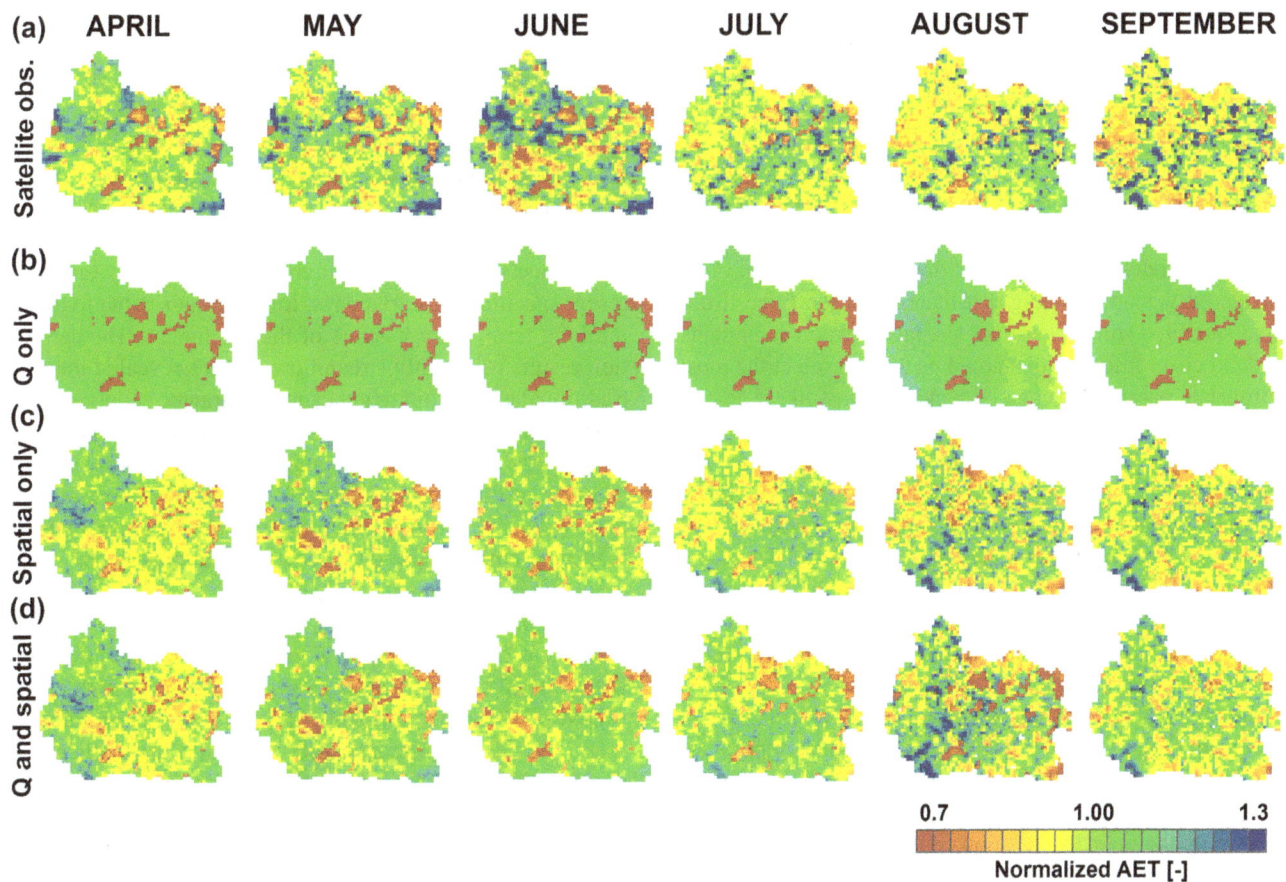

Figure 4. Three different calibration strategies: streamflow-only **(a)**, spatial-only **(b)**, and streamflow and spatial together **(c)** are compared with monthly TSEB estimates **(d)**. Calibrations are evaluated for monthly averages from April to September using cloud-free days. Note that these maps are normalised with their mean to use one representative colour bar and highlight the pattern information.

ficient of variation, Goodman and Kruskal's lambda (Goodman and Kruskal, 1954), agreement coefficient (Ji and Gallo, 2006), Theil's uncertainty, EOF, and Cramér's V (Cramér, 1946; Koch et al., 2015; Rees, 2008) proved to be inadequate in a calibration framework, since undesired visual patterns were achieved, e.g. with high correlation, but too-low standard deviation or highly separate clusters. Therefore, we developed the SPAEF metric which proved to be very efficient for calibrating the model to a satisfying spatial pattern by combining correlation coefficient, coefficient of variation ratio and histogram overlap in a robust metric that guides the model calibration well. It is our experience and recommendation that incorporating the spatial dimension in all aspects of the distributed hydrological model development from model structure, parameterisation, metric selection, sensitivity analysis and calibration is essential in order to achieve significant improvement in the spatial pattern performance of a model. We believe that traditional downstream discharge measurements contain much more accurate and robust information on the overall water balance compared to the non-continuous remotely sensed estimates, and therefore, the model constraint on biases should only originate from these streamflow observations. Conversely, it is well-known that aggregated streamflow measurements contain no information on spatial patterns upstream of the measurement (Stisen et al., 2011b). Therefore, the combination of satellite-derived patterns and aggregated streamflow measurements are an ideal way of constraining distributed hydrological models. In fact, spatial patterns should always be considered when evaluating distributed models. Even if detailed satellite estimates are not available, expert judgments and land cover information should be used to select the most appropriate parameter set (producing the most likely spatial patterns) among equally likely solutions obtained through discharge-only calibration. When a distributed model is applied, ideally it should not only produce satisfying discharge simulations, but at the same time it should also produce realistic spatial patterns of states and fluxes such as AET and soil moisture. White et al. (2017) also highlighted the importance of getting the spatial patterns right in their study since constraining the model against streamflow alone did not secure robust land cover change scenario modelling.

The monthly spatial maps are built based on the AET patterns from cloud-free days. Here, we ignore the temporal aspect and focus only on the consistent spatial patterns for each month of the growing season. The advantage of this approach is that only the main information content of the satellite data, their spatial patterns, are utilised while the uncertainty associated with the absolute values of the AET estimates are not influencing the calibrations. In addition, the simulated monthly mean AET maps reflect mainly the model parameterisation and to a lesser degree the day-to-day variation in climate forcing. This is desirable since the aim of the model calibration is to optimise the model parameterisation with a given climate forcing dataset. The current calibration framework builds on the assumption that the satellite-based estimate of AET patterns approximate an observed pattern that is suitable for model optimisation. In general, the calibration approach is deterministic by nature and does not consider error or uncertainties in either observed discharge or AET patterns. Future work could add this component to the approach. However, assessment of the uncertainties in the observed spatial patterns are far from straightforward, since the uncertainties of interest with the proposed approach are solely related to the uncertainties related to the spatial patterns and not to biases. Therefore, quantification of pattern uncertainties would require a very dense network of actual evapotranspiration measurements.

The calibration results obtained in the current study, where three strategies were tested with varying combinations of objective functions, showed that with an appropriate metric design, limited trade-offs can be achieved when combining streamflow and spatial pattern metrics in a joint calibration framework. This is largely attributed to the nature of the metric, as the spatial performance metric is bias-insensitive whereas the streamflow metrics have very little sensitivity to spatial redistribution of AET patterns as long as the spatial averages remain unchanged. Bias and temporal variability of satellite-derived AET estimates could also be useful for model optimisation; however, in this study, we deliberately limited the information content of the satellite data to address the spatial patterns. This was done because even though the satellite-based AET estimate is validated against eddy-covariance stations (Mendiguren et al., 2017) they only represent specific cloud-free days, limiting their value to assess the long term water balance of the catchment. The calibration results using only streamflow metrics revealed that this traditional calibration target cannot guarantee satisfying spatial pattern performance even though the model structure and parameterisation framework enables this without much compromise, as illustrated by the performance of the combined Q and Spatial calibrations which resulted in very similar performance of both streamflow and spatial patterns as the single objective calibrations individually.

The spatial model parameterisation applied in Skjern catchment can be site-specific due to the uniform land use (agricultural cropland) across soils ranging from very coarse sandy soil to more loamy soils whereas the calibration framework and SPAEF metric can be applied to any other river basin in the world. Regarding the dynamic scaling function, developed for incorporating remotely sensed LAI in ET_{ref} scaling, it should be noted that the use of LAI to describe the deviation of each grid cell from the assumed reference grass is a simplification. Albedo could also have been included in the dynamic scaling function; however, one could argue that albedo and LAI are somewhat correlated and including one of them is already contributing the information about the other (Chen et al., 2005; Liu et al., 2017; Stisen et al., 2008). Moreover, we limit this study to temporally averaged spatial patterns of AET and deliberately choose to ignore the day-to-day dynamics of AET. In this study, spatially varying but temporally constant field-capacity-dependent root fraction is utilised; however, it would be more elegant and physically more sound to represent the seasonality in root-growth dynamics more realistically by implementing a seasonally varying root fraction coefficient (beta) that is similar to the concept of LAI-based PET correction using the DSF module.

7   Conclusions

Our study aimed at parameterising a distributed hydrologic model for simulating distributed actual evapotranspiration patterns before an ensemble calibration using satellite-based data. This order is crucial for progressive hydrologic modelling with flexible model structure based on open-source philosophy. All these steps should be suitable for the catchment to give the model enough flexibility to adjust to pattern observations. The calibration efforts will have a limited effect on spatial patterns if the model parameterisation has not been investigated with pattern performance in mind. Ideally, the models should offer different parameterisation schemes or at least have room for development based on open-source philosophy so that we can test different spatial parameterisations for a particular calibration goal. Here, we implemented a field-capacity-dependent root fraction coefficient determining the root profile over depth for different soil and vegetation types. We introduced a dynamic scaling function which imprints the leaf area index in the potential evapotranspiration. After organising the spatial parameterisation of the model in a parsimonious manner, we also reduced the number of parameters using sequential screening. Only the informative parameters from the sequential screening are used in the subsequent ensemble calibration exercise. We then assessed the effect of different calibration strategies including monthly spatial patterns of actual evapotranspiration in combination with traditional streamflow observations. In the spatial calibration, the agreement between observed and simulated spatial patterns is added as a part of the objective function used for model optimisation. For that a multi-component bias-insensitive spatial efficiency metric is used to evaluate the simulated AET maps. The following conclusions can be drawn from our results:

- Preparing the model parameterisation for spatial calibration is a key element for achieving the calibration objectives. More specifically, the model parameterisation needs to be designed to allow the spatial parameter distribution to be optimized through calibration.

- The newly proposed spatial efficiency metric (SPAEF) has proven to be robust and easy to interpret due to its three distinct and complementary components of correlation, variance and histogram matching.

- Based on the multi-component calibration results, including spatial pattern information in calibration sig-

nificantly improves the spatial model simulations while maintaining similar streamflow performance. For the combined calibration, there is a limited trade-off between streamflow and spatial patterns for the best-performing calibration ensemble compared to the Q-only calibration. However, this trade-off disappears in the validation test, indicating that a more robust parameter set is achieved during the combined Q and Spatial calibration.

Overall, the hydrological modelling community can benefit from building familiarity with several aspects of spatial model evaluation, including spatial parameterisation and multi-component spatial performance metrics.

Code availability. Pre-processing ET with crop coefficient type dynamic scaling function is available in the mHM v5.7 and later versions (www.ufz.de/mhm/ and https://github.com/mhm-ufz/mhm). The Python and MATLAB scripts for spatial efficiency (SPAEF) and a tutorial are available in the SPACE project website (http://www.space.geus.dk/), GitHub (https://github.com/cuneyd/spaef/) and via a Researchgate repository (Demirel et al., 2017).

Competing interests. The authors declare that they have no conflict of interest.

Acknowledgements. We would like to thank the three reviewers for their useful and constructive feedback. We acknowledge the financial support for the SPACE project by the Villum Foundation (http://villumfonden.dk/) through their Young Investigator Programme (grant VKR023443). The TSEB code is retrieved from https://github.com/hectornieto/pyTSEB. All MODIS data were retrieved from the online Data Pool, courtesy of the NASA Land Processes Distributed Active Archive Center (LP DAAC), USGS/Earth Resources Observation and Science (EROS) Center, Sioux Falls, South Dakota, https://lpdaac.usgs.gov/data_access/data_pool.

Edited by: Florian Pappenberger

References

Allen, R. G., Pereira, L. S., Raes, D., and Smith, M.: Crop Evapotranspiration – Guidelines for Computing Crop Water Requirements, FAO Irrigation and drainage paper 56, http://www.fao.org/docrep/x0490e/x0490e00.htm (last access: 16 February 2018), 1998.

Berezowski, T., Nossent, J., Chormański, J., and Batelaan, O.: Spatial sensitivity analysis of snow cover data in a distributed rainfall-runoff model, Hydrol. Earth Syst. Sci., 19, 1887–1904, https://doi.org/10.5194/hess-19-1887-2015, 2015.

Beven, K. and Freer, J.: Equifinality, data assimilation, and uncertainty estimation in mechanistic modelling of complex environmental systems using the GLUE methodology, J. Hydrol., 249, 11–29, https://doi.org/10.1016/S0022-1694(01)00421-8, 2001.

Campolongo, F., Cariboni, J., and Saltelli, A.: An effective screening design for sensitivity analysis of large models, Environ. Model. Softw., 22, 1509–1518, https://doi.org/10.1016/j.envsoft.2006.10.004, 2007.

Chen, J. M., Chen, X., Ju, W., and Geng, X.: Distributed hydrological model for mapping evapotranspiration using remote sensing inputs, J. Hydrol., 305, 15–39, https://doi.org/10.1016/j.jhydrol.2004.08.029, 2005.

Conradt, T., Wechsung, F., and Bronstert, A.: Three perceptions of the evapotranspiration landscape: comparing spatial patterns from a distributed hydrological model, remotely sensed surface temperatures, and sub-basin water balances, Hydrol. Earth Syst. Sci., 17, 2947–2966, https://doi.org/10.5194/hess-17-2947-2013, 2013.

Corbari, C., Ravazzani, G., Ceppi, A., and Mancini, M.: Multipixel Calibration of a Distributed Energy Water Balance Model Using Satellite Data of Land Surface Temperature and Eddy Covariance Data, Proced. Environ. Sci., 19, 285–292, https://doi.org/10.1016/j.proenv.2013.06.033, 2013.

Cornelissen, T., Diekkrüger, B., and Bogena, H.: Using High-Resolution Data to Test Parameter Sensitivity of the Distributed Hydrological Model HydroGeoSphere, Water, 8, 202, https://doi.org/10.3390/w8050202, 2016.

Cramér, H.: Mathematical Methods of Statistics, Princeton University Press, Princeton, 1946.

Crow, W. T., Wood, E. F., Pan, M., de Wit, M., Stankiewicz, J., Crow, W. T., Coe, M. T., and Birkett, C. M.: Multiobjective calibration of land surface model evapotranspiration predictions using streamflow observations and spaceborne surface radiometric temperature retrievals, J. Geophys. Res., 311, 1917–1921, https://doi.org/10.1029/2003WR002543, 2003.

Cuntz, M., Mai, J., Zink, M., Thober, S., Kumar, R., Schäfer, D., Schrön, M., Craven, J., Rakovec, O., Spieler, D., Prykhodko, V., Dalmasso, G., Musuuza, J., Langenberg, B., Attinger, S., and Samaniego, L.: Computationally inexpensive identification of noninformative model parameters by sequential screening, Water Resour. Res., 51, 6417–6441, https://doi.org/10.1002/2015WR016907, 2015.

Dee, D. P., Uppala, S. M., Simmons, A. J., Berrisford, P., Poli, P., Kobayashi, S., Andrae, U., Balmaseda, M. A., Balsamo, G., Bauer, P., Bechtold, P., Beljaars, A. C. M., van de Berg, L., Bidlot, J., Bormann, N., Delsol, C., Dragani, R., Fuentes, M., Geer, A. J., Haimberger, L., Healy, S. B., Hersbach, H., Hólm, E. V., Isaksen, L., Kållberg, P., Köhler, M., Matricardi, M., McNally, A. P., Monge-Sanz, B. M., Morcrette, J.-J., Park, B.-K., Peubey, C., de Rosnay, P., Tavolato, C., Thépaut, J.-N., and Vitart, F.: The ERA-Interim reanalysis: configuration and performance of the data assimilation system, Q. J. Roy. Meteorol. Soc., 137, 553–597, https://doi.org/10.1002/qj.828, 2011.

Demirel, M. C., Booij, M. J., and Hoekstra, A. Y.: Effect of different uncertainty sources on the skill of 10 day ensemble low flow forecasts for two hydrological models, Water Resour. Res., 49, 4035–4053, https://doi.org/10.1002/wrcr.20294, 2013.

Demirel, M. C., Koch, J., and Stisen, S.: SPAEF: SPAtial EFficiency, Researchgate, https://doi.org/10.13140/RG.2.2.18400.58884, 2017.

Duan, Q.-Y. Y., Sorooshian, S., and Gupta, V.: Effective and efficient global optimization for conceptual rainfall-runoff models, Water Resour. Res., 28, 1015–1031, https://doi.org/10.1029/91WR02985, 1992.

Githui, F., Thayalakumaran, T., and Selle, B.: Estimating irrigation inputs for distributed hydrological modelling: a case study from an irrigated catchment in southeast Australia, Hydrol. Process., 30, 1824–1835, https://doi.org/10.1002/hyp.10757, 2016.

Goodman, L. A. and Kruskal, W. H.: Measures of Association for Cross Classifications, J. Am. Stat. Assoc., 49, 732–764, https://doi.org/10.1080/01621459.1954.10501231, 1954.

Greve, M. H., Greve, M. B., Bøcher, P. K., Balstrøm, T., Breuning-Madsen, H., and Krogh, L.: Generating a Danish raster-based topsoil property map combining choropleth maps and point information, Geogr. Tidsskr. J. Geogr., 107, 1–12, https://doi.org/10.1080/00167223.2007.10649565, 2007.

Gupta, H. V, Kling, H., Yilmaz, K. K., and Martinez, G. F.: Decomposition of the mean squared error and NSE performance criteria: Implications for improving hydrological modelling, J. Hydrol., 377, 80–91, https://doi.org/10.1016/j.jhydrol.2009.08.003, 2009.

Guzinski, R., Anderson, M. C., Kustas, W. P., Nieto, H., and Sandholt, I.: Using a thermal-based two source energy balance model with time-differencing to estimate surface energy fluxes with day–night MODIS observations, Hydrol. Earth Syst. Sci., 17, 2809–2825, https://doi.org/10.5194/hess-17-2809-2013, 2013.

Hargrove, W. W., Hoffman, F. M., and Hessburg, P. F.: Mapcurves: a quantitative method for comparing categorical maps, J. Geogr. Syst., 8, 187–208, https://doi.org/10.1007/s10109-006-0025-x, 2006.

Hendricks Franssen, H. J., Brunner, P., Makobo, P., and Kinzelbach, W.: Equally likely inverse solutions to a groundwater flow problem including pattern information from remote sensing images, Water Resour. Res., 44, 224–240, https://doi.org/10.1029/2007WR006097, 2008.

Hunink, J. E., Eekhout, J. P. C., de Vente, J., Contreras, S., Droogers, P., and Baille, A.: Hydrological Modelling using Satellite-based Crop Coefficients: a Comparison of Methods at the Basin Scale, Remote Sens., 9, 174, https://doi.org/10.3390/rs9020174, 2017.

Immerzeel, W. W. and Droogers, P.: Calibration of a distributed hydrological model based on satellite evapotranspiration, J. Hydrol., 349, 411–424, https://doi.org/10.1016/j.jhydrol.2007.11.017, 2008.

Jackson, R. B., Canadell, J., Ehleringer, J. R., Mooney, H. A., Sala, O. E., and Schulze, E. D.: A global analysis of root distributions for terrestrial biomes, Oecologia, 108, 389–411, https://doi.org/10.1007/BF00333714, 1996.

Jensen, H. E., Jensen, S. E., Jensen, C. R., Mogensen, V. O. and Hansen, S.: Jordfysik og jordbrugsmeteorologi, Jordbrugsforlaget., 2001.

Jensen, K. H. and Illangasekare, T. H.: HOBE: A Hydrological Observatory, Vadose Zone J., 10, 1–7, https://doi.org/10.2136/vzj2011.0006, 2011.

Ji, L. and Gallo, K.: An agreement coefficient for image comparison, Photogramm. Eng. Remote Sens., 72, 823–833, 2006.

Kalma, J. D., McVicar, T. R., and McCabe, M. F.: Estimating Land Surface Evaporation: A Review of Methods Using Remotely Sensed Surface Temperature Data, Surv. Geophys., 29, 421–469, https://doi.org/10.1007/s10712-008-9037-z, 2008.

Kling, H. and Gupta, H.: On the development of regionalization relationships for lumped watershed models: The impact of ignoring sub-basin scale variability, J. Hydrol., 373, 337–351, https://doi.org/10.1016/j.jhydrol.2009.04.031, 2009.

Koch, J., Jensen, K. H., and Stisen, S.: Toward a true spatial model evaluation in distributed hydrological modeling: Kappa statistics, Fuzzy theory, and EOF-analysis benchmarked by the human perception and evaluated against a modeling case study, Water Resour. Res., 51, 1225–1246, https://doi.org/10.1002/2014WR016607, 2015.

Koch, J., Mendiguren, G., Mariethoz, G., and Stisen, S.: Spatial Sensitivity Analysis of Simulated Land Surface Patterns in a Catchment Model Using a Set of Innovative Spatial Performance Metrics, J. Hydrometeorol., 18, 1121–1142, https://doi.org/10.1175/JHM-D-16-0148.1, 2017.

Kumar, R., Samaniego, L., and Attinger, S.: Implications of distributed hydrologic model parameterization on water fluxes at multiple scales and locations, Water Resour. Res., 49, 360–379, https://doi.org/10.1029/2012WR012195, 2013.

Larsen, M. A. D., Refsgaard, J. C., Jensen, K. H., Butts, M. B., Stisen, S., and Mollerup, M.: Calibration of a distributed hydrology and land surface model using energy flux measurements, Agr. Forest Meteorol., 217, 74–88, https://doi.org/10.1016/j.agrformet.2015.11.012, 2016.

Li, H. T., Brunner, P., Kinzelbach, W., Li, W. P., and Dong, X. G.: Calibration of a groundwater model using pattern information from remote sensing data, J. Hydrol., 377, 120–130, https://doi.org/10.1016/j.jhydrol.2009.08.012, 2009.

Liu, C., Sun, G., McNulty, S. G., Noormets, A., and Fang, Y.: Environmental controls on seasonal ecosystem evapotranspiration / potential evapotranspiration ratio as determined by the global eddy flux measurements, Hydrol. Earth Syst. Sci., 21, 311–322, https://doi.org/10.5194/hess-21-311-2017, 2017.

Loosvelt, L., Vernieuwe, H., Pauwels, V. R. N., De Baets, B., and Verhoest, N. E. C.: Local sensitivity analysis for compositional data with application to soil texture in hydrologic modelling, Hydrol. Earth Syst. Sci., 17, 461–478, https://doi.org/10.5194/hess-17-461-2013, 2013.

Madsen, H. B.: Distribution of spring barley roots in Danish soils, of different texture and under different climatic conditions, Plant Soil, 88, 31–43, https://doi.org/10.1007/BF02140664, 1985.

Madsen, H. B.: Computerized soil data used in agricultural water planning, Denmark, Soil Use Manage., 2, 134–139, https://doi.org/10.1111/j.1475-2743.1986.tb00697.x, 1986.

Madsen, H. B. and Platou, S. W.: Land use planning in Denmark: the use of soil physical data in irrigation planning, Hydrol. Res., 14, 267–276, 1983.

Melsen, L., Teuling, A., Torfs, P., Zappa, M., Mizukami, N., Clark, M., and Uijlenhoet, R.: Representation of spatial and temporal variability in large-domain hydrological models: case study for a mesoscale pre-Alpine basin, Hydrol. Earth Syst. Sci., 20, 2207–2226, https://doi.org/10.5194/hess-20-2207-2016, 2016.

Mendiguren, G., Koch, J., and Stisen, S.: Spatial pattern evaluation of a calibrated national hydrological model – a remote-sensing-based diagnostic approach, Hydrol. Earth Syst. Sci., 21, 5987–6005, https://doi.org/10.5194/hess-21-5987-2017, 2017.

Mizukami, N., Clark, M., Newman, A. J., Wood, A. W., Gutmann, E., Nijssen, B., Rakovec, O., and Samaniego, L.: Toward seamless large domain parameter estimation for hydrologic models, Water Resour. Res., 53, 8020–8040, https://doi.org/10.1002/2017WR020401, 2017.

Morris, M. D.: Factorial Sampling Plans for Preliminary Computational Experiments, Technometrics, 33, 161–174, https://doi.org/10.2307/1269043, 1991.

Norman, J. M., Kustas, W. P., and Humes, K. S.: Source approach for estimating soil and vegetation energy fluxes in observations of directional radiometric surface temperature, Agr. Forest Meteorol., 7, 263–293, https://doi.org/10.1016/0168-1923(95)02265-Y, 1995.

Priestley, C. H. B. and Taylor, R. J.: On the Assessment of Surface Heat Flux and Evaporation Using Large-Scale Parameters, Mon. Weather Rev., 100, 81–92, https://doi.org/10.1175/1520-0493(1972)100<0081:OTAOSH>2.3.CO;2, 1972.

Rakovec, O., Kumar, R., Attinger, S., and Samaniego, L.: Improving the realism of hydrologic model functioning through multivariate parameter estimation, Water Resour. Res., 52, 7779–7792, https://doi.org/10.1002/2016WR019430, 2016.

Rees, W. G.: Comparing the spatial content of thematic maps, Int. J. Remote Sens., 29, 3833–3844, https://doi.org/10.1080/01431160701852088, 2008.

Samaniego, L., Kumar, R., and Attinger, S.: Multiscale parameter regionalization of a grid-based hydrologic model at the mesoscale, Water Resour. Res., 46, W05523, https://doi.org/10.1029/2008WR007327, 2010.

Samaniego, L., Kumar, R., Mai, J., Zink, M., Thober, S., Cuntz, M., Rakovec, O., Schäfer, D., Schrön, M., Brenner, J., Demirel, M. C., Kaluza, M., Langenberg, B., Stisen, S., and Attinger, S.: Mesoscale Hydrologic Model, https://doi.org/10.5281/zenodo.1069203, 2017.

Schumann, G. J. P., Neal, J. C., Voisin, N., Andreadis, K. M., Pappenberger, F., Phanthuwongpakdee, N., Hall, A. C., and Bates, P. D.: A first large scale flood inundation forecasting model, Water Resour. Res., 49, 6248–6257, https://doi.org/10.1002/wrcr.20521, 2013.

Schuurmans, J. M., van Geer, F. C., and Bierkens, M. F. P.: Remotely sensed latent heat fluxes for model error diagnosis: a case study, Hydrol. Earth Syst. Sci., 15, 759–769, https://doi.org/10.5194/hess-15-759-2011, 2011.

Shin, M.-J., Guillaume, J. H. A., Croke, B. F. W., and Jakeman, A. J.: Addressing ten questions about conceptual rainfall–runoff models with global sensitivity analyses in R, J. Hydrol., 503, 135–152, https://doi.org/10.1016/j.jhydrol.2013.08.047, 2013.

Stisen, S., Jensen, K. H., Sandholt, I., and Grimes, D. I. F.: A remote sensing driven distributed hydrological model of the Senegal River basin, J. Hydrol., 354, 131–148, https://doi.org/10.1016/j.jhydrol.2008.03.006, 2008.

Stisen, S., Sonnenborg, T. O., Højberg, A. L., Troldborg, L., and Refsgaard, J. C.: Evaluation of Climate Input Biases and Water Balance Issues Using a Coupled Surface–Subsurface Model, Vadose Zone J., 10, 37–53, https://doi.org/10.2136/vzj2010.0001, 2011a.

Stisen, S., McCabe, M. F., Refsgaard, J. C., Lerer, S., and Butts, M. B.: Model parameter analysis using remotely sensed pattern information in a multi-constraint framework, J. Hydrol., 409, 337–349, https://doi.org/10.1016/j.jhydrol.2011.08.030, 2011b.

Stisen, S., Koch, J., Sonnenborg, T. O., Refsgaard, J. C., Bircher, S., Ringgaard, R., and Jensen, K. H.: Moving beyond runoff calibration – Multi-constraint optimization of a surface-subsurface-atmosphere model, Hydrol. Process., submitted, 2017.

Su, Z.: The Surface Energy Balance System (SEBS) for estimation of turbulent heat fluxes, Hydrol. Earth Syst. Sci., 6, 85–100, https://doi.org/10.5194/hess-6-85-2002, 2002.

Swain, M. J. and Ballard, D. H.: Color indexing, Int. J. Comput. Vis., 7, 11–32, https://doi.org/10.1007/BF00130487, 1991.

Vazquez, J. A., Anctil, F., Ramos, M. H., Perrin, C., and Velázquez, J. A.: Can a multi-model approach improve hydrological ensemble forecasting? A study on 29 French catchments using 16 hydrological model structures, Adv. Geosci., 29, 33–42, https://doi.org/10.5194/adgeo-29-33-2011, 2011.

Velázquez, J. A., Anctil, F., Perrin, C., and Vazquez, J. A.: Performance and reliability of multimodel hydrological ensemble simulations based on seventeen lumped models and a thousand catchments, Hydrol. Earth Syst. Sci., 14, 2303–2317, https://doi.org/10.5194/hess-14-2303-2010, 2010.

Wei, Z., Yoshimura, K., Wang, L., Miralles, D. G., Jasechko, S., and Lee, X.: Revisiting the contribution of transpiration to global terrestrial evapotranspiration, Geophys. Res. Lett., 44, 2792–2801, https://doi.org/10.1002/2016GL072235, 2017.

White, J., Stengel, V., Rendon, S., and Banta, J.: The importance of parameterization when simulating the hydrologic response of vegetative land-cover change, Hydrol. Earth Syst. Sci., 21, 3975–3989, https://doi.org/10.5194/hess-21-3975-2017, 2017.

Zhang, Y., Chiew, F. H. S., Zhang, L., and Li, H.: Use of Remotely Sensed Actual Evapotranspiration to Improve Rainfall–Runoff Modeling in Southeast Australia, J. Hydrometeorol., 10, 969–980, https://doi.org/10.1175/2009JHM1061.1, 2009.

Impact of remotely sensed soil moisture and precipitation on soil moisture prediction in a data assimilation system with the JULES land surface model

Ewan Pinnington[1,2], Tristan Quaife[1,2], and Emily Black[1,3]

[1]Department of Meteorology, University of Reading, Reading, UK
[2]National Centre for Earth Observation, University of Reading, Reading, UK
[3]National Centre for Atmospheric Science, University of Reading, Reading, UK

Correspondence: Ewan Pinnington (e.pinnington@reading.ac.uk)

Abstract. We show that satellite-derived estimates of shallow soil moisture can be used to calibrate a land surface model at the regional scale in Ghana, using data assimilation techniques. The modified calibration significantly improves model estimation of soil moisture. Specifically, we find an 18 % reduction in unbiased root-mean-squared differences in the north of Ghana and a 21 % reduction in the south of Ghana for a 5-year hindcast after assimilating a single year of soil moisture observations to update model parameters. The use of an improved remotely sensed rainfall dataset contributes to 6 % of this reduction in deviation for northern Ghana and 10 % for southern Ghana. Improved rainfall data have the greatest impact on model estimates during the seasonal wetting-up of soil, with the assimilation of remotely sensed soil moisture having greatest impact during drying-down. In the north of Ghana we are able to recover improved estimates of soil texture after data assimilation. However, we are unable to do so for the south. The significant reduction in unbiased root-mean-squared difference we find after assimilating a single year of observations bodes well for the production of improved land surface model soil moisture estimates over sub-Saharan Africa.

1 Introduction

In regions where the population relies on subsistence farming it is soil moisture, rather than precipitation per se, that is the critical factor in growing crops. The production of improved soil moisture forecasts should therefore enhance the drought resilience of these regions through improved capacity for early warning agricultural drought (Brown et al., 2017). Soil moisture is also an important variable for weather and climate prediction (Seneviratne et al., 2010), playing a key role in controlling land surface energy partitioning (Beljaars et al., 1996; Bateni and Entekhabi, 2012) and in the carbon cycle (McDowell, 2011). However, modelling soil moisture is complex and exhibits large sensitivities to meteorological forcing data and land surface model parameterisations (Pitman et al., 1999).

Globally, precipitation is the most influential meteorological driver in the estimation of soil moisture (Guo et al., 2006). However there is considerable variability in available precipitation data, which in turn has impacts on modelled predictions of soil moisture. When forcing a global land data assimilation system with different precipitation products Gottschalck et al. (2005) showed that the percentage difference in estimates of volumetric soil water content ranged between −75 and +100 %. Similarly Liu et al. (2011) showed that driving a catchment land surface model with an improved precipitation product (merged gauge and satellite observations vs. a reanalysis product) increased the model soil moisture skill by 14 %, when compared to in situ observations.

There are now a variety of remotely sensed surface soil moisture observational products from both active and passive microwave sensors. Data assimilation (DA) has been used to combine information from these observations with land surface models to improve surface soil moisture estimates (Liu et al., 2011; De Lannoy and Reichle, 2016; Yang et al.,

2016). DA refers to the suite of mathematical techniques used to combine models and observations, combining available knowledge about their respective uncertainties. These techniques are typically derived from a Bayesian standpoint and can be broadly classified as sequential and variational. Sequential methods adjust the model state and/or parameters at the time when observations are available, whereas variational methods adjust state and/or parameters at the beginning of some time window considering all observations within that window.

It has been shown by Bolten et al. (2010) that assimilation of remotely sensed surface soil moisture can significantly improve the prediction of root-zone soil moisture and drought modelling, in which a sequential DA technique is used for soil moisture state estimation. Many other recent studies also use sequential assimilation methods to update the model soil moisture state at each time step when an observation is available (Liu et al., 2011; Draper et al., 2012; De Lannoy and Reichle, 2016; Kolassa et al., 2017). In addition some studies employing sequential methods estimate the model parameters as well as the state (Moradkhani et al., 2005; Qin et al., 2009; Montzka et al., 2011). Using sequential methods in this way will likely result in parameters that vary over time, which will not be optimal when using land surface models to run forecasts because the time-varying nature of the parameters will not be carried forward. An alternative is to use variational assimilation methods for parameter estimation (Navon, 1998). Variational methods will yield time-invariant parameter estimates over the assimilation time window. For a suitably chosen length of assimilation window (i.e. over one or more whole years) this allows us to avoid seasonally varying parameters. Using variational methods to assimilate remotely sensed observations for land surface model parameter estimation has previously been shown to improve soil moisture estimates in several studies (Yang et al., 2007, 2009; Rasmy et al., 2011; Sawada and Koike, 2014; Yang et al., 2016). These studies all optimise both model parameters and state. Here we propose an alternative, which is to include the model spin-up within the data assimilation routine so that the initial soil moisture state is consistent with the updated parameters at each optimisation step.

The work in this paper forms part of the Enhancing Resilience to Agricultural Drought in Africa through Improved Communication of Seasonal Forecasts (ERADACS) project. Part of ERADACS is the development of TAMSAT-ALERT (Tropical Applications of Meteorology using SATellite data and ground-based observations AgricuLtural dEcision suppoRT), a light-weight system for prediction of agricultural drought in northern Ghana. Previous work (Brown et al., 2017) has shown that TAMSAT-ALERT's skill for predicting root-zone soil moisture in Ghana ensues largely from accurate knowledge of antecedent soil moisture conditions. In this paper we describe a method for improving soil moisture estimates for the Joint UK Land Environment Simulator (JULES; see Sect. 2.1) over Ghana through the assimilation of remotely sensed soil moisture and use of improved satellite-observed rainfall. Ultimately, we expect that the improved soil moisture estimates will increase the prediction skill of TAMSAT-ALERT, and hence the quality of drought early warning issued to farmers. We use the technique of four-dimensional variational (4D-Var) data assimilation to estimate the soil thermal and hydraulic parameters of JULES by assimilating European Space Agency Climate Change Initiative (ESA CCI) merged active and passive microwave surface soil moisture observations (Dorigo et al., 2015). We also drive the JULES model with two successive versions of the TAMSAT rainfall dataset (see Sect. 2.2) to investigate the effect of improved precipitation on soil moisture estimates. We assimilate a single year of soil moisture observations (2009), then perform a 5-year hindcast (2010–2014), driving the model with reanalysis meteorology, to judge the impact of both the precipitation products and data assimilation on the model's representation of soil moisture when compared to independent observations.

2   Method

2.1   JULES land surface model

The Joint UK Land Environment Simulator (JULES) is a process-based land surface model developed at the UK Met Office (Best et al., 2011; Clark et al., 2011). We used the global land configuration 4.0 of JULES designed for use across weather and climate modelling timescales and systems (Walters et al., 2014). JULES is typically run with 4 soil layers, with the top layer being 10 cm deep. In this paper we have updated JULES to run with a top layer of 5 cm to be more representative of the ESA CCI soil moisture observations. Another option to deal with the issue of representativity would be to use an exponential filter (Albergel et al., 2008) which has been used in sequential data assimilation studies previously (Massari et al., 2015; Alvarez-Garreton et al., 2016). The model is forced with WFDEI data (WATCH Forcing Data methodology applied to ERA-Interim reanalysis data), described by Weedon et al. (2014), for radiation, wind, temperature, pressure and humidity values. The WFDEI data have a 0.5° spatial resolution and a 3-hourly temporal resolution. The JULES model was run at a half-hourly timestep, with a soil map taken from the harmonised world soil database (Nachtergaele et al., 2008). Previously JULES has been used in sequential DA experiments (Ghent et al., 2010), and has been implemented in a variational framework with focus on the carbon cycle (Raoult et al., 2016).

2.2   TAMSAT rainfall observations

We replaced the precipitation in the WFDEI data with Tropical Applications of Meteorology using SATellite data and ground-based observations (TAMSAT) rainfall monitoring products (Maidment et al., 2014; Tarnavsky et al., 2014). TAMSAT produces daily rainfall estimates over Africa at a 4 km resolution with data ranging back to 1983. The rainfall estimates are derived from Meteosat thermal infrared images calibrated against an extensive network of African rain gauges. When aggregated over time and space, TAMSAT has been shown to have good skill over much of Africa, in comparison to ground-based observations (Maidment et al., 2013, 2017). On daily timescales, occurrence is better represented than amount (Greatrex et al., 2014), with the magnitude of high intensity rainfall events not captured. For these reasons, TAMSAT tends to be used to monitor drought rather than to provide real-time early warning of floods. Data are available from https://www.tamsat.org.uk (last access: 20 April 2018).

We ran JULES with WFDEI 3-hourly meteorological forcing data (Weedon et al., 2014) and TAMSAT daily rainfall estimates. Therefore we had to disaggregate the TAMSAT daily estimates to 3-hourly estimates. We did this by merging the TAMSAT data with the WFDEI precipitation data. We divided the WFDEI 3-hourly precipitation values by the corresponding WFDEI daily precipitation and then multiplied these values by the corresponding TAMSAT daily precipitation values. This spreads the daily TAMSAT estimates over the diurnal cycle.

In this study, we drive the JULES model with two different TAMSAT products (v2.0 and v3.0). The difference between JULES model outputs when forced with these two distinct products will help us to understand the impact of improved precipitation forcing on our estimation of soil moisture. TAMSAT v3.0 differs from TAMSAT v2.0 in that it uses an updated calibration against in situ data that is more representative of local scales. It has been shown that TAMSAT v3.0 has greatly reduced the dry bias present in TAMSAT v2.0 (Maidment et al., 2017) and has eliminated the spatial artefacts. Despite this there are still areas where both products struggle, with coastal regions subject to large amounts of warm rain and sharp topographic contrasts being an example of this. For this reason, inter-annual rainfall variability is less well represented over the south of Ghana than the north. For more information on the differences between the two TAMSAT products see Maidment et al. (2017). In Fig. 1 we show yearly cumulative rainfall averaged over 2009–2014 for TAMSAT v2.0 and v3.0; we can see the different spatial distributions of rain with v3.0 being wetter in the south and v2.0 wetter in the east. To illustrate the difference in the amount of rainfall for the two products we show cumulative rainfall for the period 2009–2014 averaged spatially over Ghana in Fig. 2. This shows TAMSAT v2.0 to be the drier of the two products, as expected.

2.3 ESA CCI soil moisture observations

In this study we use the ESA CCI level 3 version 03.2 combined active and passive soil moisture observations. This product merges data from 11 different sensors, using an algorithm described in Dorigo et al. (2017) to give an estimate of surface soil moisture together with its associated uncertainty. These estimates are assumed to represent the top 2–5 cm of soil. However, observations based on different microwave frequencies and soil moisture conditions may be representative of deeper layers (Ulaby et al., 1982). It has been previously shown that it is best to use both active and passive retrievals together (Draper et al., 2012) and that the ESA CCI merged product performs better than either the active or passive product alone (Dorigo et al., 2015). Dorigo et al. (2015) also show that the ESA CCI product performs well over western Africa when judged against in situ soil moisture observations from the African Monsoon Multidisciplinary Analyses (AMMA) network (Cappelaere et al., 2009), with stations in Benin, Mali and Niger. When judged against the AMMA network, CCI soil moisture was shown to have a high correlation (~ 0.7) and one of the lowest unbiased root-mean-squared differences (~ 0.04) of the 28 worldwide networks used in the study. This bodes well for our comparison over Ghana, which has a similar climate regime in the north to the sites in the AMMA network. Figure 3 shows the number of available daily soil moisture observations in the experiment period (2009–2014) over Ghana, with the maximum number of possible observations being 2190. We can see that there is higher data availability in the north of Ghana than in the south. There are some pixels in the south for which we have no data; this is due to high vegetation cover.

2.4 4D-Variational data assimilation

We use the method of four-dimensional variational data assimilation (4D-Var) to estimate the soil thermal and hydraulic parameters of the JULES land surface model for each grid cell over Ghana (Pinnington, 2017). 4D-Var aims to find the initial state that minimises the weighted least squares distance to the prior guess while minimising the weighted least squares distance of the model trajectory to the observations over the time window. This is done by minimising a cost function at each grid cell:

$$J(\mathbf{x}_0) = \frac{1}{2}(\mathbf{x}_0 - \mathbf{x}_b)^T \mathbf{B}^{-1}(\mathbf{x}_0 - \mathbf{x}_b)$$
$$+ \frac{1}{2}\sum_{i=0}^{N}\left(\mathbf{y}_i - \mathbf{h}_i(\mathbf{x}_0)\right)^T \mathbf{R}_{i,i}^{-1}\left(\mathbf{y}_i - \mathbf{h}_i(\mathbf{x}_0)\right), \qquad (1)$$

where $\mathbf{x}$ is the vector of model parameters, $\mathbf{x}_b$ is a prior guess and $\mathbf{x}_0$ the current update, $\mathbf{B}$ is the prior error covariance matrix, $\mathbf{y}_i$ is the observation at time t_i, $\mathbf{h}_i$ is the observation operator (here the JULES model) mapping the model parameters ($\mathbf{x}_0$) to the observation $\mathbf{y}_i$ at time t_i, $\mathbf{R}_i$ is the observation error covariance matrix and N is the number of observations. We chose a variational DA method for parameter estimation over a sequential method because variational methods ensure that the retrieved model parameters are time-invariant over the assimilation window and will hence fit seasonal model

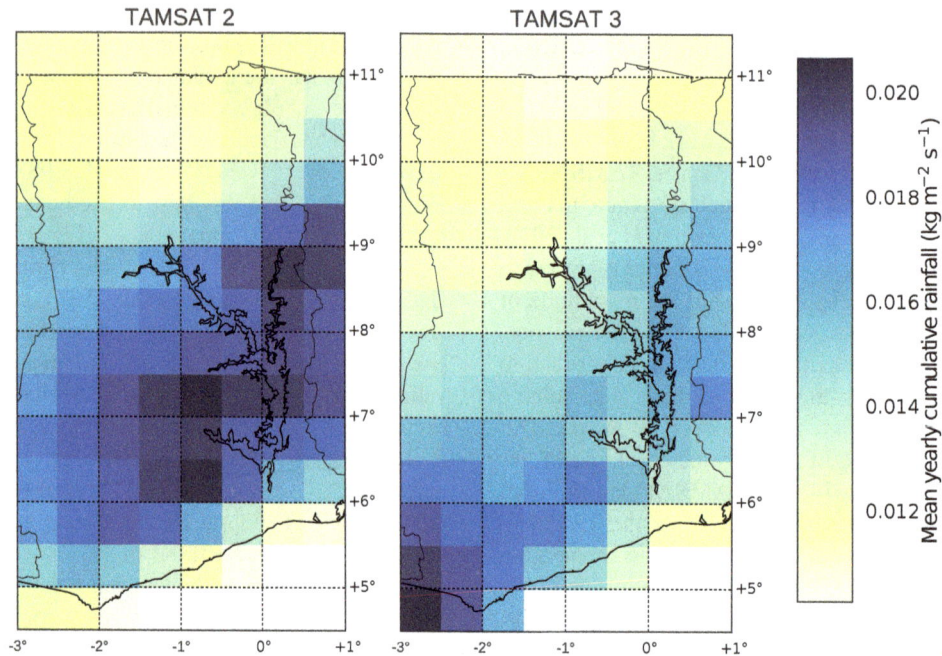

Figure 1. TAMSATv2.0 and v3.0 yearly cumulative rainfall averaged over the 6 years in our experiments (2009–2014).

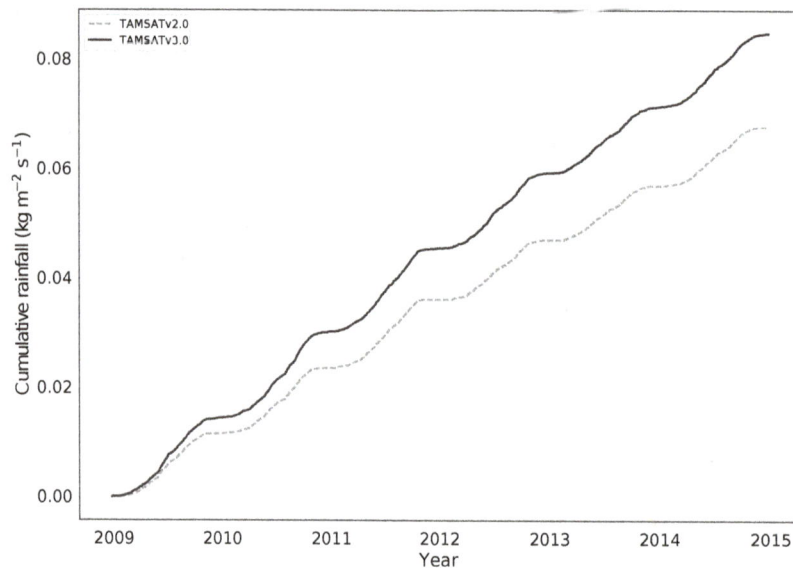

Figure 2. TAMSATv2.0 and v3.0 cumulative rainfall averaged over the whole of Ghana.

dynamics when the window is sufficiently large. As we do not have a good estimate of the error in the prior estimates of model parameters we chose a conservative 5 % standard deviation (SD) for the prior error covariance matrix **B**. This ensures we do not retrieve unrealistic estimates of soil texture after data assimilation. For the observational error covariance matrix **R** we have a diagonal matrix with variances estimated from the SDs included in the ESA CCI soil moisture product.

In this study, we updated the percentage of sand and silt in the soil at each minimisation step (with clay being updated implicitly) and then used a set of pedo-transfer functions (Cosby et al., 1984) to relate the new sand, silt and clay proportions to the eight soil parameters in JULES. This is a similar framework to that introduced in Yang et al. (2007, 2009) for data assimilation with the Simplified Biosphere model 2 (SiB2) (Sellers et al., 1996), with the exception that the soil porosity parameter of JULES is updated implicitly

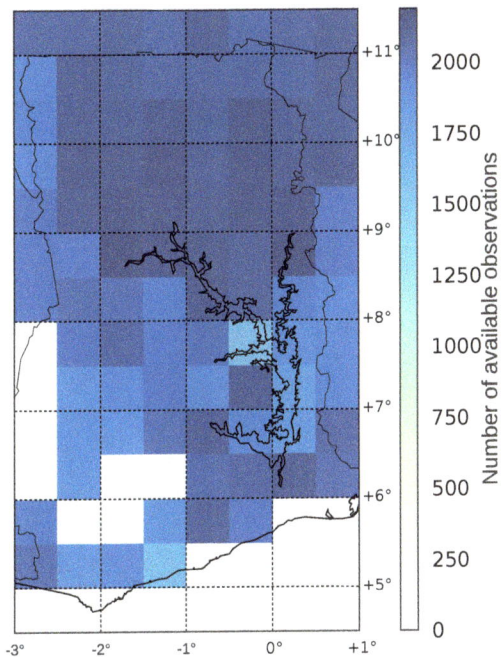

Figure 3. Number of available days of ESA CCI soil moisture observations in the experiment period (2009–2014) out of a maximum of 2190 days.

within the pedo-transfer functions rather than explicitly included in the optimisation. Parameterising the model in this way reduces the issue of equifinality (which potentially arises from minimising eight related parameters) and decreases the convergence time of the minimisation. Our prior guess for the sand, silt and clay values at each grid cell comes from the harmonised world soil database. At each minimisation step after updating the parameters of JULES we included a model spin-up to ensure that the initial soil moisture state is consistent with the updated parameters. We used the Nelder–Mead simplex algorithm (Nelder and Mead, 1965) to minimise the cost function in Eq. (1) without the use of a model adjoint. Whilst an adjoint facilitates efficient calculation of gradients in the cost function it is costly to maintain and keep up-to-date with the latest model version. The only example of an adjoint of JULES for which we are aware is provided by Raoult et al. (2016) and is implemented for version 2.2 of the model, several major versions behind the current release. In future work a 4D-Ensemble-Var (Liu et al., 2008, 2009) approach could prove a useful compromise as it allows for the use of a gradient-based descent algorithm, reducing the total number of function calls required to reach a solution without the use of an adjoint.

2.5 Experimental design

For each data assimilation experiment with JULES (driven with TAMSAT v2.0 or v3.0 rainfall) we assimilate a single

year of ESA CCI soil moisture observations (2009) and then run a 5-year hindcast (2010–2014). The hindcast allows us to evaluate the performance of each experiment against independent soil moisture observations. In our results, we consider four different model runs:

1. JULES model "free-run", driven with TAMSAT v2.0 rainfall ("prior")

2. JULES model after calibration with DA, driven with TAMSAT v2.0 rainfall ("posterior")

3. JULES model "free-run", driven with TAMSAT v3.0 rainfall ("prior")

4. JULES model after calibration with DA, driven with TAMSAT v3.0 rainfall ("posterior").

From these four distinct experiments we can interrogate the impact of both the DA and use of the updated rainfall product.

3 Results

We split our analysis over northern and southern Ghana (above and below 9° N respectively) due to the issues of data quality between the two regions. The data quality of both precipitation and soil moisture is higher in the north than the south and also much of the subsistence agriculture in Ghana takes place in the northern regions, with a higher percentage of cash crops grown in the south (Martey et al., 2013). In Fig. 4 we show the results of a data assimilation and forecast for a single grid cell in the north of Ghana; here both the prior (light grey line) and posterior (dark grey line) are forced with TAMSAT v3.0 precipitation (experiments 3 and 4 respectively, described in Sect. 2.5). From Fig. 4 we can see that the data assimilation has greatly improved the fit to the observations in the assimilation window (2009), which is to be expected, since these observations are what the model is calibrated against. However, the improved fit continues into the forecast (2010–2014) when comparing against the unassimilated observations. We can see a distinct seasonal pattern for soil moisture in northern Ghana, where there is a rainy season and corresponding "wetting-up" of soil moisture from approximately March–May and a dry season with "drying-down" of soil moisture from approximately November–January. The model skill for predicting this seasonal cycle is markedly improved after data assimilation, with a root-mean-squared difference (RMSD) of 0.035 after data assimilation compared to a RMSD of 0.094 before, for 2009. In Fig. 4 we can also see the amplitude of this seasonal cycle slightly decreasing; this is a pattern also seen in both TAMSAT products which exhibit a drying over the period 2010–2014 for this grid cell. In Fig. 5 we show the same model runs for a grid cell in the south of Ghana. The season in the south of Ghana is much less pronounced

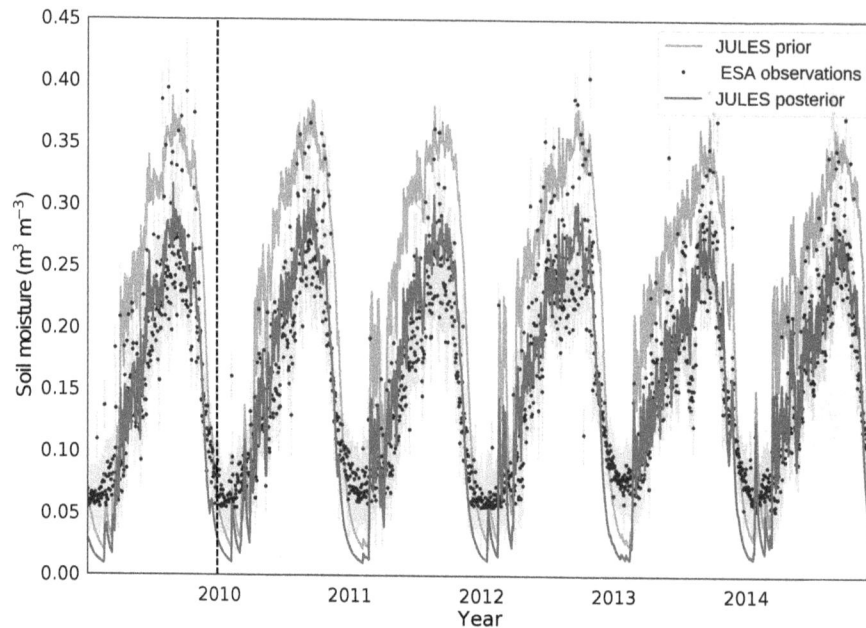

Figure 4. Soil moisture data assimilation results for a northern Ghana grid cell using TAMSAT v3.0 driving data. Light grey line: prior JULES trajectory. Dark grey line: posterior JULES trajectory. Black dots: ESA CCI level 3 soil moisture observations. Faint grey vertical lines: error bars for observations. The vertical dashed line represents the end of the assimilation window.

and this is seen in both the model runs and the observations. However, the observations are of poorer quality in the south due to the higher vegetation cover and cloud cover, adding to the noise seen in Fig. 5. Although we do improve the fit to the observations after data assimilation in Fig. 5 (RMSD of 0.059 after data assimilation compared to RMSD of 0.102 before, for 2009) we do not see the same scale of improvement as for the northern Ghana grid cell in Fig. 4. This is most likely due primarily to the higher error in both the precipitation and soil moisture observations. In addition, the less pronounced seasonal cycle is more difficult to forecast after just assimilating a single year of data. The lower layer soil moisture in JULES responds in a similar way to the top layers shown in Fig. 4 and 5, becoming slightly dried compared to our prior estimates after data assimilation. Without independent observations of these deeper layers it is difficult to know if this is realistic or not.

Figures 4 and 5 show results from experiments 3 and 4 when forcing the JULES model with TAMSAT v3.0 rainfall. In Fig. 6 we show model mean relative error (MRE) (judged against ESA CCI observations in the forecast period, 2010–2014 and calculated as the mean absolute deviation) for wet and dry seasons and experiments 1 to 4. Without DA (top row) we can see that for both wet and dry seasons there is a larger dry MRE in soil moisture in northern Ghana for TAMSAT v2.0 than v3.0 and a larger wet MRE in southern Ghana for TAMSAT v3.0 than v2.0. This finding is consistent with the comparisons of precipitation between v3.0 and v2.0 presented by Maidment et al. (2017), where TAM-

SATv3.0 was shown to reduce a dry bias present in TAMSATv2.0 when compared to ground station data. After DA (bottom row) we can see that the wet MRE in southern Ghana is largely reduced for both TAMSAT v2.0 and v3.0. However, in northern Ghana a dry MRE still remains, with this being slightly drier for TAMSAT v2.0, compared to v3.0.

Figure 7 and 8 show experiment monthly RMSDs for north and south Ghana respectively. For Fig. 7 this shows that the most accurate model run overall is experiment 4 (TAMSAT v3.0 with DA). We see in the majority of years that towards the start of the season as soils are wetting up it is experiment 3 and 4 (TAMSAT v3.0 no DA and with DA respectively) that have the lowest RMSD, suggesting that it is precipitation, as opposed to the assimilation of soil moisture, that is most important for improving soil moisture estimates during this period. This relationship changes towards the end of the rainy season, with experiment 2 and 4 being the most accurate (TAMSAT v2.0 with DA and TAMSAT v3.0 with DA respectively), suggesting that assimilation of soil moisture estimates is most important in this period. In Fig. 8 the most accurate model run is again experiment 4 (TAMSAT v3.0 with DA), although experiment 2 (TAMSAT v2.0 with DA) is much closer in accuracy than for the north. This suggests that both rainfall products are poor in the south compared to the north. We also note that experiment 1 (TAMSAT v2.0 no DA) is markedly more accurate than experiment 3 (TAMSAT v3.0 no DA) in the south. However, considering the results after DA (experiment 4 outperforming experiment 2) this can be explained by an incorrect specification of the

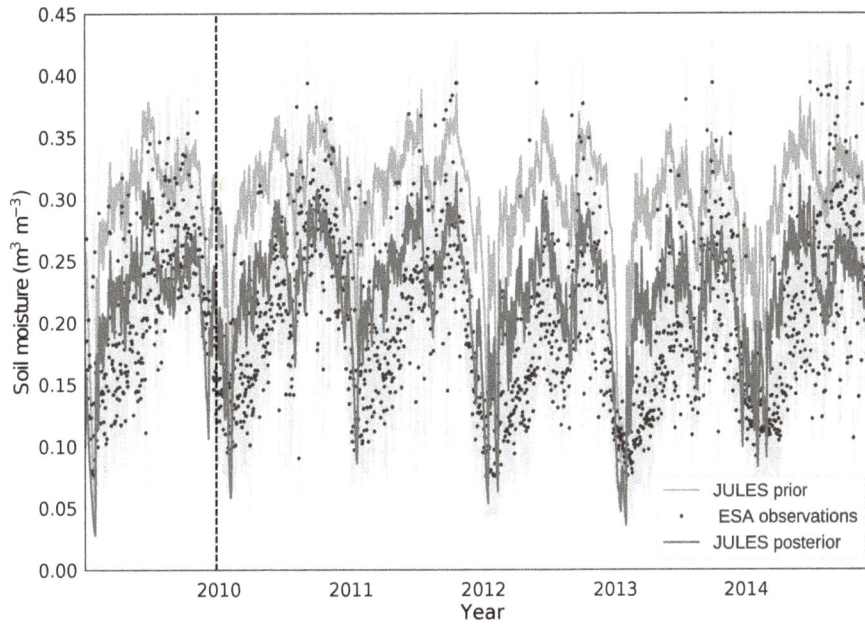

Figure 5. As Fig. 4, except for southern Ghana.

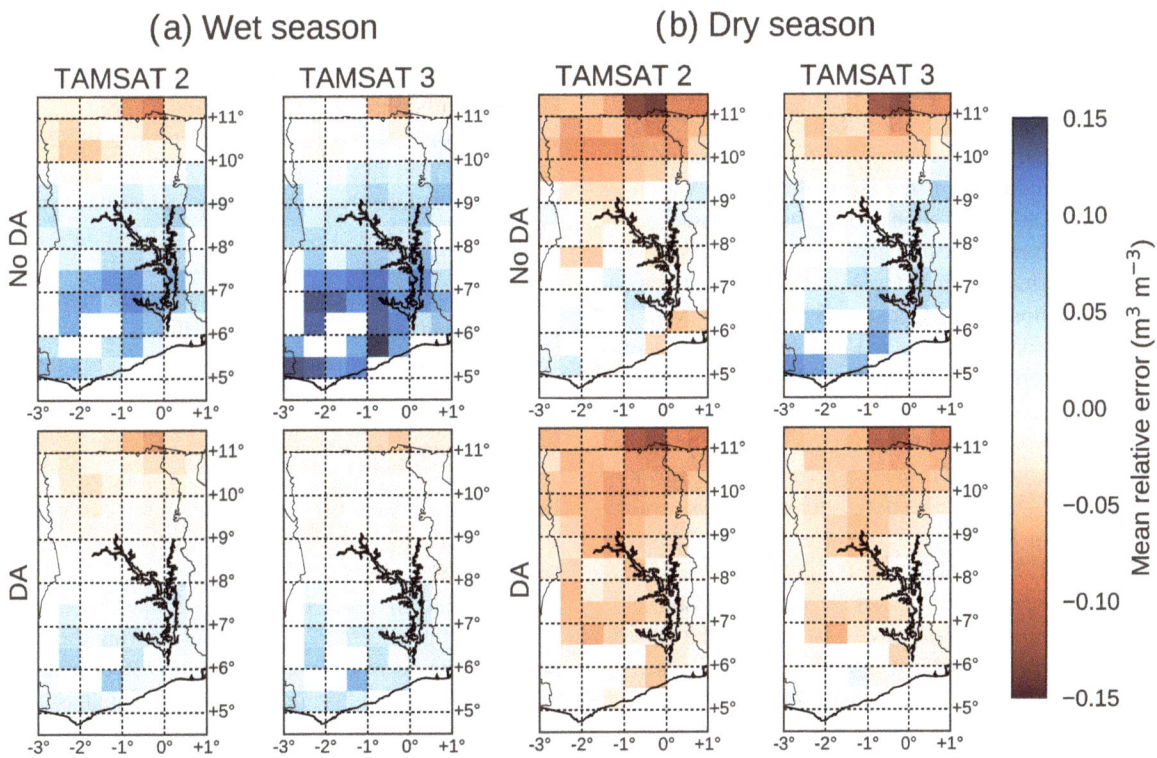

Figure 6. Soil moisture model minus observations for the 5-year JULES forecast (2010–2014) driven with TAMSAT v2.0 and v3.0 precipitation and before and after data assimilation. Subplot **(a)** shows statistics calculated over March to May for the wet period; subplot **(b)** shows statistics calculated over November to January for the dry period. White pixels indicate areas where there are no data to calculate statistics (mainly due to high vegetation cover in the south).

prior soil map in the south rather than TAMSAT v2.0 rainfall outperforming TAMSAT v3.0 (it is expected that both products perform poorly in coastal regions; Maidment et al., 2017). Experiments 2 and 4 have a lower RMSD in the south (Fig. 8) compared to the north (Fig. 7); this seems surprising given that we consider the quality of the data to be poorer in the south. However, this is in part due to the much more pronounced seasonal cycle in the north leading to peaks in RMSD when the seasonal cycle is even slightly mistimed by the model. We also have less confidence in the CCI soil moisture observations in the south so a lower RMSD in comparison to this product over this region is perhaps not indicative of a better soil moisture estimate overall.

Figure 9 compares the prior soil map used as the initial guess in the DA (i.e. from the Harmonised World Soil Data Base) with the posterior soil map retrieved by DA. The posterior soil map shown is the soil map retrieved when forcing JULES with TAMSAT v3.0 rainfall. It can be seen that after DA, the percentage clay is greatly reduced with increased percentages in silt and sand for the majority of grid cells. This change is reasonable for some grid cells, particularly in northern Ghana where soils are often much more sandy/silty in texture (Braimoh and Vlek, 2004). Comparing estimates of soil texture derived from CCI soil moisture to in situ observations is inevitably problematic due to issues of representativity in the spatial domain. However, independent sources of verification are difficult to find over Ghana. We therefore compare our soil maps to in situ observations from the Africa Soil Profiles Database (Leenaars et al., 2014). This database is compiled by the International Soil Reference and Information Centre (ISRIC), with the quality of the data being rated from 1 (highest quality) to 4 (lowest quality); here we only compare our maps to observations with a quality flag of 1 or 2. In table 1 we show the root-mean-squared error (RMSE) for our soil maps when compared to 21 in situ observations of soil texture in the north of Ghana and 36 in situ observations in the south (locations shown as red dots in Fig. 9). For the north of Ghana where we have most confidence in our results we find a reduction in RMSE for both sand and clay (almost halving the RMSE in clay). However, the RMSE for silt is increased. In the south of Ghana we do not manage to recover a better estimate of soil texture after data assimilation, with an increase in RMSE for silt and clay but a decrease in RMSE for sand. The inability of the data assimilation to improve soil texture estimates at certain points is most likely due to issues of spatial representativity between the modelled soil map and the in situ data. It is also possibly impacted by errors in our pedo-transfer functions, which may perform better if they were specifically calibrated for Ghanaian soils (Patil and Singh, 2016).

Satellite soil moisture products can be subject to larger errors and biases associated with data processing. This is particularly true for the CCI level 3 combined active and passive product used in this paper, as in order to merge information from 11 different sensors, data are matched using cumulative

Table 1. RMSE between JULES model soil maps (prior and posterior) and in situ observations of soil texture from the Africa Soil Profiles Database (Leenaars et al., 2014).

	North Ghana		
	Sand RMSE	Silt RMSE	Clay RMSE
Prior soil map	0.43	0.25	0.30
Posterior soil map	0.38	0.29	0.16
	South Ghana		
	Sand RMSE	Silt RMSE	Clay RMSE
Prior soil map	0.35	0.27	0.16
Posterior soil map	0.27	0.35	0.20

distribution functions to the GLDAS-Noah v1 model (Rodell et al., 2004). Therefore, any bias within the GLDAS-Noah model will be included in the level 3 soil moisture product used here. To make sure we are not just correcting the bias of the JULES model to that of GLDAS-Noah we include summary statistics of unbiased root-mean-squared difference (ubRMSD) and temporal correlation in Table 2. In every case we find that after data assimilation we improve both ubRMSD and correlation and in the majority of cases find the best results for experiment 4 (TAMSAT v3.0 with DA). For the north of Ghana, we reduce the ubRMSD by 18 % from experiment 3 ($0.0622 \, \text{m}^3 \, \text{m}^{-3}$) to experiment 4 ($0.0508 \, \text{m}^3 \, \text{m}^{-3}$). From experiment 2 to 4 we can see that, after data assimilation, using TAMSAT v3.0 rainfall over v2.0 has contributed to a 6 % reduction in ubRMSD when calculating statistics over the whole period. In the south of Ghana, we reduce the ubRMSD by 21 % from experiment 3 ($0.0590 \, \text{m}^3 \, \text{m}^{-3}$) to experiment 4 ($0.0467 \, \text{m}^3 \, \text{m}^{-3}$); here improved rainfall data have contributed to 10 % of this reduction. We find the highest correlations in the north of Ghana for the whole period (2010–2014); this is mainly due to the seasonal cycle being much more pronounced in this region.

4 Discussion

For northern Ghana there is a prominent seasonal cycle for soil moisture, with observations of higher quality than in the south for both TAMSAT rainfall and ESA CCI soil moisture. We find that soil moisture estimates based on TAMSAT v3.0 outperform v2.0, especially during the wetting-up phase of the seasonal cycle, with the effect of the rainfall dataset less marked during the drying-down phase. This is to be expected as little or no rain occurs during drying-down so that it is model dynamics that are the dominant factor in the estimation of soil moisture. Therefore, it is the updating of soil parameters via data assimilation and not improved precipitation that has the greatest impact on soil moisture estimates during drying-down. Conversely, improved rainfall data have the greatest impact for estimating wetting-up

Impact of remotely sensed soil moisture and precipitation on soil moisture prediction in a data assimilation...

59

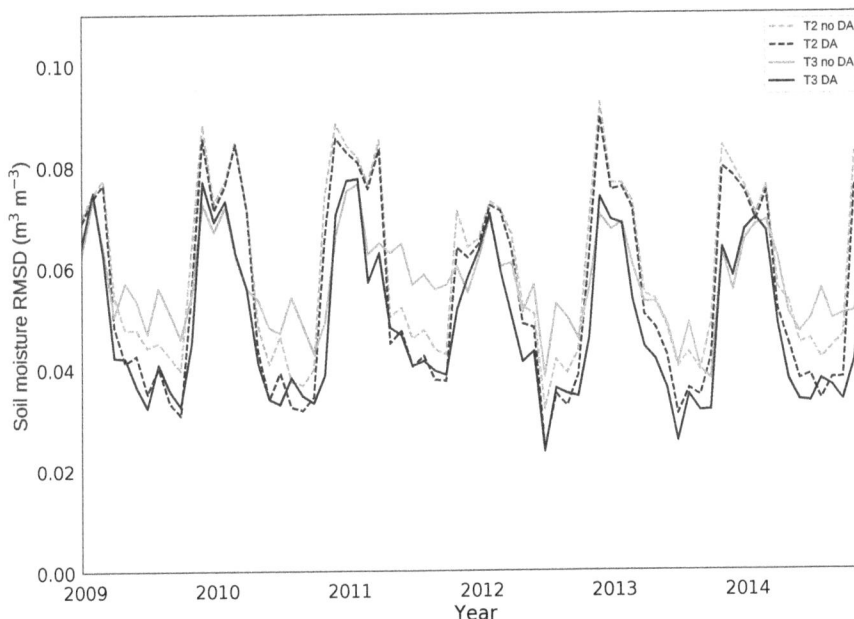

Figure 7. Monthly root-mean-squared difference (RMSD) of JULES soil moisture estimate compared to ESA CCI for northern Ghana. Light grey dashed line: prior JULES estimate, driven with TAMSAT v2.0 precipitation (exp. 1). Dark grey dashed line: prior JULES estimate, driven with TAMSAT v3.0 precipitation (exp. 3). Light grey solid line: posterior JULES estimate, driven with TAMSAT v2.0 precipitation (exp. 2). Dark grey solid line: posterior JULES estimate, driven with TAMSAT v3.0 precipitation (exp. 4).

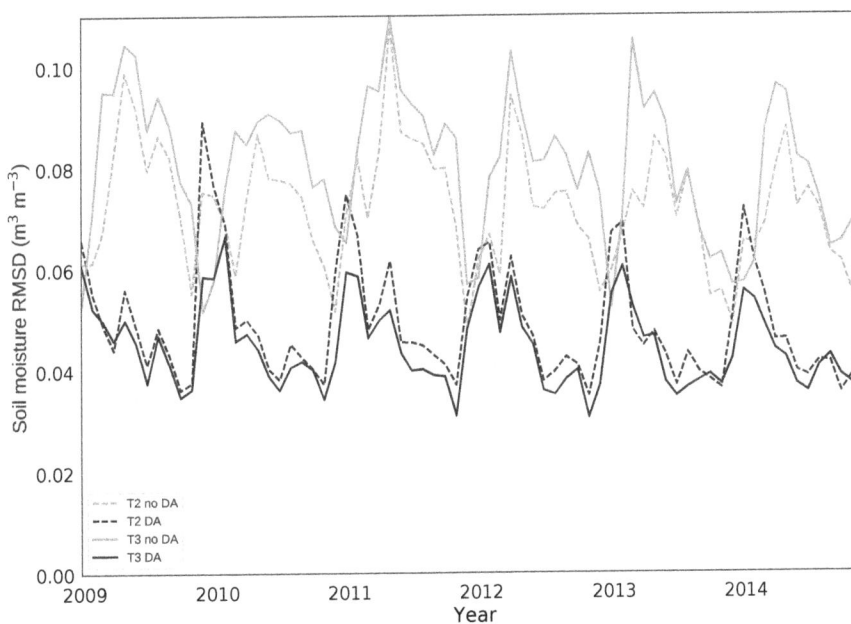

Figure 8. As Fig. 7, except for southern Ghana.

and constraining the start of the growing season. This can be seen in Fig. 7 where TAMSAT v3.0 without DA outperforms TAMSAT v2.0 with DA at certain times in the season. This is because at these times the data assimilation system is not able to overcome the errors in the precipitation forcing data to improve the estimates further. If there is too little rain-

fall, there is a point where the DA system cannot make the soil any wetter because we are not changing the model soil moisture state – only the soil texture. Assimilation of CCI soil moisture estimates in the north of Ghana allows us to re-cover improved estimates of soil texture when judged against

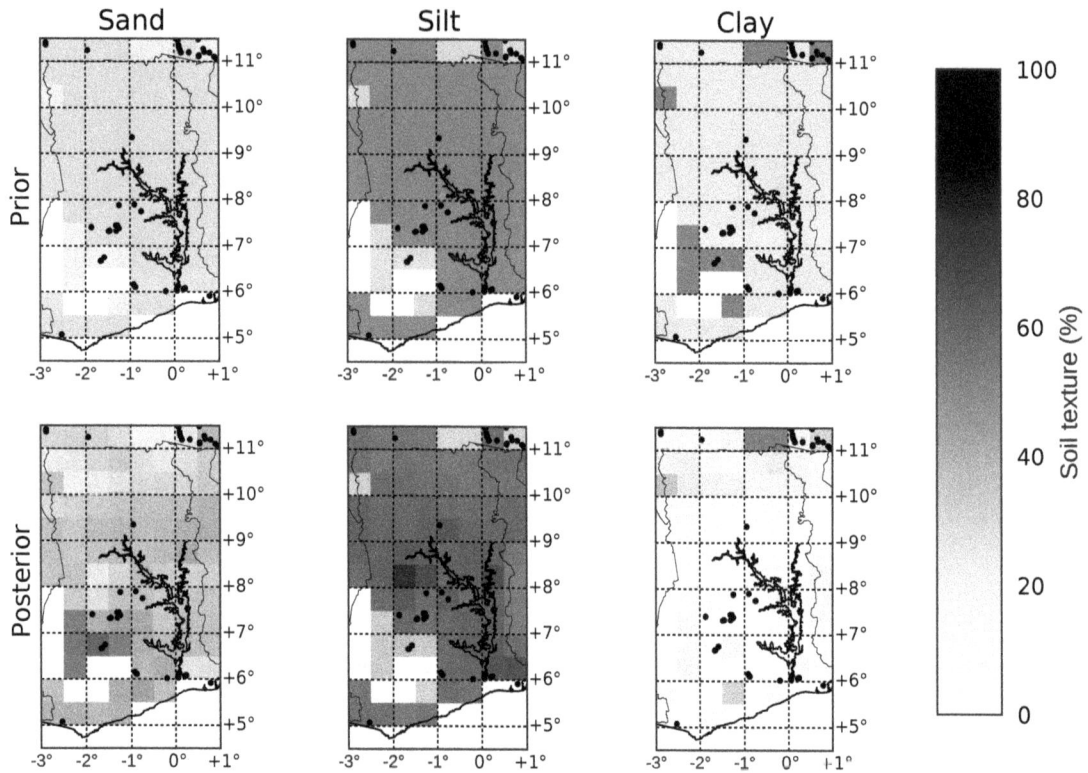

Figure 9. Prior and posterior soil maps over Ghana showing percentage of sand, silt and clay. Red dots represent locations where in situ observations of soil texture are available from the Africa Soil Profiles Database (Leenaars et al., 2014).

Table 2. Experiment statistics calculated over the north and south of Ghana in the hindcast period (2010–2014), for the whole period, wetting-up (March–May) and drying-down (November–January). The ubRMSD is calculated as $\sqrt{\frac{1}{N}\sum_{i=0}^{N}((\theta_{\text{mod}_i} - \overline{\theta}_{\text{mod}}) - (\theta_{\text{obs}_i} - \overline{\theta}_{\text{obs}}))^2}$, where N is the number of observations, θ_{mod_i} the model estimate at time i, $\overline{\theta}_{\text{mod}}$ the mean model estimate over the time window, θ_{obs_i} the observation estimate at time i and $\overline{\theta}_{\text{obs}}$ the mean observation estimate over the time window. The units of ubRMSD are $\text{m}^3\,\text{m}^{-3}$.

	North Ghana							
	(1) TAMSAT 2 no DA		(2) TAMSAT 2 DA		(3) TAMSAT 3 no DA		(4) TAMSAT 3 DA	
	ubRMSD	Correlation	ubRMSD	Correlation	ubRMSD	Correlation	ubRMSD	Correlation
Whole period	0.0605	0.86	0.0541	0.89	0.0622	0.86	0.0508	0.90
Wet	0.0643	0.59	0.0592	0.64	0.0626	0.58	0.0529	0.65
Dry	0.0396	0.77	0.0332	0.83	0.0486	0.78	0.0365	0.84
	South Ghana							
	(1) TAMSAT 2 no DA		(2) TAMSAT 2 DA		(3) TAMSAT 3 no DA		(4) TAMSAT 3 DA	
	ubRMSD	Correlation	ubRMSD	Correlation	ubRMSD	Correlation	ubRMSD	Correlation
Whole period	0.0651	0.77	0.0519	0.82	0.0590	0.76	0.0467	0.82
Wet	0.0629	0.57	0.0515	0.67	0.0571	0.55	0.0472	0.66
Dry	0.0642	0.82	0.0492	0.85	0.0604	0.83	0.0432	0.87

in situ data from the Africa Soil Profiles Database (Leenaars et al., 2014).

For southern Ghana, there is a much less prominent seasonal cycle than in the north, with poorer quality observa-

tions for both TAMSAT rainfall and ESA CCI soil moisture. This is due to large amounts of coastal convective cloud and higher vegetation cover. We find that, after assimilating soil moisture data, runs forced with TAMSAT v3.0 outperform those forced with TAMSAT v2.0. Although we do not have reliable precipitation observations in the south we can still greatly improve our forecast skill for soil moisture through DA. This bodes well for other regions with unreliable precipitation observations (Crow, 2003). In the south we find larger reductions in ubRMSD than in the north after data assimilation. However, we also have less confidence in the CCI soil moisture product to which we are comparing in the south. It is therefore unlikely that we have improved estimates more than in the north in comparison to the truth. This is backed up by the inability of our data assimilation system to recover an improved soil map when compared to in situ observations in the south.

There is likely an issue of representativity between the satellite-derived soil moisture observations and the JULES modelled soil moisture in our DA system. We make the pragmatic assumption that satellite soil moisture is representative of the top 5 cm layer of soil in JULES. However, during intense dry periods the satellite will become more sensitive to greater depths (Ulaby et al., 1982) and hence less representative of the JULES top level soil moisture. This can be seen in Fig. 4 where the model fails to capture the satellite observations during the driest periods, with the JULES model predicting a lower soil moisture than the ESA CCI observations; this same phenomenon appears at a number of grid cells during dry periods. We can also see this consistent dry bias in the bottom row of Fig. 6b. More work is needed to understand how best to address this issue between satellite-observed and modelled soil moisture. One option could be to create a multi-layer observation operator for land surface models. Previous DA studies have opted to assimilate satellite-retrieved brightness temperature and then use a radiative transfer model on top of their chosen land surface model (Moradkhani et al., 2005; Qin et al., 2009; Montzka et al., 2011; Rasmy et al., 2011; Sawada and Koike, 2014; Yang et al., 2016).

Our results highlight the importance of having quality observations of both precipitation and soil moisture. TAMSAT rainfall observations and the ESA CCI soil moisture data are available as daily products but at different spatial resolutions and different observation times. TAMSAT data are produced at 4 km spatial resolution by calculating cold cloud duration over a 5-day period of 15 min thermal infrared observations. The ESA CCI soil moisture data on the other hand are merged from various passive and active microwave observations and available in various spatial resolutions that are typically in the order of 0.25°. The core observations that make up the daily product are, in effect, instantaneous but then merged into a harmonised product. The ideal situation would be to have precipitation measurements and soil moisture observations that are representative of the same time periods and spatial domains, but there are no such current missions.

5 Conclusions

Previous studies at the grid cell level have shown that calibrating land surface models with satellite observations improves performance when judged against in situ observations (Moradkhani et al., 2005; Qin et al., 2009; Montzka et al., 2011; Rasmy et al., 2011; Sawada and Koike, 2014; Yang et al., 2016). In this study we calibrated the JULES land surface model at the regional scale (over Ghana) and show that this reduces ubRMSD and correlation when judged against independent observations in a set of hindcast experiments. From the results, it is clear that both improved rainfall estimates and the implementation of data assimilation are required in order to improve modelled estimates and forecasts of soil moisture. We have split our analysis between north and south Ghana due to the hydrological regimes varying considerably between these two regions. In the north of Ghana, where the observations are of highest quality due to lower cloud and vegetation cover, we find that improved precipitation estimates are of greatest importance for accurate representation of the start of season soil moisture. In contrast, the assimilation of relevant soil moisture observations with our land surface model gives the largest benefit for improving estimates during drying-down. This makes physical sense as when no rain is occurring it will be model dynamics that are the dominant factor in the estimation of soil moisture. After data assimilation we are able to improve our estimates of soil texture in the north, judged against in situ observations. After assimilation of a single year of soil moisture observations (2009) we reduce the ubRMSD of a 5-year model hindcast (2010–2014) by 18 % in northern Ghana and 21 % in the south, with the improved rainfall product contributing a 6 and 10 % reduction in ubRMSD respectively. The higher reduction in ubRMSD in the south is not necessarily indicative of a soil moisture estimate closer to the truth as we also have less faith in the ESA CCI soil moisture product in this region due to higher amounts of convective cloud and vegetation cover. This is supported by the fact that in the south we are unable to recover an improved estimate of soil texture after data assimilation, when judged against in situ observations. However, in the north we do recover improved soil texture estimates despite the lower reduction in ubRMSD.

Competing interests. The authors declare that they have no conflict of interest.

Acknowledgements. This work was funded by the UK Natural Environment Research Council (NE/P015352/1). This work was also partly funded by the National Centre for Earth Observation. Emily Black is supported by the Natural Environment Research Council/Global Challenges Research Fund programme ACREW (NE/R000034/1).

Edited by: Louise Slater

References

Albergel, C., Rüdiger, C., Pellarin, T., Calvet, J.-C., Fritz, N., Froissard, F., Suquia, D., Petitpa, A., Piguet, B., and Martin, E.: From near-surface to root-zone soil moisture using an exponential filter: an assessment of the method based on in-situ observations and model simulations, Hydrol. Earth Syst. Sci., 12, 1323–1337, https://doi.org/10.5194/hess-12-1323-2008, 2008.

Alvarez-Garreton, C., Ryu, D., Western, A. W., Crow, W. T., Su, C., and Robertson, D. R.: Dual assimilation of satellite soil moisture to improve streamflow prediction in data-scarce catchments, Water Resour. Res., 52, 5357–5375, https://doi.org/10.1002/2015WR018429, 2016.

Bateni, S. M. and Entekhabi, D.: Relative efficiency of land surface energy balance components, Water Resour. Res., 48, W04510, https://doi.org/10.1029/2011WR011357, 2012.

Beljaars, A. C. M., Viterbo, P., Miller, M. J., and Betts, A. K.: The Anomalous Rainfall over the United States during July 1993: Sensitivity to Land Surface Parameterization and Soil Moisture Anomalies, Mon. Weather Rev., 124, 362–383, https://doi.org/10.1175/1520-0493(1996)124<0362:TAROTU>2.0.CO;2, 1996.

Best, M. J., Pryor, M., Clark, D. B., Rooney, G. G., Essery, R. L. H., Ménard, C. B., Edwards, J. M., Hendry, M. A., Porson, A., Gedney, N., Mercado, L. M., Sitch, S., Blyth, E., Boucher, O., Cox, P. M., Grimmond, C. S. B., and Harding, R. J.: The Joint UK Land Environment Simulator (JULES), model description – Part 1: Energy and water fluxes, Geosci. Model Dev., 4, 677–699, https://doi.org/10.5194/gmd-4-677-2011, 2011.

Bolten, J. D., Crow, W. T., Zhan, X., Jackson, T. J., and Reynolds, C. A.: Evaluating the utility of remotely sensed soil moisture retrievals for operational agricultural drought monitoring, IEEE J. Sel. Top. Appl., 3, 57–66, 2010.

Braimoh, A. K. and Vlek, P. L. G.: The impact of land-cover change on soil properties in northern Ghana, Land Degrad. Dev., 15, 65–74, https://doi.org/10.1002/ldr.590, 2004.

Brown, M., Black, E., Asfaw, D., and Otu-Larbi, F.: Monitoring drought in Ghana using TAMSAT-ALERT: a new decision support system, Weather, 72, 201–205, https://doi.org/10.1002/wea.3033, 2017.

Cappelaere, B., Descroix, L., Lebel, T., Boulain, N., Ramier, D., Laurent, J.-P., Favreau, G., Boubkraoui, S., Boucher, M., Moussa, I. B., Chaffard, V., Hiernaux, P., Issoufou, H., Breton, E. L., Mamadou, I., Nazoumou, Y., Oi, M., Ottlé, C., and Quantin, G.: The AMMA-CATCH experiment in the cultivated Sahelian area of south-west Niger – Investigating water cycle response to a fluctuating climate and changing environment, J. Hydrol., 375, 34–51, https://doi.org/10.1016/j.jhydrol.2009.06.021, 2009.

Clark, D. B., Mercado, L. M., Sitch, S., Jones, C. D., Gedney, N., Best, M. J., Pryor, M., Rooney, G. G., Essery, R. L. H., Blyth, E., Boucher, O., Harding, R. J., Huntingford, C., and Cox, P. M.: The Joint UK Land Environment Simulator (JULES), model description – Part 2: Carbon fluxes and vegetation dynamics, Geosci. Model Dev., 4, 701–722, https://doi.org/10.5194/gmd-4-701-2011, 2011.

Cosby, B. J., Hornberger, G. M., Clapp, R. B., and Ginn, T. R.: A Statistical Exploration of the Relationships of Soil Moisture Characteristics to the Physical Properties of Soils, Water Resour. Res., 20, 682–690, https://doi.org/10.1029/WR020i006p00682, 1984.

Crow, W. T.: Correcting Land Surface Model Predictions for the Impact of Temporally Sparse Rainfall Rate Measurements Using an Ensemble Kalman Filter and Surface Brightness Temperature Observations, J. Hydrometeorol., 4, 960–973, https://doi.org/10.1175/1525-7541(2003)004<0960:CLSMPF>2.0.CO;2, 2003.

De Lannoy, G. J. M. and Reichle, R. H.: Assimilation of SMOS brightness temperatures or soil moisture retrievals into a land surface model, Hydrol. Earth Syst. Sci., 20, 4895–4911, https://doi.org/10.5194/hess-20-4895-2016, 2016.

Dorigo, W., Gruber, A., Jeu, R. D., Wagner, W., Stacke, T., Loew, A., Albergel, C., Brocca, L., Chung, D., Parinussa, R., and Kidd, R.: Evaluation of the ESA CCI soil moisture product using ground-based observations, Remote Sens. Environ., 162, 380–395, https://doi.org/10.1016/j.rsc.2014.07.023, 2015.

Dorigo, W., Wagner, W., Albergel, C., Albrecht, F., Balsamo, G., Brocca, L., Chung, D., Ertl, M., Forkel, M., Gruber, A., Haas, E., Hamer, P. D., Hirschi, M., Ikonen, J., de Jeu, R., Kidd, R., Lahoz, W., Liu, Y. Y., Miralles, D., Mistelbauer, T., Nicolai-Shaw, N., Parinussa, R., Pratola, C., Reimer, C., van der Schalie, R., Seneviratne, S. I., Smolander, T., and Lecomte, P.: ESA CCI Soil Moisture for improved Earth system understanding: State-of-the art and future directions, Remote Sens. Environ., 203, 185–215, https://doi.org/10.1016/j.rse.2017.07.001, 2017.

Draper, C. S., Reichle, R. H., De Lannoy, G. J. M., and Liu, Q.: Assimilation of passive and active microwave soil moisture retrievals, Geophys. Res. Lett., 39, L04401, https://doi.org/10.1029/2011GL050655, 2012.

Ghent, D., Kaduk, J., Remedios, J., Ardö, J., and Balzter, H.: Assimilation of land surface temperature into the land surface model JULES with an ensemble Kalman filter, J. Geophys. Res.-Atmos., 115, D19112, https://doi.org/10.1029/2010JD014392, 2010.

Gottschalck, J., Meng, J., Rodell, M., and Houser, P.: Analysis of Multiple Precipitation Products and Preliminary Assessment of Their Impact on Global Land Data Assimilation System Land Surface States, J. Hydrometeorol., 6, 573–598, https://doi.org/10.1175/JHM437.1, 2005.

Greatrex, H., Grimes, D., and Wheeler, T.: Advances in the Stochastic Modeling of Satellite-Derived Rainfall Estimates Using a Sparse Calibration Dataset, J. Hydrometeorol., 15, 1810–1831, https://doi.org/10.1175/JHM-D-13-0145.1, 2014.

Guo, Z., Dirmeyer, P. A., Hu, Z.-Z., Gao, X., and Zhao, M.: Evaluation of the Second Global Soil Wetness Project soil moisture simulations: 2. Sensitivity to external meteorological forcing, J. Geophys. Res.-Atmos., 111, D22S03, https://doi.org/10.1029/2006JD007845, 2006.

Kolassa, J., Reichle, R., and Draper, C.: Merging active and passive microwave observations in soil moisture data assimilation, Remote Sens. Environ., 191, 117–130, 2017.

Leenaars, J., van Oostrum, A., and Ruiperez Gonzalez, M.: Africa Soil Profiles Database, Version 1.2. A compilation of georeferenced and standardised legacy soil profile data for Sub-Saharan Africa (with dataset). Africa Soil Information Service (AfSIS) project., Tech. rep., ISRIC-World Soil Information, Wageningen, the Netherlands, 2014.

Liu, C., Xiao, Q., and Wang, B.: An Ensemble-Based Four-Dimensional Variational Data Assimilation Scheme. Part I: Technical Formulation and Preliminary Test, Mon. Weather Rev., 136, 3363–3373, https://doi.org/10.1175/2008MWR2312.1, 2008.

Liu, C., Xiao, Q., and Wang, B.: An Ensemble-Based Four-Dimensional Variational Data Assimilation Scheme. Part II: Observing System Simulation Experiments with Advanced Research WRF (ARW), Mon. Weather Rev., 137, 1687–1704, https://doi.org/10.1175/2008MWR2699.1, 2009.

Liu, Q., Reichle, R. H., Bindlish, R., Cosh, M. H., Crow, W. T., de Jeu, R., De Lannoy, G. J. M., Huffman, G. J., and Jackson, T. J.: The Contributions of Precipitation and Soil Moisture Observations to the Skill of Soil Moisture Estimates in a Land Data Assimilation System, J. Hydrometeorol., 12, 750–765, https://doi.org/10.1175/JHM-D-10-05000.1, 2011.

Maidment, R. I., Grimes, D. I. F., Allan, R. P., Greatrex, H., Rojas, O., and Leo, O.: Evaluation of satellite-based and model reanalysis rainfall estimates for Uganda, Meteorol. Appl., 20, 308–317, https://doi.org/10.1002/met.1283, 2013.

Maidment, R. I., Grimes, D., Allan, R. P., Tarnavsky, E., Stringer, M., Hewison, T., Roebeling, R., and Black, E.: The 30 year TAMSAT African Rainfall Climatology And Time series (TARCAT) data set, J. Geophys. Res.-Atmos., 119, 10619–10644, https://doi.org/10.1002/2014JD021927, 2014.

Maidment, R. I., Grimes, D., Black, E., Tarnavsky, E., Young, M., Greatrex, H., Allan, R. P., Stein, T., Nkonde, E., Senkunda, S., and Alcántara, E. M. U.: A new, long-term daily satellite-based rainfall dataset for operational monitoring in Africa, Sci. Data, 4, 170063, https://doi.org/10.1038/sdata.2017.63;, 2017.

Martey, E., Wiredu, A., Etwire, P., Fosu, M., Buah, S. S., Bidzakin, J., Ahiabor, B., and Kusi, F.: Fertilizer Adoption and Use Intensity Among Smallholder Farmers in Northern Ghana: A Case Study of the AGRA Soil Health Project, Sustainable Agriculture Research, 3, p. 24, https://doi.org/10.5539/sar.v3n1p24, 2013.

Massari, C., Brocca, L., Tarpanelli, A., and Moramarco, T.: Data Assimilation of Satellite Soil Moisture into Rainfall-Runoff Modelling: A Complex Recipe?, Remote Sensing, 7, 11403–11433, https://doi.org/10.3390/rs70911403, 2015.

McDowell, N. G.: Mechanisms Linking Drought, Hydraulics, Carbon Metabolism, and Vegetation Mortality, Plant Physiol., 155, 1051–1059, https://doi.org/10.1104/pp.110.170704, 2011.

Montzka, C., Moradkhani, H., Weihermüller, L., Franssen, H.-J. H., Canty, M., and Vereecken, H.: Hydraulic parameter estimation by remotely-sensed top soil moisture observations with the particle filter, J. Hydrol., 399, 410–421, https://doi.org/10.1016/j.jhydrol.2011.01.020, 2011.

Moradkhani, H., Sorooshian, S., Gupta, H. V., and Houser, P. R.: Dual state–parameter estimation of hydrological models us-

ing ensemble Kalman filter, Adv. Water Resour., 28, 135–147, https://doi.org/10.1016/j.advwatres.2004.09.002, 2005.

Nachtergaele, F., van Velthuizen, H., Verelst, L., Batjes, N., Dijkshoorn, K., van Engelen, V., Fischer, G., Jones, A., Montanarella, L., Petri, M., and Prieler, S.: Harmonized world soil database (version 1.0), Tech. rep., Food and Agric Organization of the UN (FAO); International Inst. for Applied Systems Analysis (IIASA); ISRIC-World Soil Information; Institute of Soil Sciences – Chinese Acadamy of Sciences (ISS-CAS); EC-Joint Research Centre (JRC), 2008.

Navon, I.: Practical and theoretical aspects of adjoint parameter estimation and identifiability in meteorology and oceanography, Dynam. Atmos. Oceans, 27, 55–79, https://doi.org/10.1016/S0377-0265(97)00032-8, 1998.

Nelder, J. A. and Mead, R.: A Simplex Method for Function Minimization, Comput. J., 7, 308–313, https://doi.org/10.1093/comjnl/7.4.308, 1965.

Patil, N. G. P. and Singh, S. K.: Pedotransfer Functions for Estimating Soil Hydraulic Properties: A Review, Pedosphere, 26, 417–430, https://doi.org/10.1016/S1002-0160(15)60054-6, 2016.

Pinnington, E. M.: JULES data assimilation Ghana, GitHub repository, available at: https://github.com/Ewan82/JULES_DA_Ghana (last access: 23 April 2018), 2017.

Pitman, A. J., Henderson-Sellers, A., Desborough, C. E., Yang, Z.-L., Abramopoulos, F., Boone, A., Dickinson, R. E., Gedney, N., Koster, R., Kowalczyk, E., Lettenmaier, D., Liang, X., Mahfouf, J.-F., Noilhan, J., Polcher, J., Qu, W., Robock, A., Rosenzweig, C., Schlosser, C. A., Shmakin, A. B., Smith, J., Suarez, M., Verseghy, D., Wetzel, P., Wood, E., and Xue, Y.: Key results and implications from phase 1(c) of the Project for Intercomparison of Land-surface Parametrization Schemes, Clim. Dynam., 15, 673–684, https://doi.org/10.1007/s003820050309, 1999.

Qin, J., Liang, S., Yang, K., Kaihotsu, I., Liu, R., and Koike, T.: Simultaneous estimation of both soil moisture and model parameters using particle filtering method through the assimilation of microwave signal, J. Geophys. Res.-Atmos., 114, D15103, https://doi.org/10.1029/2008JD011358, 2009.

Raoult, N. M., Jupp, T. E., Cox, P. M., and Luke, C. M.: Land-surface parameter optimisation using data assimilation techniques: the adJULES system V1.0, Geosci. Model Dev., 9, 2833–2852, https://doi.org/10.5194/gmd-9-2833-2016, 2016.

Rasmy, M., Koike, T., Boussetta, S., Lu, H., and Li, X.: Development of a satellite land data assimilation system coupled with a mesoscale model in the Tibetan Plateau, IEEE T. Geosci. Remote, 49, 2847–2862, 2011.

Rodell, M., Houser, P. R., Jambor, U., Gottschalck, J., Mitchell, K., Meng, C.-J., Arsenault, K., Cosgrove, B., Radakovich, J., Bosilovich, M., Entin, J. K., Walker, J. P., Lohmann, D., and Toll, D.: The Global Land Data Assimilation System, B. Am. Meteorol. Soc., 85, 381–394, https://doi.org/10.1175/BAMS-85-3-381, 2004.

Sawada, Y. and Koike, T.: Simultaneous estimation of both hydrological and ecological parameters in an ecohydrological model by assimilating microwave signal, J. Geophys. Res.-Atmos., 119, 8839–8857, https://doi.org/10.1002/2014JD021536, 2014.

Sellers, P., Randall, D., Collatz, G., Berry, J., Field, C., Dazlich, D., Zhang, C., Collelo, G., and Bounoua, L.: A Revised Land Surface Parameterization (SiB2) for

Atmospheric GCM S. Part I: Model Formulation, J. Climate, 9, 676–705, https://doi.org/10.1175/1520-0442(1996)009<0676:ARLSPF>2.0.CO;2, 1996.

Seneviratne, S. I., Corti, T., Davin, E. L., Hirschi, M., Jaeger, E. B., Lehner, I., Orlowsky, B., and Teuling, A. J.: Investigating soil moisture-climate interactions in a changing climate: A review, Earth-Sci. Rev., 99, 125–161, https://doi.org/10.1016/j.earscirev.2010.02.004, 2010.

Tarnavsky, E., Grimes, D., Maidment, R., Black, E., Allan, R. P., Stringer, M., Chadwick, R., and Kayitakire, F.: Extension of the TAMSAT Satellite-Based Rainfall Monitoring over Africa and from 1983 to Present, J. Appl. Meteorol. Clim., 53, 2805–2822, https://doi.org/10.1175/JAMC-D-14-0016.1, 2014.

Ulaby, F., Moore, R., and Fung, A.: Microwave Remote Sensing, Active and Passive: Radar Remote Sensing and Surface Scattering and Emission Theory, Vol. 2, Wesley Publishing Company, Inc., Reading, Massachusetts, 1982.

Walters, D. N., Williams, K. D., Boutle, I. A., Bushell, A. C., Edwards, J. M., Field, P. R., Lock, A. P., Morcrette, C. J., Stratton, R. A., Wilkinson, J. M., Willett, M. R., Bellouin, N., Bodas-Salcedo, A., Brooks, M. E., Copsey, D., Earnshaw, P. D., Hardiman, S. C., Harris, C. M., Levine, R. C., MacLachlan, C., Manners, J. C., Martin, G. M., Milton, S. F., Palmer, M. D., Roberts, M. J., Rodríguez, J. M., Tennant, W. J., and Vidale, P. L.: The Met Office Unified Model Global Atmosphere 4.0 and JULES Global Land 4.0 configurations, Geosci. Model Dev., 7, 361–386, https://doi.org/10.5194/gmd-7-361-2014, 2014.

Weedon, G. P., Balsamo, G., Bellouin, N., Gomes, S., Best, M. J., and Viterbo, P.: The WFDEI meteorological forcing data set: WATCH Forcing Data methodology applied to ERA-Interim reanalysis data, Water Resour. Res., 50, 7505–7514, https://doi.org/10.1002/2014WR015638, 2014.

Yang, K., Watanabe, T., Koike, T., Li, X., Fujii, H., Tamagawa, K., and Ishikawa, H.: Auto-calibration system developed to assimilate AMSR-E data into a land surface model for estimating soil moisture and the surface energy budget, J. Meteorol. Soc. Jpn., 85, 229–242, 2007.

Yang, K., Koike, T., Kaihotsu, I., and Qin, J.: Validation of a dual-pass microwave land data assimilation system for estimating surface soil moisture in semiarid regions, J. Hydrometeorol., 10, 780–793, 2009.

Yang, K., Zhu, L., Chen, Y., Zhao, L., Qin, J., Lu, H., Tang, W., Han, M., Ding, B., and Fang, N.: Land surface model calibration through microwave data assimilation for improving soil moisture simulations, J. Hydrol., 533, 266–276, https://doi.org/10.1016/j.jhydrol.2015.12.018, 2016.

ERA-5 and ERA-Interim driven ISBA land surface model simulations: which one performs better?

Clement Albergel[1], **Emanuel Dutra**[2], **Simon Munier**[1], **Jean-Christophe Calvet**[1], **Joaquin Munoz-Sabater**[3], **Patricia de Rosnay**[3], **and Gianpaolo Balsamo**[3]

[1]CNRM UMR 3589, Météo-France/CNRS, Toulouse, France
[2]Instituto Dom Luiz, IDL, Faculty of Sciences, University of Lisbon, Lisbon, Portugal
[3]ECMWF, Reading, UK

Correspondence: Clement Albergel (clement.albergel@meteo.fr)

Abstract. The European Centre for Medium-Range Weather Forecasts (ECMWF) recently released the first 7-year segment of its latest atmospheric reanalysis: ERA-5 over the period 2010–2016. ERA-5 has important changes relative to the former ERA-Interim atmospheric reanalysis including higher spatial and temporal resolutions as well as a more recent model and data assimilation system. ERA-5 is foreseen to replace ERA-Interim reanalysis and one of the main goals of this study is to assess whether ERA-5 can enhance the simulation performances with respect to ERA-Interim when it is used to force a land surface model (LSM). To that end, both ERA-5 and ERA-Interim are used to force the ISBA (Interactions between Soil, Biosphere, and Atmosphere) LSM fully coupled with the Total Runoff Integrating Pathways (TRIP) scheme adapted for the CNRM (Centre National de Recherches Météorologiques) continental hydrological system within the SURFEX (SURFace Externalisée) modelling platform of Météo-France. Simulations cover the 2010–2016 period at half a degree spatial resolution.

The ERA-5 impact on ISBA LSM relative to ERA-Interim is evaluated using remote sensing and in situ observations covering a substantial part of the land surface storage and fluxes over the continental US domain. The remote sensing observations include (i) satellite-driven model estimates of land evapotranspiration, (ii) upscaled ground-based observations of gross primary production, (iii) satellite-derived estimates of surface soil moisture and (iv) satellite-derived estimates of leaf area index (LAI). The in situ observations cover (i) soil moisture, (ii) turbulent heat fluxes, (iii) river discharges and (iv) snow depth. ERA-5 leads to a consistent improvement over ERA-Interim as verified by the use of these eight independent observations of different land status and of the model simulations forced by ERA-5 when compared with ERA-Interim. This is particularly evident for the land surface variables linked to the terrestrial hydrological cycle, while variables linked to vegetation are less impacted. Results also indicate that while precipitation provides, to a large extent, improvements in surface fields (e.g. large improvement in the representation of river discharge and snow depth), the other atmospheric variables play an important role, contributing to the overall improvements. These results highlight the importance of enhanced meteorological forcing quality provided by the new ERA-5 reanalysis, which will pave the way for a new generation of land-surface developments and applications.

1 Introduction

Observing and simulating the response of land biophysical variables to extreme events is a major scientific challenge in relation to the adaptation to climate change. To that end, land surface models (LSMs) constrained by high-quality gridded atmospheric variables and coupled with river-routing models are essential (Schellekens et al., 2017; Dirmeyer et al., 2006). Such LSMs should represent land surface biogeophysical variables like surface and root zone soil moisture (SSM and RZSM, respectively), biomass, and leaf area index (LAI) in a way that is fully consistent with the representation of surface and energy flux as well as river discharge simulations. Land surface simulations, such as those from the Global Soil Wetness Project (GSWP, Dirmeyer et al., 2002, 2006; Dirmeyer, 2011), combined with seasonal forecasting systems have been of paramount importance in triggering progress in land-related predictability as documented in the Global Land–Atmosphere Coupling Experiments (GLACE; Koster et al., 2009a, 2011). The land surface state estimates used in those studies were generally obtained with offline (or stand-alone) model simulations, forced by 3-hourly meteorological fields from atmospheric reanalysis. In the past decade, several improved global atmospheric reanalyses of the satellite era (1979–onwards) have been produced that enable new applications of offline land surface simulations. Amongst them are NASA's Modern Era Retrospective Analysis for Research and Applications (MERRA; Rienecker et al., 2011, and MERRA2; Gelaro et al., 2017) as well as ECMWF's (European Centre for Medium-Range Weather Forecasts) Interim reanalysis (ERA-Interim; Dee et al., 2011). Their offline use in either LSMs or land data assimilation system (LDAS), with or without meteorological corrections (e.g. precipitations), led to global land surface variables (LSVs) reanalysis data sets that can support, for example water resources analysis (Schellekens et al., 2017), like MERRA-Land and MERRA2-Land (Reichle et al., 2011, 2017), ERA-Interim/Land (Balsamo et al., 2015), the forthcoming ERA5-Land (Muñoz-Sabater et al., 2018), the North American LDAS (NLDAS; Mitchell et al., 2004), the Global LDAS (GLDAS; Rodell et al., 2004) and LDAS-Monde (Albergel et al., 2017). The quality of those offline land surface simulations relies on the accuracy of the forcing and of the realism of the LSM itself (Balsamo et al., 2015).

ECMWF recently released the first 7-year segment of its latest atmospheric reanalysis: ERA-5 over the period 2010–2016. ERA-5 has important changes relative to the former ERA-Interim atmospheric reanalysis including higher spatial and temporal resolutions as well as a better global balance of precipitation and evaporation. As ERA-5 will eventually replace the ERA-Interim reanalysis assessing its ability to force a LSM with respect to ERA-Interim is highly relevant. In this study, ERA-5, ERA-Interim and a combination of both (ERA-5 with precipitation of ERA-Interim) are used to constrain the CO_2-responsive version of the Interactions between Soil, Biosphere, and Atmosphere (ISBA; Noilhan and Mahfouf, 1996; Calvet et al., 1998, 2004; Gibelin et al., 2006) LSM fully coupled with the CNRM (Centre National de Recherches Météorologiques) version of the Total Runoff Integrating Pathways (TRIP; Oki et al., 1998) continental hydrological system (CTRIP hereafter; Decharme et al., 2010) within the SURFEX (SURFace Externalisée; Masson et al., 2013) modelling system of Météo-France. The ISBA models leaf-scale physiological processes and plant growth, with transfer of water and heat through the soil relying on a multilayer diffusion scheme.

In this study, SURFEX is applied over a data-rich area: North America (latitudes from 20.0 to 55.0° N, longitudes from 130.0 to 60.0° W) for the period 2010–2016. ERA-5 added values with respect to ERA-Interim are assessed by providing verification and diagnostics comparing ISBA LSV outputs when forced by either ERA-5, ERA-Interim, ERA-5 with ERA-Interim precipitations to several in situ measurement data sets or satellite-derived estimates of Earth observations. Specifically, in situ measurements of (i) soil moisture from the USCRN (US Climate Reference Network; Bell et al., 2013) spanning the United States of America and (ii) turbulent heat fluxes from FLUXNET-2015 (http://fluxnet.fluxdata.org/data/fluxnet2015-dataset/, last access: June 2018) are used in the evaluation, together with (iii) river discharges from the United States Geophysical Survey (USGS; https://waterwatch.usgs.gov/, last access: June 2018) and (iv) snow depth measurements from the Global Historical Climatology Network (GHCN; Menne et al., 2012a, b). The following are also used: (i) satellite-driven model estimates of land evapotranspiration from the Global Land Evaporation Amsterdam Model (GLEAM; Martens et al., 2017), (ii) upscaled ground-based observations of gross primary production (GPP) from the FLUXCOM project (Jung et al., 2017), (iii) satellite-derived estimates of surface soil moisture (SSM) from the Climate Change Initiative (CCI) of the European Space Agency (ESA CCI SSM v4; Dorigo et al., 2015, 2017) and (iv) satellite-derived estimates of LAI from the Copernicus Global Land Service program (CGLS; http://land.copernicus.eu/global/, last access: June 2018).

Section 2 presents the details of two atmospheric reanalyses data sets (ERA-Interim and ERA-5), the SURFEX model configuration and the evaluation strategy with the observational data sets. Section 3 provides a set of statistical diagnostics to assess and evaluate the impact of ERA-5 on ISBA with respect to ERA-Interim. Finally, Sect. 4 provides perspectives and future research directions.

2 Methodology

2.1 ERA-Interim and ERA-5 reanalyses

ERA-Interim is a global atmospheric reanalysis produced by ECMWF (Dee et al., 2011). It uses the integrated forecast system (IFS) version 31r1 (more information at https://www.ecmwf.int/en/forecasts/documentation-and-support/changes-ecmwf-model/ifs-documentation, last access: June 2018) with a spatial resolution of about 80 km (T255) and with analyses available for 00:00, 06:00, 12:00 and 18:00 UTC. It covers the period from 1 January 1979 onward and continues to be extended forward in near-real time (with a delay of approximately 1 month). Reanalyses merge observations and model forecasts in data assimilation methods to provide an accurate and reliable description of the climate over the last few decades. Berrisford et al. (2009) provide a detailed description of the ERA-Interim product archive. ERA-5 (Hersbach and Dee, 2016) is the latest and fifth generation of European reanalyses produced by the ECMWF and a key element of the EU-funded Copernicus Climate Change Service (C3S). It is expected that ERA-5 will replace the production of the current ERA-Interim reanalysis (Dee et al., 2011) before the end of 2018, from 1979 to close to the Near Real Time (NRT) period, i.e. in ERA-5 regular routine updates will be conducted to keep close to NRT. In a second phase, an extension back to 1950 is also expected. ERA-5 adds different characteristics to ERA-Interim reanalysis, which makes it richer in term of climate information.

ERA-5 uses one of the most recent versions of the Earth system model and data assimilation methods applied at ECMWF, which makes it able to use modern parameterizations of Earth processes compared to older versions used in ERA-Interim. For instance, developments were done at ECMWF which allows the reanalysis to use a variational bias scheme not only for satellite observations but also for ozone, aircraft and surface pressure data. ERA-5 also benefits from reprocessed data sets that were not ready yet during the production of ERA-Interim. Two other important features of ERA-5 are the improved temporal and spatial resolutions: from 6-hourly in ERA-Interim to hourly in ERA-5, and from 79 km in the horizontal dimension and 60 levels in the vertical to 31 km and 137 levels in ERA-5. Finally, ERA-5 also provides an estimate of uncertainty through the use of a 10-member ensemble of data assimilations (EDA) at a coarser resolution (63 km horizontal resolution) and 3-hourly frequency.

2.2　SURFEX modelling system

2.2.1　The ISBA land surface model

This study makes use of the CO_2-responsive version of the ISBA LSM included in the open-access SURFEX modelling platform of Météo-France (Masson et al., 2013). The most recent version of SURFEX (version 8.1) is used with the "NIT" biomass option for ISBA. The latter simulates the diurnal cycle of water and carbon fluxes, plant growth, and key vegetation variables like LAI and above-ground biomass on a daily basis. It can be coupled to the CTRIP river-routing model in order to simulate streamflow. In this version of ISBA, a single-source energy budget of a soil–vegetation composite is computed. Also, the ISBA parameters are defined for 12 generic land surface patches, which include nine plant functional types (needle leaf trees, evergreen broadleaf trees, deciduous broadleaf trees, C_3 crops, C_4 crops, C_4 irrigated crops, herbaceous, tropical herbaceous and wetlands), bare soil, rocks, and permanent snow and ice surfaces. A more comprehensive model description can be found in Masson et al. (2013).

ISBA accounts for the atmospheric CO_2 concentration on stomatal aperture (Calvet et al., 1998, 2004; Gibelin et al., 2006). Also, photosynthesis and its coupling with stomatal conductance on a leaf level are accounted for. The vegetation net assimilation of CO_2 is estimated and used as an input to a simple vegetation growth submodel able to predict LAI: photosynthesis drives the dynamic evolution of the vegetation biomass and LAI variables in response to atmospheric and climate conditions. During the growing phase, enhanced photosynthesis corresponds to a CO_2 uptake, which leads to vegetation growth. In contrast, lack of photosynthesis leads to higher mortality rates. The GPP is defined as the carbon uptake while the ecosystem respiration (RECO) is the release of CO_2, the difference between these two quantities being the net ecosystem CO_2 exchange (NEE). Evaporation due to (i) plant transpiration, (ii) liquid water intercepted by leaves, (iii) liquid water contained in top soil layers and (iv) the sublimation of snow and soil ice are combined to represent the total evaporative flux.

The ISBA 12-layer explicit snow scheme (Boon and Etchevers, 2001; Decharme et al., 2016) and its multilayer soil diffusion scheme (ISBA-Dif) are used. The later is based on the mixed form of the Richards equation (Richards, 1931) and explicitly solves the one-dimensional Fourier law. It also incorporates soil freezing processes developed by Boone et al. (2000) and Decharme et al. (2013). The total soil profile is vertically discretized; both the temperature and moisture of each soil layer are computed according to their textural and hydrological characteristics. The Brookes and Corey model (Brooks and Corey, 1966) determines the closed-form equations between the soil moisture and the soil hydrodynamic parameters, including the hydraulic conductivity and the soil matrix potential (Decharme et al., 2013). The default discretization with 14 layers over 12 m depth is used. The lower boundary of each layer being: 0.01, 0.04, 0.1, 0.2, 0.4, 0.6, 0.8, 1, 1.5, 2, 3, 5, 8 and 12 m deep (see Fig. 1 of Decharme et al., 2011). Amounts of clay, sand and organic carbon in the soil determine the thermal and hydrodynamic soil properties (Decharme et al., 2016). They are taken from the Harmonized World Soil Database (HWSD; Wieder et al., 2014). As for hydrology, the infiltration, surface evaporation and total runoff are accounted for in the soil water balance. The infiltration rate defines the discrepancy between the surface runoff and the throughfall rate. The later being defined as the sum of rainfall not intercepted by the canopy, dripping from the canopy (i.e. interception reservoir) and snow melt water. The soil evaporation affects only the superficial layer (top 1 cm) and is proportional to its relative humidity. Transpiration water from the root zone (the region where the roots are asymptotically distributed) follows the equations in Jackson et al. (1996). Canal et al. (2014) provide more information on the root density profile.

Both the surface runoff (the lateral subsurface flow in the topsoil) and a free drainage condition at the bottom soil layer contribute to ISBA total runoff. The Dunne runoff (i.e. when no further soil moisture storage is available) and lateral subsurface flow from a subgrid distribution of the topography are computed using a basic TOPMODEL approach. The Horton runoff (i.e. when rainfall has exceeded infiltration capacity) is estimated from the maximum soil infiltration capacity and a subgrid exponential distribution of the rainfall intensity.

2.2.2 The CTRIP hydrological system

CTRIP is driven by three prognostic equations corresponding to (i) the groundwater, (ii) the surface stream water and (iii) the seasonal floodplains. Streamflow velocity is computed using the Manning formula as described in Decharme et al. (2010). When the river water level overtops the riverbank, it fills up the floodplain reservoir which empties when the water level drops below this threshold (Decharme et al., 2012). Occurrence of flooding impacts the ISBA soil hydrology through infiltration, and it also influences the overlying atmosphere via free surface-water evaporation and precipitation interception. The groundwater scheme is based on the two-dimensional groundwater flow equation for the piezometric head (Vergnes and Decharme, 2012). Its coupling with ISBA enables accounting for the presence of a water table under the soil moisture column. It allows for upward capillary fluxes into the soil (Vergnes et al., 2014). CTRIP is coupled to ISBA through OASIS-MCT (Voldoire et al., 2017). Once a day, ISBA provides CTRIP with updates on runoff, drainage, groundwater and floodplain recharges, and CTRIP feedbacks to ISBA the water table fall or rise, floodplain fraction, and flood potential infiltration. The current CTRIP version consists of a global streamflow network at $0.5° \times 0.5°$ spatial resolution.

2.3 Evaluation strategy and data sets

Three experiments are considered for the evaluation: (i) SURFEX forced by ERA-Interim, all atmospheric variables interpolated to $0.5° \times 0.5°$ spatial resolution (referred as ei_S hereafter, the benchmark experiment); (ii) SURFEX forced by ERA-5, all atmospheric variables interpolated at $0.5° \times 0.5°$ spatial resolution except precipitation (rain and snow interpolated to hourly time steps assuming a constant flux) that comes from ERA-Interim (referred as e5ei_S hereafter); and (iii) SURFEX forced by ERA-5, all atmospheric variables interpolated at $0.5° \times 0.5°$ spatial resolution (referred as e5_S hereafter). A bilinear interpolation from the native reanalysis grid to the regular grid has been used. For all three experiments, the first year (2010) was spun up 20 times to allow the model to reach equilibrium. Comparing e5_S to ei_S provides the overall improvements from ERA-Interim to ERA-5. The idealized e5ei_S simulation was carried out to assess the role of precipitation changes from ERA-Interim to ERA-5.

This study makes use of several in situ measurement data sets as well as satellite-derived estimates of Earth observations that are described in the next two sections. The different performance metrics used for the evaluation are also described. Their choice is of crucial interest; it is governed by the nature of the variable itself and is influenced by the purpose of the investigation and its sensitivity to the considered variables (Stanski et al., 1989). No single metric or statistic can capture all the attributes of environmental variables; some are robust with respect to some attributes while insensitive to others (Entekhabi et al., 2010). While performance metrics like the correlation coefficient (R), unbiased root mean squared differences (ubRMSD), root mean squared differences (RMSDs) and efficiency score (depending on the considered variable) are first applied to the three simulations independently, metrics like the normalized information contribution (NIC; e.g. Kumar et al., 2009) are then used to quantify improvement or degradation from one data set to another. Table 1 summarizes the different data sets used for the evaluation and the performance metrics used.

2.3.1 In situ measurement of soil moisture, river discharges, snow depth and fluxes

USCRN is a network of climate-monitoring stations maintained and operated by the National Oceanic and Atmospheric Administration (NOAA). It aims at providing climate-science-quality measurements of air temperature and precipitation. To increase the network's capability of monitoring soil processes and drought, soil observations were added to USCRN instrumentation. At each USCRN station in the conterminous United States in 2011, the USCRN team completed the installation of triplicate-configuration soil moisture and soil temperature probes at five standard depths (5, 10, 20, 50 and 100 cm) as prescribed by the World Meteorological Organization. The 111 stations present data between 2009 and 2016. Stations provide data at an hourly time step. Similar to a prior study, data sets potentially affected by frozen conditions were masked out using an observed temperature threshold of 4 °C (e.g. Albergel et al., 2013a). The second layer of soil of ISBA between 1 and 4 cm depth (the diffusion scheme is used in this study) is compared to in situ measurements at 5 cm depth at a 3-hourly time step (model output) between April and September in order to avoid frozen conditions as much as possible . The ability of ei_S, e5ei_S and e5_S to reproduce surface soil moisture variability is first assessed using the correlation coefficient (R) and unbiased root mean square differences (ubRMSD). Climatology differences between model and in situ observations make a direct comparison difficult (Koster et al., 2009b). Soil moisture time series usually show a strong seasonal pattern possibly increasing the skill values between modelled and observed data sets. To avoid seasonal effects, monthly anomaly time series are calculated. At each grid and observation point, the difference from the mean is produced for a sliding window of 5 weeks, and the difference is scaled to the standard deviation as in Albergel et al. (2013b). For each surface soil moisture estimate at day i, a period F is defined, with $F = [i - 17, i + 17]$ (corresponding to a 5-week window). If at least five measurements are

Table 1. Evaluation data sets and associated metrics used in this study.

Data sets used for the evaluation	Source	Associated metrics
In situ measurements of soil moisture (USCRN; Bell et al., 2013)	https://www.ncdc.noaa.gov/crn	R (on both volumetric and anomaly time series) ubRMSD
In situ measurements of streamflow (USGS)	https://nwis.waterdata.usgs.gov/nwis	Nash–Sutcliffe efficiency (NSE), normalized information contribution (NIC) based on NSE, ratio of simulated and observed streamflow (Q)
In situ measurements of snow depth (GHCN; Menne et al., 2012a, b)	https://www.ncdc.noaa.gov/climate-monitoring/	R, bias and ubRMSD
In situ measurements of sensible and latent heat fluxes (FLUXNET-2015)	http://fluxnet.fluxdata.org/data/fluxnet2015-dataset/	R, RMSD
Satellite-derived surface soil moisture (ESA CCI SSM v4, Dorigo et al., 2015, 2017)	http://www.esa-soilmoisture-cci.org	R (on both volumetric and anomaly time series)
Satellite-derived leaf area index (GEOV1; Baret et al., 2013)	http://land.copernicus.eu/global/	R and RMSD
Satellite-driven model estimates of land evapotranspiration (GLEAM; Martens et al., 2017)	http://www.gleam.eu	R and RMSD
Upscaled estimates of gross primary production (GPP; Jung et al., 2017)	https://www.bgc-jenna.mpg.de/geodb/projects/Home.php	R and RMSD

available in this period, the average soil moisture value and the standard deviation are calculated. Anomaly time series reflect the time-integrated impact of antecedent meteorological forcing. The latter is mainly reflected in the upper layer of soil. The correlation coefficient is also computed for anomaly time series (R_{ano}). For correlations, the p value (a measure of the correlation significance) is also calculated indicating the significance of the test (as in Albergel et al., 2010), and only cases where the p value is below 0.05 (i.e. the correlation is not a coincidence) are retained. Stations with nonsignificant R values can be considered suspect and are excluded from the computation of the network average metrics. This process may remove some reliable stations too (e.g. in areas where the model might not realistically represent soil moisture).

Over the period 2010–2016, river discharge from ei_S, e5ei_S and e5_S are compared to daily streamflow data from the USGS http://nwis.waterdata.usgs.gov/nwis, last access: June 2018). Data are chosen for subbasins with large drainage areas (10 000 km^2 or greater) and with a long observation time series (4 years or more). Smaller basins are excluded due to the low resolution of CTRIP ($0.5° \times 0.5°$). It is common to express observed and simulated river discharge (Q) data in m^3 s^{-1}. Given that the observed drainage areas may differ slightly from the simulated ones, specific discharge in mm d^{-1} (the ratio of Q to the drainage area) is used in this study, similarly to Albergel et al. (2017). Stations with drainage areas differing by more than 20 % from the simulated ones are also discarded. This criterion aims to

ensure a meaningful comparison between observed and simulated values. It is necessary for coping with the significant distortions in the model representation of the river network that are caused by the coarse spatial resolution of the CTRIP global river network ($0.5° \times 0.5°$). Impact on Q is evaluated using the efficiency score (NSE; Nash and Sutcliffe, 1970). NSE evaluates the model ability to represent the monthly discharge dynamics and is given by

$$\text{NSE} = 1 - \frac{\sum\limits_{t=1}^{T} \left(Q_s^t - Q_o^t \right)^2}{\sum\limits_{t=1}^{T} \left(Q_o^t - \overline{Q_o^t} \right)^2}, \tag{1}$$

where Q_s^t is the simulated river discharge (by either ei_S, e5ei_S or e5_S) at time t and Q_o^t is observed river discharge at time t, T is the total number of days and $\overline{Q_o^t}$ is the average observed discharge. NSE can vary between $-\infty$ and 1. A value of 1 corresponds to identical model predictions and observed data. A value of 0 implies that the model predictions have the same accuracy as the mean of the observed data. Negative values indicate that the observed mean is a more accurate predictor than the model simulation. Only stations with a NSE greater than -1 for the benchmark experiment, ei_S, are considered, leading to 172 stations over the considered domain. A normalized information contribution (NIC; as in Kumar et al., 2009) measure is then computed to quantify the improvement or degradation due to the specific atmospheric reanalysis used to force ISBA. The NIC$_{NSE}$ values are computed for both e5_S and e5ei_S with respect to ei_S as

$$NIC_{NSE(e5;5ei)} = \frac{NSE_{(e5;e5ei)} - NSE_{(ei)}}{1 - NSE_{(ei)}}. \qquad (2)$$

The NIC_{NSE} metric provides a normalized measure of the improvement through the use of either NSE_{e5ei} or NSE_{e5} as a fraction of the maximum possible skill improvement $(1 - NSE_{ei})$. Positive and negative NIC_{NSE} values indicate improvements and degradations in either e5_S or e5ei_S relative to ei_S river discharge estimates, respectively. NICs along with their 95 % confidence interval of the median derived from a 10 000 samples bootstrapping are provided for e5_S and e5ei_S. The ratio of simulated and observed river discharges is also computed $\left(Q_s^t / Q_o^t\right)$; the closer to 1 it is, the better the simulated river discharges are.

The Global Historical Climatology Network (GHCN) daily data set, developed to meet the needs of climate analysis and monitoring studies that require data at a daily time resolution, contains records from over 75 000 stations in 179 countries and territories (Menne et al., 2012a, b). Numerous daily variables are provided, including maximum and minimum temperature, total daily precipitation, snowfall and snow depth. In this study, over North America, stations with daily snow depth data from 2010 to 2016, with less than 10 % missing and at least 15 days of snow presences per year on average (to avoid using stations always reporting zero snow depth) are used, it results in 1901 stations out of 2056. The ability of ei_S, e5ei_S and e5_S to reproduce snow depth and its variability is assessed using the bias, correlation coefficient (R) and unbiased root mean square difference (ubRMSD).

Daily observations of sensible and latent heat fluxes from the FLUXNET-2015 data set with at least 2 years of data are used over the period 2010–2016 to evaluate the ability of e5_S, e5ei_S and ei_S to reproduce flux variability. The FLUXNET-2015 data set includes data collected at sites from multiple regional flux networks as well as several improvements to the data quality control protocols and the data processing pipeline (http://fluxnet.fluxdata.org/data/fluxnet2015-dataset/). The 37 stations are retained for the evaluations and two metrics are considered: R and RMSD.

Performance metrics are applied to each individual station of each network; thereafter, network metrics are computed by providing the median values of the statistics from the individual stations within each network. For each metric, the 95 % confidence interval of the median derived from a 10 000 samples bootstrapping is provided.

2.3.2 Satellite-derived estimates of surface soil moisture, leaf area index, land evapotranspiration and gross primary production

In response to the GCOS (Global Climate Observing System) endorsement of soil moisture as an essential climate variable,

the European Space Agency Water Cycle Multimission Observation Strategy (WACMOS) project and Climate Change Initiative (CCI; http://www.esa-soilmoisture-cci.org, last access: June 2018) have supported the generation of a surface soil moisture product based on multiple microwave sources (ESA CCI SSM hereafter). The first version of the combined product was released in June 2012 by the Vienna University of Technology (Liu et al., 2011, 2012; Wagner et al., 2012). Several authors (e.g. Albergel et al., 2013a, b; Dorigo et al., 2015, 2017) have highlighted the quality and stability over time of the product. Despite some limitations, this data set has already shown potential in assessing model performance (e.g. Szczypta et al., 2014; van der Schrier et al., 2013). In this study the combined ESA CCI SSM latest version of the product (v4) is used. It merges SSM observations from seven microwave radiometers (SMMR, SSM/I, TMI, ASMR-E, WindSat, AMSR2, SMOS) and four scatterometers (ERS-1, 2 AMI, MetOp-A and B ASCAT) into a combined data set covering the period November 1978 to December 2016. Data are in volumetric ($m^3\, m^{-3}$) units and quality flags (snow coverage, temperature below 0°C or dense vegetation) are provided. For a more comprehensive overview of the product, see Dorigo et al. (2015, 2017). As topographic relief is known to negatively affect remote sensing estimates of soil moisture (Mätzler and Standley, 2000), the time series for pixels whose average altitude exceeded 1500 m above sea level were discarded. Data on pixels with urban land cover fractions larger than 15 % were also discarded, to limit the effects of artificial surfaces. The altitude and urban area thresholds were set according to Draper et al. (2011) and Barbu et al. (2014), who processed satellite-based SSM retrievals for data assimilation studies with the ISBA LSM. As for in situ measurements of soil moisture, correlation is applied to both the volumetric and anomaly time series.

The GEOV1 LAI used in this study is produced by the European CGLS (http://land.copernicus.eu/global/) as evaluated in Boussetta et al. (2015). The LAI observations are retrieved from the SPOT-VGT and then PROBA-V (from 1999 to present) satellite data according to the methodology proposed by Baret et al. (2013). As in Barbu et al. (2014), the 1 km spatial resolution observations are interpolated by an arithmetic average to the 0.5° model grid points, if at least 50 % of the observation grid points are observed (i.e. half the maximum amount). LAI observations have a temporal frequency of 10 days at best (in presence of clouds, no observations are available). Correlation and root mean squared differences are used to assess the ability of ei_S, e5ei_S and e5_S to reproduce LAI variability.

The GLEAM product uses a set of algorithms to estimate both terrestrial evaporation and RZSM based on satellite data (Miralles et al., 2011). It is a useful validation tool to assess model performance given that such quantities are difficult to measure directly on large scales. Potential evaporation

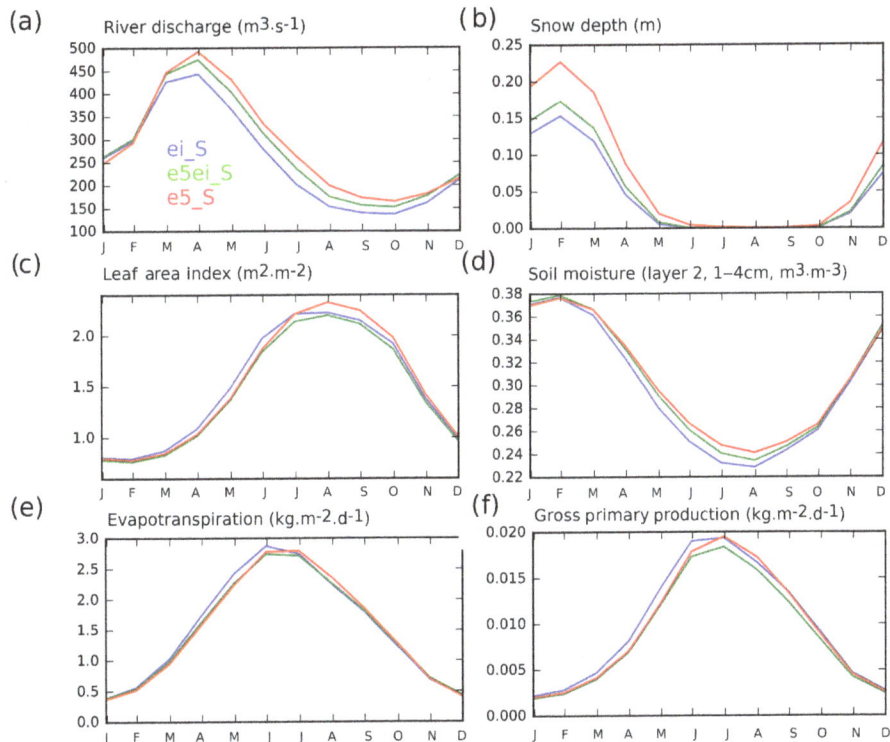

Figure 1. Seasonal time series of the six main land surface variables (LSVs) evaluated in this study over the whole domain for 2010–2016: **(a)** river discharge, **(b)** snow depth, **(c)** leaf area index, **(d)** liquid soil moisture in the second layer of soil (1–4 cm depth), **(e)** evapotranspiration and **(f)** gross primary production. LSVs simulated with SURFEX forced by ERA-Interim (ei_S) are in blue, by ERA-5 (e5_S) with precipitation from ERA-Interim (e5ei_S) in green and by ERA-5 (e5_S) in red.

rates are constrained by satellite-derived SSM data, while the global evaporation model in GLEAM is mainly driven by various microwave remote-sensing observations. It is now a well established data set that has been widely used to study land–atmosphere feedbacks (e.g. Miralles et al., 2014b; Guillod et al., 2015), as well as trends and spatial variability in the hydrological cycle (e.g. Jasechko et al., 2013; Greve et al., 2014; Miralles et al., 2014a; Zhang et al., 2016). This study makes use of the latest version available, v4.0. It is a 37-year data set spanning from 1980 to 2016 and is derived from a variety of sources, such as vegetation optical depth and snow water equivalent, satellite-derived SSM estimates, reanalysis air temperature and radiation, and a multi-source precipitation product (Martens et al., 2017). It is available at a spatial resolution of $0.25° \times 0.25°$. A full description of the data set, including an extensive validation using measurements from 64 eddy-covariance towers worldwide is provided by Martens et al. (2017). As for LAI, the correlation and root mean squared differences are the two performance metrics used to evaluate the representation of evapotranspiration from the three data sets.

The final product used in this study is a daily GPP estimate from the FLUXCOM project (Jung et al., 2017). It is an upscaled product derived from the FLUXNET. In FLUXCOM,

selected machine-learning-based regression tools that span the full range of commonly applied algorithms (from model tree ensembles to multiple adaptive regression splines, to artificial neural networks, and to kernel methods), and several representatives of each family are used to provide a spatial upscaling of GPP at regional to global scales. It is limited to a $0.5° \times 0.5°$ spatial resolution and a daily temporal resolution over the period 1982–2013 (Tramontana et al., 2016). FLUXCOM fluxes can be used as a way of benchmarking LSMs on large scales (Jung et al., 2009, 2010, 2011; Beer et al., 2010; Bonan et al., 2011; Slevin et al., 2017). The product can be found at the Max Planck Institute for Biogeochemistry data portal (https://www.bgc-jena.mpg.de/geodb/projects/Home.php, last access: June 2018). Correlation and root mean squared differences are the two performance metrics used to evaluate the representation of carbon uptake from the three data sets.

3   Results

Seasonal time series of the six main LSVs evaluated in this study over the whole domain for 2010–2016 are illustrated on Fig. 1. They are (Fig. 1a) river discharge (although averag-

Table 2. Comparison of surface soil moisture with in situ observations for ei_S, e5ei_S and e5_S over the period 2010–2016 (April to September months are considered). Median correlations R (on volumetric and anomaly time series) and ubRMSD are given for the USCRN. Scores are given for significant correlations with p values < 0.05.

	Median R^1 on volumetric time series, 95 % confidence interval[2] (% of stations for which this configuration is the best)	Median R^3 on anomalies time series, 95 % confidence interval[2] (% of stations for which this configuration is the best)	Median ubRMSD[1] ($m^3\,m^{-3}$), 95 % confidence interval[2] (% of stations for which this configuration is the best)
ei_S	0.66 ± 0.02 (20 %)	0.53 ± 0.02 (15 %)	0.052 ± 0.003 (19 %)
e5ei_S	0.69 ± 0.02 (20 %)	0.54 ± 0.04 (10 %)	0.052 ± 0.002 (24 %)
e5_S	0.71 ± 0.02 (60 %)	0.58 ± 0.03 (75 %)	0.050 ± 0.003 (57 %)

[1] Only for stations presenting significant R values on volumetric time series (p value < 0.05): 110 stations; [2] 95% confidence interval of the median derived from a 10 000 samples bootstrapping; [3] Only for stations presenting significant R values on anomaly time series (p value < 0.05): 107 stations

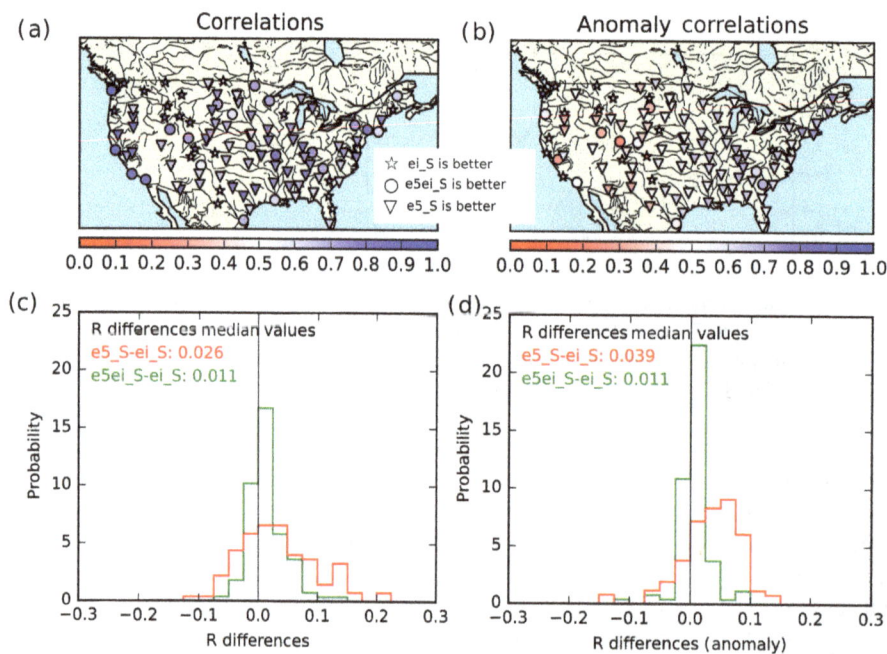

Figure 2. Maps of correlation (R) on volumetric time series (**a**) and anomaly time series (**b**) between in situ measurements at 5 cm depth from the USCRN and the ISBA LSM within the SURFEX modelling platform forced by either ERA-Interim (ei_S), ERA-5 with ERA-Interim precipitations (e5ei_S) or ERA-5 (e5_S). For each station presenting significant R (p values < 0.05), the simulation that presents the better R values is represented. Star symbols are when ei_S presents the best value, circles when it is e5ei_S and downward pointing triangles when it is e5_S. Panel (**c**) shows a histogram of R differences on volumetric time series, $R(e5\_S) - R(ei\_S)$ in red and $R(e5ei\_S) - R(ei\_S)$ in green, median values of the differences are also reported. (**d**) Same as (**c**) for R values on anomaly time series.

ing this variable over the whole domain has no real meaning, it is certainly useful to appreciate the differences between the three data sets), (Fig. 1b) snow depth, (Fig. 1c) leaf area index, (Fig. 1d) liquid soil moisture in the second layer of soil (1–4 cm depth), (Fig. 1e) evapotranspiration and (Fig. 1f) gross primary production. LSVs simulated with the ISBA LSM forced by ERA-Interim (ei_S) are in blue, by ERA-5 with precipitation from ERA-Interim (e5ei_S) in green and by ERA-5 (e5_S) in red. From Fig. 1, one can see that river discharge, snow depth and surface soil moisture are the most impacted by the use of ERA-5, suggesting that precipitation is the main driver of the differences.

3.1 Evaluations using in situ measurements

This section presents the results of the comparison versus in situ observations of LSVs from model simulations using either ei_S, e5ei_S or e5_S starting with soil moisture. The statistical scores for 2010–2016 surface soil moisture from ei_S, e5ei_S and e5_S are presented in Table 2. Median R values on volumetric time series (anomaly time series) along with their 95 % confidence intervals are 0.66 ± 0.02 (0.53 ± 0.02), 0.69 ± 0.02 (0.54 ± 0.04) and 0.71 ± 0.02 (0.58 ± 0.03), while median ubRMSD are 0.052 ± 0.003, 0.052 ± 0.002 and 0.050 ± 0.003 for ei_S, e5ei_S and e5_S,

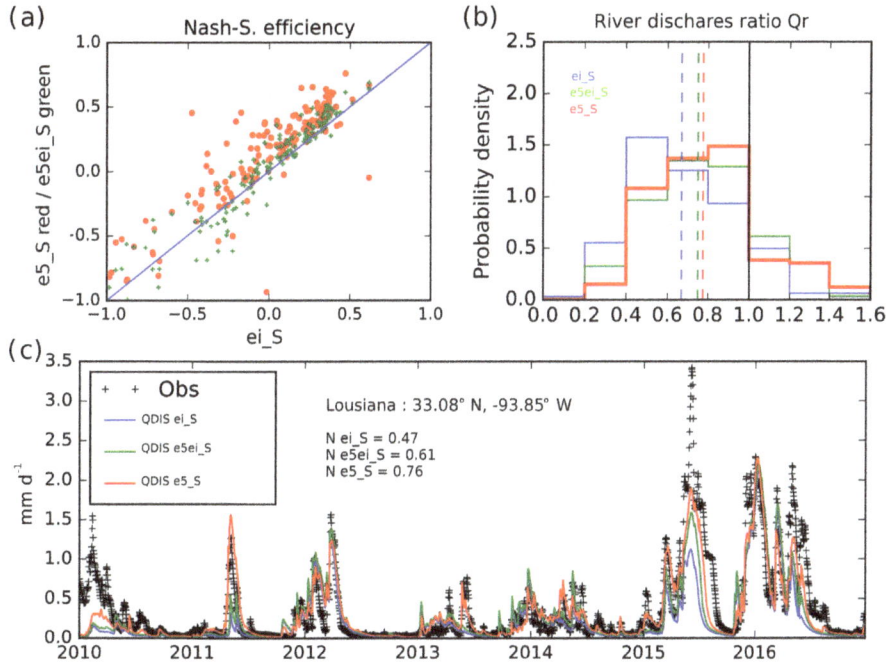

Figure 3. (a) Scatter plot of efficiency scores between in situ and simulated river discharges Q; efficiency scores for Q simulated with SURFEX forced either by ERA-5 but ERA-Interim precipitations (e5ei_S, green crosses) or ERA-5 (e5_S, red dots) as a function of efficiency scores for Q simulated using ERA-Interim (ei_S). **(b)** Histograms of river discharge ratio for ei_S (Qr_ei, in blue), e5ei_S (Qr_e5ei, in green) and e5_S (Qr_e5, in red). **(c)** Hydrograph for a river station in Louisiana (33.08° N, 1.52° W) representing scaled Q (using either observed or simulated drainage areas), in situ data (black crosses), simulated river discharges from ei_S (blue solid line), e5ei_S (green solid line) and e5_S (red solid line).

respectively. These results underline the better capability of the ISBA LSM to represent surface soil moisture variability when forced by the ERA-5 reanalysis. Also, the latest configuration (e5_S) presents more stations with better R values on volumetric time series (anomaly time series) than both ei_S and e5ei, respectively 60 and 75 % (out of 110 and 107 stations, respectively). This is also reflected in Fig. 2, illustrating correlation values on volumetric time series (Fig. 2a) and anomaly time series (Fig. 2b) on maps. Star symbols represent stations for which ISBA LSM performs best when forced by ERA-Interim, circles when it is forced by ERA-5 with ERA-Interim precipitations and downward pointing triangles when it is forced by all ERA-5 atmospheric variables. Both maps in Fig. 2 are dominated by downward pointing triangles. Figure 2c and d show histograms of R differences on volumetric (anomaly) time series for soil moisture from e5_S (in red) e5ei_S (in green) with respect to ei_S, median values of the differences are also reported.

The 172 out of 344 gauging stations retained for the evaluation according to the criteria described in the methodology section present NSE scores in the $[-1, 1]$ interval. Figure 3 presents the performance of each data set for this pool of stations. Figure 3a is a scatter plot of NSE scores between in situ and simulated river discharges Q; NSE scores for Q simulated with either ERA-5 but ERA-Interim precipitations (e5ei_S, green crosses) or ERA-5 (e5_S, red dots)

as a function of NSE scores for Q simulated using ERA-Interim (ei_S). When considering e5_S, almost all the red dots are above the 1 : 1 diagonal, suggesting a general improvement from the use of e5_S. For a large part, e5ei_S green crosses are above this diagonal, suggesting that the improvement in e5_S does not only come from precipitation but also from other variables. Median NSE values are 0.06 ± 0.06, 0.12 ± 0.07 and 0.24 ± 0.05 for ei_S, e5ei_S and e5_S, respectively. Figure 3b shows an histogram of river discharge ratio for ei_S (Qr_ei in blue), e5ei_S (Qr_e5ei in green) and e5_S (Qr_e5 in red), median values are 0.67, 075 and 0.77, respectively. While all three experiments underestimate Q (a value of 1 being a perfect match), the use of e5ei_S and e5_S leads to better results. Finally, Fig. 3c illustrates hydrographs for a river station in Louisiana (33.08° N, $-93.85°$ W) representing scaled Q (using either observed or simulated drainage areas), in situ data (black crosses), simulated river discharges from ei_S (blue solid line), e5ei_S (green solid line) and e5_S (red solid line). From this hydrograph, the added value of e5_S is clear, particularly for the 2011 and 2015 main events. NSE scores are 0.47, 0.61 and 0.76 for ei_S, e5ei_S and e5_S, respectively. Figure 4 illustrates the added value of using e5_S (panel **a**) or e5ei_S (panel **b**) with respect to ei_S. For 156 out of the pool of 172 stations, NIC_{NSE} values computed using e5_S with respect to ei_S are positive (large blue circles) showing a gen-

(a) NIC : e5_S vs ei_S   (b) NIC : e5ei_S vs ei_S

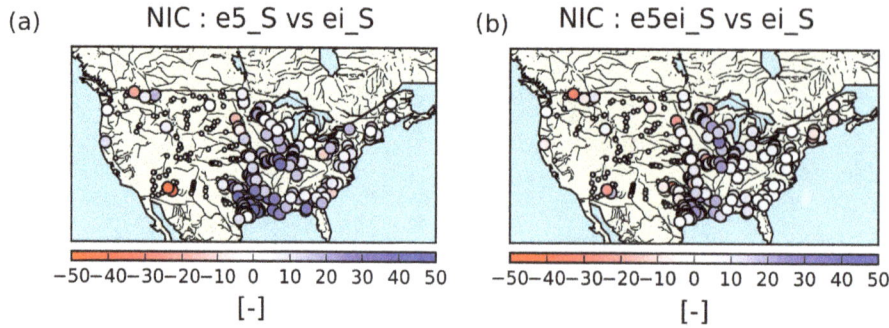

−50 −40 −30 −20 −10 0 10 20 30 40 50   −50 −40 −30 −20 −10 0 10 20 30 40 50
[-]   [-]

Figure 4. Normalized information contribution scores based on efficiency scores (NIC$_{NSE}$) **(a)** e5_S with respect to ei_S and **(b)** e5ei_S with respect to ei_S. Small dots represent stations for which the benchmark experiment (ei_S) present efficiency scores less than −1, large circles when it presents values more than −1. Positive values (blue large circles) suggest an improvement over ei_S, negative values (red large circles) a degradation. For sack of clarity, a factor of 100 has been applied to NIC.

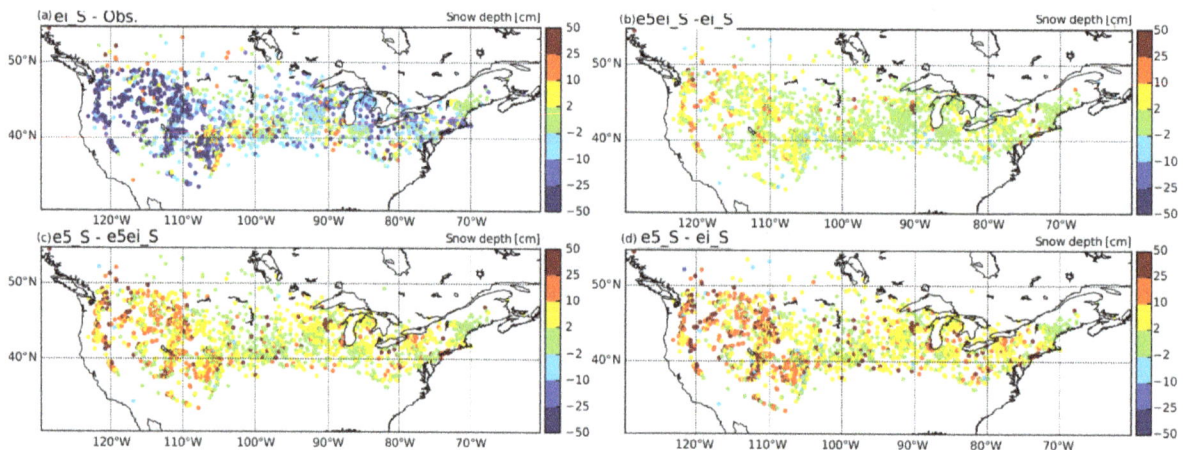

Figure 5. Mean snow depth bias for December–January–February in ei_S **(a)** and differences between e5ei_S and ei_S **(b)**, e5_S and e5ei_S **(c)**, and e5_S and ei_5 **(d)**.

eral improvement from the use of e5_S (representing 91 % of the stations) with a median NIC$_{NSE}$ value of 14 % ± 0.05. When considering e5ei_S versus ei_S, they are still 118 (69 %) with a median NIC$_{NSE}$ value of 4 % ± 0.02 suggesting that the improvement in e5_S does not only come from precipitation but also from other variables. It is also worthnoticing that stations where a score degradation is observed (large red circles) are located in areas known for irrigation, which is not represented in ISBA. All scores computed for seasons (December–January–February, March–April–May, June–July–August, September–October–November) suggest the same ranking (not shown).

The mean snow depth bias of ei_S (see Fig. 5) highlights a clear underestimation of winter snow depth accumulation mainly over the Rocky Mountains. This is likely a result of the underestimation of snowfall by ei_S associated with an overestimation of snow melt due to the coarse resolution of the ei_S reflected in a smooth topography. The replacement of all forcing variables by e5_S but keeping ei_S precipitation (e5ei_S, Fig. 5b) shows a slight increase in snow

depth. This result justifies the above hypothesis that part of the snow underestimation is also due to temperature issues linked with a coarse model orography. Moving to the full e5_S forcing, there is a clear increase in snow depth when compared with both ei_S and e5ei_S forced simulations resulting from an increase in snowfall in e5_S. Figure 6 presents the mean seasonal cycle of bias and ubRMSD (Fig. 6a) and correlations (Fig. 6b) over the period 2010–2016. In addition to the added values of e5_S in terms of the mean snow depth already presented in Fig. 5, the temporal variability and random errors are also improved. Comparably with what was discussed for the mean bias, e5ei_S shows some benefits when compared with ei_S in terms of ubRMSD and correlation (median bias, ubRMSD and R values of e5ei_S over the whole period are −1.70 ± 0.33, 7.40 ± 0.65 and 0.77 ± 0.01 cm, respectively, for ei_S they are −2.11 ± 0.33, 7.58 ± 0.65 and 0.75 ± 0.01 cm, respectively), while e5_S has a clear improvement in ubRMSD and correlation (median bias, ubRMSD and R values of e5_ei over the whole period are −0.64 ± 0.19, 7.00 ± 0.65

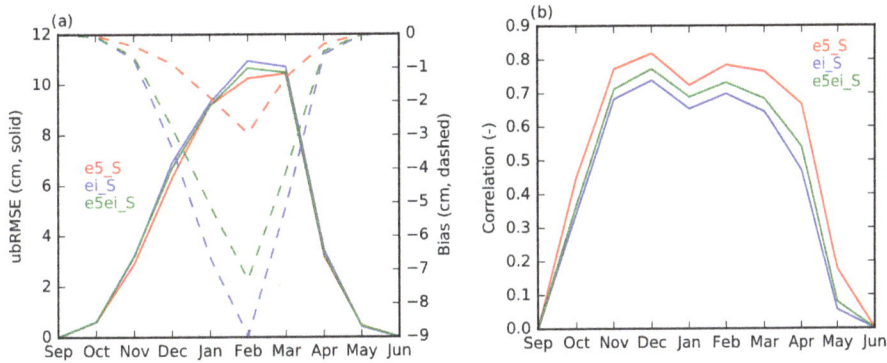

Figure 6. (a) Mean seasonal cycle of the bias (dashed lines) and ubRMSD (solid lines) averaged over all stations and **(b)** the mean seasonal cycle of the correlations for ei_S (in blue), e5ei_S (in green) and e5_S (in red).

Table 3. Comparison of snow depth with in situ measurements, median Bias, ubRMSD and R values are given for the three seasons affected by snow (SON, DJF, MAM) and for the whole period (All). SON, DJF and MAM stand for September–October–November, December–January–February and March–April–May, respectively.

		Median bias (cm)[1], 95 % confidence interval[2] (% of stations for which this configuration is the best)	Median ubRMSD (cm)[1], 95 % confidence interval[2] (% of stations for which this configuration is the best)	Median R[1], 95 % confidence interval[2] (% of stations for which this configuration is the best)
ei_S	SON	-0.27 ± 0.04 (13 %)	2.05 ± 0.17 (13 %)	0.70 ± 0.01 (21 %)
	DJF	-6.28 ± 0.86 (11 %)	10.34 ± 0.63 (17 %)	0.72 ± 0.01 (20 %)
	MAM	-1.90 ± 0.33 (15 %)	7.82 ± 0.79 (17 %)	0.65 ± 0.01 (18 %)
	All	-2.11 ± 0.33 (11 %)	7.58 ± 0.65 (14 %)	0.75 ± 0.01 (19 %)
e5ei_S	SON	-0.25 ± 0.04 (12 %)	2.03 ± 0.15 (10 %)	0.74 ± 0.01 (23 %)
	DJF	-4.84 ± 0.80 (14 %)	9.98 ± 0.50 (14 %)	0.75 ± 0.01 (21 %)
	MAM	-1.49 ± 0.33 (14 %)	7.61 ± 0.76 (13 %)	0.69 ± 0.02 (22 %)
	All	-1.70 ± 0.33 (14 %)	7.40 ± 0.65 (14 %)	0.77 ± 0.01 (20 %)
e5_S	SON	-0.14 ± 0.03 (76 %)	1.83 ± 0.14 (77 %)	0.79 ± 0.01 (56 %)
	DJF	-1.70 ± 0.44 (75 %)	9.64 ± 0.46 (69 %)	0.80 ± 0.01 (59 %)
	MAM	-0.57 ± 0.22 (71 %)	7.43 ± 0.79 (70 %)	0.76 ± 0.01 (60 %)
	All	-0.64 ± 0.19 (75 %)	7.00 ± 0.65 (72 %)	0.82 ± 0.01 (61 %)

[1] only for stations presenting more than 80 % of (daily) data; 1901 out of 2056 stations. [2] 95 % confidence interval of the median derived from a 10 000 samples bootstrapping.

and 0.82 ± 0.01 cm, respectively). The improvements on the snow depth simulations are consistent throughout the entire snow-cover season (see Fig. 6a and b) with a maximum improvement from January to March. These results highlight the cumulative effect of the forcing quality on the snow depth simulation. Finally Table 3 presents scores from the comparison of snow depth with in situ measurements; median bias, ubRMSD and R values are given for the three seasons affected by snow (September–October–November, December–January–February and March–April–May) and for the whole period. e5_S always presents better scores when compared to ei_S and it is always the configuration presenting the highest percentage of stations with the best scores. Looking at the 95 % confidence interval, for the correlation and bias, it is clear that the changes are significant.

Results from the comparisons between ei_S, e5ei_S, e5_S and in situ sensible and latent flux measurements are presented in Table 4 and illustrated by Fig. 7. The 37 stations present significant correlation values (at p value < 0.05). For sensible heat flux, median correlation and RMSD values are 0.62 ± 0.11, 0.62 ± 0.11 and 0.65 ± 0.11 and 39.58 ± 3.71, 32.89 ± 3.86 and 32.73 ± 2.61 W m^{-2} for ei_S, e5ei_S and e5_S, respectively. For latent heat flux, they are 0.63 ± 0.05, 0.62 ± 0.07 and 0.70 ± 0.04 and 39.00 ± 5.38, 37.12 ± 4.37 and 36.66 ± 4.94 W m^{-2}, respectively. As for surface soil moisture, river discharge and snow depth, e5_S presents better results than e5ei_S and ei_S. At the station level, Fig. 7 illustrates scatter plots of correlations and RMSD for sensible and latent heat flux from ei_S, e5ei_S, e5_S against in situ measurements of sensible (Fig. 7a for correlation, Fig. 7c

Table 4. Comparison of sensible (H) and latent (LE) heat flux with in situ observations for ei_S, e5ei_S and e5_S. Median correlations (R) and median RMSD are given for the FLUXNET stations. Scores are given for significant correlations with p values < 0.05.

	H median R^1, 95 % confidence interval[2] (% of stations for which this configuration is the best)	H median RMSD[1] $\mathrm{W\,m^{-2}}$, 95 % confidence interval[2] (% of stations for which this configuration is the best)	LE median R^1, 95 % confidence interval[2] (% of stations for which this configuration is the best)	LE median RMSD[1] $\mathrm{W\,m^{-2}}$, 95 % confidence interval[2] (% of stations for which this configuration is the best)
ei_S	0.62 ± 0.11 (8 %)	39.58 ± 3.71 (5 %)	0.63 ± 0.05 (8 %)	39.00 ± 5.38 (16 %)
e5ei_S	0.62 ± 0.11 (27 %)	32.89 ± 3.86 (27%)	0.62 ± 0.07 (11 %)	37.12 ± 4.37 (22 %)
e5_S	0.65 ± 0.11 (65 %)	32.73 ± 2.61 (68 %)	0.70 ± 0.04 (81 %)	36.66 ± 4.94 (62 %)

[1] only for stations presenting significant R values (p value < 0.05): 37 stations; [2] 95 % confidence interval of the median derived from a 10 000 samples bootstrapping

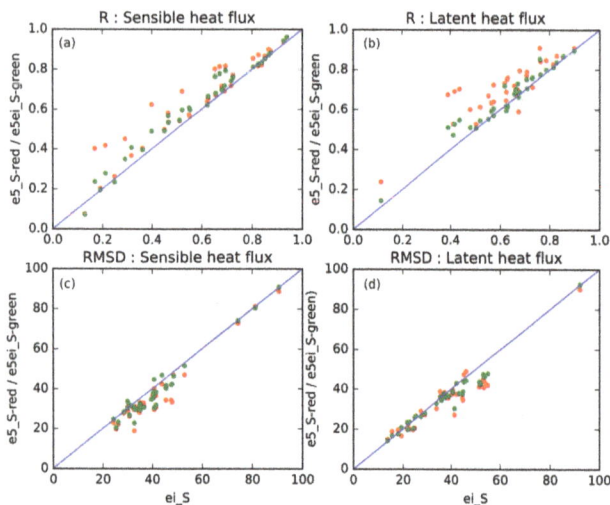

Figure 7. Scatter plots illustrating evaluation of ei_S, e5ei_S and e5_S against in situ measurements of sensible (**a** for correlation, **c** for RMSD) and latent (**b** for correlation, **d** for RMSD) heat flux. Scores for either e5ei_S (green dots) or e5_S (in red) are presented as a function of those for ei_S.

for RMSD) and latent (Fig. 7b for correlation, Fig. 7d for RMSD) heat flux. Scores for either e5ei_S (green dots) or e5_S (in red) are presented as a function of those for ei_S. When looking at the correlations, almost all of e5_S and e5ei_S symbols (in red and green, respectively, in Fig. 7a, c) are above the 1 : 1 diagonal indicating that e5_S and e5ei_S better represent sensible and latent heat flux than ei_S. The same tendency is observed for RMSD with most of the symbols below the 1 : 1 diagonal. If RMSD values are comparable for e5_S and e5ei_S, R values are clearly higher for e5_S.

3.2 Evaluations using satellite-derived estimates

Figure 8 illustrates the comparison between ESA CCI SSM v4 and soil moisture from the ISBA second layer of soil over 2010–2016. Figure 8a shows seasonal correlations on volumetric time series and Fig. 8b on anomaly time series. Scores for ISBA LSM forced by ERA-Interim (ei_S) are in blue,

ERA-5 but with precipitation from ERA-Interim (e5ei_S) in green and ERA-5 (e5_S) in red. From Fig. 8a, one can appreciate the added value of using ERA-5 atmospheric forcing particularly from April to September. It is also interesting to notice that when using all ERA-5 atmospheric fields except for the precipitation, a similar added value is noticeable suggesting that all improvements from ERA-5 do not only come from precipitation. However, when evaluating the short-term variability of soil moisture (i.e. removing the seasonal effect), it is really ERA-5 that provides the best results. Correlation on volumetric (anomaly) time series for all grid points put together over 2010–2016 are 0.668 (0.464), 0.682 (0.468) and 0.689 (0.490) for ei_S, e5ei_S and e5_S, respectively. Additionally to the global seasonal scores, Fig. 8c and d present maps of correlation differences between soil moisture from e5_S and ei_S on volumetric time series and anomaly time series, respectively. Grey areas represent areas that were flagged out for elevation greater than 1500 m above sea level. As visible on Fig. 8c and d, the use of ERA-5 mainly leads to improvements all over the considered domain. Focusing on correlation differences, ($R_{e5} - R_{ei}$) on volumetric (or anomaly) time series, 68 % (77 %) of the values are positive – indicating an improvement from e5_S – with median values of 4.5 % (4.11 %) and include values up to 40 % (45 %). It shows the added value of using ERA-5 to force ISBA LSM compared to ERA-Interim.

Figure 9 illustrates seasonal scores between ISBA LSM forced by either ERA-Interim (ei_S in blue), ERA-5 but ERA-Interim precipitation (e5ei_S in green) or ERA-5 (e5_S in red) for the following variables: (Fig. 9a, b) evapotranspiration estimates from the GLEAM project over 2010–2016, (Fig. 9c, d) upscaled GPP from the FLUXCOM project over 2010–2013 and (Fig. 9e, f) LAI estimates from the CGLS project over 2010–2016. The left column (Fig. 9a, c and e) are for RMSDs and the right column (Fig. 9b, d and e) for correlations. For evapotranspiration, and to a lesser extend GPP, one can notice a decrease in RMSD when using ERA-5 atmospheric reanalysis compared to ERA-Interim atmospheric reanalysis; however, it fails at improving LAI. Considering evapotranspiration, correlation (RMSD) values for all grid points put together over 2010–2016

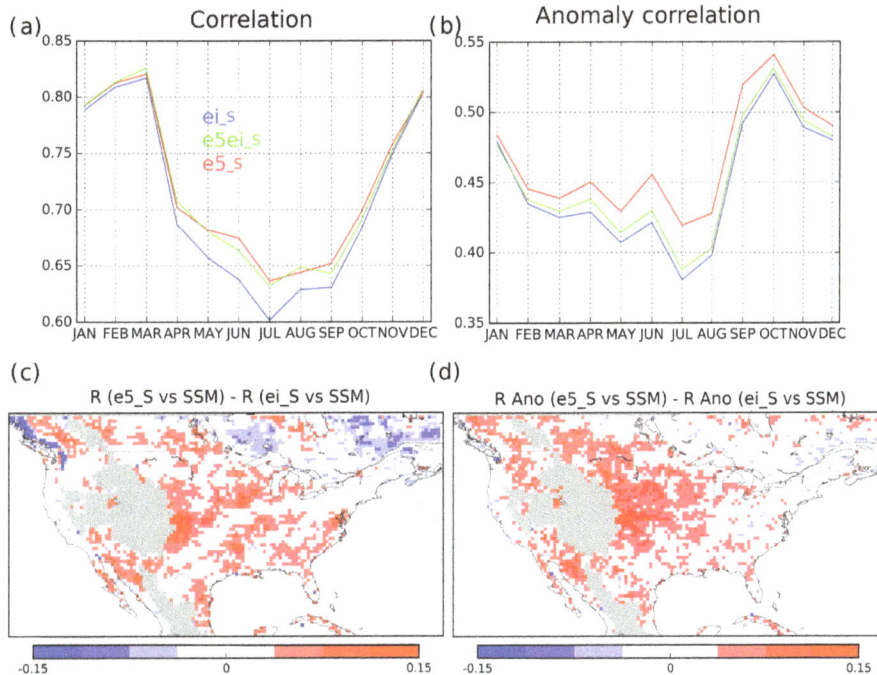

Figure 8. Seasonal correlations for **(a)** volumetric time series and **(b)** anomaly time series between surface soil moisture (SSM) estimates from the ESA CCI project (ESA CCI SSM v4) and soil moisture from the second layer of soil of the ISBA LSM forced by ERA-Interim (ei_S, in blue), ERA-5 but with precipitation from ERA-Interim (e5ei_S, in green) and ERA-5 (e5_S, in red) over the period 2010–2016. Maps of correlation differences between soil moisture from e5_S and ei_S for volumetric time series **(c)** and anomaly time series **(d)** are shown, areas in red represent an improvement from the use of ERA-5. Grey areas represent areas that were flagged out for elevation greater than 1500 m above sea level.

are 0.786 (0.927 $\mathrm{kg\,m^{-2}\,d^{-1}}$), 0.778 (0.917 $\mathrm{kg\,m^{-2}\,d^{-1}}$) and 0.795 (0.889 $\mathrm{kg\,m^{-2}\,d^{-1}}$) for ei_S, e5ei_S and e5_S, respectively. They are 0.726 (2.429 $\mathrm{kg\,m^{-2}\,d^{-1}}$), 0.733 (2.167 $\mathrm{kg\,m^{-2}\,d^{-1}}$) and 0.734 (2.227 $\mathrm{kg\,m^{-2}\,d^{-1}}$) for GPP and 0.715 (1.050 $\mathrm{m^2\,m^{-2}}$), 0.710 (1.026 $\mathrm{m^2\,m^{-2}}$) and 0.697 (1.079 $\mathrm{m^2\,m^{-2}}$) for LAI, respectively.

Improvements (in red) and degradations (in blue) from the use of ERA-5 in the ISBA LSM with respect to ERA-Interim for evapotranspiration, GPP and LAI are illustrated by Fig. 10 (respectively from top to bottom). Figure 10a, c and e show RMSD differences while Fig. 10b, d and f show R differences. Both differences in RMSD and R values suggest an improvement from the use of ERA-5 as the two figures are mainly dominated by red colours, RMSD and R represent 56 and 53 % of the domain, respectively for evapotranspiration (Fig. 10a, b), 60 and 69 % for GPP (Fig. 10c, d), but only 47 and 44 % for LAI (Fig. 10e, f).

4   Discussion and conclusions

This study assesses the ability of the recently released ERA-5 atmospheric reanalysis to force the ISBA land surface model (LSM) with respect to ERA-Interim reanalysis over North America for 2010–2016. The results presented above using the three atmospheric reanalysis data

sets (ERA-Interim, ei_S; ERA-5 but with precipitation from ERA-Interim, e5ei_S; and ERA-5, e5_S, with all meteorological variables) to force the ISBA LSM provide two important insights: (i) firstly the use of ERA-5 leads to significant improvements in the representation of the land surface variables (LSVs) linked to the terrestrial water cycle assessed in this study (surface soil moisture, river discharges, snow depth and turbulent fluxes) but failed impacting LSVs linked to the vegetation cycle (carbon uptake and LAI). Even when they are small, improvements are systematic when using ERA-5. (ii) Secondly, if most of the improvements seem to come from a better representation of the precipitation in ERA-5, the e5ei_S experiment also presents improvements with respect to the ei_S experiment and suggests that the other meteorological forcing from ERA-5 are better represented too. However, it is acknowledged that the use of 3-hourly ERA-Interim liquid and solid precipitations rescaled at an hourly time step in ERA-5 might have sometimes led to inconsistent configurations (e.g. precipitations while having a very strong net radiation).

ERA-5 has a great potential to further improve the representation of LSVs if used to force offline LDAS. In recent years, several LDAS have emerged at different spatial scales, (i) regional like the Coupled Land Vegetation LDAS (CLVLDAS; Sawada and Koike, 2014, Sawada et al., 2015)

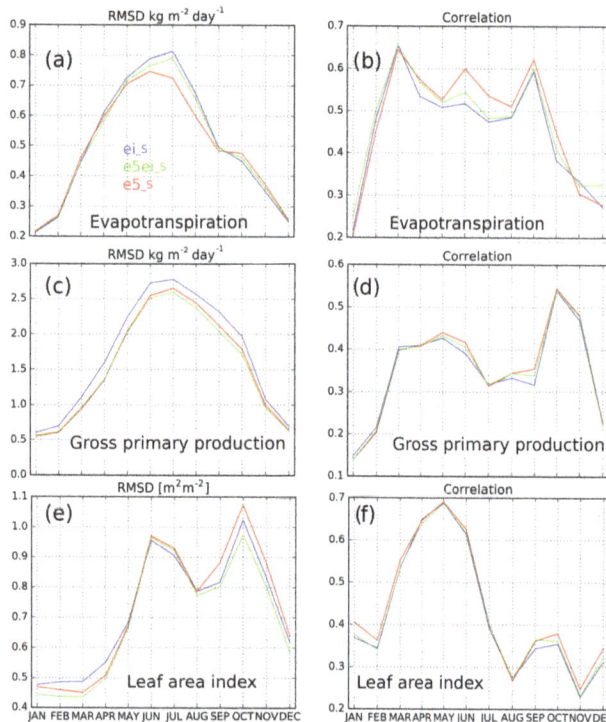

Figure 9. Seasonal scores between ISBA LSM within SURFEX forced by either ERA-Interim (ei_S, in blue), ERA-5 but ERA-Interim precipitation (e5ei_S, in green) or ERA-5 (e5_S, in red) and (**a, b**) evapotranspiration estimates from the GLEAM project over the period 2010–2016, (**c, d**) upscaled GPP from the FLUXCOM project over 2010–2013 and (**e, f**) LAI estimates from the CGLS project over 2010–2016. The left column (**a, c, e**) are for RMSD and the right column (**b, d, e**) are for correlations.

and the Famine Early Warning Systems Network (FEWS-NET) LDAS (FLDAS; McNally et al., 2017), (ii) continental like the North American LDAS (NLDAS; Mitchell et al., 2004; Xia et al., 2012) and the National Climate Assessment LDAS (NCA-LDAS; Kumar et al., 2018), and (iii) global like the Global Land Data assimilation (GLDAS; Rodell et al., 2004) and more recently LDAS-Monde (Albergel et al., 2017, 2018). LDAS-Monde is a global capacity system able to sequentially assimilate satellite-derived estimates of surface soil moisture and LAI. Albergel et al. (2017) found that the main improvements of their analysis (i.e. with assimilation) when compared to an open-loop experiment (simple model run) were linked to vegetation variables and the assimilation of vegetation estimates. They have also proposed further advances on a better use of satellite-based microwave data in the assimilation system. Having LDAS-Monde analysis forced by ERA-5 atmospheric forcing should both combine the strengths of an improved atmospheric reanalysis on the terrestrial water cycle and of the assimilation of satellite-derived products on the vegetation cycle. Effort will now be concentrated on the use of ERA-5 and strengthening LDAS-

Monde through the direct assimilation of satellite-based soil moisture and vegetation properties from microwave remote sensing. It will enable fostering links with potential applications like climate reanalysis of the LSVs as well as going from a monitoring system of the LSVs and extreme events (like agricultural drought) to a forecasting system. Preliminary results suggest that a LSV forecast initialized by an analysis is more robust than one initialized by a simple model run (Albergel et al., 2018). Preliminary tests over Europe also indicate similar benefits from the use of ERA-5 (not shown). When the whole ERA-5 period will be available (1979–present), in addition to the availability of the ERA-5 10-member ensemble of data assimilation (at lower spatial and temporal resolutions though), it will be possible to develop a global long-term ensemble of LSV reanalysis forced by high quality atmospheric data. It will make it possible providing uncertainties in the representation of the atmospheric forcing, while LSVs may require special considerations and perturbation methods. Capturing those uncertainties coming from the simplifications and assumptions in the LSM is of paramount interest for many applications from monitoring to forecasting.

Data availability. The ERA-Interim (ERA-I) and ERA-5 datasets are distributed by ECMWF (http://apps.ecmwf.int/datasets/, ECMWF, last access: June 2018). The ECOCLIMAP dataset is distributed by CNRM (https://opensource.umr-cnrm.fr/projects/ecoclimap, CNRM, 2013). The SURFEX model code is distributed by CNRM (http://www.umr-cnrm.fr/surfex/, CNRM, 2016). The satellite-derived LAI GEOV1 observations are freely accessible from the Copernicus Global Land Service (http://land.copernicus.eu/global/; last access: June 2018). The ESA CCI surface soil moisture dataset is distributed by ESA (http://www.esa-soilmoisture-cci.org/, last access: June 2018, Dorigo et al., 2017). The satellite-driven model estimates of land evapotranspiration are freely accessible at http://www.gleam.eu (last access: June 2018; Martens et al., 2017). The upscaled estimates of gross primary production are freely accessible at https://www.bgc-jena.mpg.de/geodb/projects/Home.php (last access: June 2018; Jung et al., 2017). In situ measurements of soil moisture are freely available at https://www.ncdc.noaa.gov/crn (last access: June 2018; Bell et al., 2013). In situ measurements of streamflow are freely available at https://nwis.waterdata.usgs.gov/nwis (last access: June 2018, USGS). In situ measurements of snow depth are freely available at https://www.ncdc.noaa.gov/climate-monitoring/ (last access: June 2018; Menne et al., 2012a, b). In situ measurements of sensible and latent heat fluxes (FLUXNET-2015) are freely available at http://fluxnet.fluxdata.org/data/fluxnet2015-dataset/ (last access: June 2018).

Author contributions. CA and ED conceived and designed the experiments; CA performed the experiments; all the authors analysed the results; CA wrote the paper.

Competing interests. The authors declare that they have no conflict of interest.

Special issue statement. This article is part of the special issue "Integration of Earth observations and models for global water resource assessment". It is not associated with a conference.

Acknowledgements. Results were generated using the Copernicus Climate Change Service Information 2017. Emanuel Dutra's work was supported by the Portuguese Science Foundation (FCT) under project IF/00817/2015.

Edited by: Frederiek Sperna Weiland

References

Albergel, C., Calvet, J.-C., de Rosnay, P., Balsamo, G., Wagner, W., Hasenauer, S., Naeimi, V., Martin, E., Bazile, E., Bouyssel, F., and Mahfouf, J.-F.: Cross-evaluation of modelled and remotely sensed surface soil moisture with in situ data in southwestern France, Hydrol. Earth Syst. Sci., 14, 2177–2191, https://doi.org/10.5194/hess-14-2177-2010, 2010.

Albergel, C., Dorigo, W., Balsamo, G., Muñoz-Sabater, J., de Rosnay, P., L. Isaksen, Brocca, L., de Jeu, R., and Wagner, W.: Monitoring multi-decadal satellite earth observation of soil moisture products through land surface reanalyses, Remote Sens. Environ., 138, 77–89, https://doi.org/10.1016/j.rse.2013.07.009, 2013a.

Albergel, C., Dorigo, W., Reichle, R. H., Balsamo, G., de Rosnay, P., Munoz-Sabater, J., Isaksen, L., de Jeu, R., and Wagner, W.: Skill and global trend analysis of soil moisture from reanalyses and microwave remote sensing, J. Hydrometeorol., 14, 1259–1277, https://doi.org/10.1175/JHM-D-12-0161.1, 2013b.

Albergel, C., Munier, S., Leroux, D. J., Dewaele, H., Fairbairn, D., Barbu, A. L., Gelati, E., Dorigo, W., Faroux, S., Meurey, C., Le Moigne, P., Decharme, B., Mahfouf, J.-F., and Calvet, J.-C.: Sequential assimilation of satellite-derived vegetation and soil moisture products using SURFEX_v8.0: LDAS-Monde assessment over the Euro-Mediterranean area, Geosci. Model Dev., 10, 3889–3912, https://doi.org/10.5194/gmd-10-3889-2017, 2017.

Albergel, C., Munier, S., Bocher, A., Draper, C., Leroux, D. J., Barbu, A. L., and Calvet, J.-C.: LDAS-Monde global capacity integration of satellite derived observations applied over North America: assessment, limitations and perspectives. to be sumitted to Remote Sensing, Special Issue "Assimilation of Remote Sensing Data into Earth System Models", in preparation, 2018.

Balsamo, G., Albergel, C., Beljaars, A., Boussetta, S., Brun, E., Cloke, H., Dee, D., Dutra, E., Muñoz-Sabater, J., Pappenberger, F., de Rosnay, P., Stockdale, T., and Vitart, F.: ERA-Interim/Land: a global land surface reanalysis data set, Hydrol. Earth Syst. Sci., 19, 389–407, https://doi.org/10.5194/hess-19-389-2015, 2015.

Barbu, A. L., Calvet, J.-C., Mahfouf, J.-F., and Lafont, S.: Integrating ASCAT surface soil moisture and GEOV1 leaf area index into the SURFEX modelling platform: a land data assimilation application over France, Hydrol. Earth Syst. Sci., 18, 173–192, https://doi.org/10.5194/hess-18-173-2014, 2014.

Baret, F., Weiss, M., Lacaze, R., Camacho, F., Makhmared, H., Pacholczyk, P., and Smetse, B.: GEOV1: LAI and FAPAR essential climate variables and FCOVER global time series capitalizing over existing products, Part 1: Principles of development and production, Remote Sens. Environ., 137, 299–309, 2013.

Beer, C., Reichstein, M., Tomelleri, E., Ciais, P., Jung, M., Carvalhais, N., Rödenbeck, C., Arain, M. A., Baldocchi, D., Bonan, G. B., Bondeau, A., Cescatti, A., Lasslop, G., Lindroth, A., Lomas, M., Luyssaert, S., Margolis, H., Oleson, K. W., Roupsard, O., Veenendaal, E., Viovy, N., Williams, C., Woodward, F. I., and Papale, D.: Terrestrial gross carbon dioxide uptake: global distribution and covariation with climate, Science, 329, 834–838, https://doi.org/10.1126/science.1184984, 2010.

Bell, J. E., Palecki, M. A., Collins, W. G., Lawrimore, J. H., Leeper, R. D., Hall, M. E., Kochendorfer, J., Meyers, T. P., Wilson, T., Baker, B., and Diamond, H. J.: U.S. Climate Reference Network soil moisture and temperature observatons, J. Hydrometeorol., 14, 977–988, https://doi.org/10.1175/JHM-D-12-0146.1, 2013.

Berrisford, P., Dee, D. P., Fielding, K., Fuentes, M., Kallberg, P., Kobayashi, S., and Uppala, S. M.: The ERA-Interim archive, ERA Rep. 1, 16 pp. available at: https://www.ecmwf.int/en/elibrary/8173-era-interim-archive (last access: June 2018), 2009.

Bonan, G. B., Lawrence, P. J., Oleson, K. W., Levis, S., Jung, M., Reichstein, M., Lawrence, D. M., and Swenson, S. C.: Improving canopy processes in the Community Land Model version 4 (CLM4) using global flux fields empirically inferred from FLUXNET data, J. Geophys. Res., 116, G02014, https://doi.org/10.1029/2010JG001593, 2011.

Boone, A. and Etchevers, P.: An intercomparison of three snow schemes of varying complexity coupled to the same land-surface model: local scale evaluation at an Alpine site, J. Hydrometeorol., 2, 374–394, 2001.

Boone, A., Masson, V., Meyers, T., and Noilhan, J.: The influence of the inclusion of soil freezing on simulations by a soil vegetation-atmosphere transfer scheme, J. Appl. Meteorol., 39, 1544–1569, 2000.

Boussetta, S., Balsamo, G., Dutra, E., Beljaars, A., and Albergel, C.: Assimilation of surface albedo and vegetation states from satellite observations and their impact on numerical weather prediction, Remote Sens. Environ., 163, 111–126, https://doi.org/10.1016/j.rse..2015.03.009, 2015.

Brooks, R. H. and Corey, A. T.: Properties of porous media affecting fluid flow, J. Irrig. Drain. Div. Am. Soc. Civ. Eng., 17, 187–208, 1966.

Calvet, J.-C., Noilhan, J., Roujean, J.-L., Bessemoulin, P., Cabelguenne, M., Olioso, A., and Wigneron, J.-P.: An interactive vegetation SVAT model tested against data from six contrasting sites, Agr. Forest Meteorol., 92, 73–95, 1998.

Calvet, J.-C., Rivalland, V., Picon-Cochard, C., and Guehl, J.-M.: Modelling forest transpiration and CO_2 fluxes – response to soil moisture stress, Agr. Forest Meteorol., 124, 143–156, https://doi.org/10.1016/j.agrformet.2004.01.007, 2004.

Canal, N., Calvet, J.-C., Decharme, B., Carrer, D., Lafont, S., and Pigeon, G.: Evaluation of root water uptake in the ISBA-A-gs land surface model using agricultural yield statistics over France, Hydrol. Earth Syst. Sci., 18, 4979–4999, https://doi.org/10.5194/hess-18-4979-2014, 2014.

Decharme, B., Alkama, R., Douville, H., Becker, M., and Cazenave, A.: Global evaluation of the ISBA-TRIP continental hydrologic system, Part 2: Uncertainties in river routing simulation related to flow velocity and groundwater storage, J. Hydrometeorol., 11, 601–617, 2010.

Decharme, B., Boone, A., Delire, C., and Noilhan, J.: Local evaluation of the Interaction between soil biosphere atmosphere soil multilayer diffusion scheme using four pedotransfer functions, J. Geophys. Res., 116, D20126, https://doi.org/10.1029/2011JD016002, 2011.

Decharme, B., Alkama, R., Papa, F., Faroux, S., Douville, H., and Prigent, C.: Global offline evaluation of the ISBA-TRIP flood model, Clim. Dynam., 38, 1389–1412, https://doi.org/10.1007/s00382-011-1054-9, 2012.

Decharme, B., Martin, E., and Faroux, S.: Reconciling soil thermal and hydrological lower boundary conditions in land surface models, J. Geophys. Res.-Atmos., 118, 7819–7834, https://doi.org/10.1002/jgrd.50631, 2013.

Decharme, B., Brun, E., Boone, A., Delire, C., Le Moigne, P., and Morin, S.: Impacts of snow and organic soils parameterization on northern Eurasian soil temperature profiles simulated by the ISBA land surface model, The Cryosphere, 10, 853–877, https://doi.org/10.5194/tc-10-853-2016, 2016.

Dee, D. P., Uppala, S. M., Simmons, A. J., Berrisford, P., Poli, P., Kobayashi, S., Andrae, U., Balmaseda, M. A., Balsamo, G., Bauer, P., Bechtold, P., Beljaars, A. C. M., van de Berg, I., Biblot, J., Bormann, N., Delsol, C., Dragani, R., Fuentes, M., Greer, A. J., Haimberger, L., Healy, S. B., Hersbach, H., Holm, E. V., Isaksen, L., Kallberg, P., Kohler, M., Matricardi, M., McNally, A. P., Mong-Sanz, B. M., Morcette, J.-J., Park, B.-K., Peubey, C., de Rosnay, P., Tavolato, C., Thepaut, J. N., and Vitart, F.: The ERA-Interim reanalysis: Configuration and performance of the data assimilation system, Q. J. Roy. Meteorol. Soc., 137, 553–597, https://doi.org/10.1002/qj.828, 2011.

Dirmeyer, P. A.: A history and review of the Global Soil-Wetness Project (GSWP), J. Hydrometeorol., 12, 729–749, https://doi.org/10.1175/JHM-D-10-05010.1, 2011.

Dirmeyer, P. A., Gao, X., and Oki, T.: The second Global Soil Wetness Project – Science and implementation plan, IGPO Int. GEWEX Project Office Publ. Series 37, Global Energy and Water Cycle Exp. (GEWEX) Proj. Off., Silver Spring, MD, 65 pp., 2002.

Dirmeyer, P. A., Gao, X., Zhao, M., Guo, Z., Oki, T., and Hanasaki N.: The Second Global Soil Wetness Project (GSWP-2): Multi-model analysis and implications for our perception of the land surface, B. Am. Meteorol. Soc., 87, 1381–1397, https://doi.org/10.1175/BAMS-87-10-1381, 2006.

Dorigo, W. A., Gruber, A., De Jeu, R. A. M., Wagner, W., Stacke, T., Loew, A., Albergel, C., Brocca, L., Chung, D., Parinussa, R. M., and Kidd, R.: Evaluation of the ESA CCI soil moisture product using ground-based observations, Remote Sens. Environ., 162, 380–395, https://doi.org/10.1016/j.rse.2014.07.023, 2015.

Dorigo, W., Wagner, W., Albergel, C. Albrecht, F., Balsamo, G., Brocca, L., Chung, D., Ertl, M., Forkel, M., Gruber, A., Haas, E., Hamer, P. D., Hirschi, M., Ikonen, J., de Jeu, R., Kidd, R., William Lahoz g, Liu, Y. Y., Miralles, D., Mistelbauer, T., Nicolai-Shaw, N., Parinussa, R., Pratola, C., Reimer, C., van der Schalie, R., Seneviratne, S. I., Smolander, T., and Lecomte, P.: ESA CCI soil moisture for improved Earth system understanding: state-of-the art and future directions, Remote Sens. Environ., 201, 185–215, https://doi.org/10.1016/j.rse.2017.07.001, 2017.

Draper, C., Mahfouf, J.-F., Calvet, J.-C., Martin, E., and Wagner, W.: Assimilation of ASCAT near-surface soil moisture into the SIM hydrological model over France, Hydrol. Earth Syst. Sci., 15, 3829–3841, https://doi.org/10.5194/hess-15-3829-2011, 2011.

Gelaro, R., McCarty, W., Suárez, M. J., Todling, R., Molod, A., Takacs, L., Randles, C. A., Darmenov, A., Bosilovich, M. G., Reichle, R., Wargan, K., Coy, L., Cullather, R., Draper, C., Akella, S., Buchard, V., Conaty, A., da Silva, A. M., Gu, W., Kim, G., Koster, R., Lucchesi, R., Merkova, D., Nielsen, J. E., Partyka, G., Pawson, S., Putman, W., Rienecker, M., Schubert, S. D., Sienkiewicz, M., and Zhao, B.: The Modern-Era Retrospective Analysis for Research and Applications, Version 2 (MERRA-2), J. Climate, 30, 5419–5454, https://doi.org/10.1175/JCLI-D-16-0758.1, 2017.

Gibelin, A.-L., Calvet, J.-C., Roujean, J.-L., Jarlan, L., and Los, S. O.: Ability of the land surface model ISBA-A-gs to simulate leaf area index at global scale: comparison with satellite products, J. Geophys. Res., 111, 1–16, https://doi.org/10.1029/2005JD006691, 2006.

Greve, P., Orlowsky, B., Mueller, B., Sheffield, J., Reichstein, M., and Seneviratne, S. I.: Global assessment of trends in wetting and drying over land, Nat. Geosci., 7, 716–721, https://doi.org/10.1038/ngeo2247, 2014.

Guillod, B. P., Orlowsky, B., Miralles, D. G., Teuling, A. J., and Seneviratne, S. I.: Reconciling spatial and temporal soil moisture effects on afternoon rainfall, Nat. Comm., 6, 6443, https://doi.org/10.1038/ncomms7443, 2015.

Entekhabi, D., Reichle, R. H., Koster, R. D., and Crow, W. T.: Performance metrics for soil moisture retrieval and application requirements, J. Hydrometeor., 11, 832–840, 2010.

Hersbach, H. and Dee, D.: "ERA-5 reanalysis is in production", ECMWF newsletter, number 147, Spring 2016, p. 7, 2016.

Jackson, R. B., Canadell, J., Ehleringer, J. R., Mooney, H. A., Sala, O. E., and Schulze, E. D.: A global analysis of root distributions for terrestrial biomes, Oecologia, 108, 389–411, https://doi.org/10.1007/BF00333714, 1996.

Jasechko, S., Sharp, Z. D., Gibson, J. J., Birks, S. J., Yi, Y., and Fawcett, P. J.: Terrestrial water fluxes dominated by transpiration, Nature, 496, 347–350, https://doi.org/10.1038/nature11983, 2013.

Jung, M., Reichstein, M., and Bondeau, A.: Towards global empirical upscaling of FLUXNET eddy covariance observations: validation of a model tree ensemble approach using a biosphere model, Biogeosciences, 6, 2001–2013, https://doi.org/10.5194/bg-6-2001-2009, 2009.

Jung, M., Reichstein, M., Ciais, P., Seneviratne, S. I., Sheffield, J., Goulden, M. L., Bonan, G., Cescatti, A., Chen, J., de Jeu, R., Dolman, A. J., Eugster, W., Gerten, D., Gianelle, D., Gobron, N., Heinke, J., Kimball, J., Law, B. E., Montagnani, L., Mu, Q., Mueller, B., Oleson, K., Papale, D., Richardson, A. D., Roupsard, O., Running, S., Tomelleri, E., Viovy, N., Weber, U., Williams, C., Wood, E., Zaehle, S., and Zhang, K.: Recent decline in the global land evapotranspiration trend due to limited moisture supply, Nature, 467, 951–954, https://doi.org/10.1038/nature09396, 2010.

Jung, M., Reichstein, M., Margolis, H. A., Cescatti, A., Richardson, A. D., Arain, M. A., Arneth, A., Bernhofer, C., Bonal, D., Chen, J., Gianelle, D., Gobron, N., Kiely, G., Kutsch, W., Lasslop, G., Law, B. E., Lindroth, A., Merbold, L., Montagnani, L., Moors, E. J., Papale, D., Sottocornola, M., Vaccari, F., and Williams, C.: Global patterns of land–atmosphere fluxes of carbon dioxide, latent heat, and sensible heat derived from eddy covariance, satellite, and meteorological observations, J. Geophys. Res., 116, G00J07, https://doi.org/10.1029/2010JG001566, 2011.

Jung, M., Reichstein, M., Schwalm, C. R., Huntingford, C., Sitch, S., Ahlström, A., Arneth, A., Camps-Valls, G., Ciais, P., Friedlingstein, P., Gans, F., Ichii, K., Jain, A. K., Kato, E., Papale, D., Poulter, B., Raduly, B., Rödenbeck, C., Tramontana, G., Viovy, N., Wang, Y.-P., Weber, U., Zaehle, S., and Zeng, N.: Compensatory water effects link yearly global land CO_2 sink changes to temperature, Nature, 541, 516–520, https://doi.org/10.1038/nature20780, 2017.

Koster, R., Mahanama, S., Yamada, T., Balsamo, G., Boisserie, M., Dirmeyer, P., Doblas-Reyes, F., Gordon, T., Guo, Z., Jeong, J.-H., Lawrence, D., Li, Z., Luo, L., Malyshev, S., Merryfield, W., Seneviratne, S. I., Stanelle, T., van den Hurk, B., Vitart, F., and Wood, E. F.: The contribution of land surface initialization to sub-seasonal forecast skill: First results from the GLACE-2 Project, Geophys. Res. Lett., 37, L02402, https://doi.org/10.1029/2009GL041677, 2009a.

Koster, R., Guo, Z., Yang, R., Dirmeyer, P., Mitchell, K., and Puma, M.: On the nature of soil moisture in land surface models, J. Climate, 22, 4322–4335, https://doi.org/10.1175/2009JCLI2832.1, 2009b.

Koster, R., Mahanama, S. P. P., Yamada, T. J., Balsamo, G., Berg, A. A., Boisserie, M., Dirmeyer, P. A., Doblas-Reyes, F. J., Drewitt, G., Gordon, C. T., Guo, Z., Jeong, J.-H., Lee, W.-S., Li, Z., Luo, L., Malyshev, S., Merryfield, W. J., Seneviratne, S. I., Stanelle, T., van den Hurk, B. J. J. M., Vitart, F., and Wood, E. F.: The second phase of the global land-atmosphere coupling experiment: soil moisture contributions to sub-seasonal forecast skill, J. Hydrometeorol., 12, 805–822, https://doi.org/10.1175/2011JHM1365.1, 2011.

Kumar, S., Reichle, R. H., Koster, R. D., Crow, W. T., and Peters-Lidard, C.: Role of Subsurface Physics in the Assimilation of Surface Soil Moisture Observations, J. Hydrometeor., 10, 1534–1547, https://doi.org/10.1175/2009JHM1134.1, 2009.

Kumar, S. V., Jasinski, M., Mocko, D., Rodell, M., Borak, J., Li, B., Kato Beaudoing, H., and Peters-Lidard, C. D.: NCA-LDAS land analysis: Development and performance of a multisensor, multivariate land data assimilation system for the National Climate Assessment, J. Hydrometeor., https://doi.org/10.1175/JHM-D-17-0125.1, online first, 2018.

Liu, Y. Y., Parinussa, R. M., Dorigo, W. A., De Jeu, R. A. M., Wagner, W., van Dijk, A. I. J. M., McCabe, M. F., and Evans, J. P.: Developing an improved soil moisture dataset by blending passive and active microwave satellite-based retrievals, Hydrol. Earth Syst. Sci., 15, 425–436, https://doi.org/10.5194/hess-15-425-2011, 2011.

Liu, Y. Y., Dorigo, W. A., Parinussa, R. M., De Jeu, R. A. M., Wagner, W., McCabe, M. F., Evans, J. P., and van Dijk, A. I. J. M.: Trend-preserving blending of passive and active microwave soil moisture retrievals, Remote Sens. Environ., 123, 280–297, https://doi.org/10.1016/j.rse.2012.03.014, 2012.

Martens, B., Miralles, D. G., Lievens, H., van der Schalie, R., de Jeu, R. A. M., Fernández-Prieto, D., Beck, H. E., Dorigo, W. A., and Verhoest, N. E. C.: GLEAM v3: satellite-based land evaporation and root-zone soil moisture, Geosci. Model Dev., 10, 1903–1925, https://doi.org/10.5194/gmd-10-1903-2017, 2017.

Mätzler, C. and Standley, A.: Relief effects for passive microwave remote sensing, Int. J. Remote Sens., 21, 2403–2412, https://doi.org/10.1080/01431160050030538, 2000.

Masson, V., Le Moigne, P., Martin, E., Faroux, S., Alias, A., Alkama, R., Belamari, S., Barbu, A., Boone, A., Bouyssel, F., Brousseau, P., Brun, E., Calvet, J.-C., Carrer, D., Decharme, B., Delire, C., Donier, S., Essaouini, K., Gibelin, A.-L., Giordani, H., Habets, F., Jidane, M., Kerdraon, G., Kourzeneva, E., Lafaysse, M., Lafont, S., Lebeaupin Brossier, C., Lemonsu, A., Mahfouf, J.-F., Marguinaud, P., Mokhtari, M., Morin, S., Pigeon, G., Salgado, R., Seity, Y., Taillefer, F., Tanguy, G., Tulet, P., Vincendon, B., Vionnet, V., and Voldoire, A.: The SURFEXv7.2 land and ocean surface platform for coupled or offline simulation of earth surface variables and fluxes, Geosci. Model Dev., 6, 929–960, https://doi.org/10.5194/gmd-6-929-2013, 2013.

McNally, A., Arsenault, K., Kumar, S., Shukla, S., Peterson, P., Wang, S., Funk, C., Peters-Lidard, C. D., and Verdin, J. P.: A land data assimilation system for sub-Saharan Africa food and water security applications, Scientific Data, 4, 170012, https://doi.org/10.1038/sdata.2017.12, 2017.

Menne, M. J., Durre, I., Vose, R. S., Gleason, B. E., and Houston, T. G.: An overview of the Global Historical Climatology Network-Daily Database, J. Atmos. Ocean. Tech., 29, 897–910, doi.10.1175/JTECH-D-11-00103.1, 2012a.

Menne, M. J., Durre, I., Korzeniewski, B., McNeal, S., Thomas, K., Yin, X., Anthony, S., Ray, R., Vose, R. S., Gleason, B. E., and Houston, T. G.: Global Historical Climatology Network – Daily (GHCN-Daily), Version 3, snow depth, NOAA National Climatic Data Center, https://doi.org/10.7289/V5D21VHZ, 2012b.

Miralles, D. G., Holmes, T. R. H., De Jeu, R. A. M., Gash, J. H., Meesters, A. G. C. A., and Dolman, A. J.: Global land-surface evaporation estimated from satellite-based observations, Hydrol. Earth Syst. Sci., 15, 453–469, https://doi.org/10.5194/hess-15-453-2011, 2011.

Miralles, D. G., Teuling, A. J., van Heerwaarden, C. C., and de Arellano, J. V.-G.: Mega-heatwave temperatures due to combined soil desiccation and atmospheric heat accumulation, Nat. Geosci., 7, 345–349, https://doi.org/10.1038/NGEO2141, 2014a.

Miralles, D. G., van den Berg, M. J., Gash, J. H., Parinussa, R. M., de Jeu, R. A. M., Beck, H. E., Holmes, T. R. H., Jiménez, C., Verhoest, N. E. C., Dorigo, W. A., Teuling, A. J., and Dolman, A. J.: El Niño-La Niña cycle and recent trends in continental evaporation, Nat. Clim. Change, 4, 122–126, https://doi.org/10.1038/NCLIMATE2068, 2014b.

Mitchell, K., Lohman, D., Houser, P., Wood, E., Schaake, J., Robock, A., Cosgrove, B., Sheffield, J., Duan, Q., Luo, L., Higgins, R., Pinker, R., Tarpley, J., Lettenmaier, D., Marshall, C., Entin, J., Pan, M., Shi, W., Koren, V., Meng, J., Ramsay, B., and Bailey, A.: The multi-institution North American Land Data Assimilation System (NLDAS): Utilizing multiple GCIP products and partners in a continental distributed hydrological modeling system, J. Geophys. Res., 109, D07S90, https://doi.org/10.1029/2003JD003823, 2004.

Muñoz-Sabater J., Dutra E., Balsamo G., Boussetta S., Zsoter E., Albergel C., and Agusti-Panareda A.: ERA5-Land: an improved version of the ERA5 reanalysis land component, Joint ISWG and LSA-SAF Workshop IPMA, Lisbon, 26–28 June 2018.

Nash, J. E. and Sutcliffe, V.: River forecasting through conceptual models, J. Hydrol., 10, 282–290, 1970.

Noilhan, J. and Mahfouf, J.-F.: The ISBA land surface parameterisation scheme, Global Planet. Change, 13, 145–159, 1996.

Oki, T. and Sud, Y. C.: Design of Total Runoff Integrating Pathways (TRIP), a global river chanel network, Earth Interact., 2, 1–36, 1998.

Reichle, R. H., Koster, R. D., De Lannoy, G. J. M., Forman, B. A., Liu, Q., Mahanama, S. P. P., and Toure, A.: Assessment and enhancement of MERRA land surface hydrology estimates, J. Climate, 24, 6322–6338, https://doi.org/10.1175/JCLI-D-10-05033.1, 2011.

Reichle, R. H., Draper, C. S., Liu, Q., Girotto, M., Mahanama, S. P., Koster, R. D., and De Lannoy, G. J.: Assessment of MERRA-2 Land Surface Hydrology Estimates, J. Climate, 30, 2937–2960, https://doi.org/10.1175/JCLI-D-16-0720.1, 2017.

Richards, L. A.: Capillary conduction of liquids in porous mediums, Physics, 1, 318–333, 1931.

Rienecker, M. M., Suarez, M. J., Gelaro, R., Todling, R., Julio, B., Liu, E., Bosilovich, M. G., Schubert, S. D., Takacs, L., Kim, G.-K., Bloom, S., Chen, J., Collins, D., Conaty, A., da Silva, A., Gu, W., Joiner, J., Koster, R. D., Lucchesi, R., Molod, A., Owens, T., Pawson, S., Pegion, P., Redder, C. R., Reichle, R., Robertson, F. R., Ruddick, A. G., Sienkiewicz, M., and Woollen, J.: MERRA– NASA's modern-era retrospective analysis for research and applications, J. Climate, 24, 3624–3648, https://doi.org/10.1175/JCLI-D-11-00015.1, 2011.

Rodell, M., Houser, P. R., Jambor, U., Gottschalck, J., Mitchell, K., Meng, C.-J., Arsenault, K., Cosgrove, B., Radakovich, J., Bosilovich, M., Entin, J. K., Walker, J. P., Lohmann, D., and Toll, D.: The Global Land Data Assimilation System, B. Am. Meteorol. Soc., 85, 381–394, 2004.

Sawada, Y. and Koike, T.: Simultaneous estimation of both hydrological and ecological parameters in an ecohydrological model by assimilating microwave signal, J. Geophys. Res.-Atmos., 119, 8839–8857, https://doi.org/10.1002/2014JD021536, 2014.

Sawada, Y., Koike, T., and Walker, J. P.: A land data assimilation system for simultaneous simulation of soil moisture and vegetation dynamics, J. Geophys. Res.-Atmos., 120, 5910–5930, https://doi.org/10.1002/2014JD022895, 2015.

Schellekens, J., Dutra, E., Martínez-de la Torre, A., Balsamo, G., van Dijk, A., Sperna Weiland, F., Minvielle, M., Calvet, J.-C., Decharme, B., Eisner, S., Fink, G., Flörke, M., Peßenteiner, S., van Beek, R., Polcher, J., Beck, H., Orth, R., Calton, B., Burke, S., Dorigo, W., and Weedon, G. P.: A global water resources ensemble of hydrological models: the eartH2Observe Tier-1 dataset, Earth Syst. Sci. Data, 9, 389–413, https://doi.org/10.5194/essd-9-389-2017, 2017.

Slevin, D., Tett, S. F. B., Exbrayat, J.-F., Bloom, A. A., and Williams, M.: Global evaluation of gross primary productivity in the JULES land surface model v3.4.1, Geosci. Model Dev., 10, 2651–2670, https://doi.org/10.5194/gmd-10-2651-2017, 2017.

Stanski, H. R., Wilson, L. J., and Burrows, W. R.: Survey of common verifications methods in meteorology, WMO World Weather Watch Tech. Rep. 8/WMO/TD-358, 114 pp., 1989.

Szczypta, C., Calvet, J.-C., Maignan, F., Dorigo, W., Baret, F., and Ciais, P.: Suitability of modelled and remotely sensed essential climate variables for monitoring Euro-Mediterranean droughts, Geosci. Model Dev., 7, 931–946, https://doi.org/10.5194/gmd-7-931-2014, 2014.

Tramontana, G., Jung, M., Schwalm, C. R., Ichii, K., Camps-Valls, G., Ráduly, B., Reichstein, M., Arain, M. A., Cescatti, A., Kiely, G., Merbold, L., Serrano-Ortiz, P., Sickert, S., Wolf, S., and Papale, D.: Predicting carbon dioxide and energy fluxes across global FLUXNET sites with regression algorithms, Biogeosciences, 13, 4291–4313, https://doi.org/10.5194/bg-13-4291-2016, 2016.

van der Schrier, G., Barichivich, J., Briffa, K. R., and Jones, P. D.: A scPDSI-based global data set of dry and wet spells for 1901–2009, J. Geophys. Res.-Atmos., 118, 4025–4048, 2013.

Vergnes, J.-P. and Decharme, B.: A simple groundwater scheme in the TRIP river routing model: global off-line evaluation against GRACE terrestrial water storage estimates and observed river discharges, Hydrol. Earth Syst. Sci., 16, 3889–3908, https://doi.org/10.5194/hess-16-3889-2012, 2012.

Vergnes, J.-P., Decharme, B., and Habets, F.: Introduction of groundwater capillary rises using subgrid spatial variability of topography into the ISBA land surface model, J. Geophys. Res.-Atmos., 119, 11065–11086, https://doi.org/10.1002/2014JD021573, 2014.

Voldoire, A., Decharme, B., Pianezze, J., Lebeaupin Brossier, C., Sevault, F., Seyfried, L., Garnier, V., Bielli, S., Valcke, S., Alias, A., Accensi, M., Ardhuin, F., Bouin, M.-N., Ducrocq, V., Faroux, S., Giordani, H., Léger, F., Marsaleix, P., Rainaud, R., Redelsperger, J.-L., Richard, E., and Riette, S.: SURFEX v8.0 interface with OASIS3-MCT to couple atmosphere with hydrology, ocean, waves and sea-ice models, from coastal to global scales, Geosci. Model Dev., 10, 4207–4227, https://doi.org/10.5194/gmd-10-4207-2017, 2017.

Wagner, W., Dorigo, W., de Jeu, R., Fernandez, D., Benveniste, J., Haas, E., and Ertl, M.: Fusion of active and passive microwave observations to create an Essential Climate Variable data record on soil moisture, Proc. XXII ISPRS Congress, Melbourne, Australia, ISPRS, 315–321, 2012.

Wieder, W. R., Boehnert, J., Bonan, G. B., and Langseth, M.: Re-gridded Harmonized World Soil Database v1.2, Data set, OakRidge National Laboratory Distributed Active Archive Center, Oak Ridge, Tennessee, USA, available at: http://daac.ornl.gov, 2014.

Xia, T., Mitchell K., Ek, M., Sheffield, J., Cosgrove, B., Wood, E., Luo, L., Alonge, C., Wei, H., Meng, J., Livneh, B., Lettenmaier, D., Koren, V. Duan, Q., Mo, K., Fan, Y., and Mocko, D.: Continental-scale water and energy flux analysis and validation for the North American Land Data Assimilation System project phase 2 (NLDAS-2): 1. Intercomparison and application of model products, J. Geophys. Res., 117, D03109, https://doi.org/10.1029/2011JD016048, 2012.

Zhang, Y., Peña-Arancibia, J. L., McVicar, T. R., Chiew, F. H. S., Vaze, J., Liu, C., Lu, X., Zheng, H., Wang, Y., Liu, Y. Y., Miralles, D. G., and Pan, M.: Multi-decadal trends in global terrestrial evapotranspiration and its components, Sci. Rep.-UK, 6, 19124, https://doi.org/10.1038/srep19124, 2016.

The challenge of forecasting impacts of flash floods: test of a simplified hydraulic approach and validation based on insurance claim data

Guillaume Le Bihan[1], Olivier Payrastre[1], Eric Gaume[1], David Moncoulon[2], and Frédéric Pons[3]

[1]IFSTTAR, GERS, EE, F-44344 Bouguenais, France

[2]CCR, 157 boulevard Haussmann, 75008 Paris, France

[3]Cerema, Direction Méditerranée, 30 rue Albert Einstein, F-13593 Aix-en-Provence, France

Correspondence to: Olivier Payrastre (olivier.payrastre@ifsttar.fr)

Abstract. Up to now, flash flood monitoring and forecasting systems, based on rainfall radar measurements and distributed rainfall–runoff models, generally aimed at estimating flood magnitudes – typically discharges or return periods – at selected river cross sections. The approach presented here goes one step further by proposing an integrated forecasting chain for the direct assessment of flash flood possible impacts on inhabited areas (number of buildings at risk in the presented case studies). The proposed approach includes, in addition to a distributed rainfall–runoff model, an automatic hydraulic method suited for the computation of flood extent maps on a dense river network and over large territories. The resulting catalogue of flood extent maps is then combined with land use data to build a flood impact curve for each considered river reach, i.e. the number of inundated buildings versus discharge. These curves are finally used to compute estimated impacts based on forecasted discharges. The approach has been extensively tested in the regions of Alès and Draguignan, located in the south of France, where well-documented major flash floods recently occurred. The article presents two types of validation results. First, the automatically computed flood extent maps and corresponding water levels are tested against rating curves at available river gauging stations as well as against local reference or observed flood extent maps. Second, a rich and comprehensive insurance claim database is used to evaluate the relevance of the estimated impacts for some recent major floods.

1 Introduction

Hydro-meteorological forecasts are essential for efficient real-time flood management, especially when the situation is evolving rapidly. Forecasts provide crucial information to emergency managers for the anticipation and appraisal of the forthcoming floods that may affect areas at risk. In the particular case of flash floods, often affecting simultaneously a large number of small ungauged streams, suitable forecasting systems are still currently under development around the world. The first approaches developed, namely the flash flood guidance, were based on a preliminary analysis of rainfall volumes generating bankfull flow, for several durations and initial soil moisture conditions (Georgakakos, 2006; Norbiato et al., 2008). More recent approaches aimed to directly forecast peak discharges at ungauged locations based on highly distributed hydrological models and radar based quantitative precipitation estimates (QPEs) or nowcasts (Cole and Moore, 2009; Rozalis et al., 2010; Wang et al., 2011; Javelle et al., 2014; Gourley et al., 2014, 2017; Naulin et al., 2013; Versini et al., 2014). Such models provide indications of possible flood magnitudes, but are still rarely designed to directly evaluate the possible associated impacts. A large number of simultaneous alarms may be generated in the case of a significant rainfall event by such highly distributed flash flood forecasting systems. And it is now recognised that end users, such as emergency managers, who have little time for situation analysis and decision making during flash floods, crucially need rapid assessment of the possible field consequences and damage severity (Schroeder et al., 2016; Cole et al., 2016). Moreover, a direct forecast of possible field consequences opens the possibility for assessing the performance of flash flood forecasting systems in ungauged areas, based on reported consequences, as a surrogate for measured flood discharges (Versini et al., 2010a; Naulin et al., 2013; Javelle et al., 2014; Moncoulon et al., 2014; Saint-Martin et al., 2016; Le Bihan et al., 2016). In the near future, real-time assimilation of proxy data for flood magnitude such as information contained in reports of rescue services or social networks can be envisaged. This article presents a proposal of such an integrated flash flood impact forecasting chain and illustrates its validation against insurance claims. If successful, such an approach may be of great help for both emergency managers to better appraise the expected flash flood impacts, and for hydrologists to improve their modelling approaches in ungauged situations.

Translating discharges into local possible impacts requires an estimation of the corresponding flood extent, as well as the knowledge of the level of exposure (location) of the considered assets and possibly of the vulnerability of these assets. This information may be difficult to assess and incorporate at the large scale at which flash flood forecasting systems are implemented to monitor a large number of small streams. Large-scale flood mapping approaches based on digital terrain models (DTMs) were recently proposed and tested (Yamazaki et al., 2011; Pappenberger et al., 2012; Sampson et al., 2015). These works may offer an interesting way for automatic treatment of DTM and flood mapping. But they were not initially designed for the simulation of a large range of flood magnitudes and were generally applied at relatively large spatial resolutions: computation square grids from 100 m up to 1 km. Such resolutions are not suited to the representation of floodplains of small streams. On the other hand, detailed flood inundation mapping approaches are available at higher resolutions (Bradbrook et al., 2005; Sanders, 2007; Nguyen et al., 2015), but require large computational resources which limit the implementation possibility at a large scale. In both cases, most of the proposed mapping approaches would not be compatible with application in real time.

The approach proposed hereafter combines applicability at a large scale (computational efficiency), the possibility to be integrated in a real-time forecasting chain, and high spatial resolution for an appropriate representation of floodplains of small ungauged streams. It is proposed to compute automatically (i.e. without manual corrections) a series of flood extent maps for each river reach (a river reach being defined as the portion of river located between two confluences), covering a large spectrum of discharge values (i.e. discharge return period values). A DTM treatment method is proposed for the extraction of river cross sections, which are used in one-dimensional (1-D) steady-state hydraulic numerical models

for the computation of water stages and flood extent maps. Land use databases are then analysed to compute the number of buildings in the estimated flooded areas for each discharge value and each river reach. Based on this preliminary analysis, a relationship between the discharge and the number of affected buildings is adjusted at the river reach scale and is then used as impact model of an integrated rainfall–runoff–impact simulation chain.

Even if the proposed procedure may appear relatively straightforward, the main challenge lies in its automatic application and validation over extended territories on a dense stream network – typically streams with upstream watershed areas larger than 5 km^2. The validation is an essential step which should reveal if such a forecasting chain is able to provide reasonably accurate results, despite the necessary simplifications of such large-scale applications (standard roughness coefficient values, 1-D steady-state hydraulic models, missing bathymetric data, etc.), and sources of uncertainty (DTM accuracy, unknown vulnerabilities, etc.).

This article presents the proposed method and its application on two well-documented test case studies. Two types of evaluations are conducted. First, the automatically computed flood maps and corresponding water levels are tested against the rating curves at river gauging stations as well as against local reference or observed flood maps. Second, a rich and comprehensive insurance claim database provided by the largest French reinsurance company (Caisse Centrale de Réassurance) is used to evaluate the relevance of the estimated impacts (number of possibly inundated buildings) for some recent major floods. The article is organised as follows: the first section presents the methodology developed for both the implementation of the impact model, including the computation of a catalogue of flood extent maps, and for the implementation of the rainfall–runoff–impact simulation chain; Sect. 3 presents the two application case studies and the data sets used for the validation; Sect. 4 examines and discusses the obtained results.

2   The proposed rainfall–runoff–impact simulation chain

2.1   Simplified automatic implementation of 1-D steady-state hydraulic models

The Cartino method (Pons et al., 2014), has recently been proposed to automatically build the input files and run one-dimensional (1-D) hydraulic models based on data extracted from high-resolution DTMs. This method has been used and adapted herein to derive catalogues of flood extent maps for a wide range of discharge values. The software proceeds in three steps (Fig. 1). First, the locations of the cross sections are selected (Fig. 1a) and their shapes extracted from the DTM (Fig. 1b). Second, the corresponding input files are built and the selected 1-D hydraulic model is run to compute longitudinal water level profiles corresponding to each

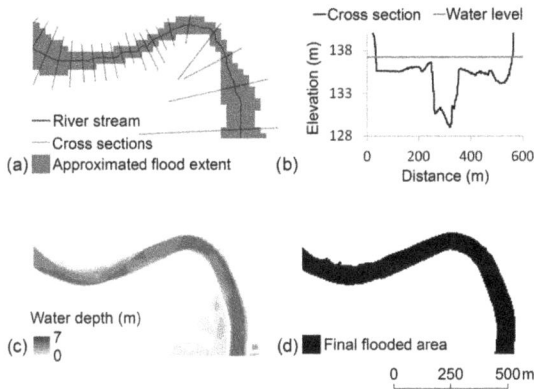

Figure 1. Overall principle of the computation of flood maps based on CartinoPC software: **(a)** input information (position of river streams and approximated extent of flooded area) and position of cross sections, **(b)** example of one cross section including the computed water level (1-D hydraulic model), **(c)** map of flooded areas and water depths obtained after post-treatment, **(d)** final map of flooded areas after removal of disconnected areas.

selected discharge value (Fig. 1b): the Mascaret 1-D model (Goutal et al., 2012) has been used in the present case study. Third, the estimated water levels are interpolated between successive cross sections and compared to the DTM elevations to compute the flood extent and water depth maps (Fig. 1c).

The first two steps are run in an iterative way to adjust the width of the cross sections and their inter-distances for each considered discharge value. The cross sections should be wide enough to contain the simulated flow and successive cross sections should not overlap. The procedure is initiated based on a first estimation of the possible extent of the flooded area (provided as input as well as the position of the river bed; see Fig. 1a), which is used to define the initial width of each section. The distances between profiles are then defined as a proportion of each cross-section width (proportion defined as input parameter). After each run, it is checked if the computed water level does not exceed the altitude of the borders of the cross section. If it is the case, the cross section is enlarged in a proportion also defined as input parameter. Distances between profiles are adapted consequently. Note that the final set of cross-sectional profiles and their locations vary between the runs and depend in particular on the considered discharge value.

To ensure automatic computation, important simplifications are introduced in the structure of the hydraulic model: cross-section shapes are estimated based on a simple extraction from the available DTM, without additional information on topography or bathymetry; specific sections such as weirs or bridges are not represented; a unique roughness coefficient is used for all stream reaches ($n = 0.05$ hereafter); no distinction is made between river bed and floodplain. Of course all these assumptions, even if necessary for the sake of simplic-

ity, may have an impact on the accuracy of the results. This point will be evaluated and discussed in the next sections.

An automatic verification is performed after each hydraulic computation to eliminate the main errors in the shapes of cross sections, mainly associated with the limits of DTM information used as input: bridges still appearing in the DTM, remaining noise due to dense vegetation. This verification is based on the comparison between the wetted areas of the successive cross sections: automatic removal of cross sections appearing as inconsistent with the immediate upstream cross section (ratio between successive wetted areas exceeding 3), before running the hydraulic model again.

One crucial aspect for the computation is the delimitation of the active river bed in presence of depressions in the floodplain (perched rivers). In this case, the choice of the input information (default extent of flooded areas and parameter values) highly influences the result and the computation time. For instance, too wide an initial flooded area or too fast an increase in cross-section widths may lead to incorporating depressions that are not connected to the river bed and hence not active for the considered discharge (Figs. 1b and c). This may lead to both overestimation of the local extent of the flooded area and to a decrease in the simulated water level, with an impact on the upstream cross sections due to backwater effects. On the other hand, too narrow an initial flooded area and/or too slow increase in the widths of the profiles highly increases the computation time. To cope with these difficulties, the computations are first conducted for the smallest discharge (2-year flood), using a narrow initial extent of flooded area as input of the Cartino method. The width of the cross sections are then progressively increased to detect as accurately as possible the limits of the river bed. If necessary, the remaining flooded areas disconnected from the river bed are finally removed from the final result (Fig. 1d). The computed flooded area is then used as initial cross-section extent for the next computation (next larger discharge value; see Sect. 2.2), and so on.

2.2 Definition of the impact model based on a catalogue of flood extent maps

The hydraulic computations are conducted for each river reach of the considered stream network, and for a range of discharge values corresponding to return periods of 2, 5, 10, 20, 30, 50, 100, 300, 500, and 1000 years. The discharge quantiles are estimated based on the French SHYREG flood frequency database (Aubert et al., 2014). Nevertheless, the accuracy of this information on flow frequency is not crucial for the implementation of the impact model; it just enables the derivation of flood maps for defined discharge values of relatively homogeneous magnitude for all the considered river reaches.

All the computations are made in steady-state regime to simplify the procedure and facilitate its fast application at a large scale. The underlying assumption is that the extent of

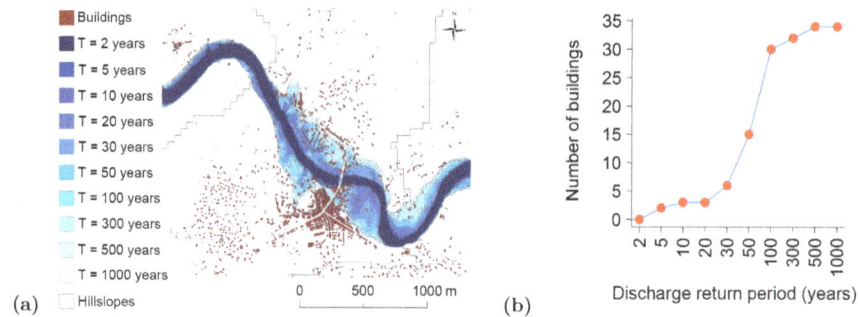

Figure 2. Illustration of the implementation of the impact model on one river reach: (**a**) catalogue of flooded areas, (**b**) interpolated discharge–impact relationship.

flooded areas is mainly influenced by the peak discharges, rather than the flow volumes. This assumption is consistent with the application domain of the method (flash floods occurring in small upstream watersheds). The discharge is assumed to be homogeneous on each river reach (no lateral flow introduced) as the length of the considered river reaches is limited (2 km in average; see Sect. 3). The roughness coefficient is fixed to $n = 0.05$ according to the conclusions of Lumbroso and Gaume (2012) about the relevant range of roughness values in the case of flash floods. The downstream limit condition for each reach is the water level computed for the downstream river reach for the discharge of the same return period. For the last downstream river reach, the boundary condition corresponds to the normal depth for a uniform flow regime. Finally, a post-treatment procedure is applied to ensure better overall consistency of the catalogue of flood extent maps: each map is corrected to systematically include the flood extents computed for the lower discharge values.

Based on this catalogue, an impact model is derived for each river reach. The hillslope limits of the rainfall–runoff model are used to delineate the river reaches and the corresponding flooded areas (Fig. 2a). The assets present in the 10 estimated flooded areas are counted. The number of buildings may be used as an indicator of human asset exposure. More precisely, the number of private houses with a single and georeferenced insurance policy in the available insurance database has been considered herein to enable comparisons with the reported number of insurance claims (Sect. 3.6). The values computed for the 10 flood extent maps for each river reach are then linearly interpolated to build a continuous relationship between discharge and number of flooded houses, i.e. number of impacted insurance policies (Fig. 2b).

2.3 Main limits of the impact model obtained

Despite all of the aforementioned precautions, some sources of errors affecting the quality of results remain. These errors are mainly related to the simplification of the procedure necessary for fast and automatic implementation of hydraulic computations over a large river network. These errors are mainly due to

- errors in the retrieval of the shapes of the river cross sections due to the automatic extraction and to the limits of topographic information used (DTM);

- absence of representation of friction losses due to bridges and other hydraulic singularities;

- choice of a fixed Manning roughness coefficient, equal to 0.05,

- steady-state regime computations;

- remaining difficulties to determine the active river cross section in cases of perched river bed despite the previously described precautions.

For these reasons, the information obtained should not be considered as a highly accurate estimation of flooded areas and related impacts. It should nevertheless give an order of magnitude of the level of flooding and enable some comparisons of impacts at a regional scale.

2.4 The rainfall–runoff–impact simulation chain

The impact model is finally incorporated in a full simulation chain combining radar-based quantitative precipitation estimates (QPEs; see Sect. 3) as input, and a distributed rainfall–runoff model for the simulation of discharges over the stream network. Quantitative precipitation forecasts or nowcasts may also be used as input of the chain to increase anticipation lead times. However, QPEs have been considered herein to focus the analysis on the accuracy of the proposed rainfall–runoff–impact simulation chain.

The CINECAR hydrological model (Gaume et al., 2004; Naulin et al., 2013; Versini et al., 2010b) was selected to build the simulation chain. CINECAR is a distributed rainfall–runoff model based on a representation of the catchment as a ramified series of stream reaches, to which both left and right-hand hillslopes are connected. The spatial resolution of the model (areas of hillslopes and associated length of river reaches) can easily be adapted and has been defined herein to match the resolution of the impact model. This means that the river reaches considered are identical in the rainfall–runoff model and in the impact model: they correspond to the same stream network, defined here from an elementary catchment surface of $5\,\mathrm{km}^2$.

CINECAR only simulates the rapid component of the runoff and is suited for modelling the rising limb and peak phases of significant flash floods. The Soil Conservation Service–curve number model (SCS-CN) is used to compute runoff rates and the corresponding effective rainfall at each computation time step (USDA, 1986). A temporal resolution of 15 min has been used here for the computations. The effective rainfall is then propagated onto both the hillslopes and the river network using either the kinematic wave model (Borah et al., 1980), or the Hayami solution for diffusive wave model (Moussa, 1996) to represent the flood wave attenuation in river reaches with slopes less than 0.6 %.

Since CINECAR was developed for the purpose of forecasting flood hydrographs in ungauged catchments, it includes a limited number of calibration parameters. For sake of simplicity, the hillslopes are represented by schematic rectangles of the same area as the actual hillslopes, and the river reaches are assumed to have a rectangular cross section. The width of the cross section is the main parameter controlling the transfer function and is estimated based on the Strahler order of river reaches and the discharge return period (Naulin et al., 2013). The curve number (CN) value is the second key parameter which controls the temporal evolution of runoff rates. The USDA method is used for the estimation of the CN (USDA, 1986), depending on the bedrock type, land use, and 5-day antecedent rainfall for initialisation. This model was applied in 2013 to the entire Cévennes Region (Sect. 3), and was validated with respect to measured data (Naulin et al., 2013). It provided satisfactory results for large flood events, similar to the ones obtained with locally calibrated standard conceptual rainfall–runoff models, with an average Nash criterion computed for single flood events equal to 0.49.

The same model version has been used herein without any further adjustments for the Argens watershed (second case study), whose characteristics are relatively similar to the watersheds of the Cévennes area. The performances of the model on this new application case study could be verified according to stream gauge and peak discharge data available for the June 2010 event.

The discharge values computed with the CINECAR model are finally converted into estimated impacts according to the continuous discharge–impact relationships defined for each river reach (Sect. 2.2 and Fig. 2b).

3 Presentation of the two case studies

3.1 The region of Alès in the Cévennes area, south-eastern France, and the September 2002 flood

The region of Alès is located in the core of the Cévennes region, well known to be prone to frequent and intense flash floods (Gaume et al., 2009). Moreover, it has been identified during the implementation of the European Union (EU) Floods Directive as one of the areas most exposed to flood risk in France (areas with potential significant flood risk – APSFR – selected for the implementation of risk management plans). The region is presented in Fig. 3, indicating the exact limits of the APSFR of Alès. Its vulnerability to floods is mainly related to the presence of the town of Alès, but also to other highly vulnerable smaller towns such as Anduze (Fig. 3).

This territory is part of three main watersheds: the Gardon d'Anduze, the Gardon d'Alès, and the Cèze rivers. These rivers have their upstream course in the Cévennes relief, and reach in their downstream part a plateau area with limited slopes. The APSFR of Alès is located in the transition zone between the mountainous and plateau areas. Therefore, this case study includes a large variety of river bed configurations including steep and narrow V-shaped valleys as well as flat and wide floodplains. Some statistics about the river bed characteristics are provided in Table 1.

The region was hit on September 2002 by a catastrophic flash-flood event (Delrieu et al., 2005). A maximum rainfall accumulation of 680 mm in 24 h was recorded in the town of Anduze, which is among the highest daily records in the region (estimated return period widely exceeding 100 years). The estimated peak discharges reached 900 (800–1000) $\mathrm{m}^3\,\mathrm{s}^{-1}$ at Mialet and 3500 (3000–4000) $\mathrm{m}^3\,\mathrm{s}^{-1}$ at Anduze on the Gardon d'Anduze river, and 2500 (2100–2900) $\mathrm{m}^3\,\mathrm{s}^{-1}$ at Alès on the Gardon d'Alès river. Twenty-three casualties and EUR 1.2 billion in damages were reported. This event induced a large number of insurance claims. It has therefore been selected here for the evaluation of the impact simulation chain.

The area selected for the study corresponds to the exact limits of the APSFR of Alès (area of about $1000\,\mathrm{km}^2$). It includes 400 km of river streams having at least a $5\,\mathrm{km}^2$ upstream catchment area. The stream network has been divided into 192 river reaches, among which only 70 reaches (132 km) are covered by the current operational flood forecasting service (forecasts published on the Vigicrues website; see main river network in Fig. 3c).

3.2 The region of Draguignan in the Argens watershed, south-eastern France, and the June 2010 flood

The Argens river watershed ($2700\,\mathrm{km}^2$) is located in the eastern part of the French Mediterranean region. It has been hit

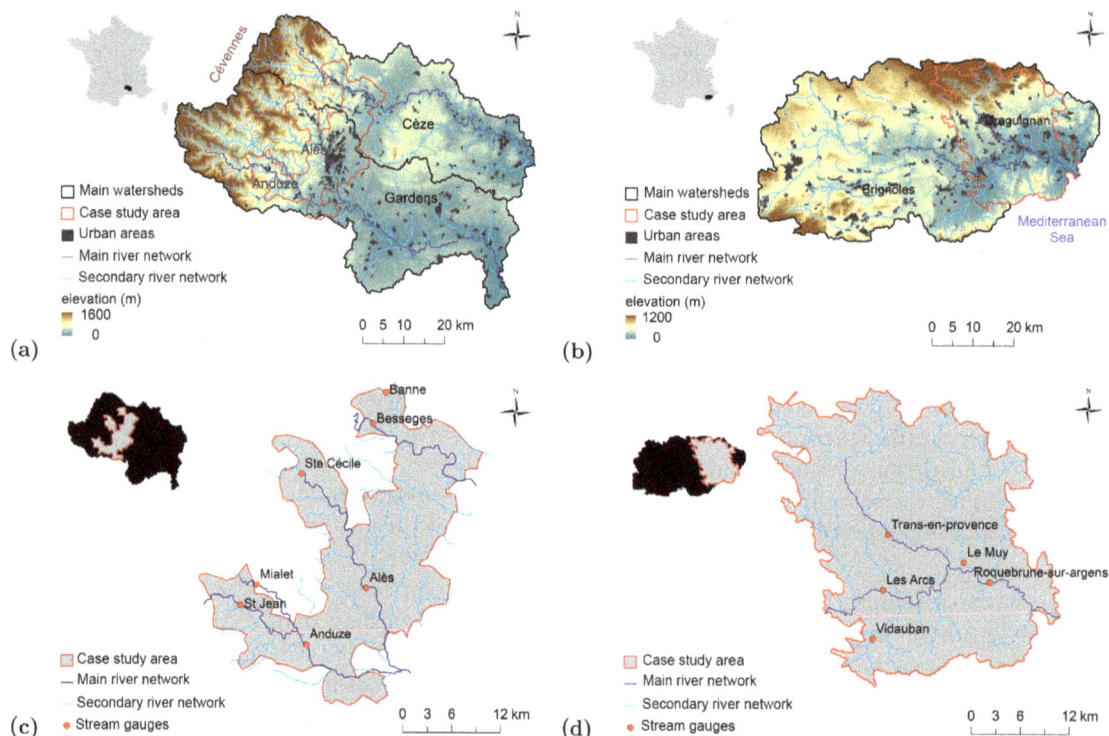

Figure 3. Location of the case studies: **(a)** APSFR of Alès territory in the Gardon and Cèze watersheds, **(b)** region of Draguignan in the Argens watershed, and river networks considered in the impact models ($5 \, \text{km}^2$ upstream catchment surface): **(c)** APSFR of Alès, and **(d)** region of Draguignan.

Table 1. Characteristics of the river networks considered in the two case studies (extracted from SYRAH database).

Case study		Bed slope (percent)	River bed width (m)	Floodplain width (m)
		avg./min–max	avg./min–max	avg./min–max
Alès	main network	0.5/0.2–1	37/16–84	470/120–1670
	secondary network	3.4/0.2–20	7/2–34	430/60–3130
Draguignan	main network	0.6/0.1–1.2	25/5.5–41	730/130–3200
	secondary network	2.1/0.05–21	4/1–11	380/3–4800

by several severe flash floods in recent years. This watershed was also selected as an APSFR for the implementation of the EU Floods Directive.

Among recent flash floods observed in this region, the June 2010 event is certainly the most catastrophic one. It particularly affected the region of Draguignan located in the eastern part of the Argens watershed (Fig. 3), where the maximum rainfall accumulation exceeded locally 400 mm in 24 h (estimated return period exceeding 100 years). The peak discharges were estimated to 440 (360–520) $\text{m}^3 \, \text{s}^{-1}$ at Trans-en Provence on the Nartuby river (tributary of the Argens river; see Fig. 3), and respectively 1000 (800–1200) $\text{m}^3 \, \text{s}^{-1}$ and 2500 (2200–2900) $\text{m}^3 \, \text{s}^{-1}$ on the Argens

river at Les Arcs and Roquebrune-sur-Argens. Twenty-five casualties and EUR 1 billion in estimated damages were reported after this event as well as a large number of insurance claims.

As for the Alès case study, the region of Draguignan presents a varied topography, and consequently a wide variety of river bed configurations, from narrow valleys in the upstream part of the studied watersheds, to wide floodplains in their downstream part. The area includes 345 km of rivers with at least $5 \, \text{km}^2$ upstream catchment areas, which were divided into 173 river reaches. Only 42 of these river reaches are covered by the operational flood forecasting service (Fig. 3d). This illustrates the possible added value of the proposed integrated flash flood impact simulation chain in France.

3.3 Available digital terrain models

The implementation of the impact models is based on commonly available high-resolution DTMs in both case studies. Nevertheless, the characteristics of these DTMs significantly differ:

- The DTM produced in 2007 by the Conseil Départemental du Gard was available in the case of the Alès case study; it has a 20 m initial resolution and was interpolated at 5 m; its altimetric accuracy was estimated to less than 20 cm in non-vegetated areas, and less than 1 m in vegetated areas.

- The 5 m resolution DTM extracted from the Institut Géographique National (IGN) RGE Alti database was available in the case of the Draguignan region; its altimetric precision ranges from 20 cm (main rivers covered with lidar data) to 70 cm (other areas covered with photogrammetry products for instance), and artificial structures such as bridges are removed in this DTM. This second product can be considered as standard DTM data that will be available over the whole French territory by the end of 2017.

It should finally be noted that if the terrain information used herein is now commonly available in France, its accuracy remains limited and may affect the quality of results: DTMs extracted from lidar measurements are currently limited to the main rivers which are not the scope of this study, and would probably lead to results of better quality. The geographic coverage of lidar data is, however, evolving very fast, and these data should become available also for small rivers in the near future.

3.4 Rainfall and discharge data

Both regions are equipped with relatively dense stream gauge and rain gauge networks, complemented with weather radars. In the case of the APSFR of Alès, the whole data set was carefully checked in the framework of the Observatoire Hydro-Météorologique Cévennes-Vivarais (OHM-CV) research observatory. It can therefore be considered as exceeding conventional quality standards.

The locations of stream gauges are shown in Fig. 3. Given the limited possibilities to conduct direct flow measurements during intense flash floods, the rating curves are often extrapolated, with consequently a reduced accuracy of estimated discharges for high water levels and large floods.

Radar-based QPEs are available for both case studies, at 1 km^2 spatial resolution and 5 min temporal resolution. They correspond to the operational Meteo France Panthere QPEs in the case of the Draguignan June 2010 event, and to a radar

QPE reanalysis provided by the OHM-CV research observatory in the case of the September 2002 event in the region of Alès (Delrieu et al., 2009).

3.5 Reference flood maps from previous studies

Thanks to the recent application of the EU Floods Directive in both considered areas, great effort was put into mapping flooded areas, leading to detailed inundation maps available for three reference events: the 30-year return period flood ("common" event), 300-year return period flood ("medium" event), and 1000-year return period flood ("large" event). These maps were generally obtained based on 1-D hydraulic modelling, carefully implemented by experts in hydraulics. These maps were criticised and validated using all available information, including the observed extents of inundation during past floods such as the September 2002 and June 2010 floods. These maps were used as a reference here for the evaluation of the computed catalogues of flooded areas. Unfortunately these maps were produced on only part of the considered river networks. This limits the validation possibilities. In the case of the Alès case study for instance, the river network covered by reference maps represents 192 km (out of 400 km included in the case study) and includes 84 river reaches (out of 192).

3.6 Insurance claim database

Over the last 15 years, an insurance claim database has been developed by the Caisse Centrale de Réassurance (CCR) within the framework of its reinsurance contracts with its clients (Moncoulon et al., 2014). This database covers the entire French territory, and the quality of data is considered as acceptable for the period since 1997. It includes information on both nature and location of insurance policies and claims for all events classified as so-called "CATNAT" events. "CATNAT" events are flood events with an estimated return period exceeding 10 years and consequently for which the natural disaster insurance compensation is activated, in accordance with the compensation scheme in force in France since 1982.

The CCR database is certainly the most comprehensive available database on flood field consequences in France. It has nevertheless some limits, and the content and accuracy of the insurance policy and claim data incorporated have evolved over the years. Consequently the data had to be carefully selected to enable an objective comparison with the modelling results:

- With regard to the policies, the database currently includes more than 80 % of the policies of the French insurance market, insurance against natural hazards being mandatory in France. Great effort has also been put into the accurate georeferencing of the policies in recent years: approximatively 70 % of the policies are geocoded at the street number, about 15 % geocoded at the street centre, and 14 % geocoded at the town centre. Only the policies accurately located (i.e. georeferenced at the street number or centre) were considered here for incorporation in the impact model. Note that the address of the policy is generally the address of the owner of the asset. It does often, but not always, correspond to the address of the insured asset. Note also that in the case of flats, several policies can be located at the same address, but are generally not all exposed to flooding. Therefore, only insurance policies and claims corresponding to private houses were considered here to ensure a more direct correspondence between houses, policies exposed to flooding, and claims.

- As far as the claims are concerned, their collation in the database is less systematic: between 30 and 50 % of the total number of CATNAT insurance claims for the French market are documented in the database depending on the year. This depends on the comprehensiveness of the data provided by the insurance companies to the CCR. To base the comparison on faithful and robust data, the policies of the insurance companies documenting more than 80 % of their claims in the CCR database were selected here for the validation of the forecasting chain. The claims with a null compensation amount were also removed, since there is no certainty in this case that real damage occurred, or at least the amount of damage did not exceed the insurance excess in these cases.

Finally, only part of the database could be considered for the comparisons, after application of the above-mentioned selection rules: georeferenced policies, private houses, comprehensiveness of the information on claims provided by the insurance companies. A comparison with the IGN BD Topo database (buildings of < 7 m height corresponding mostly to individual houses) shows that the number of selected policies represents about 20 % of the total number of buildings in the considered floodplains (Table 2). The proportion is higher in the Draguignan 2010 case, a sign of the improved quality of the CCR database over 2002. To enable a direct comparison of the forecasted impacts with the number of reported claims, the impact model (Fig. 2) was finally built based on the selected policies: the forecasted information corresponds in this case to a number of possibly impacted individual houses, for which almost comprehensive claim data are available. As the details of the insurance data of individuals are confidential, the validation was based on claim data aggregated at the river reach scale. Moreover, to ensure a total confidentiality, the analyses and comparisons were only conducted on river reaches with at least 20 recorded policies in the database.

The analysis of the available claim data reveals some additional surprises. First, despite the comprehensiveness of the selected claim data, the ratio between reported claims and policies rarely exceeds 50 % even in areas which are likely to have been flooded (Fig. 4). Some houses with raised basements may be out of water even in the flooded areas, but not at such a high proportion. This ratio is also explained by the remaining proportion of claims not documented in the database (up to ~ 20 % according to the data selection), by the significant proportion of claims with a null compensation amount, and maybe also by some non-declaration of flood damages to the insurance companies. The large difference between the number of policies and the number of reported claims is a common feature for all floods in the CCR database (Moncoulon et al., 2014). In total, the combination of the limited proportion of selected insurance policies (policy / building ratio) and of the partial documentation of the claims and damages (claim / policy ratio) leads to a relatively low average ratio between the number of buildings and the number of reported claims in the floodplains: 6 to 9 % (Table 2). This explains why, despite the richness of the CCR database, the evaluation of the proposed forecasting chain based on insurance claims, with the ambition to provide results at the stream reach level, had to be limited to extreme flood events with large numbers of reported claims.

The second major surprise is the significant proportion of reported claims located outside the estimated 1000-year flood envelop (over 10 % for the two considered events; see Fig. 4). This is also a general feature observed in the CCR database. Several explanations can be put forward to explain the presence of claims outside the identified flood areas: (i) damages may be induced by small watercourses not represented in the flood model, (ii) they may also be triggered by other local processes (runoff accumulation in low points, sewer saturation, cellar flooding due to groundwater raising, etc.), and (iii) the address of the owners of the insurance policies may not correspond to the location of the affected assets. Information in the database does not allow for differentiation between damages induced by direct stream flooding and other processes. The existence of a significant proportion of claims not directly related to river overflows represented by the model adds to the complexity of the validation exercise. For the purpose of this validation, and considering that the contours of the flooded area are only available for a limited number of the considered river reaches, every claim located in the maximum possible flood envelop (estimated 1000-year flood envelop) for each stream reach has been considered for the computation of the reference number of observed claims per reach. This could lead to overestimating the reference number of claims for reaches with actual limited flooding and will be discussed in the section presenting the results.

Despite all these constraints which limit the information content of the data, this CCR database is still a rich and unique source of information to measure the impacts of flash floods in small rivers. Until now, it has been used to assess economic losses at the event scale (Moncoulon et al., 2014). We tested here whether it could provide a number of private houses affected by the floods for each river reach, to be compared to the outputs of the proposed forecasting chain.

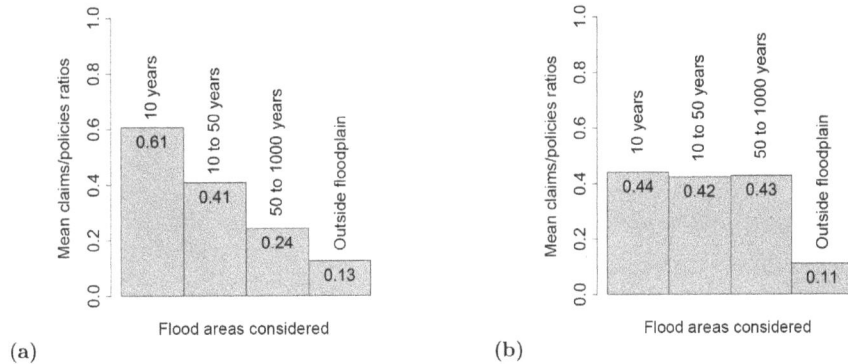

Figure 4. Spatial distribution of mean claims / policies ratios, inside and outside the floodplains: **(a)** Alès 2002 event, **(b)** Draguignan 2010 event.

Table 2. Mean claims / policies and policies / buildings ratios computed inside the 1000-year floodplains for the September 2002 and June 2010 events. The claims and policies are those selected in the CCR database according to the rules described in Sect. 3.6. The buildings considered are those of less than 7 m height according to the IGN BD Topo database.

	Mean claims / policies ratio	Mean policies / buildings ratio	Global claim / buildings ratio
September 2002	0.37	0.17	0.06
June 2010	0.43	0.21	0.09

4　Results and discussion

As mentioned in the introduction, the results are presented hereafter in two steps:

- First, an evaluation of the accuracy of the catalogue of flooded areas is presented based on the Alès case study. Two different types of evaluations results are examined: the water levels estimated at stream gauges are compared to existing stage–discharge relationships (rating curves), and the estimated flooded areas are compared to reference areas computed for the purpose of the implementation of the EU Floods Directive.

- Second, the results of the whole rainfall–runoff–impact simulation chain are presented for the both case studies, i.e. the September 2002 and June 2010 floods, and evaluated against real observed flooded areas and insurance claim data.

4.1　Comparison of water levels at stream gauges locations (Alès case study)

A first evaluation of the results of hydraulic computations is proposed here based on information available at stream gauges. These gauges indeed offer locally the opportunity to compare the rating curves based on expert know-how with the results of the 10 steady-state hydraulic computations used for the implementation of the impact model. Considering that the distance between cross sections may reach up to 100 m in the proposed method and that their locations is variable,

additional cross sections corresponding to the exact locations of the stream gauges were manually added for the hydraulic computations to enable comparisons.

The results are presented in Fig. 5 for three different stream gauges: Mialet, Banne, and Alès. This figure illustrates contrasted situations, which are detailed hereafter. It has first to be mentioned that only the Mialet and Alès stations are used for flood forecasting purposes: for these two stations, the rating curves have been extrapolated based on local hydraulic models. The Banne station is mainly used for low-flow measurements and its rating curve was extrapolated based on an assumption on the hydraulic control of the gauging section; it is nevertheless an interesting case since the station is located on a small tributary.

The case of the Mialet station (Fig. 5a) appears as an ideal situation, where the shape of the cross section is well retrieved from the DTM and the computed water levels are very close to the existing rating curve. Note that the real measured discharge values are low for the three stream gauges. The comparisons are focussed on the extrapolation range of the rating curve; i.e. the automatically implemented hydraulic model is essentially compared with local expert know-how generally based on a detailed hydraulic computation. The very satisfactory result obtained in Mialet may be explained by the simple shape of the cross section (deep and relatively narrow in this case), the limited presence of vegetation in the river bed that could affect local roughness, and the significant slope (i.e. limited risk of backwater influences). The selected roughness coefficient value of 0.05, an average value based

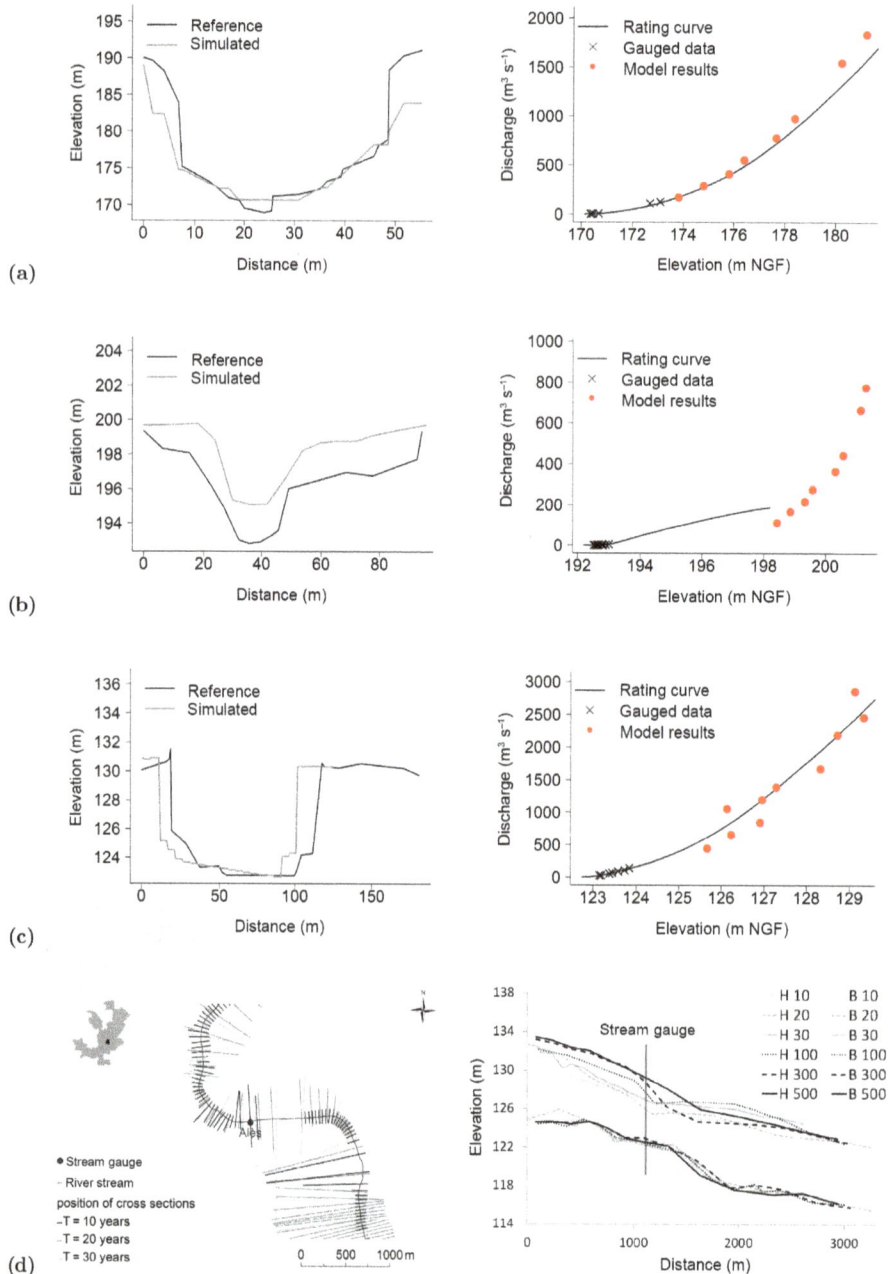

Figure 5. Examples of comparison of cross sections and water levels at three stream gauges: **(a)** Mialet station, **(b)** Banne station, **(c)** Alès station, **(d)** Alès station: position of cross sections and longitudinal profile of river bed (B) and water levels (H).

on post-event studies (Lumbroso and Gaume, 2012), corresponds well to the locally adjusted one.

The comparison for the Banne station is less satisfactory (Fig. 5b). The two cross sections have similar shapes but do not seem to have the same reference altitude. A difference of about 2 m exists for an unknown reason. This may nevertheless have little influence on the relative water levels and corresponding computed flooded areas. However, if the computed water levels are reduced by 2 m, the computed discharges still appear much larger than the corresponding

discharge estimates based on the local rating curve for the larger stage or discharge values. The slope of the local rating curve appears very low and does not even follow the evolution of cross-sectional areas with the water stages. Such a rating curve shape could result from a local backwater effect and could illustrate the limits of the hydraulic model used, which does not account for such phenomena. In this case, there is nevertheless no hydraulic singularity immediately downstream the gauge that could generate such an effect. The reference extrapolated rating curve is questionable.

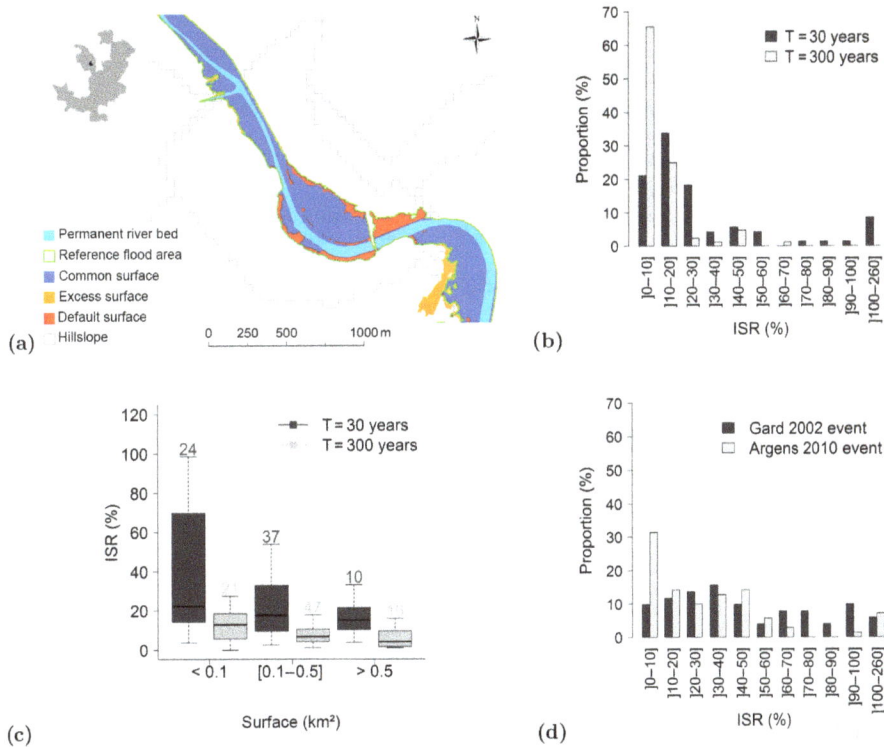

Figure 6. Comparison between estimated and reference flooded areas: **(a)** definition of common surface (S_c), excess surface (S_e), and default surface (S_d) for the computation of ISRs, **(b)** distributions of ISR scores for the 30-year and 300-year discharge quantiles on the APSFR of Alès, **(c)** distributions of ISR scores depending on the extent of the reference flooded area, and **(d)** distributions of ISR scores for the full simulation chain for the Alès 2002 and Draguignan 2010 events.

The case of the Alès station illustrates other sources of difficulties (Fig. 5c). Again in this case, the topography of the river bed appears well retrieved from the DTM even if a horizontal displacement is noticeable. However, the shape of the computed stage–discharge relationship, even if on average close to the rating curve of the station, appears chaotic and non-monotonous. The stream gauge is located just upstream a large meandering stretch where the valley is perched with a large flood plain on the right bank (Fig. 5d). This flood plain is only inundated during extreme floods. Depending on the run (i.e. on the discharge value), it can be included or excluded from the modelled cross sections located just downstream the gauge and may generate an inundated area not connected to the river bed, finally eliminated in the post-treatment as described previously. However, this artificial inclusion of the flood plain in the cross sections, for some intermediate discharge values, leads to underestimating the computed water level at the gauged cross section due to the backwater propagation (see longitudinal profiles in Fig. 5d). Clearly, the proposed procedure could not eliminate all the problems encountered when modelling perched rivers with 1-D hydraulic models, despite the precautions taken. The use of 2-D hydraulic models could help solve the problems encountered in the future but at the cost of a large increase in computation time. A detailed analysis of the results obtained

for the two case studies nevertheless reveals that the number of remaining problematic river reaches is limited. Moreover, these problematic configurations mainly correspond to relatively large rivers (315 km^2 of drainage area at the Alès station), which correspond to the limit of the target application domain of the proposed method.

Similar results were obtained for the other available gauging stations. Overall, the results are extremely satisfactory, almost exceeding the initial expectancies. The cross-sectional shapes can be correctly retrieved from the existing DTM despite its limits, at least sufficiently accurately for the reconstruction of local stage–discharge relationships. It is important to note that the low-flow water heights are limited in these Mediterranean rivers. This explains why bathymetric data are not crucial to obtain a relevant estimation of the cross-sectional shapes: aerial topographic surveys are often sufficient to get an accurate representation of the river beds in these regions. The selected average roughness coefficient value appears to be suited to the local expert know-how. In the future, checking stage–discharge relationships at gauging stations could help to adjust the values of roughness coefficients in the proposed approach extrapolated to other areas. It is important nevertheless to keep in mind that hydraulic singularities such as bridges, which cannot be characterised through the DTM, may locally largely influence the stage–

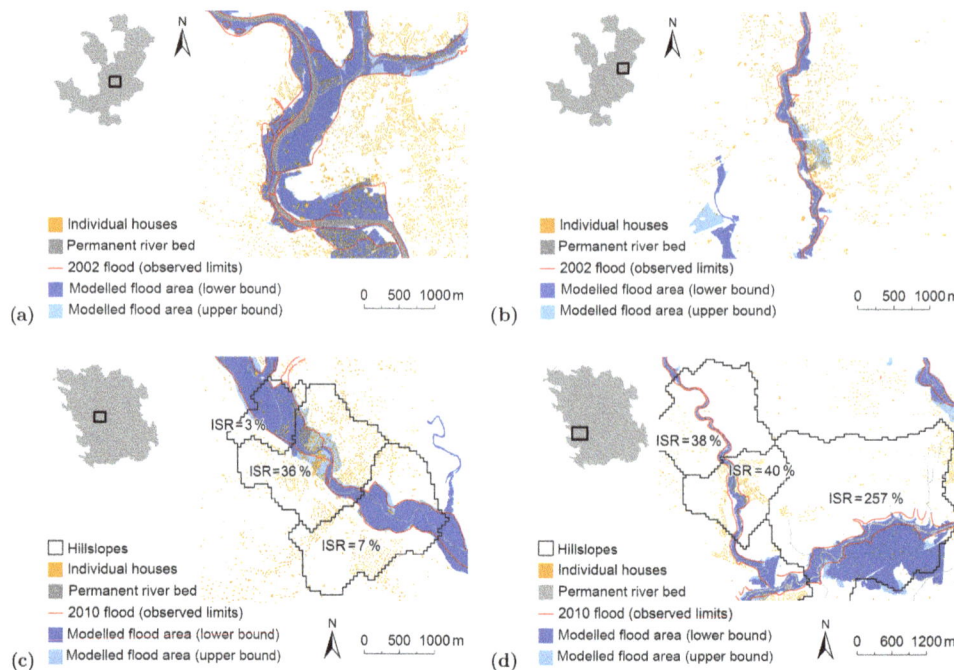

Figure 7. Examples of comparison between observed and estimated flood extents, and related ISR values (computed for the lower bounds of estimated flood extents): **(a)** Alès 2002 event for the Grabieux and Gardons rivers, **(b)** Alès 2002 event for the Avène river, **(c)** Draguignan 2010 event for the Nartuby river, and **(d)** Draguignan 2010 event for the Florieye and Argens rivers.

discharge relationship, even if this could not really be illustrated in the presented case study.

4.2 Comparison with reference flooded areas (Alès case study)

Reference flood maps have been produced for the purpose of the implementation of the EU Floods Directive for discharge values corresponding to the 30- and 300-year flood events for almost half of the considered stream reaches. The automatically computed maps could be compared to these reference maps for the same discharge values. For each river reach, the estimated surface (ES) and observed or reference surface (RS) are compared. The surface area in common (S_c), the excess surface area (S_e: computed but not observed), and the default surface area (S_d: observed but not computed) are evaluated (Fig. 6a). Note that $ES = S_c + S_e$ and $RS = S_c + S_d$. A synthetic incoherent surface ratio (ISR) is then computed. It represents the extent of excess and default surfaces, expressed as a proportion of the reference surface (Eq. 1):

$$ISR = \frac{S_e + S_d}{RS}. \qquad (1)$$

The permanent river bed (represented in Fig. 6a) which is not affected by estimation uncertainties, is not considered in the computation of the surfaces ES and RS and hence in the computation of ISR.

Figure 6b presents the distributions of ISRs computed for the 84 river reaches for which reference inundation maps

were available (71 reaches for the 30-year flood) for the Alès case study. The results appear overall satisfactory: the ISR rarely exceeds 30 %. This ratio includes both default and excess surfaces: the real difference between ES and RS is in fact more limited. This suggests that the errors in the estimation of impacts will also be more limited. The ISR is sensitive to small differences between computed and reference maps as illustrated by the examples shown in Fig. 7 and low values are difficult to reach and suggest a quasi perfect agreement. Figure 6c also shows that the ISR values depend on the magnitude of the simulated floods. The results obtained for the 300-year flood appear much more accurate, with ISRs almost never exceeding 50 %. This can be explained by the fact that the floodplains are largely flooded in the case of high return period discharges, with limited possibilities for large errors. The estimation of the flooded areas for the 30-year discharges appear less accurate, with a significant proportion of stream reaches with large relative errors: ISRs exceeding 100 % for almost 10 % of the reaches. These relative errors are nevertheless essentially related to observed flooded surfaces less than 0.1 km^2 (Fig. 6c). This corresponds to the very beginning of the river overflow that is hardly captured accurately with a hydraulic model based on automatic extractions from a DTM, whose roughness parameters are averaged over large areas and with no description of local hydraulic singularities. However, large relative errors in small flooded areas will have limited influence on the impact model. Overall, the simulated flooded areas correspond pretty well to the refer-

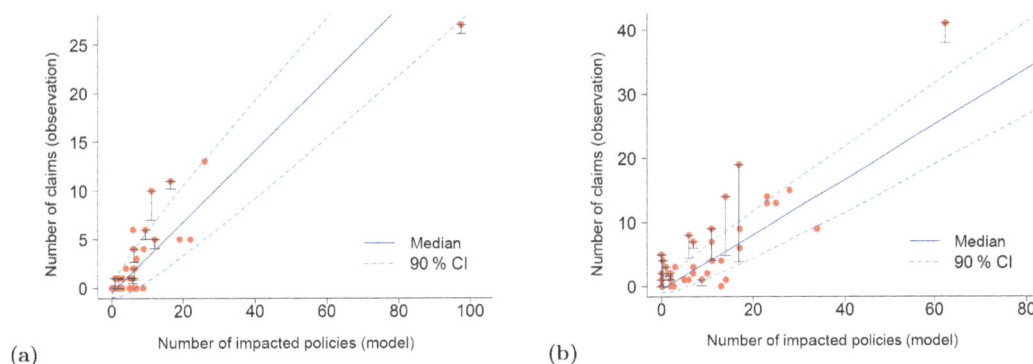

Figure 8. Comparison of estimated number policies affected by the flood (model), and associated number of claims for each river reach: (a) Alès 2002 event and (b) Draguignan 2010 event. Blue lines correspond to the mean claims / policies binomial relationship and associated 90 % confidence interval, vertical bars represent uncertainty in the number of claims related to river flooding.

ence areas. These results confirm the validation results obtained from the rating curves: the proposed 1-D hydraulic model, based on automatic extractions from relatively accurate DTM and on a regionally averaged roughness coefficient, stands overall reasonably well compared with local expert-based hydraulic models, at least in the test region. Some reasons can be put forward to explain this result: the limited need of bathymetric data in the Mediterranean context; the river slopes – typically 0.5 % or more – limiting the distance over which backwater effects propagate and therefore the local influence of hydraulic singularities not taken into account in the model; the reduced sensitivity of water stages (h) to variations of discharges (Q) due to the typical shape of stage–discharge relationships ($h \propto Q^{3/5}$); finally, the reduced sensitivity of the flooded area to the water stages, except at the beginning of the river overflow, due to the cross-sectional shapes of the river beds, with generally narrow valleys and well-delimited flat flood plains.

4.3 Accuracy of forecasted flood extents for the Alès 2002 and Draguignan 2010 floods

For both considered validation flood events, observed inundated areas were carefully mapped after the floods on part of the affected streams. These observed flood extent maps were compared for each river reach to the forecasted flood extents. More precisely, the ISRs described in the previous section were computed based on the map of the flood catalogue corresponding to a discharge value immediately lower or equal to the simulated peak discharge value. There is indeed not necessarily a map in the catalogue corresponding to the exact value of the simulated peak discharge. This choice should lead to a slight underestimation of the flooded areas if the hydrological and hydraulic models were perfect. This is illustrated in Fig. 7, which shows for some river reaches, the observed flood extent and both modelled extents corresponding to the discharge values immediately lower and higher than

the forecasted peak discharge. This figure also shows to what range of differences the ISR values correspond.

The ISRs obtained for both events are summarised in Fig. 6d. Without surprise, the ISR values are significantly increased when actual flood events are considered, if compared to the initial evaluation of the catalogue of flooded areas presented in the previous section. When actual floods are simulated, additional sources of uncertainties affect the computed flooded areas. The simulated peak discharge on which the forecasted maps are based may differ significantly from the observed ones. Moreover, the observed flooded areas may be the result of local processes (dike breaches, blockages) particularly during extreme flood events. These processes are not represented in the hydraulic models, either the proposed simplified regional model or the local models used to elaborate the reference flood maps. In fact, Fig. 7 shows that the differences between observed and simulated flood extents are explained not only by uncertainties in the simulated peak discharge values which would result in systematic over or underestimations of flood extents but also by local processes imperfectly accounted for in the hydraulic models.

Finally, a large proportion of the relative ISRs remains less than 50 %. Consequently, the computed flood extent maps may be sufficiently realistic to provide an approximation of the local field consequences of floods and of their spatial distribution. This is verified in the next section. It is noteworthy that the ISRs are lower in the case of the Draguignan 2010 flood despite that the models (rainfall–runoff and hydraulic models) were extrapolated to this event and area without any further calibration. This can be explained by the higher accuracy of the DTM available in this area. In any case, it is a promising result that seems to reveal that uncertainties related to the extrapolation of the models may be of secondary importance compared to the other sources of uncertainties of the proposed approach. Nevertheless general conclusions cannot be drawn from one single example.

4.4 Validation of forecasted impacts based on insurance claims

Even after a careful selection of the appropriate validation data, the observed claim / policy ratio is significantly less than 1 (Table 2 and Fig. 4), and varies between river reaches. As shown in Sect. 3.6, this is explained by claims with no compensation and/or non-declared claims, buildings with raised basements, inaccurate location of insured buildings, and also the imperfect filling of the claim database. Clearly, the number of reported claims per river reach has a random component due both to the claim triggering processes and to the limits of the claim database. To account for randomness in the validation process, the number of reported claims can for instance be considered as the result of a random binomial process $B(n, p)$, with n being the number of policies in the considered flooded area and p the probability that a corresponding claim with non-null compensation is observed. If p is considered to be the same for all reaches and equal to the average claim / policy ratio for the selected sample of insurance policies, a confidence interval (90 % binomial confidence interval herein) can be estimated for the number of claims corresponding to every computed number of impacted policies (policies located in the estimated flooded area). Ideally, the average claim / policy ratio should be estimated inside the actually observed flooded area. However, this information is clearly not available everywhere on the considered stream networks. By default, the claim / policy ratios have been computed here within the 1000-year flood extent (Table 2). The result is presented in Fig. 8 for the two case studies.

This figure shows a relatively good agreement between the forecasted number of impacted policies and the number of reported claims per river reach. The observed spread of the results nevertheless exceeds the width of the 90 % intervals, especially in the Draguignan 2010 case study: more than 10 % of the dots lie outside the confidence limits with predominantly underestimations by the model. This indicates that some other sources of errors affect the relationship between forecasted and observed number of claims. First, the reported claim / policy ratio is significantly affected by water depth and flood duration, two variables that have not been considered here. Second, the claim / policy ratio and the reference number of claims have been computed based on the maximum possible estimated flood envelop (1000-year flood envelop), i.e. without using the observed flooded area (not available for a large part of the stream network), or the computed flooded area (which has to be evaluated and should therefore not be used in the validation process). This may lead to slight underestimation of the claim / policy ratio since some non-flooded areas are incorporated in the estimation. This choice may also lead to overestimating the reference claim number for stream reaches where limited overflow occurs, especially in densely urbanised areas (see points A and B in Fig. 8b), since a significant proportion of claims are not

related to the streams represented in the model (Sect. 3.6). The number of reported claims in the computed flooded areas has also been estimated and the difference with the reference number of claims is indicated by the dotted lines in Fig. 8. This difference is modest in the Alès 2002 case, where the floods have been extreme over the entire considered area. The correction is much more significant in the Draguignan case, especially for almost all the points located over the 95 % confidence limit (Fig. 8b). Several of these reaches, particularly points A and B, are located in the upper south-eastern part of the studied area, which has not been affected by the most intense rainfalls. The observed flood extent was only partially mapped in this area after the 2010 flood, but the mapped extent indicates moderate overflow in this sector and is in good agreement with the modelled flooded area. The high number of claims located outside areas flooded by the streams modelled, the impossibility to separate these claims, and the absence of observed flood extent maps for stream affected by moderate floods clearly limit the use of the insurance claim database for the validation of flood impact models.

Finally, Fig. 9 compares the spatial distributions of the simulated peak discharges, of the maximum forecasted impacts (i.e. number of flooded private houses with georeferenced policy), and of the number of associated claims according to the CCR database. It should be noted that information on claims is provided only for river reaches with at least 20 policies recorded in the database. This explains why no values have been provided for several stream reaches, essentially non-urbanised reaches with limited exposed assets. This figure illustrates the large differences in the outputs of the hydrological rainfall–runoff model (Fig. 9a and b), and of the integrated rainfall–runoff–impact modelling chain (Fig. 9c and d) which provides a much contrasted analysis. It also shows good overall consistency between the forecasted impacts (Fig. 9c and d) and observed claims (Fig. 9e and f) for the location of the main hotspots. Apart from some exceptions, the ranking and magnitudes of the field impacts appears to be well captured by the proposed forecasting chain. That is mainly the information needed by rescue services to adapt and dispatch their rescue means during flood event management. Such information could also help targeting more effectively alert messages.

5 Conclusions

Flood event managers need to assess, in real time, the severity of possible field consequences associated with hydro-meteorological forecasts, to be able to take appropriate decisions. Automatic assessment methods are necessary in the case of fast-evolving events such as flash floods, when little time is available for information processing and analysis. Moreover, the direct estimation of field consequences opens the possibility of testing the performance of forecast-

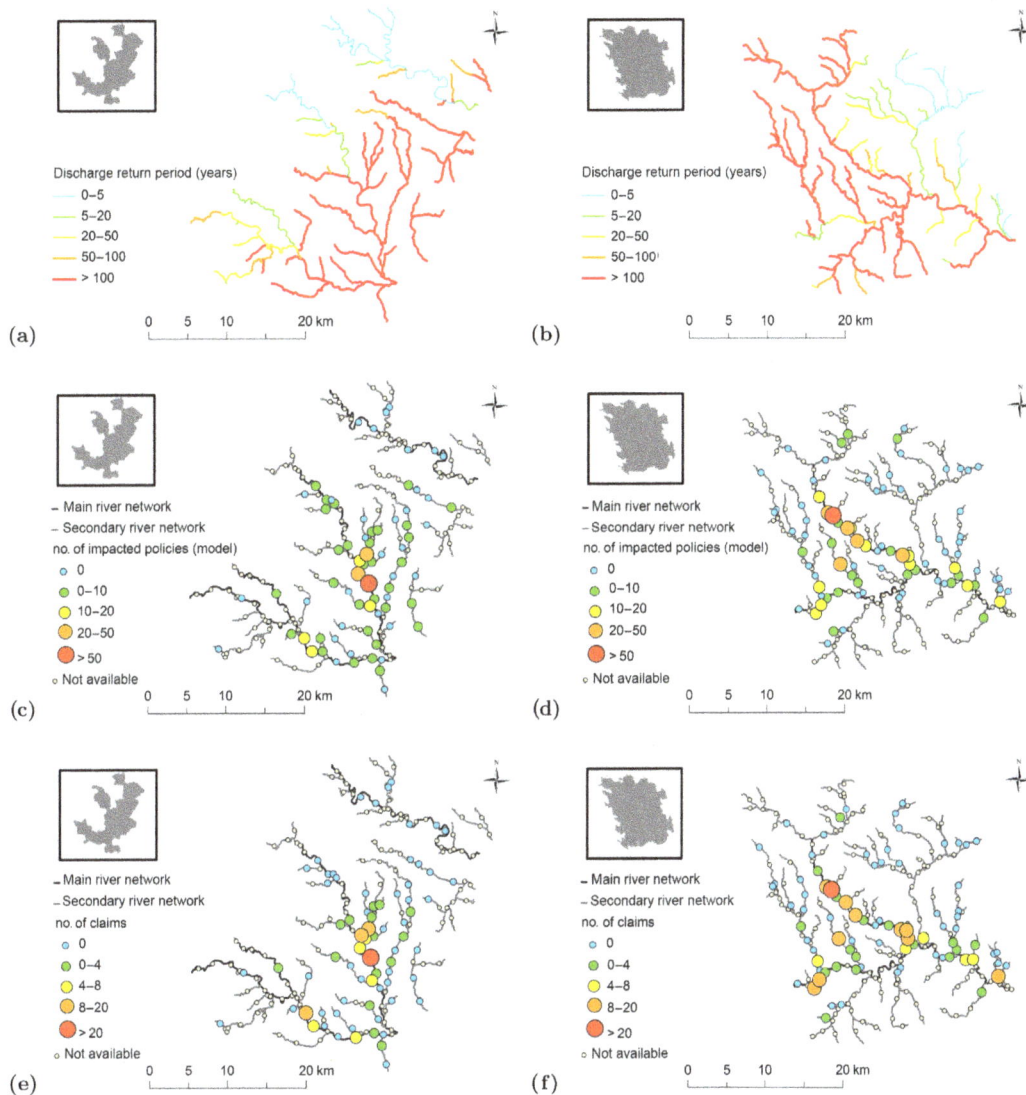

Figure 9. Maps of the peak discharges (return periods) and related impacts (number of flooded policies) simulated by the model, and of the number of claims extracted from the CCR database: **(a)** peak discharges, Alès 2002; **(b)** peak discharges, Draguignan 2010; **(c)** estimated impacts, Alès 2002; **(d)** estimated impacts, Draguignan 2010; **(e)** claims, Alès 2002; **(f)** claims, Draguignan 2010.

ing chains in ungauged areas, where various observations related to damages may be available.

This paper has tested the potential of simple approaches for the estimation of the magnitudes of possible field consequences within flash-flood forecasting chains. The proposed methods have been selected to be implemented with limited calibration effort, over extended areas, and at detailed spatial scales. Particular attention has been paid to the performance evaluation of the proposed chain. An original and particularly rich and comprehensive insurance claim database has therefore been used. It is to our knowledge the first time an insurance claim database is used for such a purpose.

The proposed approaches certainly deserve further validation, but the results presented herein on two case studies appear extremely satisfactory and promising. The flood map-

ping based on 1-D steady-state hydraulic modelling, automatically implemented over a large river network with an average roughness value, stands the comparison, for most of the considered river reaches, with local expert-based hydraulic simulations: stage–discharge relationships at gauging stations and flood maps computed for the implementation of the EU Floods Directive. The whole hydrological and hydraulic simulation chain provides estimates of maximum flooded areas that also appear generally close to the flood extents mapped after the two test events when such maps were available. The typical configuration of the streams affected by flash floods can be put forward to explain such satisfactory results: (i) narrow valleys and well-delimited floodplains, (ii) limited need for bathymetric data for Mediterranean streams, (iii) steep stream slopes implying limited spatial influence of hydraulic singularities (bridges) and their induced backwater effects, and (iv) huge discharge contrasts between moderate, large, and extreme floods that are well captured by rainfall–runoff models despite the inevitable modelling uncertainties. The slightly better results obtained in the Draguignan case suggest that some improvements could still be achieved with more accurate topographic data, especially lidar data enabling better retrieval of the cross-sectional shape of the main stream bed. The influence of the DTM accuracy on the results of the proposed approach will have to be tested. Some other difficulties nevertheless remain. The proposed modelling approach has to be further improved to properly handle complex hydraulic configurations such as perched river beds (Fig. 5). And the observed flood extents may also locally differ from the estimated flooded area even if the discharge value has been well forecasted due to local effects difficult to anticipate such as blockages and breaches. A perfect fit is out of reach. This suggests that we should not put too much confidence in theoretical flood maps computed either a priori or in real time. Such maps, if provided to the flood event managers, should be presented as indications of possible flood scenarios close but not identical to the actual flood.

The validation of the estimated damages based on insurance claims faced some difficulties related to the specificity of insurance data. The CCR database used is probably the most comprehensive source of information about flood insurance losses in France. However, the validation process requires both accurate geocoding of insurance policies and comprehensive information on claims, which limits the amount of available information. A high proportion of claims is also not related to the streams included in the model, limiting the possibility of using this data as a reference for moderate floods if the actual flood extent is unknown (Fig. 8). The validation exercise could nevertheless be successfully achieved for the two extreme floods studied herein, providing interesting additional information on the accuracy of the whole simulation chain. It should also be considered that the quality of insurance data is continuously increasing and that some of the limits identified here (e.g. geocoding, comprehensiveness of claim information) should be significantly reduced in the future. Moreover, better representation of the claims triggering processes should be possible based on the inundation water depths obtained from the hydraulic computations; direct relationships between the claim ratios and water depths are indeed commonly used by insurance companies. Therefore, insurance claim data should be considered a relevant option for the validation of flood forecasting results, particularly in the case of flash floods affecting ungauged rivers. Although data of this sort are generally confidential, they may be accessible through partnerships with insurance companies. Other sources of information on flash flood impacts could also be used, such as the logs of emergency services, emergency calls, information shared on social networks (USDHS, 2012; Jongman et al., 2015; Tkachenko et al., 2017), or information gathered in the field after or during the event (Ortega et al., 2009; Ruin et al., 2014). This information is also affected by uncertainties and severe biases, especially in flash flood situations: absence of information due to local breakdowns of communication networks, reduction of social network activity, and partial filling of emergency logs in strongly affected areas during the turmoil of the event. Some of this information has nevertheless the benefit of being available in real time (digitised logs, emergency calls, social networks) and could help validate and improve forecasted impacts. Finally, the combined use of flood impact forecasting models and field data mining and processing methods is without doubts a promising avenue for the development of innovative flood forecasting and warning services.

Data availability. Most of the data used in this study are not under open licence and cannot be provided for direct download. The main data providers were:

- the Institut Géographique National (IGN; http://www.ign.fr) and the Conseil Départemental du Gard (http://www.gard.fr) for DTM data.
- the Direction Régionale de l'Environnement, de l'Aménagement et du Logement Auvergne Rhone Alpes, Service de Prévision des Crues du Grand Delta (DREAL Auvergne Rhone Alpes, SPC Grand Delta; http://www.auvergne-rhone-alpes.developpement-durable.gouv.fr) for stream gauge data and rating curves.
- the OHM-CV observatory (http://www.ohmcv.fr) and Hymex program (http://www.hymex.org) for QPEs. The radar and rain gauge input data were provided by Météo-France, the SPC Grand Delta, and Electricité de France (EDF).
- the CCR for insurance claim data (www.ccr.fr).

Competing interests. The authors declare that they have no conflict of interest.

Acknowledgements. The authors would like to express their gratitude to François Bourgin for his careful reading of the manuscript. They also want to thank the following:

- The French Ministry of Environment, Direction Générale de la Prévention des Risques, Service Central D'Hydrométéorologie et d'Appui à la Prévision des Inondations (DGPR/SCHAPI, http://vigicrues.gouv.fr), for the financial support of this work.
- The DREAL Auvergne Rhone Alpes, SPC Grand Delta, the Conseil Départemental du Gard, and the IGN, for supplying part of the data.
- The OHM-CV observatory and the HyMeX program for their help in accessing the rainfall data (QPEs). OHM-CV is an observation service supported by the Institut National des Sciences de l'Univers, section Surface et Interfaces Continentales and the Observatoire des Sciences de l'Univers de Grenoble. OHM-CV is a key observation system of the HyMeX program.

Edited by: Matjaz Mikos

References

Aubert, Y., Arnaud, P., Ribstein, P., and Fine, J.-A.: La méthode SHYREG – application sur 1605 bassins versants en France métropolitaine, Hydrolog. Sci. J., 59, 993–1005, https://doi.org/10.1080/02626667.2014.902061, 2014.

Borah, D., Prasad, S., and Alonso, C.: Kinematic wave routing incorporating shok fitting, Water Resour. Res., 3, 529–541, 1980.

Bradbrook, K., Waller, S., and Morris, D.: National Floodplain Mapping: Datasets and Methods – 160.000 km in 12 months, Natural Hazards, 36, 103–123, https://doi.org/10.1007/s11069-004-4544-9, 2005.

Cole, J., Moore, R., Wells, S., and Mattingley, P.: Realtime forecasts of flood hazard and impact: some UK experiences, E3S Web of Conferences, p. 18015, https://doi.org/10.1051/e3sconf/20160718015, 2016.

Cole, S. J. and Moore, R. J.: Distributed hydrological modelling using weather radar in gauged and ungauged basins, Adv. Water Resour., 32, 1107–1120, https://doi.org/10.1016/j.advwatres.2009.01.006, 2009.

Delrieu, G., Nicol, J., Yates, E., Kirstetter, P.-E., Creutin, J.-D., Anquetin, S., Obled, C., Saulnier, G.-M., Ducrocq, V., Gaume, E., Payrastre, O., Andrieu, H., Ayral, P.-A., Bouvier, C., Neppel, L., Livet, M., Lang, M., du Chatelet, J. P., Walpersdorf, A., and Wobrock, W.: The Catastrophic Flash-Flood Event of 8–9 September 2002 in the Gard Region, France: A First Case Study for the Cévennes-Vivarais Mediterranean Hydrometeorological Observatory, J. Hydrometeorol., 6, 34–52, 2005.

Delrieu, G., Boudevillain, B., Nicol, J., Chapon, B., Kirstetter, P.-E., Andrieu, H., and Faure, D.: Bollène – 2002 Experiment: Radar Quantitative Precipitation Estimation in the Cévennes-Vivarais Region, France, J. Appl. Meteorol. Clim., 48, 1422–1447, https://doi.org/10.1175/2008JAMC1987.1, 2009.

Gaume, E., Livet, M., Desbordes, M., and Villeneuve, J. P.: Hydrologic analysis of the Aude, France, flash flood 12 and 13 november 1999, J. Hydrol., 286, 135–154, https://doi.org/10.1016/j.jhydrol.2003.09.015, 2004.

Gaume, E., Bain, V., Bernardara, P., Newinger, O., Barbuc, M., Bateman, A., Blaškovičová, L., Blöschl, G., Borga, M., Dumitrescu, A., Daliakopoulos, I., Garcia, J., Irimescu, A., Kohnova, S., Koutroulis, A., Marchi, L., Matreata, S., Medina, V., Preciso, E., Sempere-Torres, D., Stancalie, G., Szolgay, J., Tsanis, I., Velasco, D., and Viglione, A.: A compilation of data on European flash floods, J. Hydrol., 367, 70–78, https://doi.org/10.1016/j.jhydrol.2008.12.028, 2009.

Georgakakos, K. P.: Analytical results for operational flash flood guidance, J. Hydrol., 317, 81–103, https://doi.org/10.1016/j.jhydrol.2005.05.009, 2006.

Gourley, J., Flamig, Z., Vergara, H., Kirstetter, P.-E., Clark, R., Argyle, E., Arthur, A., Martinaitis, S., Terti, G., Erlingis, J., Hong, Y., and Howard, K.: The Flash project: improving the tools for flash flood monitoring and prediction across the United States, B. Am. Meteorol. Soc., 98, 361–372, 2017.

Gourley, J. J., Flamig, Z. L., Hong, Y., and Howard, K. W.: Evaluation of past, present and future tools for radar-based flash flood prediction in the USA, Hydrolog. Sci. J., 59, 1377–1389, https://doi.org/10.1080/02626667.2014.919391, 2014.

Goutal, N., Lacombe, J.-M., Zaoui, F., and El-Kadi-Abderrezzak, K.: MASCARET: A 1-D open-source software for flow hydrodynamic and water quality in open channel networks, in: River Flow 2012 vols 1 and 2, edited by: Munoz, R., 1169–1174, CRC Press – Tailor and Francis Group, Boca Raton, USA, 2012.

Javelle, P., Demargne, J., Defrance, D., Pansu, J., and Arnaud, P.: Evaluating flash-flood warnings at ungauged locations using post-event surveys: a case study with the AIGA warning system, Hydrolog. Sci. J., 59, 1390–1402, https://doi.org/10.1080/02626667.2014.923970, 2014.

Jongman, B., Wagemaker, J., Romero, B., and de Perez, E.: Early Flood Detection for Rapid Humanitarian Response: Harnessing Near Real-Time Satellite and Twitter Signals, ISPRS Int. J. Geo-Inf., 4, 2246–2266, https://doi.org/10.3390/ijgi4042246, 2015.

Le Bihan, G., Payrastre, O., Gaume, E., Moncoulon, D., and Pons, F.: Regional models for distributed flash-flood nowcasting: towards an estimation of potential impacts and damages, E3S Web of Conferences, p. 18013, https://doi.org/10.1051/e3sconf/20160718013, 2016.

Lumbroso, D. and Gaume, E.: Reducing the uncertainty in indirect estimates of extreme flash flood discharges, J. Hydrol., 414–415, 16–30, https://doi.org/10.1016/j.jhydrol.2011.08.048, 2012.

Moncoulon, D., Labat, D., Ardon, J., Leblois, E., Onfroy, T., Poulard, C., Aji, S., Rémy, A., and Quantin, A.: Analysis of the French insurance market exposure to floods: a stochastic model combining river overflow and surface runoff, Nat. Hazards Earth Syst. Sci., 14, 2469–2485, https://doi.org/10.5194/nhess-14-2469-2014, 2014.

Moussa, R.: Analytical Hayami solution for the diffusive wave flood routing problem with lateral inflow, Hydrol. Process., 10, 1209–1227, 1996.

Naulin, J.-P., Payrastre, O., and Gaume, E.: Spatially distributed flood forecasting in flash flood prone areas: Application to road network supervision in Southern France, J. Hydrol., 486, 88–99, https://doi.org/10.1016/j.jhydrol.2013.01.044, 2013.

Nguyen, P., Thorstensen, A., Sorooshian, S., Hsu, K., and AghaKouchak, A.: Flood Forecasting and Inundation Mapping Using HiResFlood-UCI and Near-Real-Time Satellite Precipitation Data: The 2008 Iowa Flood, J. Hydrometeorol., 16, 1171–1183, https://doi.org/10.1175/JHM-D-14-0212.1, 2015.

Norbiato, D., Borga, M., Esposti, S. D., Gaume, E., and Anquetin, S.: Flash flood warning based on rainfall thresholds and soil moisture conditions: An assessment for gauged and ungauged basins, J. Hydrol., 362, 274–290, https://doi.org/10.1016/j.jhydrol.2008.08.023, 2008.

Ortega, K., Smith, T., Manross, K., Scharfenberg, K., Witt, A., Kolodziej, A., and Gourley, J.: The severe hazards analysis and verification experiment, B. Am. Meteorol. Soc., 90, 1519–1530, https://doi.org/10.1175/2009BAMS2815.1, 2009.

Pappenberger, F., Dutra, E., Wetterhall, F., and Cloke, H. L.: Deriving global flood hazard maps of fluvial floods through a physical model cascade, Hydrol. Earth Syst. Sci., 16, 4143–4156, https://doi.org/10.5194/hess-16-4143-2012, 2012.

Pons, F., Laroche, C., Fourmigue, P., and Alquier, M.: Cartographie des surfaces inondables extrêmes pour la directive inondation: cas de la Nartuby, La Houille Blanche, 2, 34–41, https://doi.org/10.1051/lhb/2014014, 2014.

Rozalis, S., Morin, E., Yair, Y., and Price, C.: Flash flood prediction using an uncalibrated hydrological model and radar rainfall data in a Mediterranean watershed under changing hydrological conditions, J. Hydrol., 394, 245–255, https://doi.org/10.1016/j.jhydrol.2010.03.021, 2010.

Ruin, I., Lutoff, C., Boudevillain, B., Creutin, J., Anquetin, S., Rojo, M., Boissier, L., Bonnifait, L., Borga, M., Colbeau-Justin, L., Creton-Cazanave, L., Delrieu, G., Douvinet, J., Gaume, E., Gruntfest, E., Naulin, J., Payrastre, O., and Vannier, O.: Social and hydrological hesponses to extreme precipitations: an interdisciplinary strategy for postflood investigation, Weather and Climate Society, 6, 135–153, 2014.

Saint-Martin, C., Fouchier, C., Javelle, P., Douvinent, J., and Vinet, F.: Assessing the exposure to floods to estimate the risk of flood-related damage in French Mediterranean basins, E3S Web of Conferences, p. 04013, https://doi.org/10.1051/e3sconf/20160704013, 2016.

Sampson, C., Smith, A., Bates, P., Neal, J., Alfieri, L., and Freer, J.: A high-resolution global flood hazard model, Water Resour. Res., 51, 7358, https://doi.org/10.1002/2015WR016954, 2015.

Sanders, B. F.: Evaluation of on-line {DEMs} for flood inundation modeling, Adv. Water Resour., 30, 1831–1843, https://doi.org/10.1016/j.advwatres.2007.02.005, 2007.

Schroeder, A., Gourley, J., Henderson, J., Parhi, P., Rahmani, V., Reed, K., Schumacher, R., B.K., S., and Taraldsen, M.: The development of a flash flood severity index, J. Hydrol., 541, 523–532, https://doi.org/10.1016/j.jhydrol.2016.04.005, 2016.

Tkachenko, N., Jarvis, S., and Procter, R.: Predicting floods with Flickr tags, PLoS One, 12, e0172870, https://doi.org/10.1371/journal.pone.0172870, 2017.

USDA: Urban hydrology for small watersheds, Technical Release N. 55, Tech. rep., United States Department of Agriculture, Natural Resources Conservation Service, Washington DC, 1986.

USDHS: First responder communities of practice, virtual social media working group, community engagement guidance and best practice, Tech. rep., United States Department of Homeland Security, Science and technology directorate, Washington DC, 2012.

Versini, P.-A., Gaume, E., and Andrieu, H.: Application of a distributed hydrological model to the design of a road inundation warning system for flash flood prone areas, Nat. Hazards Earth Syst. Sci., 10, 805–817, https://doi.org/10.5194/nhess-10-805-2010, 2010a.

Versini, P.-A., Gaume, E., and Andrieu, H.: Assessment of the susceptibility of roads to flooding based on geographical information – test in a flash flood prone area (the Gard region, France), Nat. Hazards Earth Syst. Sci., 10, 793–803, https://doi.org/10.5194/nhess-10-793-2010, 2010b.

Versini, P.-A., Berenguer, M., Corral, C., and Sempere-Torres, D.: An operational flood warning system for poorly gauged basins: demonstration in the Guadalhorce basin (Spain), Natural Hazards, 71, 1355–1378, https://doi.org/10.1007/s11069-013-0949-7, 2014.

Wang, J., Yang, H., Li, L., Gourley, J., Khan, S., Yilmaz, K., Adler, R., Policelli, F., Habib, S., Irwn, D., Limaye, A., Korme, T., and Okello, L.: The coupled routing and excess storage (CREST) distributed hydrological model, Hydrolog. Sci. J., 56, 84–98, https://doi.org/10.1080/02626667.2010.543087, 2011.

Yamazaki, D., Kanae, S., Kim, H., and Oki, T.: A physically based description of floodplain inundation dynamics in a global river routing model, Water Resour. Res., 47, w04501, https://doi.org/10.1029/2010WR009726, 2011.

A geostatistical data-assimilation technique for enhancing macro-scale rainfall–runoff simulations

Alessio Pugliese[1], **Simone Persiano**[1], **Stefano Bagli**[2], **Paolo Mazzoli**[2], **Juraj Parajka**[3], **Berit Arheimer**[4], **René Capell**[4], **Alberto Montanari**[1], **Günter Blöschl**[3], and **Attilio Castellarin**[1]

[1]Department DICAM, University of Bologna, Bologna, Italy
[2]GECOsistema srl, Cesena, Italy
[3]Institute for Hydraulic and Water Resources Engineering, TU Wien, Vienna, Austria
[4]Swedish Meteorological and Hydrological Institute (SMHI), Norrköping, Sweden

Correspondence: Attilio Castellarin (attilio.castellarin@unibo.it)

Abstract. Our study develops and tests a geostatistical technique for locally enhancing macro-scale rainfall–runoff simulations on the basis of observed streamflow data that were not used in calibration. We consider Tyrol (Austria and Italy) and two different types of daily streamflow data: macro-scale rainfall–runoff simulations at 11 prediction nodes and observations at 46 gauged catchments. The technique consists of three main steps: (1) period-of-record flow–duration curves (FDCs) are geostatistically predicted at target ungauged basins, for which macro-scale model runs are available; (2) residuals between geostatistically predicted FDCs and FDCs constructed from simulated streamflow series are computed; (3) the relationship between duration and residuals is used for enhancing simulated time series at target basins. We apply the technique in cross-validation to 11 gauged catchments, for which simulated and observed streamflow series are available over the period 1980–2010. Our results show that (1) the procedure can significantly enhance macro-scale simulations (regional LNSE increases from nearly zero to ≈ 0.7) and (2) improvements are significant for low gauging network densities (i.e. 1 gauge per $2000 \, \mathrm{km}^2$).

1 Introduction

The steady increase in computational capabilities together with the expanding accessibility of regional and global datasets (e.g. soil properties, land-cover, morphology, climate characteristics, satellite-based gridded precipitation) trigger the development of regional- to continental-scale and global-scale hydrological models (Archfield et al., 2015), hereafter referred to as macro-scale models.

During the last decade, several of these macro-scale models have become operational and thus continuously provide data automatically for decision-making. For instance, the distributed rainfall–runoff-routing model LISFLOOD (De Roo et al., 2000) provides daily forecast for operational warning services through the systems of EFAS (Pappenberger et al., 2013) and GLOFAS (Alfieri et al., 2013); the LAR-SIM models (Haag and Luce, 2008) are used operationally for simulating streamflow at large areas in southern Germany, Luxembourg, Austria, Switzerland, and the eastern part of France; the WATFLOOD, developed at the University of Waterloo, is used operationally in Canada (Kouwen et al., 1993); and the S-HYPE model (Strömqvist et al., 2012) is running operationally for flood or drought forecasting and water quality assessments for the Swedish landmass, providing high-resolution information to authorities and citizens (Hjerdt et al., 2011).

Other macro-scale models are used for off-line water assessments and research purposes. For instance, the global WaterGAP Global Hydrological model (Alcamo et al., 2003) assists in water accounting; the SAFRAN-ISBA-MODCOU model (Habets et al., 2008) has been applied over the entire French territory to combine a meteorological analysis system, a land surface model, and a hydrogeological model; the PGB-IPH model (Pontes et al., 2017) has been applied to many South American basins; and the SWIM model (Krysanova et al., 1998) couples water balance simulations with water quality for small to mid-size watersheds, i.e. regional meso-scale.

The macro-scale hydrological models are getting more and more popular due to three main reasons: (1) they can provide users with a large-scale representation of hydrological behaviour, which is fundamental information for effectively addressing several water resources planning and management problems (e.g. surface water availability assessment, instream water quality studies, ecohydrological studies); (2) they can be used to compute a variety of hydrological signatures everywhere along the stream network at the resolution of the model; (3) model outputs in some cases are open-access and freely distributed, so that their regional runs represent a wealth of information for addressing the problem of hydrological predictions in data-scarce regions of the world (Pechlivanidis and Arheimer, 2015; Donnelly et al., 2016; Beck et al., 2016). Accurate regional hydrological simulations undoubtedly foster and support the implementation of improved large-scale and trans-boundary policies for water resources system management and flood-risk mitigation or climate change adaptation (de Paiva et al., 2013; Sampson et al., 2015; Falter et al., 2016; Arheimer et al., 2017).

However, improved accuracy in terms of average regional performance does not necessarily imply homogeneous improvements in local performance. In fact, due to the difficulties to perform local calibrations and validations of macro-scale models over the entire modelled regions, local performance can be rather diverse (see e.g. de Paiva et al., 2013; Donnelly et al., 2016). Factors controlling the heterogeneity of local performance may be various, for instance the quality of macro-scale input data, local water management, representativeness of model structure chosen, the influence of local geophysical and micro-climatic factors.

There is a recognized and noteworthy value in readily available and easy accessible simulated daily streamflow series for scarcely gauged, or ungauged, areas of the world to enhance awareness and decision-making (e.g. Arheimer et al., 2011; Hjerdt et al., 2011). Nevertheless, the harmonization and enhancement of local performances of macro-scale models is still a scientific challenge that is worth addressing in operational hydrology, and which raises different research questions, such as the following:

– How could we deal with locally biased simulations?

– Can we assimilate additional data to improve model performance without re-calibration?

– Is there a minimum gauging network density that makes the post-modelling data assimilation viable and effective?

Recent literature shows the significant potential of kriging-based techniques for performing regional prediction of streamflow indices in ungauged locations (Skøien et al., 2006; Castiglioni et al., 2011; Pugliese et al., 2014). Among such techniques, topological kriging, or top-kriging (see Skøien et al., 2006), has shown high prediction accuracy and excellent adaptability to a variety of water-related applications, such as prediction of low-flow indices (Castiglioni et al., 2009), interpolation of river temperatures (Laaha et al., 2013), estimation of flood quantiles (Archfield et al., 2013), regionalization of flow–duration curves (Castellarin, 2014; Castellarin et al., 2018; Pugliese et al., 2014, 2016), estimation of daily runoff in ungauged basins (Parajka et al., 2015) and reconstruction of historical daily streamflow series (Farmer, 2016).

Our study aims to develop and test a geostatistical data-assimilation procedure for better agreement between locally observed streamflow and model results from macro-scale rainfall–runoff models. The procedure employs top-kriging for geostatistically interpolating empirical period-of-record flow–duration curves (FDCs) along the stream network available at gauged basins. Interpolated FDCs are assimilated into simulated daily streamflow series at ungauged stream-network nodes, enhancing the local accuracy of simulated daily streamflow series. We test our method by improving E-HYPE model simulations (European-HYdrological Predictions for the Environment model; Donnelly et al., 2016; Hundecha et al., 2016), which provides approximately 30 years of simulated daily streamflows freely and openly accessible for 35 408 prediction nodes in Europe (mean catchment size of 215 km^2, see also http://hypeweb.smhi.se/europehype/time-series/, last access: 30 July 2018). We address the Tyrolean region, as this area gives particularly poor simulations in the E-HYPE version 3 and thus would benefit from statistical enhancement of results. For the geostatistical interpolation, we use a group of 46 gauged catchments obtained from Austrian and Italian water services, and not used when setting up E-HYPE. With the observed streamflows we construct and interpolate empirical FDCs. Then, we assess the value and potential of assimilating this streamflow information into E-HYPE simulated series. In particular, (1) we cross-validate the proposed data-assimilation procedure for 11 E-HYPE prediction nodes located nearby an existing stream gauge, and (2) we assess the enhancement of simulated series resulting from the geostatistical data assimilation under different hypotheses on the spatial density of the stream-gauging network.

The paper is structured as follows: first, methods and procedures are presented in a general way, then we illustrate the case study and application. In particular, Sect. 2 presents the proposed procedure, while Sect. 3 details the system of cross-validations and sensitivity analyses we adopted for assessing the procedure. Section 4 illustrates the study area and E-HYPE simulation data. The last three sections report results, discussion and conclusions, respectively.

2 A new geostatistical streamflow-data assimilation method

2.1 Geostatistical interpolation of empirical flow–duration curves (TNDTK)

Top-kriging is a powerful geostatistical procedure developed by Skøien et al. (2006) for the prediction of hydrological variables. Like all kriging approaches, top-kriging produces predictions of hydrological variables at ungauged sites with a linear combination of the empirical information collected at neighbouring gauging stations. Through this method, the unknown value of the streamflow index of interest at prediction location x_0, $Z(x_0)$ can be estimated as a weighted average of the regionalized variable, measured within the neighbourhood:

$$Z(x_0) = \sum_{j=1}^{n} \lambda_j Z(x_j), \quad (1)$$

where λ_j is the kriging weight for the empirical value $Z(x_j)$ at location x_j, and n is the number of neighbouring stations used for interpolation. Kriging weights λ_j can be found by solving the typical ordinary kriging linear system (Eq. 2a) with the constraint of unbiased estimation (Eq. 2b):

$$\sum_{j=1}^{n} \gamma_{i,j} \lambda_j + \theta = \gamma_{0,i} \quad i = 1, \ldots, n, \quad (2a)$$

$$\sum_{j=1}^{n} \lambda_j = 1, \quad (2b)$$

where θ is the Lagrange parameter and $\gamma_{i,j}$ is the semi-variance between catchment i and j (Isaaks and Srivastava, 1990). The semi-variance, or variogram, represents the spatial variability of the regionalized variable Z. Unique from any other method of kriging, top-kriging considers the variable defined over a non-zero support S, the catchment drainage area (Cressie, 1993; Skøien et al., 2006). The kriging system of Eqs. (2a) and (2b) remains the same, but the semi-variances between the measurements need to be obtained by regularization, i.e. smoothing the point variogram over the support area.

The point variogram can then be back-calculated by fitting aggregated variogram values to the sample variogram (Skøien et al., 2006). Pugliese et al. (2014) proposed a method for using top-kriging to predict FDCs at ungauged locations that they termed total negative deviation top-kriging (TNDTK). The authors reduce the dimensionality of the problem by seeking a unique index of site-specific FDCs. Unlike other regional approaches (e.g. regional regression of streamflow quantiles, see e.g. Castellarin et al., 2013), the kriging-based method interpolates the entire curve, therefore ensuring its monotonicity (see e.g. Pugliese et al., 2014; Castellarin, 2014). This is accomplished by first standardizing the empirical FDCs at site x, $\Psi(x, d)$, for some reference value, $Q^*(x)$, to yield a dimensionless FDC:

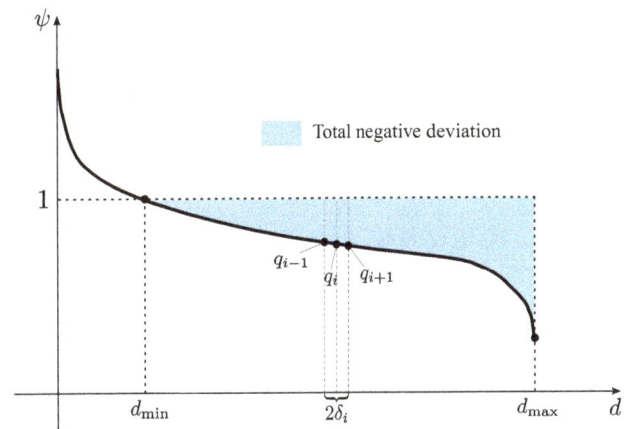

Figure 1. A sketch of the total negative deviation (TND).

$$\psi(x, d) = \frac{\Psi(x, d)}{Q^*(x)}, \quad (3)$$

where d denotes a specific duration. Pugliese et al. (2014) identified an overall point index that effectively summarizes the entire curve. This index, which the authors termed total negative deviation (TND), is derived by integrating the area between the lower limb of the FDC and the reference streamflow value Q^* (see Fig. 1). Empirical TND values are computed as follows:

$$\text{TND}(x) = \sum_{i=1}^{m} |q_i(x) - 1| \delta_i, \quad (4)$$

where $q_i = \frac{Q_i}{Q^*}$ represents the ith empirical dimensionless quantile standardized for the selected reference value Q^*, δ_i is half of the frequency interval between the $(i + 1)$th and $(i - 1)$th quantile and the summation involves only the m standardized quantiles lower than 1. The equality between a given streamflow value and the reference value Q^* is represented by a horizontal dashed line in Fig. 1, i.e. the threshold given by the equation $\frac{Q}{Q^*} = 1$. The range of the summation, m, in Eq. (4) is a function of the maximum duration $d_{\max}$, which is itself a function of that sample with minimum length across gauged sites in the study region. Having calculated empirical TNDs, Pugliese et al. (2014) propose using the TNDs as a regionalized variable to develop site-specific weighting schemes. The same weights, derived through the solution of the linear kriging system (Eqs. 2a and 2b), are used for a batch prediction of the continuous, dimensionless FDC for the ungauged site x_0:

$$\hat{\psi}(x_0, d) = \sum_{j=1}^{n} \lambda_j \psi(x_j, d) \quad \forall d \in (0, 1), \quad (5)$$

where λ_j, with $j = 1, \ldots, n$, is the weights resulting from the kriging interpolation of TNDs, $\psi(x_j, d)$ is the dimensionless empirical FDC at the donor site x_j, and $\hat{\psi}(x_0, d)$ is the predicted dimensionless FDC. It is worth highlighting that the computation of the linear kriging system (Eqs. 2a and 2b) depends on n, the number of neighbouring sites on which to base the spatial interpolation, a fact that will be explored below.

If a reliable model for predicting Q^* at the ungauged site x_0 can be developed, the prediction of the dimensional FDC, $\hat{\Psi}(x_0, d)$, is obtained as follows:

$$\hat{\Psi}(x_0, d) = \hat{Q}^*(x_0)\, \hat{\psi}(x_0, d) \quad \forall d \in (0, 1), \tag{6}$$

where $\hat{Q}^*(x_0)$ is the prediction of Q^* at the ungauged site x_0 and $\hat{\psi}(x_0, d)$ has the same meaning as in Eq. (5).

The main objective of our study is to improve rainfall–runoff model simulations, whereas an assessment of the reliability of the regional metric TND is out of the scope of this paper. Nevertheless, further details on TNDTK (i.e. cross-validations in different geomorphological and climatic regions, sensitivity analyses, comprehensive assessments and comparisons to state-of-the-art models for predicting FDCs in ungauged basins) are reported in recent studies to which an interested reader is referred (see e.g. Pugliese et al., 2014, 2016; Kim et al., 2017).

2.2 Algorithm for assimilation of local streamflow data

Following the approach proposed by Smakhtin and Masse (2000), we present a novel procedure for predicting the model residuals that may be associated with macro-scale rainfall–runoff model simulations (e.g. LISFLOOD, HYPE, PGB-IPH; see Introduction). This method relies on a regional prediction of the long-term FDC in the same site where these simulations are available.

For instance, let $\Psi(x_0, d)$ be the "true" unknown FDC for a given catchment x_0 and $\hat{\Psi}_{\mathrm{SIM}}(x_0, d)$ be its prediction constructed on the basis of the daily streamflows simulated through the macro-scale model. We can assume that a general relationship between the two curves exists and reads as follows:

$$\Psi(x_0, d) = \hat{\Psi}_{\mathrm{SIM}}(x_0, d) + \varepsilon(x_0, d) \quad \forall d \in (0, 1), \tag{7}$$

where $\varepsilon(x_0, d)$ are the model residuals defined over the duration domain d, which we may term the residual–duration curve (εDC). Evidently, although the "true" residual–duration curve is unknown at ungauged basins, one can nevertheless estimate such a curve on the basis of geostatistically interpolated flow–duration curves $\hat{\Psi}_{\mathrm{TNDTK}}(x_0, d)$ introduced in Sect. 2.1,

$$\hat{\varepsilon}(x_0, d) = \hat{\Psi}_{\mathrm{TNDTK}}(x_0, d) - \hat{\Psi}_{\mathrm{SIM}}(x_0, d) \quad \forall d \in (0, 1). \tag{8}$$

The estimated residual–duration curve obtained from the regional prediction of the long-term flow–duration curve can then be used for assimilating local streamflow information into the simulated daily streamflow series. The procedure is sketched in Fig. 2: (1) given a simulated streamflow series (red line in the top-right), select a specific day t and the corresponding discharge $Q(t)$; (2) retrieve the duration d associated with $Q(t)$ from the flow–duration curve constructed from simulated data (red line in the top-left quadrant); (3) read the estimated residual $\hat{\varepsilon}(t)$ off of the predicted residual–duration curve (blue line in the bottom-left quadrant); and (4) assimilate the residual into the simulated series as $Q(t) + \hat{\varepsilon}(t)$. The iteration of the algorithm through all time steps leads to an enhanced simulated series (blue line in the top-right quadrant).

This new assimilation procedure shares some analogies with a technique called "quantile mapping", used in the context of bias corrections for global climate model predictions (see e.g. Komma et al., 2007). The procedure we propose in this context, though, is a rather general tool that can be applied to, for example, any macro-scale rainfall–runoff model for locally enhancing long simulated streamflow series without the need to re-run computationally intensive simulations, provided that the model itself is behavioural and validated on the basis of streamflow data that were not available for model calibration. The performance of the assimilation procedure depends on a variety of drivers, e.g. the quality of the simulated streamflows, which can be severely impacted by the local quality of input data even for a behavioural and well-calibrated model, the quality of streamflow data and the density of the stream-gauging network (see Sect. 3.1.3), the accuracy of the chosen regional model for predicting FDCs (we refer to TNDTK herein, but there are other viable options, see e.g. Castellarin et al., 2013; Castellarin, 2014). Regarding the latter element, indeed, the proposed procedure deeply relies on accurate regional FDC predictions by means of an unbiased regional model, which necessarily must be validated beforehand for the area of interest. Otherwise, detriments are likely to be expected.

3 Testing the proposed algorithm: cross-validation procedures and sensitivity analysis

3.1 Structure of the analysis

The rationale of the analyses implemented in this study drives a sequence of operations, which can be summarized as follows:

1. Streamflow simulations from a given model, e.g. a rainfall–runoff macro-scale model, are available in a well-defined study area.

2. We suppose that recorded streamflow data are made available for a reasonable number of stream gauge stations within the study area, and we apply a suitable model for the regionalization of FDCs.

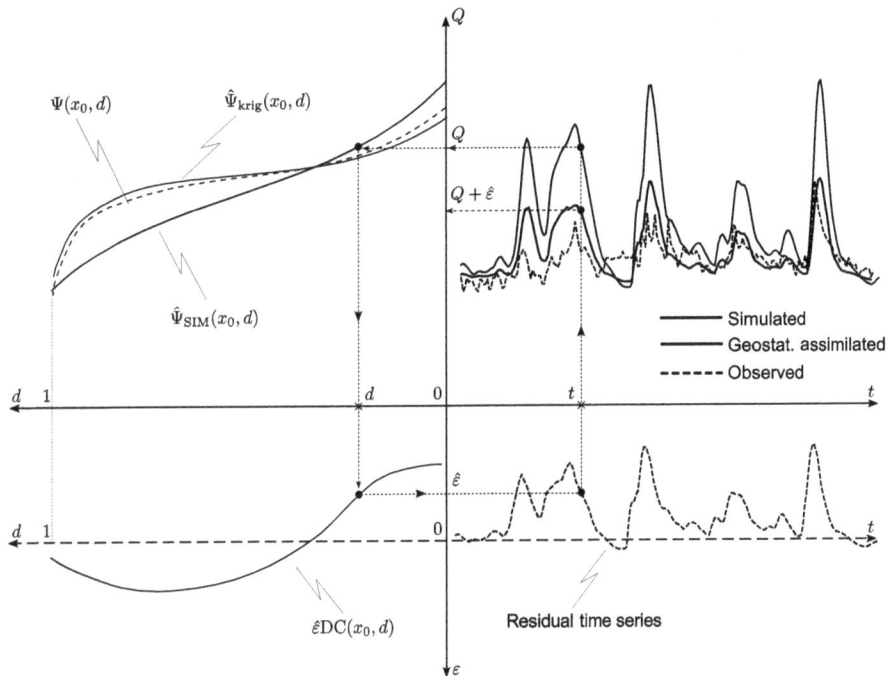

Figure 2. Illustration of the proposed data-assimilation procedure for a given simulated time series. Top-right panel: real streamflow series (unknown, since the basin is ungauged, black dashed line); macro-scale model simulation (red solid line); geostatistically enhanced streamflow series (blue solid line). Top-left panel: FDC predictions obtained from simulated streamflows (red solid line) and via geostatistical interpolation (black solid line), with real (unknown) FDC (black dashed line). Bottom-left panel: estimated residual–duration curve (blue solid line) computed as the difference between the two predicted FDCs in the top-left panel. Bottom-right panel: time series of residuals (blue dashed line).

3. We validate the regional model with respect to available streamflow observations. In this case we adopted a leave-one-out cross-validation (LOOCV; see e.g. Kroll and Song, 2013; Salinas et al., 2013; Wan Jaafar et al., 2011; Srinivas et al., 2008), even though different validation schemes might be preferred in other regions (see e.g. Pugliese et al., 2016; Castellarin et al., 2018).

4. We validate the assimilation procedure by sequentially (a) neglecting all the streamflow information at a given nodes of the river network, (b) predicting FDCs, and (c) applying the assimilation method illustrated in Sect. 2.2.

5. We evaluate the sensitivity of the assimilation procedure to the stream gauge density of the study area.

The methodologies adopted for addressing points 3–5 are illustrated in Sects. 3.1.1, 3.1.2, and 3.1.3, in this order; the accuracy of predictions (e.g. regional FDCs, assimilated and simulated streamflow series) relative to their empirical or observed counterparts is quantified through performance indices described in Sect. 3.2.

3.1.1 Cross-validation of the FDC geostatistical interpolator (point 3 above)

We proposed to assess the accuracy of the geostatistical predictor of FDCs (i.e. TNDTK, see Sect. 2.1) in cross-validation with respect to available streamflow observations from a sufficient number of gauged catchments. We chose the mean annual flow (MAF), computed as the average flow of recorded historical streamflow series, as the reference value Q^* (see details in Sect. 2.1).

TNDTK operates by first applying top-kriging to empirical TND values (see Sect. 2.1), which we performed by calculating a binned sample variogram first, and then by modelling binned empirical data with a five-parameter "modified" exponential theoretical variogram (a combination of an exponential and a fractal model; see details in Skøien et al., 2006). The fitted theoretical point variogram and its five parameters were obtained through the weighted least squares (WLS) regression method from Cressie (1993) by simultaneously fitting all regularized binned variograms that were computed for various area classes (in this case study we employed 2 variogram bins as a result of the range of drainage areas, which spans over 3 orders of magnitude; see details on binning methods in Skøien, 2018). Recent applications of TNDTK indicate $n = 6$ as an optimal number of neighbouring donor stations, and thus we chose the same value for this case study as well (see details in Pugliese et al., 2014, 2016). Then, TNDTK uses the kriging weights obtained for predicting TND values for interpolating the dimensionless FDCs at the location of interest as the weighted average of dimensionless empirical FDCs constructed from the $n = 6$ neighbouring gauged sites (see Eq. 5, in which λ_j, with $j = 1, \ldots, 6$ and $n = 6$, is the kriging weights). While the computation of TNDs does not require any specific resampling scheme of the FDCs, the prediction of the curves in ungauged locations is carried out using a fixed number of quantiles that should be selected to thoroughly represent the variability from high to low flows. Thus, observed FDCs are resampled to 20 equally spaced points, in the normal space, leading to the widest range of durations compatible with the shorter observed streamflow series in the dataset. We adopted a LOOCV procedure (see e.g. Pugliese et al., 2014, 2016) to test the accuracy and uncertainty associated with FDC predictions. This simulates the ungauged conditions at each and every gauged site in the study area by (1) removing it in turn from the dataset and (2) referring to the $n = 6$ neighbouring gauges for predicting its dimensionless FDC. Given that the geostatistical assimilation procedure uses dimensional FDCs, we also tested the suitability of standard top-kriging for predicting MAF at ungauged locations in the study area, still through a LOOCV procedure (general validity of top-kriging for predicting mean annual flows is also described in Blöschl et al., 2013). For MAF interpolation, we adopted the same settings used for predicting TND values (i.e. a five-parameter "modified" exponential theoretical variogram and $n = 6$ neighbouring sites). We then used cross-validated dimensionless FDCs and MAF predictions at each and every gauging station in the study area to obtain cross-validated predictions of dimensional FDCs for each measuring node through Eq. (6).

3.1.2 Cross-validation of the geostatistical assimilation procedure (point 4 above)

We applied the proposed assimilation procedure as outlined in Sect. 2.2, by, firstly, assessing the efficiency of the procedure through a leave-one-out cross-validation. We predicted the FDC associated with each simulation node by using TNDTK and by, also, neglecting the hydrological information coming from the closest (i.e. immediately upstream or downstream) gauged catchment, therefore assuming that no streamflow information is available near the simulation node. The workflow of the validation algorithm is as follows:

a. We select one pair among n_p possible pairs, let us term it pair i–j, where i stands for simulation node and j stands for the corresponding stream gauge.

b. We drop the daily streamflow series observed at stream gauge j from the set of observed series.

c. We interpolate FDC at simulation node i through TNDTK (as illustrated in Sect. 2.1) using the remaining $n_g - 1$ gauged sites, where n_g is total number of stream gauges in the study region.

d. We apply the assimilation procedure outlined in Sect. 2.2 and depicted in Fig. 2 to the streamflow series simulated for the simulation node i.

e. We compare the original simulated daily streamflow series and the geostatistically enhanced one at prediction node i with the daily streamflow series observed at stream gauge j times the corresponding area ratio A_i/A_j (i.e. A_i is the drainage area of simulation node i, A_j is the drainage area of stream gauge j; see also Sect. 4.1).

f. We repeat all previous steps for each one of the remaining n_p pairs.

For the sake of consistency, we anticipate here that we will refer to the procedure presented above as GAE-HYPE (i.e. geostatistically assimilated E-HYPE) in the remainder of the paper. The acronym clearly recalls the rainfall–runoff model used in this study (see Sect. 4.2); however, the procedure disregards a specific rainfall–runoff model.

Finally, it is worth highlighting here that FDCs obtained from either the geostatistical model or the rainfall–runoff simulation model are resampled to 20 equally spaced points across the normally transformed duration intervals (see details in Pugliese et al., 2014). Thus, as a result, the produced εDCs reflect the same sampling scheme of the curves. Nevertheless, the procedure does not foresee any restriction to the resolution of the resampled curve, allowing for a finer resampling scheme in other analyses.

3.1.3 Stream-gauging network density and effectiveness of geostatistical data assimilation (point 5 above)

Since the proposed geostatistical data-assimilation procedure (GAE-HYPE) relies upon the local availability of stream gauges records, understanding to what extent the performances of the assimilation method are driven by gauging network density is a fundamental of paramount importance. Therefore, we performed a sensitivity analysis and assessed the degree of enhancement of simulated daily streamflow sequences associated with different scenarios of streamflow data availability, repeating for each scenario the procedure described in Sect. 3.1.2. Thus, we randomly discarded some of the gauges available over the study area and varied the total number of available stations continuously from the lowest to the highest gauge density. At each density scenario, we performed exactly the same kriging settings and same LOOCV illustrated in detail in Sect. 3.1.1 and 3.1.2, respectively.

3.2 Performance indices

We assessed the performances for predicting regional FDCs by means of Nash–Sutcliffe efficiency (Nash and Sutcliffe, 1970) computed for log-transformed streamflows (LNSEs). These indices are computed as follows:

$$
\mathrm{LNSE}_{\mathrm{FDC},j} = 1 - \frac{\sum\limits_{k=1}^{n_\mathrm{d}} \left(\ln \Psi \left(x_j, d_k \right) - \ln \hat{\Psi} \left(x_j, d_k \right) \right)^2}{\sum\limits_{k=1}^{n_\mathrm{d}} \left(\ln \Psi \left(x_j, d_k \right) - \mu_j \right)^2},
$$
$$
j = 1, \ldots, n_\mathrm{g}, \tag{9}
$$

where $\Psi(x_j, d_k)$ and $\hat{\Psi}(x_j, d_k)$ are the empirical and the predicted kth streamflow quantiles at site x_j, respectively, μ_j is the mean of empirical log-transformed streamflow quantiles at site x_j, n_d is the number of discretization points throughout duration range, and n_g is the number of stream gauges.

Another useful metric of performance for the assessment of FDC predictions is the overall absolute curve error (see Ganora et al., 2009), which reads as follows:

$$
\delta_{\mathrm{FDC},j} = \sum_{k=1}^{n_\mathrm{d}} \left| \Psi \left(x_j, d_k \right) - \hat{\Psi} \left(x_j, d_k \right) \right| \quad j = 1, \ldots, n_\mathrm{g}, \tag{10}
$$

where $\Psi(x_j, d_k)$ and $\hat{\Psi}(x_j, d_k)$ have the same meaning as in Eq. (9).

Similarly, concerning streamflow time series, the assessment of modelled streamflows is carried out with LNSE, but, in this case, it reads as follows:

$$
\mathrm{LNSE}_{\mathrm{mod},j} = 1 - \frac{\sum\limits_{t=1}^{t_\mathrm{s}} \left(\ln Q_{\mathrm{emp},j}(t) - \ln Q_{\mathrm{mod},j}(t) \right)}{\sum\limits_{t=1}^{t_\mathrm{s}} \left(\ln Q_{\mathrm{emp},j}(t) - \omega_j \right)^2},
$$
$$
j = 1, \ldots, n_\mathrm{p}, \tag{11}
$$

where $Q_{\mathrm{emp},j}(t)$ and $Q_{\mathrm{mod},j}(t)$ are the empirical and predicted streamflow at site x_j and time t, respectively, ω_j is the mean of empirical log-transformed streamflow at site x_j, t_s and n_p is the number of selected simulation nodes.

Furthermore, we assessed the efficiency of the data-assimilation procedure through the following metric:

$$
\mathrm{LNSE}_{\mathrm{ratio},j} = \frac{\mathrm{LNSE}_{\mathrm{mod}2,j} - \mathrm{LNSE}_{\mathrm{mod}1,j}}{1 - \mathrm{LNSE}_{\mathrm{mod}1,j}},
$$
$$
j = 1, \ldots, n_\mathrm{p}. \tag{12}
$$

$\mathrm{LNSE}_{\mathrm{ratio}}$ quantifies the degree of enhancement of "model 2" relative to "model 1" standardized by the maximum possible improvement (i.e. $1 - \mathrm{LNSE}_{\mathrm{mod}1}$). An $\mathrm{LNSE}_{\mathrm{ratio}}$ close to zero means no significant enhancement (detriment of original sequences in case of negative values), whereas an $\mathrm{LNSE}_{\mathrm{ratio}}$

close to 1 indicates that no further enhancement is possible. Such an index is derived from the reciprocal root-mean-squared error ratio between the two models.

Finally, in order to verify whether or not the proposed assimilation procedure GAE-HYPE outperforms rainfall–runoff simulations, for different gauge density scenarios (see Sect. 3.1.3), we used the Wilcoxon signed-rank test with the null hypothesis that simulation model LNSEs are greater than GAE-HYPE ones at 5 % significance level (Hollander and Wolfe, 1999).

4 Data

4.1 Study region

We focus on a large alpine region located in Tyrol (Italy, Austria and, for small portion only, Switzerland). Our analyses consider two types of data, observed daily streamflows and E-HYPE simulated daily streamflows (see Sect. 4.2), representing different sets of catchments (Fig. 3). In this study, E-HYPE represents the rainfall–runoff model selected to evaluate the procedure presented in Sect. 2.2, which has shown significantly poor results in this region. Indeed, this alpine area is particularly suitable for hydro-power generation, and therefore the presence of dams along the stream network could likely alter the streamflow regime downstream, producing a significant alteration of the natural flow conditions. E-HYPE only simulates the dams present in the global database of GranD (Lehner et al., 2011), which might not be representative for hydropower production at the local scale (Arheimer et al., 2017). Thus, we removed from the initial group of gauged catchments all basins for which the streamflow regime is highly or significantly altered by upstream dams. Table 1 reports the main characteristics of streamflow regimes for 46 gauged basins and 11 selected E-HYPE prediction nodes. Among all E-HYPE prediction nodes available in Tyrol we selected only those whose catchments were the closest to gauged ones, i.e. difference in terms of drainage areas < 14 % and distance between catchment centroids < 15 km. These criteria resulted in the selection of 11 E-HYPE prediction nodes that are evenly distributed in the study region (see red lines in Fig. 3). We addressed the limited differences existing between drainage areas of E-HYPE and gauged basins by adopting the drainage-area ratio technique (DAR, see e.g. Farmer and Vogel, 2013), that is by rescaling daily streamflows according to drainage areas of the corresponding catchment. Such a method assumes the same unit daily streamflow for any pair of hydrologically similar catchments i and j, which reads as follows:

$$
\frac{Q_i(t)}{A_i} = \frac{Q_j(t)}{A_j}, \tag{13}
$$

where $Q_i(t)$ represents the daily streamflow at day t for catchment i, and $Q_j(t)$ is the daily streamflow at day t for

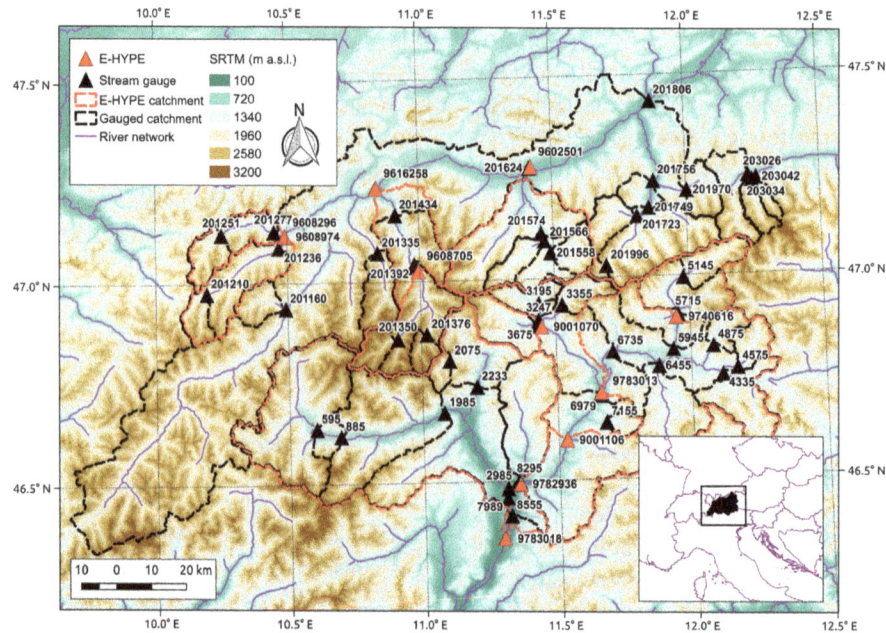

Figure 3. Study area: Tyrol. Catchment boundaries for 11 E-HYPE prediction nodes (red) and 46 stream gauges (black).

Table 1. Study catchments: streamflow properties standardized by drainage area ($m^3 s^{-1} km^{-2}$) for either gauged catchments or E-HYPE catchments, mean annual flow (q_{MAF}), and 50 % and 95 % streamflow quantiles (q_{50} and q_{95}, respectively).

	Gauged catchments (46)			E-HYPE catchments (11)		
	q_{MAF}	q_{50}	q_{95}	q_{MAF}	q_{50}	q_{95}
Min.	0.0147	0.0078	0.0023	0.0261	0.0057	0.0009
25th percentile	0.0205	0.0158	0.0046	0.0276	0.0117	0.0022
Median	0.0309	0.0188	0.007	0.0294	0.0169	0.0032
Mean	0.0315	0.0195	0.0066	0.034	0.0168	0.0028
75th percentile	0.0369	0.0221	0.008	0.0351	0.0223	0.0035
Max.	0.0588	0.043	0.0116	0.0622	0.0275	0.0047

catchment j. In our application, i and j could correspond to any given pair (stream gauge, E-HYPE prediction node), and A_i and A_j the corresponding drainage areas. Finally, it is worth pointing out that neither the observed nor the E-HYPE series present zero values in their recording (or simulation) periods for each of the 11 selected nodes.

4.2 Pan-European rainfall–runoff simulation: E-HYPE

The HYdrological Predictions for the Environment (HYPE) model is a hydrological model for small-scale and large-scale assessments of water resources and water quality, developed at the Swedish Meteorological and Hydrological Institute (SMHI) during 2005–2007 (Lindström et al., 2010). The European application, E-HYPE, has been proved to be a powerful tool for water resources managers and practitioners, addressing nutrient concentration in river flow as well as water forecasts on short or seasonal timescale. It is also

widely used to estimate snow storage and accumulated TWh (terawatt hour) of water inflow to hydropower dams and in climate change impact analysis (Donnelly et al., 2017). The website Hypeweb (http://hypeweb.smhi.se, last access: 30 July 2018) provides visualization and free downloading of 30 years of continuous and consistent daily streamflow simulations across the European river network at rather fine scale (i.e. the average size of elementary catchments is equal to 215 km^2) as well as forecasts and climate change impact analysis.

The HYPE model is open-access and can be downloaded with documentation and model set-up guidelines from the model website (http://hypecode.smhi.se/, last access: 30 July 2018). It simulates water flow and substances on their way from precipitation through soil, river and lakes to the river outlet (Lindström et al., 2010). River basins are divided into sub-basins, which in turn are divided into classes (the finest

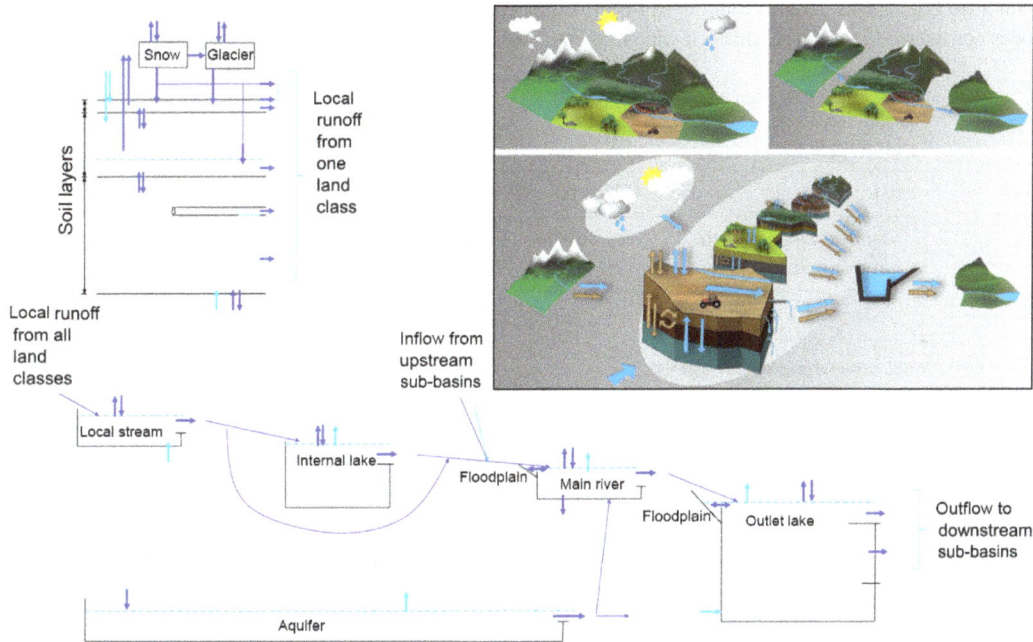

Figure 4. Schematic concept of the HYPE model (all equations are available at http://hypecode.smhi.se/, last access: 30 July 2018).

calculation units) depending on land use, soil type and elevation (Fig. 4).

The soil is modelled as several layers, which may have different thicknesses for each class. In E-HYPE, each sub-basin can have up to some 40 soil and land-use classes, which are lumped within the sub-basins, while the watercourses are routed through the river network. The model parameters can be associated with land use (e.g. evapotranspiration) or soil type (e.g. water content in soil), or be common for the whole catchment or a region with geophysical similarities (Hundecha et al., 2016). This way of coupling the parameters with geographic information makes the model better suited for simulations in ungauged catchments.

5 Results

The application of the geostatistical method TNDTK through an LOOCV procedure reveals an agreement between empirical values and predictions as shown in Figs. 5 and 6. Specifically, Fig. 5 reports empirical (x axis) against geostatistically predicted (y axis) MAF values as well as LNSEs obtained in cross-validation (i.e. 0.96). Figure 6a shows a scatter diagram between observed (x axis) and predicted streamflows (y axis) from FDCs. This agreement is also confirmed by the distributions of on-site LNSE values (see box plots in Fig. 6b); the median LNSE is equal to 0.97, while mean LNSE is ca. 0.90. The performance obtained in cross-validation legitimizes the use of TNDTK for predicting FDCs in the study area at the 11 E-HYPE prediction nodes of interest, for which TNDTK delivers high prediction capability, with LNSE val-

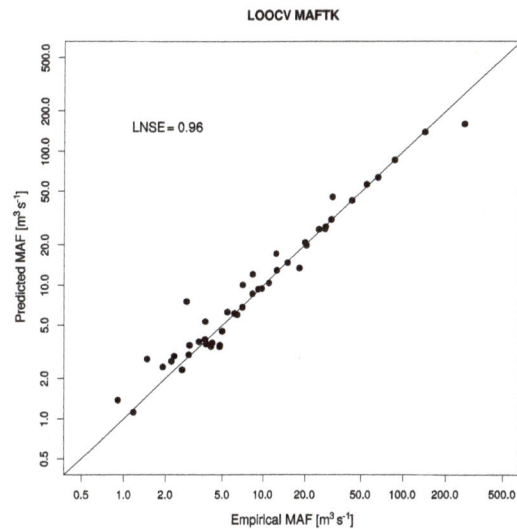

Figure 5. Top-kriging predictions of mean annual flow (MAF) in cross-validation mode.

ues above 0.97 (see the spatial distribution of efficiency values in Fig. 10a).

Figure 7 reports the results obtained by applying the aforementioned cross-validation algorithm. Circles in Fig. 7a represent the cumulative absolute error δ (see Eq. 10) computed for each catchment pair $i-j$, between empirical FDCs and predicted FDCs for either E-HYPE (δ_{EHYPE} on the y axis) or TNDTK (δ_{TNDTK} on the x axis) predictions. This figure clearly shows that for 9 out of 11 target sites the geostatistical method TNDTK outperforms E-HYPE in pre-

Table 2. Nash–Sutcliffe Efficiencies computed on log-transformed daily streamflows for E-HYPE and GAE-HYPE: median values for the 11 prediction nodes considered in the study; smallest enhancement (IDs 3675 – 9001070), largest enhancement (IDs 201236 – 9608296).

LNSE			E-HYPE	GAE-HYPE
Median			0.045	0.685
Pair	GAUGE ID	E-HYPE ID		
Smallest enhancement	3675	9001070	0.527	0.69
Largest enhancement	201236	9608296	−0.462	0.594

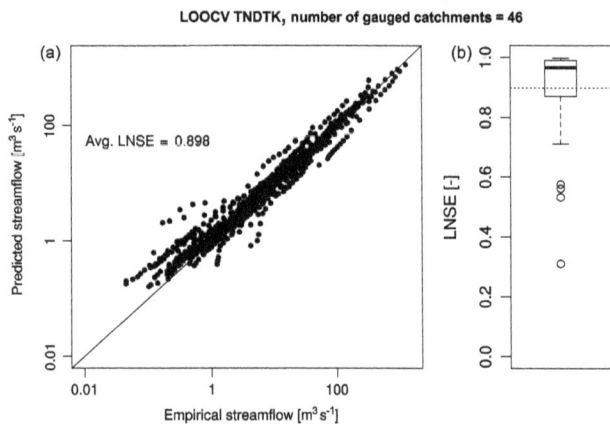

Figure 6. (a) Scatter diagrams of empirical (*x* axis) vs. predicted (*y* axis) streamflows. **(b)** Box-plot representation of on-site LNSE values, summarizing the first, second (median) and third quartiles along with whiskers extending to the most extreme non-outlying data point (outliers are highlighted as circles and lay at more than 1.5 times the interquartile-range from the nearest quartile); the average on-site LNSE value is reported in **(a)** and illustrated as a dashed line in **(b)**.

dicting FDCs. Moreover, one of the two sites for which E-HYPE outperforms TNDTK shows nearly the same performance as TNDTK (i.e. the circle is very close to the 1 : 1 line), while the other one (i.e. site 3675-9001070, highlighted with a black dot in the figure) is associated with the worst performance of TNDTK relative to E-HYPE (see also Sect. 6 on this). Figure 7b reports estimated residual–duration curves ($\hat{\varepsilon}$DCs) for the selected sites. For the sake of representation, we report standardized residuals in the *y* axis, i.e. residuals divided by the corresponding streamflow quantiles predicted via TNDTK; we referred to TNDTK quantiles for standardization since the real empirical FDC is supposed to be unknown (see cross-validation algorithm illustrated in Sect. 3.1.2). Overall, $\hat{\varepsilon}$DCs show negative values for lower durations and positive values for higher durations (see also Fig. 8). This means that, in Tyrol, E-HYPE tends to overestimate streamflow in wet periods as well as to underestimate streamflows in drier ones relative to the geostatistically predicted FDCs (i.e. TNDTK, see the left panels in Fig. 8). We eventually used the $\hat{\varepsilon}$DC curves, which are estimates of E-

HYPE residuals, to assimilate locally available streamflow data into E-HYPE simulated series as illustrated in Fig. 2, obtaining what we termed GAE-HYPE simulations (see right panels of Fig. 8).

Representativeness of simulations (i.e. E-HYPE and GAE-HYPE simulated daily streamflows) is assessed through a Nash–Sutcliffe efficiency of log-flows (LNSE) computed by referring to a recorded streamflow time period of the paired stream gauge (see pairing method adopted in Sect. 4.1). Improvements are obtained with the proposed data-assimilation procedure relative to E-HYPE (Fig. 9). Indeed, we obtained an enhancement of LNSE values of GAE-HYPE simulations relative to the original E-HYPE ones for all the 11 selected sites, which show LNSE increments from −0.462 to 0.594 in the best case (catchment pair IDs: 201236–9608296) and from 0.527 to 0.690 in the worst case (catchment pair IDs: 3675–9001070, see also Table 2). The median on-site LNSE value increases from 0.045 to 0.685, which ultimately underlines the benefits introduced with the proposed method. Figure 9, also, illustrates the impact of geostatistical data assimilation for the two E-HYPE prediction nodes mentioned above (i.e. the one characterized by the best improvements in terms of overall LNSE value, and the one associated with the most limited improvement).

Moreover, looking at the spatial distribution of LNSE values across the 11 selected prediction nodes within the study area depicted in Fig. 10, it is clear how the proposed enhancement strategy benefits from the unbiased estimations of FDCs. In fact, TNDTK shows homogeneous and rather high performance for predicting FDCs (Fig. 10a); also, Fig. 10b and c reveal that the enhancement capabilities of the assimilation procedure are lower for those catchments where E-HYPE performs better (see elementary catchments filled in yellow to green in Fig. 10), whereas the assimilation procedure proves to be powerful when E-HYPE performs worse (see elementary catchments coloured in orange to red), bringing efficiencies from negative to positive values in all cases (from green to blue).

The assessment of gauge density impacts on the proposed procedure reveals that enhancements are obtained even with the lowest gauge density scenario (i.e. seven gauging stations). Figure 11 displays a clear pattern, showing an improvement in the degree of enhancement associated

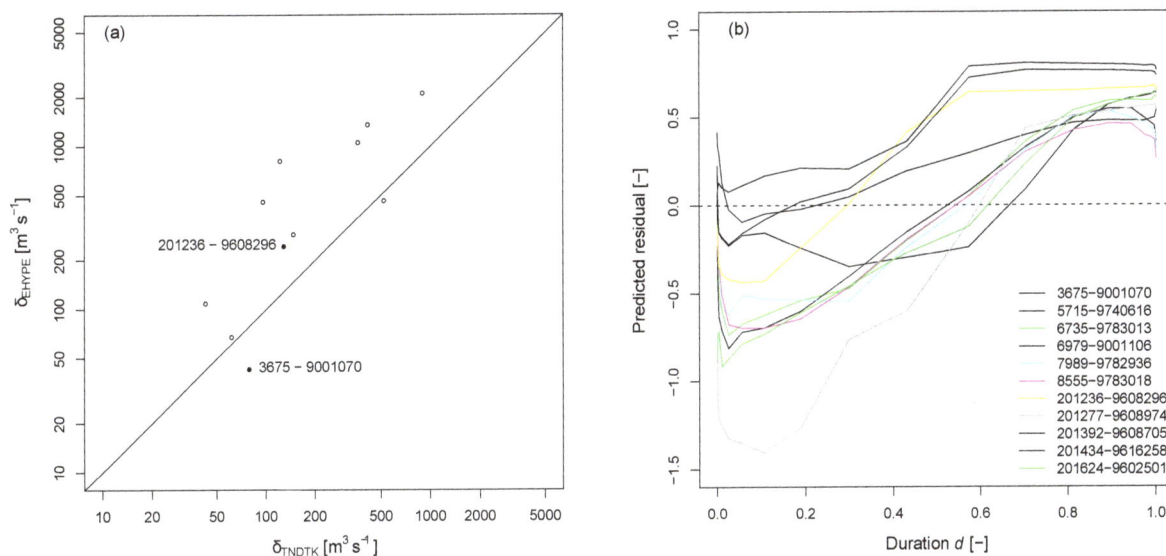

Figure 7. 11 E-HYPE prediction nodes: **(a)** comparison between TNDTK (x axis) and E-HYPE (y axis) in terms of distances δs between empirical and predicted FDCs; the 1 : 1 line represents equivalent performance for TNDTK and E-HYPE; **(b)** standardized residual–duration curves computed as illustrated in Eq. (8) (TNDTK streamflow quantiles are used for standardization).

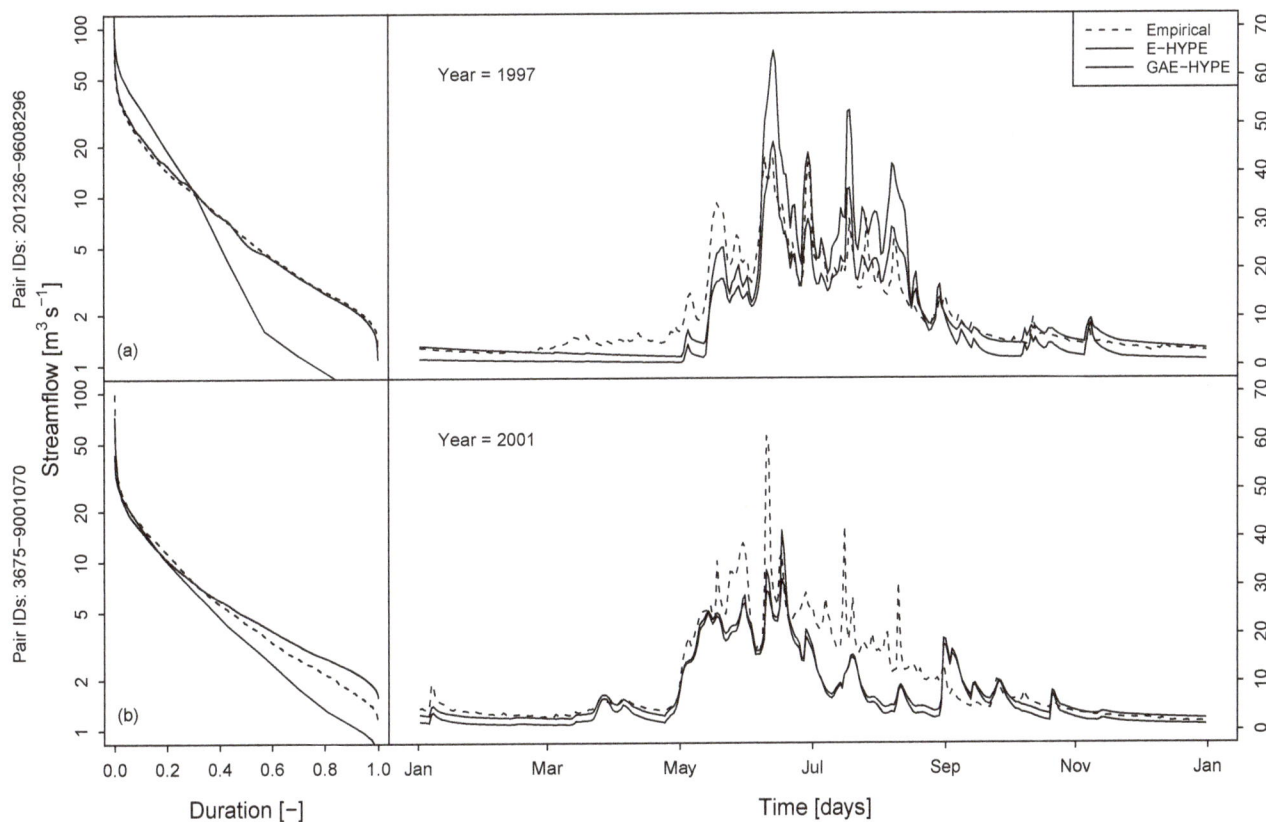

Figure 8. Examples of comparison between observed streamflow series (black dashed lines) and simulated daily streamflows via E-HYPE (red solid lines) and GAE-HYPE (blue solid lines) for two representative sites and a given year, showing two cases for which the geostatistical assimilation procedure resulted in sizeable **(a)** and limited **(b)** improvements, respectively.

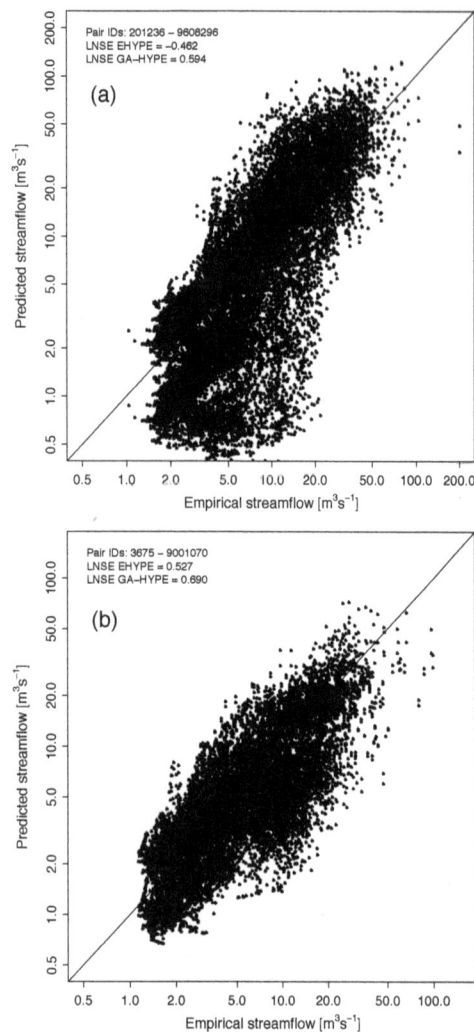

Figure 9. Scatter diagrams of empirical vs. simulated daily stream-flows for either E-HYPE (red dots) and GAE-HYPE (blue dots) for two representative sites, showing the cases in which the data-assimilation procedure respectively produced the largest **(a)** and smallest **(b)** degree of enhancement for the study area, respectively.

with an increasing gauge density. Moreover, Fig. 11 shows how the degree of enhancement flattens out in cases in which there are more than approximately 25 gauges available per $10\,000\,\mathrm{km}^2$. Finally, the p values resulting from the Wilcoxon signed-rank tests (see Sect. 3.2) highlight that the assimilation procedure outperforms E-HYPE: the null hypothesis is rejected, with p values always lower than 0.04 %, regardless of the particular stream gauge density scenario.

6   Discussion

This new geostatistical procedure enables practitioners and water resources managers and planners to profit from the

wealth of hydrological information, by adjusting open-data products with local observations. We enhanced the streamflow series simulated by macro- and continental-scale rainfall–runoff models at ungauged prediction nodes by assimilating streamflow observations, which are locally available in the region of interest, without having to redo the original hydrological model calculations. This is a recurrent condition since local streamflow data are released under different license terms and policies: some of them could be public and open-access, while some others might not be openly and freely accessible by the broad public. The E-HYPE model obviously lacks storage capacity in the Tyrol region and the proposed approach to enhance the results should be seen as temporary until a new model version accounting for this is available. We do not propose the procedure as a general fix for structurally unsuitable (or non-behavioural, see Beven and Binley, 1992) models, which have been proved to be unfit for either the area of interest or the water problem at hand. For a more sustainable solution, we suggest using another model structure or re-calibrating the model, instead of postprocessing the output. However, this procedure makes sense for making a first assessment of water issues in regions where information is otherwise missing, but only macro-scale models are readily available.

Our study shows for Tyrol that it is possible to significantly enhance rainfall–runoff simulations resulting from macro-scale, regional or continental hydrological models by geostatistically assimilating (geographically sparse) streamflow observations (see e.g. Figs. 8, 9 and 10); provided that available streamflow series are long enough to obtain a good empirical approximation of the long-term FDC for the site of interest (i.e. 5–10 years; see Castellarin et al., 2013). Indeed, series length of the observed streamflow dataset controls the magnitude of duration extremes (i.e. duration boundary interval), which, in turn, might affect the adopted resampling scheme needed for predicting FDCs at simulation nodes. Nonetheless, in some specific application, very high or very low durations might be of particular interest (e.g. studies focused on flood or drought only); therefore a preliminary investigation on the resampling scheme (e.g. duration extremes, duration intervals, resolution of the points to represent correctly the whole curve) should always be taken into account.

One of the main advantages of the proposed method is that the end user can get enhanced streamflow simulations without any further model calibration or refinement. Even though one could argue that when additional streamflow data become available at neighbouring gauges it should be used for improving the performance of the model at the site of interest, calibrating and validating macro-scale and regional model could be a time-consuming and computationally demanding task. The proposed procedure, instead, is neither computational nor data-intensive, and is implemented only using observed streamflow data and a GIS vector layer with

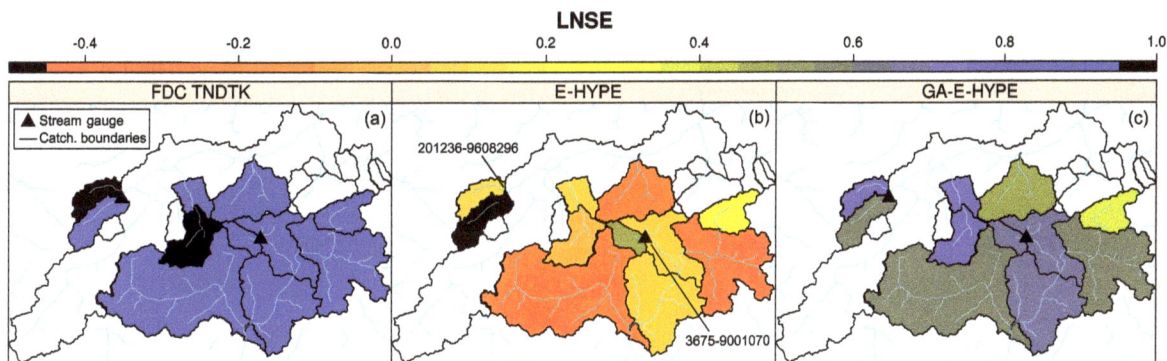

Figure 10. Spatial distribution of Nash–Sutcliffe efficiency computed for log-transformed streamflows (LNSEs) at the 11 E-HYPE prediction nodes considered in the study: geostatistically predicted flow–duration curves (FDC TNDTK, **a**); predicted daily streamflow time series (E-HYPE, **b**, and GAE-HYPE, **c**, respectively); the locations of the two sites considered in Figs. 9 and 8 are highlighted with black triangles.

catchment boundaries (see e.g. Fig. 3). The application requires the identification of a suitable regional model for predicting FDC in ungauged basis (see e.g. Fig. 6). However, it has advantages, such as (a) a regional model can be a very informative and useful tool for water resources managers and planners, and (b) the subsequent advantages obtained from the data-assimilation procedure is transferred downstream in the entire regional river network (see Fig. 11).

One important limitation of the proposed method is that, once a target prediction node is considered, any given simulated streamflow value is associated with a single duration, which corresponds to a particular estimated residual, which will be used in turn for correcting the streamflow value itself (see Figs. 2 and 8). Essentially, this means that the volumes from E-HYPE are discarded while the sequencing of E-HYPE simulations is retained. Moreover, this algorithm cannot possibly account for seasonal (or interannual) modifications in the hydrological behaviour of the catchment. Indeed, as shown in the time series comparison in Fig. 8, when the geostatistical prediction of FDCs is unreliable, the assimilation procedure reflects such low accuracy (i.e. the procedure fails to correctly capture high-flow and low-flow regimes; see e.g. the resulting FDC from GAE-HYPE simulations at the catchment pair 3675–9001070 in Fig. 8), propagating this bias throughout the whole simulated series. Finally, designing a theoretical framework that combines statistical data-driven approaches with deterministic process-driven ones is seen by many as the correct way for tackling the "prediction in ungauged basins" (PUB) problem and further advancing the scientific research in this area (see e.g. Di Prinzio et al., 2011). We believe that our geostatistical data-assimilation procedure for macro-scale hydrological models is one example in this direction. Future analyses will focus on the relaxation of the main limitation of the approach (i.e. the incorporation of seasonal patterns in a data-assimilation procedure) and on the extension of its applicability to anthropologically altered streamflow regimes.

7 Conclusions

This research work focuses on the development of an innovative method for enhancing streamflow series simulated by macro-, continental-, and global-scale rainfall–runoff models by means of a geostatistical prediction of model residuals. We focus on Tyrol as study region and E-HYPE (European-HYdrological Predictions for the Environment, from the Swedish Meteorological and Hydrological Institute, SMHI) as a macro-scale hydrological model; nevertheless, the geostatistical data-assimilation procedure is general and can be applied to simulated streamflow series coming from other macro-scale rainfall–runoff models. The proposed data-assimilation procedure utilizes streamflow data that are locally available for the area of interest, which were not considered in the implementation of the macro-scale hydrological model; it (1) adopts top-kriging for regionally interpolating empirical period-of-record flow–duration curves (FDCs) that can be constructed from locally available streamflow data; (2) constructs residual–duration relationships at any prediction node in the study region where simulated streamflow series are available, by comparing FDCs resulting from geostatistical interpolation (top-kriging) and rainfall–runoff simulation (E-HYPE); and (3) uses the residual–duration curve to enhance macro-scale simulated streamflows.

The cross-validation tests of the proposed approach with different scenarios of streamflow data availability show the significant advantages of geostatistical data assimilation even for very low stream-gauging network densities (i.e. ca. 1 gauge per 2000 km^2). It can become a stand-alone numerical tool to be used for enhancing results from macro-scale models anywhere along the stream network of a given region. Potential applications are envisaged for a variety of water resources management and planning problems that require accurate streamflow series (e.g. regional assessment of hydropower potential, habitat suitability studies, surface

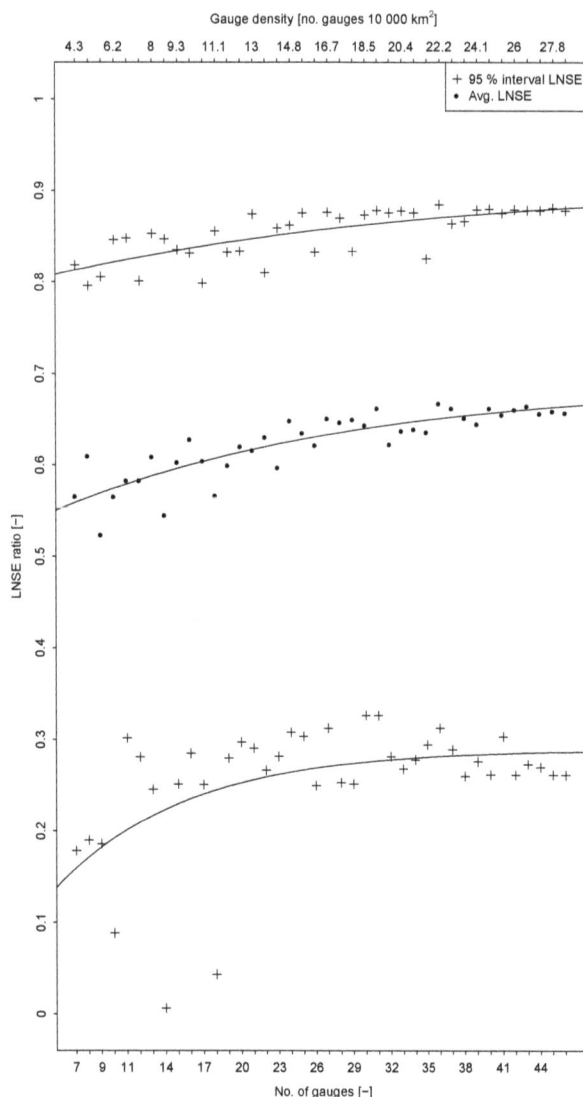

Figure 11. LNSE ratio (see Eq. 12) as a function of stream gauge availability: black dots represent the average of 11 LNSE ratio values, while crosses indicate their 95 % confidence interval.

water allocation, civil protection management strategies, climate change trends, safety of river structures). Future analyses will address the main limitation of the proposed geostatistical data-assimilation procedure, aiming to incorporate observed seasonal and inter-annual variations of the hydrological behaviour of the study region as well as other hydrological features (e.g. baseflow index or peak-flow data) into the geostatistical regionalization of model residuals.

Code and data availability. The analysis was carried out in the Virtual Water Science Lab developed within the FP7 funded research project SWITCH-ON (grant agreement no. 603587). We invite the interested reader to explore the experiment protocol here: http://dl-ng005.xtr.deltares.nl/view/462/ (Pugliese et al., 2017).

Author contributions. AP performed the numerical analyses, data handling and homogenization. SP contributed to the algorithm coding. AC, AP, AM, SP, BA, RC, GB, and JP had an essential role in designing the framework of the analysis, developing methods and testing procedures, while SB and PM contributed to large dataset pre-processing. All authors made a substantial contribution to the critical interpretation of results and provided important ideas to further improve the study. All authors actively took part in drafting, writing, and revising the paper.

Competing interests. The authors declare that they have no conflict of interest.

Acknowledgements. The analyses presented in the study are the main outcomes of the international experiment Geostatistical enhancement of European hydrological predictions (GEEHP), which was performed in the Virtual Water-Science Lab developed within the European Commission FP7 funded research project SWITCH-ON (Sharing Water-related Information to Tackle Changes in the Hydrosphere – for Operational Needs, grant agreement no. 603587). The overall aim of the project is to promote data sharing to exploit open data sources. The study also contributes to developing the framework of the "Panta Rhei" Research Initiative of the International Association of Hydrological Sciences (IAHS).

Edited by: Fuqiang Tian

References

Alcamo, J., Döll, P., Henrichs, T., Kaspar, F., Lehner, B., Rösch, T., and Siebert, S.: Development and testing of the WaterGAP 2 global model of water use and availability, Hydrolog. Sci. J., 48, 317–337, https://doi.org/10.1623/hysj.48.3.317.45290, 2003.

Alfieri, L., Burek, P., Dutra, E., Krzeminski, B., Muraro, D., Thielen, J., and Pappenberger, F.: GloFAS – global ensemble streamflow forecasting and flood early warning, Hydrol. Earth Syst. Sci., 17, 1161–1175, https://doi.org/10.5194/hess-17-1161-2013, 2013.

Archfield, S. A., Pugliese, A., Castellarin, A., Skøien, J. O., and Kiang, J. E.: Topological and canonical kriging for design flood prediction in ungauged catchments: an improvement over a traditional regional regression approach?, Hydrol. Earth Syst. Sci., 17, 1575-1588, https://doi.org/10.5194/hess-17-1575-2013, 2013.

Archfield, S. A., Clark, M., Arheimer, B., Hay, L. E., McMillan, H., Kiang, J. E., Seibert, J., Hakala, K., Bock, A., Wagener, T., Farmer, W. H., Andréassian, V., Attinger, S., Viglione, A., Knight, R., Markstrom, S., and Over, T.: Accelerating advances in continental domain hydrologic modeling, Water Resour. Res., 51, 10078–10091, https://doi.org/10.1002/2015WR017498, 2015.

Arheimer, B., Wallman, P., Donnelly, C., Nyström, K., and Pers, C.: E-HypeWeb: Service for Water and Climate Information – and Future Hydrological Collaboration across Europe?, in: Environmental Software Systems. Frameworks of eEnvironment, IFIP Advances in Information and Communication Technology, Springer, Berlin, Heidelberg, 657–666, https://doi.org/10.1007/978-3-642-22285-6_71, 2011.

Arheimer, B., Donnelly, C., and Lindström, G.: Regulation of snow-fed rivers affects flow regimes more than climate change, Nat. Commun., 8, 62, https://doi.org/10.1038/s41467-017-00092-8, 2017.

Beck, H. E., van Dijk, A. I. J. M., de Roo, A., Miralles, D. G., McVicar, T. R., Schellekens, J., and Bruijnzeel, L. A.: Global-scale regionalization of hydrologic model parameters, Water Resour. Res., 52, 3599–3622, https://doi.org/10.1002/2015WR018247, 2016.

Beven, K. and Binley, A.: The future of distributed models: Model calibration and uncertainty prediction, Hydrol. Process., 6, 279–298, https://doi.org/10.1002/hyp.3360060305, 1992.

Blöschl, G., Sivapalan, M., Thorsten, W., Viglione, A., and Savenije, H.: Runoff prediction in ungauged basins: synthesis across processes, places and scales, Cambridge University Press, Cambridge, 2013.

Castellarin, A.: Regional prediction of flow-duration curves using a three-dimensional kriging, J. Hydrol., 513, 179–191, https://doi.org/10.1016/j.jhydrol.2014.03.050, 2014.

Castellarin, A., Botter, G., Hughes, D. A., Liu, S., Ouarda, T. B. M. J., Parajka, J., Post, M., Sivapalan, M., Spence, C., Viglione, A., and Vogel, R.: Prediction of flow duration curves in ungauged basins, in: chap. 7, Cambridge University Press, Cambridge, 135–162, 2013.

Castellarin, A., Persiano, S., Pugliese, A., Aloe, A., Skøien, J. O., and Pistocchi, A.: Prediction of streamflow regimes over large geographical areas: interpolated flow–duration curves for the Danube region, Hydrolog. Sci. J., https://doi.org/10.1080/02626667.2018.1445855, 63, 845–861, 2018.

Castiglioni, S., Castellarin, A., and Montanari, A.: Prediction of low-flow indices in ungauged basins through physiographical space-based interpolation, J. Hydrol., 378, 272–280, https://doi.org/10.1016/j.jhydrol.2009.09.032, 2009.

Castiglioni, S., Castellarin, A., Montanari, A., Skøien, J. O., Laaha, G., and Blöschl, G.: Smooth regional estimation of low-flow indices: physiographical space based interpolation and top-kriging, Hydrol. Earth Syst. Sci., 15, 715–727, https://doi.org/10.5194/hess-15-715-2011, 2011.

Cressie, N. A. C.: Statistics for spatial data, Wiley series in probability and mathematical statistics: Applied probability and statistics, J. Wiley, New York, USA, ISBN: 978-0-471-00255-0, 1993.

de Paiva, R. C. D., Buarque, D. C., Collischonn, W., Bonnet, M.-P., Frappart, F., Calmant, S., and Bulhões Mendes, C. A.: Large-scale hydrologic and hydrodynamic modeling of the Amazon River basin, Water Resour. Res., 49, 1226–1243, https://doi.org/10.1002/wrcr.20067, 2013.

De Roo, A. P. J., Wesseling, C. G., and Van Deursen, W. P. A.: Physically based river basin modelling within a GIS: the LISFLOOD model, Hydrol. Process., 14, 1981–1992, https://doi.org/10.1002/1099-1085(20000815/30)14:11/12<1981::AID-HYP49>3.0.CO;2-F, 2000.

Di Prinzio, M., Castellarin, A., and Toth, E.: Data-driven catchment classification: application to the pub problem, Hydrol. Earth Syst. Sci., 15, 1921–1935, https://doi.org/10.5194/hess-15-1921-2011, 2011.

Donnelly, C., Andersson, J. C. M., and Arheimer, B.: Using flow signatures and catchment similarities to evaluate the E-HYPE multi-basin model across Europe, Hydrolog. Sci. J., 61, 255–273, https://doi.org/10.1080/02626667.2015.1027710, 2016.

Donnelly, C., Greuell, W., Andersson, J., Gerten, D., Pisacane, G., Roudier, P., and Ludwig, F.: Impacts of climate change on European hydrology at 1.5, 2 and 3 degrees mean global warming above preindustrial level, Climatic Change, 143, 13–26, https://doi.org/10.1007/s10584-017-1971-7, 2017.

Falter, D., Dung, N., Vorogushyn, S., Schröter, K., Hundecha, Y., Kreibich, H., Apel, H., Theisselmann, F., and Merz, B.: Continuous, large-scale simulation model for flood risk assessments: proof-of-concept, J. Flood Risk Manage., 9, 3–21, https://doi.org/10.1111/jfr3.12105, 2016.

Farmer, W. H.: Ordinary kriging as a tool to estimate historical daily streamflow records, Hydrol. Earth Syst. Sci., 20, 2721–2735, https://doi.org/10.5194/hess-20-2721-2016, 2016.

Farmer, W. H. and Vogel, R. M.: Performance-weighted methods for estimating monthly streamflow at ungauged sites, J. Hydrol., 477, 240–250, https://doi.org/10.1016/j.jhydrol.2012.11.032, 2013.

Ganora, D., Claps, P., Laio, F., and Viglione, A.: An approach to estimate nonparametric flow duration curves in ungauged basins, Water Resour. Res., 45, W10418, https://doi.org/10.1029/2008WR007472, 2009.

Haag, I. and Luce, A.: The integrated water balance and water temperature model LARSIM-WT, Hydrol. Process., 22, 1046–1056, https://doi.org/10.1002/hyp.6983, 2008.

Habets, F., Boone, A., Champeaux, J. L., Etchevers, P., Franchistéguy, L., Leblois, E., Ledoux, E., Le Moigne, P., Martin, E., Morel, S., Noilhan, J., Quintana Seguí, P., Rousset-Regimbeau, F., and Viennot, P.: The SAFRAN-ISBA-MODCOU hydrometeorological model applied over France, J. Geophys. Res., 113, D06113, https://doi.org/10.1029/2007JD008548, 2008.

Hjerdt, N., Arheimer, B., Lindström, G., Westman, Y., Falkenroth, E., and Hultman, M.: Going Public with Advanced Simulations, in: Environmental Software Systems. Frameworks of eEnvironment: Proceedings 9th IFIP WG 5.11 International Symposium, ISESS 2011, 27–29 June 2011, Brno, Czech Republic, edited byL Hřebíček, J., Schimak, G., and Denzer, R., Springer, Berlin, Heidelberg, 574–580, https://doi.org/10.1007/978-3-642-22285-6_62, 2011.

Hollander, M. and Wolfe, D. A.: Nonparametric Statistical Methods, Wiley, New York, USA, 1999.

Hundecha, Y., Arheimer, B., Donnelly, C., and Pechlivanidis, I.: A regional parameter estimation scheme for a pan-European multi-basin model, J. Hydrol.: Reg. Stud., 6, 90–111, https://doi.org/10.1016/j.ejrh.2016.04.002, 2016.

Isaaks, E. H. and Srivastava, R. M.: An Introduction to Applied Geostatistics, Oxford University Press, New York, USA, ISBN: 0195050134, 1990.

Kim, D., Jung, I. W., and Chun, J. A.: A comparative assessment of rainfall–runoff modelling against regional flow duration curves for ungauged catchments, Hydrol. Earth Syst. Sci., 21, 5647–5661, https://doi.org/10.5194/hess-21-5647-2017, 2017.

Komma, J., Reszler, C., Blöschl, G., and Haiden, T.: Ensemble prediction of floods – catchment non-linearity and forecast probabilities, Nat. Hazards Earth Syst. Sci., 7, 431–444, https://doi.org/10.5194/nhess-7-431-2007, 2007.

Kouwen, N., Soulis, E., Pietroniro, A., Donald, J., and Harrington, R.: Grouped Response Units for Distributed Hydrologic Modeling, J. Water Resour. Pl. Manage.-ASCE, 119, 289–305, 1993.

Kroll, C. N. and Song, P.: Impact of multicollinearity on small sample hydrologic regression models, Water Resour. Res., 49, 3756–3769, https://doi.org/10.1002/wrcr.20315, 2013.

Krysanova, V., Müller-Wohlfeil, D.-I., and Becker, A.: Development and test of a spatially distributed hydrological/water quality model for mesoscale watersheds, Ecol. Model., 106, 261–289, https://doi.org/10.1016/S0304-3800(97)00204-4, 1998.

Laaha, G., Skøien, J., Nobilis, F., and Blöschl, G.: Spatial Prediction of Stream Temperatures Using Top-Kriging with an External Drift, Environ. Model Assess., 18, 671–683, https://doi.org/10.1007/s10666-013-9373-3, 2013.

Lehner, B., Liermann, C. R., Revenga, C., Vörösmarty, C., Fekete, B., Crouzet, P., Döll, P., Endejan, M., Frenken, K., Magome, J., Nilsson, C., Robertson, J. C., Rödel, R., Sindorf, N., and Wisser, D.: High-resolution mapping of the world's reservoirs and dams for sustainable river-flow management, Front. Ecol. Environ., 9, 494–502, https://doi.org/10.1890/100125, 2011.

Lindström, G., Pers, C., Rosberg, J., Strömqvist, J., and Arheimer, B.: Development and testing of the HYPE (Hydrological Predictions for the Environment) water quality model for different spatial scales, Hydrol. Res., 41, 295–319, https://doi.org/10.2166/nh.2010.007, 2010.

Nash, J. and Sutcliffe, J.: River flow forecasting through conceptual models part I – A discussion of principles, J. Hydrol., 10, 282–290, https://doi.org/10.1016/0022-1694(70)90255-6, 1970.

Pappenberger, F., Stephens, E., Thielen, J., Salamon, P., Demeritt, D., van Andel, S. J., Wetterhall, F., and Alfieri, L.: Visualizing probabilistic flood forecast information: expert preferences and perceptions of best practice in uncertainty communication, Hydrol. Process., 27, 132–146, https://doi.org/10.1002/hyp.9253, 2013.

Parajka, J., Merz, R., Skøien, J. O., and Viglione, A.: The role of station density for predicting daily runoff by top-kriging interpolation in Austria, J. Hydrol. Hydromech., 63, 228–234, https://doi.org/10.1515/johh-2015-0024, 2015.

Pechlivanidis, I. G. and Arheimer, B.: Large-scale hydrological modelling by using modified PUB recommendations: the India-HYPE case, Hydrol. Earth Syst. Sci., 19, 4559–4579, https://doi.org/10.5194/hess-19-4559-2015, 2015.

Pontes, P. R. M., Fan, F. M., Fleischmann, A. S., de Paiva, R. C. D., Buarque, D. C., Siqueira, V. A., Jardim, P. F., Sorribas, M. V., and Collischonn, W.: MGB-IPH model for hydrological

and hydraulic simulation of large floodplain river systems coupled with open source GIS, Environ. Model. Softw., 94, 1–20, https://doi.org/10.1016/j.envsoft.2017.03.029, 2017.

Pugliese, A., Castellarin, A., and Brath, A.: Geostatistical prediction of flow–duration curves in an index-flow framework, Hydrol. Earth Syst. Sci., 18, 3801–3816, https://doi.org/10.5194/hess-18-3801-2014, 2014.

Pugliese, A., Farmer, W. H., Castellarin, A., Archfield, S. A., and Vogel, R. M.: Regional flow duration curves: Geostatistical techniques versus multivariate regression, Adv. Water Resour., 96, 11–22, https://doi.org/10.1016/j.advwatres.2016.06.008, 2016.

Pugliese, A., Bagli, S., Mazzoli, P., Parajka, J., Arheimer, B., Capell, R., and Castellarin, A.: Geostatistical Enhancement of European Hydrological Predictions (GEEHP): a SWITCH-ON experiment protocol, available at: http://dl-ng005.xtr.deltares.nl/view/462/ (last access: 30 July 2018), 2017.

Salinas, J. L., Laaha, G., Rogger, M., Parajka, J., Viglione, A., Sivapalan, M., and Blöschl, G.: Comparative assessment of predictions in ungauged basins – Part 2: Flood and low flow studies, Hydrol. Earth Syst. Sci., 17, 2637–2652, https://doi.org/10.5194/hess-17-2637-2013, 2013.

Sampson, C. C., Smith, A. M., Bates, P. D., Neal, J. C., Alfieri, L., and Freer, J. E.: A high-resolution global flood hazard model, Water Resour. Res., 51, 7358–7381, https://doi.org/10.1002/2015WR016954, 2015.

Skøien, J. O.: rtop: Interpolation of data with variable spatial support, r package version 0.5–14, http://CRAN.R-project.org/package=rtop, last access: 30 July 2018.

Skøien, J. O., Merz, R., and Blöschl, G.: Top-kriging – geostatistics on stream networks, Hydrol. Earth Syst. Sci., 10, 277–287, https://doi.org/10.5194/hess-10-277-2006, 2006.

Smakhtin, V. Y. and Masse, B.: Continuous daily hydrograph simulation using duration curves of a precipitation index, Hydrol. Process., 14, 1083–1100, https://doi.org/10.1002/(SICI)1099-1085(20000430)14:6<1083::AID-HYP998>3.0.CO;2-2, 2000.

Srinivas, V., Tripathi, S., Rao, A. R., and Govindaraju, R. S.: Regional flood frequency analysis by combining self-organizing feature map and fuzzy clustering, J. Hydrol., 348, 148–166, https://doi.org/10.1016/j.jhydrol.2007.09.046, 2008.

Strömqvist, J., Arheimer, B., Dahné, J., Donnelly, C., and Lindström, G.: Water and nutrient predictions in ungauged basins: set-up and evaluation of a model at the national scale, Hydrolog. Sci. J., 57, 229–247, https://doi.org/10.1080/02626667.2011.637497, 2012.

Wan Jaafar, W. Z., Liu, J., and Han, D.: Input variable selection for median flood regionalization, Water Resour. Res., 47, W07503, https://doi.org/10.1029/2011WR010436, 2011.

8

The benefit of seamless forecasts for hydrological predictions over Europe

Fredrik Wetterhall and Francesca Di Giuseppe

European Centre for Medium-range Weather Forecasts, Shinfield Park, Reading, UK

Correspondence: Fredrik Wetterhall (fredrik.wetterhall@ecmwf.int)

Abstract. Two different systems provide long-range forecasts at ECMWF. On the sub-seasonal timescale, ECMWF issues an extended-range ensemble prediction system (ENS-ER) which runs a 46-day forecast integration issued twice weekly. On longer timescales, the current seasonal forecasting system (SYS4) produces a 7-month outlook starting from the first of each month. SYS4 uses an older model version and has lower spatial and temporal resolution than ENS-ER, which is issued with the current operational ensemble forecasting system. Given the substantial differences between the ENS-ER and the SYS4 configurations and the difficulties of creating a seamless integration, applications that rely on weather forcing as input such as the European Flood Awareness System (EFAS) often follow the route of the creation of two separate systems for different forecast horizons. This study evaluates the benefit of a seamless integration of the two systems for hydrological applications and shows that the seamless system outperforms SYS4 in terms of skill for the first 4 weeks, but both forecasts are biased. The benefit of the new seamless system when compared to the seasonal forecast can be attributed to (1) the use of a more recent model version in the sub-seasonal range (first 46 days) and (2) the much more frequent updates of the meteorological forecast.

1 Introduction

The European Centre for Medium-range Weather Forecasts (ECMWF) produces a range of forecasts, among them a 10-day deterministic high-resolution forecast (HRES) and a lower resolution 15-day 51-member ensemble prediction system (ENS) that is extended to 46 days twice weekly (Mondays and Thursdays at 00:00 UTC; Vitart et al., 2008).

In this paper we refer to the extended ENS as ENS-ER. On longer time ranges ECMWF issues a seasonal ensemble forecast system (SYS4), operational since November 2011. SYS4 issues a 7-month prediction (extended to 13 months four times a year) once every month (Molteni et al., 2011). The ENS-ER forecast system benefits from frequent updates of the model physics and data assimilation system (Vitart et al., 2008). ECMWF releases official model updates on average 2–3 times a year which typically include improved schemes for physical processes, better use of observations and their assimilation, and sometimes an increase in model resolution. The seasonal forecast has a lower resolution, is an older model version than ENS-ER and is also updated much less frequently. This implies that the skill of the seasonal forecasting system is lower relative to ENS-ER in the overlapping first 6 weeks.

Applications that use numerical weather predictions as forcing, such as the operational European Flood Awareness System (EFAS; Thielen et al., 2009; Bartholmes et al., 2009; Smith et al., 2016) are often designed for a specific purpose. EFAS has, since the start, focused on early warning of floods in the medium-range forecast horizon, typically up to 15 days. Recently, a seasonal hydrological outlook forced by SYS4 was implemented operationally with a lead time of 7 months (Arnal et al., 2018).

This extension to the monthly and seasonal timescales is potentially very useful in order to (i) produce products which extend the previous forecast horizon, (ii) benefit from hindcasts for pre- and post-processing to produce output of higher quality (e.g., model-based return periods) and (iii) design completely new early-warning frameworks complementing the existing ones. The extended lead time provided by running EFAS forced by weather prediction across different timescales could potentially provide added benefit in terms of very early planning, for example for agriculture, energy (Bazile et al., 2017) and transport sectors (Meißner et al., 2017), as well as water resources management (Sene et al., 2018). Such a forecast system would be a first step to close the identified gap between hydrological forecasts on the medium (up to 15 days) and seasonal range (White et al., 2017). These extended-range systems may not be able to capture extremes of short-lived events like floods but they are able to detect anomalous conditions on longer lead times, such as low flows (Meißner et al., 2017) and droughts (Dutra et al., 2014).

The concept of seamless forecasts was first introduced by Palmer and Webster (1993). Palmer et al. (2008) formally expanded the idea showing how short-lived phenomena under certain conditions may persist and increase predictability at longer timescales. Since then the concept of a unified or seamless framework for weather and climate prediction has been vastly debated (Hurrell et al., 2009; Brunet et al., 2010). However, as noticed by Hoskins (2013) in his seminal paper: while "the atmosphere knows no barriers in time-scales", model implementation is often segmented for practical reasons. Still, major efforts have been made to create unified systems. Indeed, the ENS-ER was the first attempt to create a seamless extension of the ECMWF medium-range forecast to the monthly timescales (Vitart et al., 2008). Similarly, the UK Met Office has, in the past 25 years, worked to create a unified model that could work across all timescales (Brown et al., 2012). Also the climate community has moved in the same direction. For example, the EC-Earth project shows that a bridge can be made between weather, seasonal forecasting and beyond (Hazeleger et al., 2010, 2012).

The latter projects went all the way to create new systems starting from existing components and were therefore costly and time demanding. In contrast, a practical and simpler approach could be taken. The seamless idea could be translated into a concatenation of "the best" forecast at each lead time. The clear advantage of this off-the-shelf seamless prediction conversion is that it uses products that are already available and operational, thereby avoiding the complications of new developments, while at the same time generating forecast products that meet the demands of different types of users (Pappenberger et al., 2013). However, there is an underlying complexity in this simplification; the difference in design between the various forecasting systems makes the concatenation not entirely straight-forward. The forecasting systems are related since they are from different generations of the same model development; however, they have non-matching temporal and spatial resolutions, different hindcast span and different ensemble sizes. One important consequence of this is that the more frequent updates to the extended range compared to the seasonal forecasting system at ECMWF causes the model errors from the two systems to diverge over time, and only closing this gap when the seasonal system is updated to a newer model version (Di Giuseppe et al., 2013).

Then model outputs either need to be bias-corrected to be a useful forcing to drive sectoral models such as EFAS or final products should be provided in terms of anomalies calculated against the model climate, taking into consideration the bias of the seamless forecast system. In both cases the seamless system needs to account for the use of the hindcast dataset and the application of some bias correction algorithm. In return, the advantage is in the gain in skill and the extension of the lead time.

In this work the benefit of a seamless hydrometeorological system was tested for a span of time ranges from 1 week to 6 months for stream flow forecasts over the European domain using the EFAS system. The aim was to test whether integrating medium-range forecasts with seasonal prediction contributes to enhance hydrological predictability on the seasonal timescale. Specifically, the questions addressed were the following. What is the gain, in terms of hydrological forecasting, when using a more recent model version in the first 46 days provided by the use of the ENS-ER? What is the skill gain provided by having more frequent forecast updates?

2 Methods

2.1 Hydrological model system

The hydrometeorological system used in this study was the European Flood Awareness System (EFAS; Thielen et al., 2009; Bartholmes et al., 2009; Smith et al., 2016). EFAS is an operational early-warning system covering most of the European domain and has been run operationally since October 2012 as part of the COPERNICUS Emergency Management Service (CEMS). The hydrological component of EFAS is the distributed rainfall-runoff model LISFLOOD (De Roo et al., 2000; Van Der Knijff et al., 2010; Burek et al., 2013). LISFLOOD calculates the main hydrological processes on sub-daily and daily timescales that generate runoff for each grid cell. In the operational setup, EFAS covers most of Europe on a 5 km × 5 km equal-area grid. The runoff is transformed through a routing scheme to estimate the river discharge at each grid cell along the river network. The routing scheme also takes into account water retention in lakes and reservoirs. This study will concentrate on the forecast of river discharge at the outlets of the sub-basins of the river network that were used for calibration of the current EFAS system (Smith et al., 2016; Zajac et al., 2013). The total number of outlets used was 679, and they represent river basins of all sizes and characteristics across the EFAS domain.

In its operational implementation the latest calibration (referred to as "tuning" in the numerical weather prediction, NWP, nomenclature) of LISFLOOD used an observational dataset of meteorological forcing data (precipitation and temperature) and observed discharge covering the model domain over the period 1990–2013 (Smith et al., 2016;

Table 1. Technical details of the forecast and the hindcast used in this paper.

System	Time res.	Spatial res.	Horizon	Ensemble size	Issue frequency	Hindcast set	Hindcast ensemble size
ENS-ER	3 h/6 h	18/36 km*	46 days	51	Twice weekly	20 years	11 members
SYS4	6 h	80 km	7/13 months	51	Monthly	30 years	15/51 members
SEAM	6 h	5 km	6 months	51	Twice weekly	20 years	11 members

* The resolution changes to 36 km at day 16 of the forecast.

Zajac et al., 2013). The meteorological dataset comprises more than 5000 synoptic stations that have been interpolated to a 5 km × 5 km Lambert azimuthal equal-area projection (Ntegeka et al., 2013). The high-resolution gridded observations of precipitation and temperature were used for the calibration of LISFLOOD. The observational meteorological dataset was also used to generate a reference modeled climatology of discharge (hereafter called water balance, WB) which is used as (i) initial conditions for the operational forecast and hindcasts and (ii) a reference model run to assess the performance of the forecasts. Using the WB run as proxy observation simplifies the interpretation of the skill scores as it avoids the complication of having to assess the bias against observed discharge. The purpose of this paper is rather to assess the skill of the two forecasts used for forcings rather than the total skill of the forecasting system.

2.2 Seamless integration of meteorological forcing data

Every Monday and Thursday, ECMWF issues an extended-range ensemble forecast (ENS-ER) by continuing the integration time beyond day 15 up to day 46, with a lower-resolution model (Fig. 1, Table 1). Each ENS-ER integration comes with an 11-member hindcast set produced for the same dates as the forecast date over the previous 20 years. This hindcast set provides identical integrations as the current operational forecast with the difference that ERA-Interim reanalysis (ERAI; Dee et al., 2011) and ERAI land reanalysis (Balsamo et al., 2015) are used to provide the initial conditions, whereas the operational ensemble forecast uses the operational analysis. The hindcast data together with observations can be used in many applications, for example to calibrate the forecast in an operational setting (Di Giuseppe et al., 2013).

The operational seasonal forecast (SYS4) issues a new forecast at the beginning of each month with a lead time up to 7 months, four times a year extended to 13 months (Fig. 2). SYS4 has a hindcast consisting of 30 years starting at each month and consisting of 15 members. The new seamless forecasting system (hereafter called SEAM) was created by concatenating each ENS-ER ensemble member with a randomly selected SYS4 ensemble member at day 46, which is the last day of the ENS-ER (Fig. 2). SEAM benefits from the frequent updates of the ENS-ER and has the 7-month horizon of the seasonal system.

Since the two systems have different resolutions (Table 1) the horizontal resolution was homogenized to the 5 km × 5 km equal-area grid through a mass-conservative interpolation for precipitation and a bilinear interpolation for temperature before it was used as input to the hydrological model in EFAS. The mass-conservative interpolation summarizes the partial contribution of the meteorological input fields onto the LISFLOOD grid. The time step was reduced to daily by averaging (accumulating for precipitation and evapotranspiration) the three hourly outputs of the ENS-ER and the six hourly outputs of SYS4. Since the ENS-ER has a reduced hindcast (20 years) and number of members (11), SEAM has the same number of members and hindcast period. Note that in real-time mode, a full 51-member SEAM is possible. The technical details of the forecast and the hindcast used in this experiment are presented in Table 1. For simplicity, SYS4 and SEAM will, from now on, refer to the full hydro-meteorological model chain and not only the meteorological forcing.

2.3 Experimental setup

This study focuses on the performance of SYS4 and SEAM over the hindcasts of the operational forecast. The hindcasts starting from 14 May 2015 (the first available date with 11-member hindcast for ENS-ER) to 2 June 2016 were used as input to the full EFAS modeling chain. As described above, the hindcasts are the reforecasts over the previous 20 years and are produced for each individual run of the ENS-ER. This provided 13 monthly starting dates for SYS4 and 111 biweekly starting dates for SEAM with a corresponding hindcast set covering all seasons over the previous 20-year period, each with 15 and 11 ensemble members, respectively (Fig. 1). The output was averaged to weekly means before the skill score analysis. Since the starting dates of the SEAM and SYS4 were not always in sync (the starting date of the SYS4 integrations are only sometimes on a Monday or Thursday), it is impossible to do a completely like-for-like comparison since the validation periods would be slightly different. However, this error will be random and given the sample sizes (260 and 2220) it was not considered to have a big impact on the results.

SEAM was validated against the runs with SYS4 to assess the added value of the merged forecast. Further, both model systems were compared against a climatological benchmark

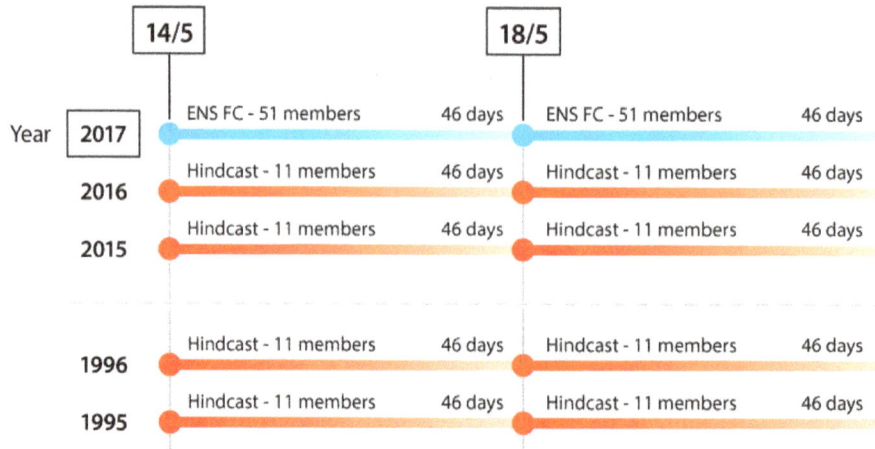

Figure 1. Schematic overview of the operational ECMWF ensemble forecast for the extended range and its associated hindcast. The hindcasts consists of a reduced ensemble forecast (11 members) with the same starting date of year as the current forecast, but run for the previous 20 years.

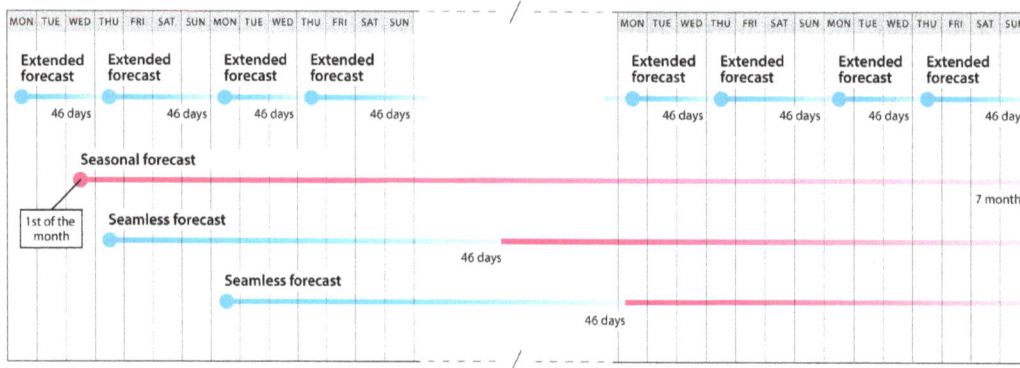

Figure 2. Schematic overview of the seasonal, extended-range forecast and merged systems. The extended forecast is issued every Monday and Thursday and extends up until 46 days, the seasonal forecast is issued on the first of each month and extends up until 7 months (13 months in February, May, August and November). The merged forecast concatenates the latest extended forecast with the latest seasonal forecast.

simulation (hereafter called CLIM). CLIM was constructed by forcing the LISFLOOD with 15 randomly selected time series of observed meteorological forcing from the period 1990–2014, excluding the modeled year. CLIM has the advantage of having the same initial conditions as SYS4 and SEAM hindcasts, but has no expected predictive skill beyond the horizon of the initial conditions. The advantage of CLIM is that in theory it has near-perfect reliability with regards to the WB runs since it is produced with the same unbiased forcing data. It should, therefore, score better or equal to the hindcasts as predictor on time ranges beyond their respective limits of predictability.

2.4   Score metrics

The performance of the two forecast systems was compared against the WB run at the 679 sub-basin outlets using deterministic and probabilistic scores. WB is treated as a proxy for

observations in the evaluation. The scores used were the continuous ranked probability score (CRPS; Hersbach, 2000), mean relative error (MRE) and forecast reliability through an attributes diagram. All scores were calculated for SYS4 and SEAM over the hindcast period. CRPS is a common tool to validate probabilistic forecasts and can been seen as generalization of the mean absolute error to the probabilistic realm of ensemble forecasts. It is defined as

$$\text{CRPS} = \frac{1}{N}\sum_{t=1}^{N}\int_{-\inf}^{+\inf}\Big[F_t(x(n)) - H_t(x(n) - x_0)^2\Big]\mathrm{d}x, \quad (1)$$

where $x(n)$ is the forecast at time step t of N number of forecasts and x_0 is the observed value (WB). The CRPS is the continuous extension of the ranked probability score (RPS), where $F(x)$ is the cumulative distribution function (CDF)

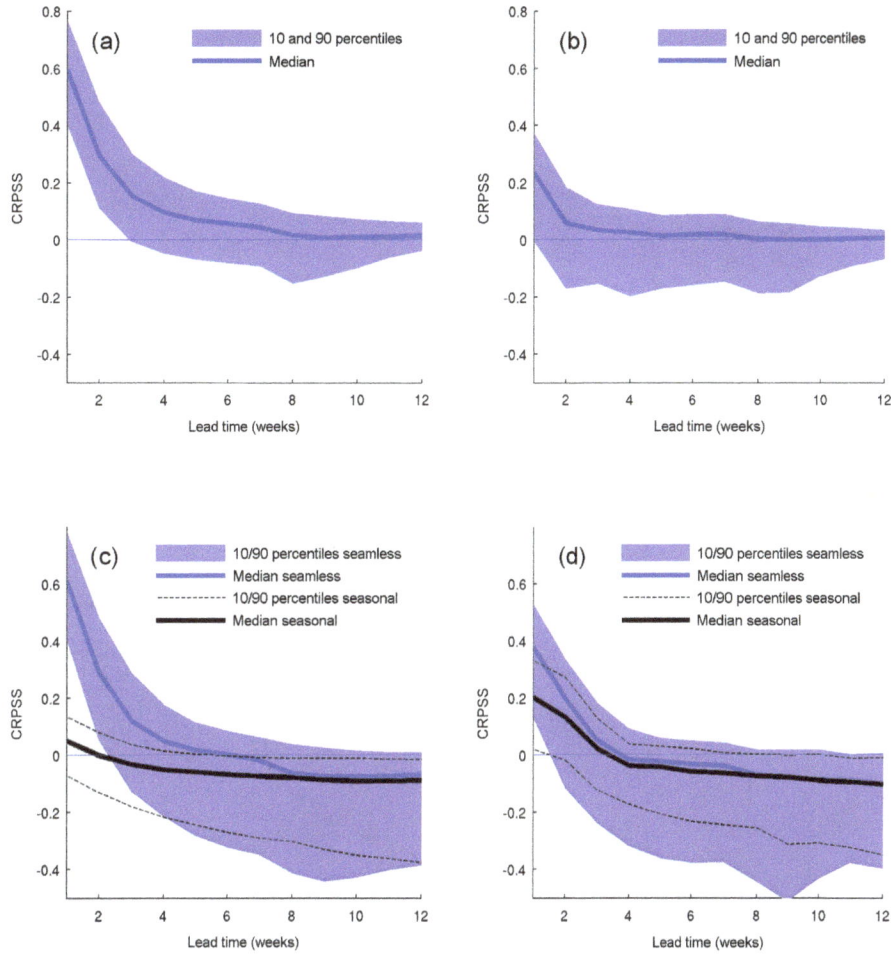

Figure 3. Continuous ranked probability skill score (CRPSS) for **(a)** merged forecast against seasonal forecast for all start dates evaluated over the 679 basin outlet points; **(b)** as in **(a)** but only for the first merged forecast of each month; **(c)** merged forecast against climatology for all lead times in blue and **(d)** as in **(c)** but for the first merged forecast in the month. The shaded blue area denotes the 10–90 percentile of the CRPSS and the blue line the median. The black solid (dotted) lines in **(c)** and **(d)** denote the mean and 10–90th percentiles of the CRPSS of the seasonal against the climatological forecast.

$F(x) = p(X - x)$ and $H(x - x_0)$ is the Heaviside function, which has the value 0 when $x - x_0 < 0$ and 1 otherwise.

The CRPS compares the cumulative probability distribution of the discharge forecasted by the ensemble forecast system to an observation. It is sensitive to the mean forecast biases as well as the spread of the ensemble. Since the SEAM has 11 members and SYS4 and CLIM have 15 members in the hindcast, the CRPS are not directly comparable. Ferro et al. (2008) showed that for two ensemble distributions with different ensemble sizes, M and m, the unbiased estimate for CRPS_M based on CRPS calculated from the ensemble size m is

where

$$\Delta_t = \frac{1}{m(m-1)} \sum_{i \neq j} |X_{t,i} - X_{t,j}| \qquad (3)$$

is Gini's mean difference of ensemble members $[X_{t,1}, \ldots, X_t, m]$ at time t. From the CRPS a skill score (CRPSS) can be derived by comparing CRPS of the verified forecast against a reference forecast.

$$\mathrm{CRPS}_M = \mathrm{CRPS}_m - \frac{M - m}{2Mmn} \sum_{t=1}^{n} \Delta_t, \qquad (2)$$

$$\mathrm{SS}_{\mathrm{CRPS}} = 1 - \frac{\mathrm{CRPS}_{\mathrm{fc}}}{\mathrm{CRPS}_{\mathrm{rf}}} \qquad (4)$$

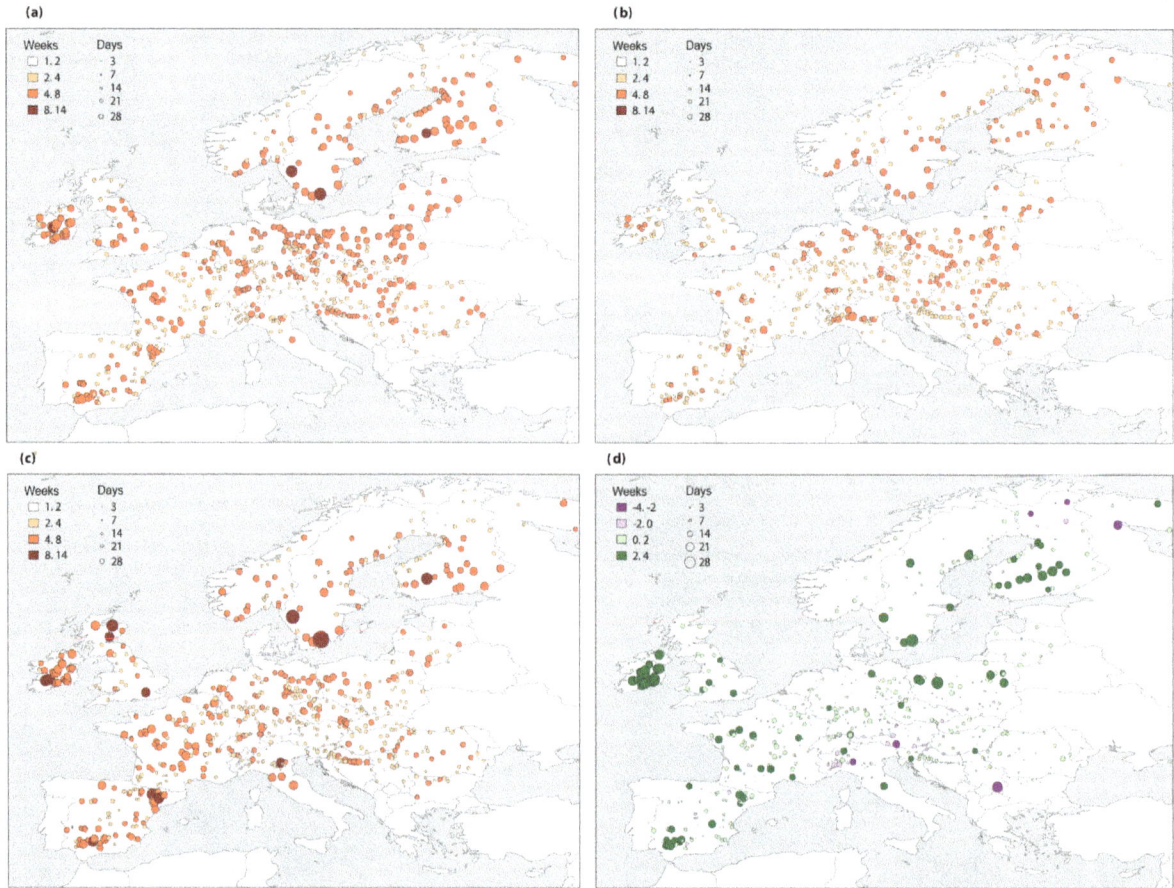

Figure 4. The number of weeks (days) before the CRPSS goes below zero using only the first forecast of the month for **(a)** SEAM against CLIM; **(b)** SYS4 against CLIM; **(c)** SEAM against SYS4; and **(d)** difference between SEAM against CLIM and SYS4 against CLIM. The dimension of the circles is proportional to the number of days while the color scale refers to the number of weeks. The size and color of the circles are therefore showing the same information, and are both added for clarity.

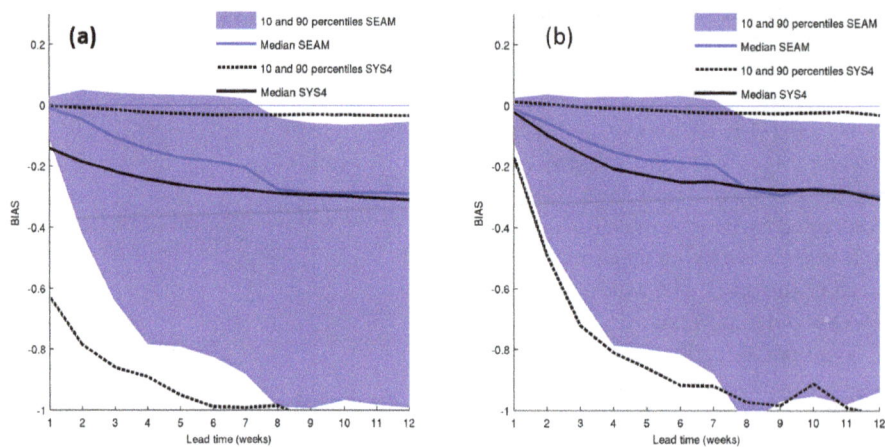

Figure 5. Mean relative error over all outlet points as a function of lead time in weeks **(a)** for all starting dates of the forecasts and **(b)** for the starting dates close to the beginning of the months. Negative values denote that the forecast is too wet in comparison with the CLIM run. The SEAM (SYS4) forecast is in blue (black) where the solid line denotes the median and the filled area (area between dotted lines) denote the 10 to 90th percentiles.

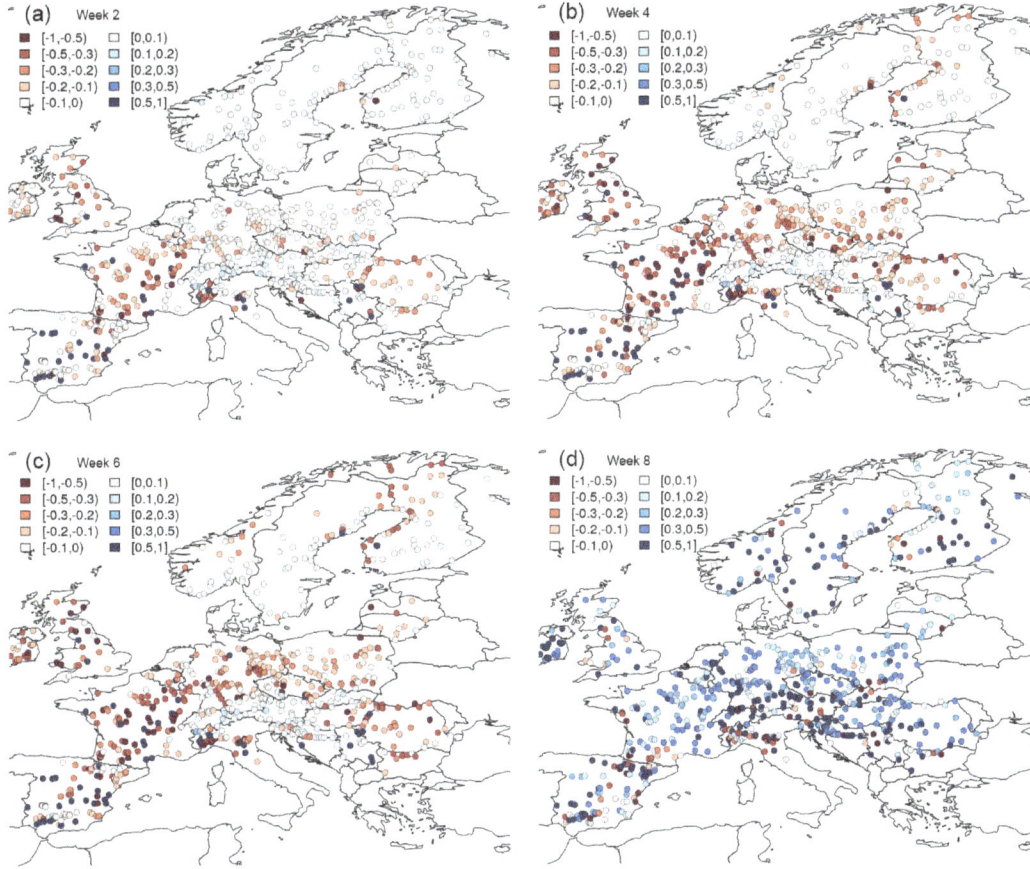

Figure 6. Mean relative error for each of the outlet points for the SEAM forecast over the outlet points for **(a)** week 2, **(b)** week 4 and **(c)** week 6. Red indicates where the forecast is too wet, and blue where it is too dry. **(d)** Shows the difference in absolute error between SEAM and SYS4, where blue (red) denotes points where SEAM has a smaller (larger) MAE than SYS4.

The mean relative error (MRE) was measured as the forecast bias in comparison with WB normalized with WB, here defined as

$$\text{MRE} = \frac{1}{n} \sum_{t=1}^{N} \frac{x_o - \overline{x}(n)}{x_o}, \qquad (5)$$

where x_o denotes the observed value (WB) and $\overline{x}(n)$ denotes the forecasted ensemble mean at time t.

The reliability was assessed through an attributes diagram, where the forecast probability of exceeding a certain threshold is compared with observed frequencies (Hsu and Murphy, 1986; Weisheimer and Palmer, 2014). The forecast reliability was evaluated for the 10, 50 and 90th percentiles of observed discharge at each outlet.

3 Results and discussion

3.1 Overall forecast skill

The forecast skill gain provided by SEAM with respect to SYS4 is mostly concentrated to the first 6 weeks (Fig. 3a)

when the forcing data are from the ENS-ER. The difference in CRPSS is 0.6 at week 1, which then decreases to 0.1 by week 6. All points used in the validation show a gain in skill up until week 3, then some points show a benefit of using the SYS4 instead of SEAM. However, in some catchments there is skill up further than 8 weeks. The overall better performance of SEAM with respect to SYS4 is partly because of the use of a more recent model version and partly because of the more frequent updates of the atmospheric and hydrological initial conditions. It is possible to disentangle the relative contributions between these two factors by only considering a reduced number of starting dates for the SEAM forecast, i.e., dates that are the closest to the SYS4 starting dates (Fig. 3b). This reduced statistic provides a measure of the expected contribution of *only* employing a newer model cycle in the first weeks while both simulations benefits from the same hydrological initialization. In this case, the skill gain in CRPS reduces to between 0 and 0.4 (median 0.2) against SYS4 for the first week, reducing to neutral around week 4. Therefore the most relevant gain comes from the more frequent initializations of the hydrological model.

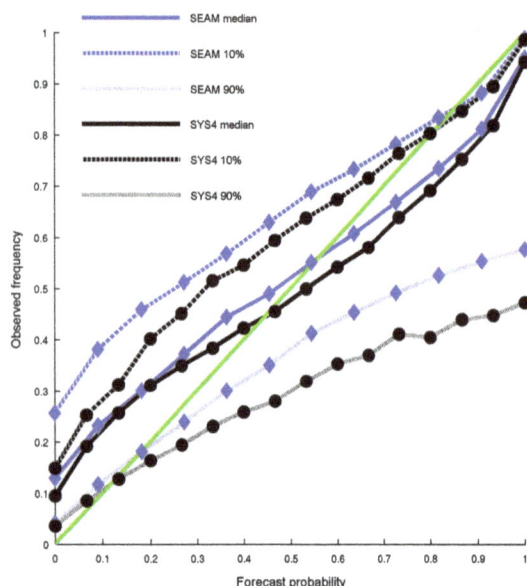

Figure 7. Reliability diagram for SEAM (blue) and SYS4 (black) for week 4 for all outlet points. Solid lines indicate the reliability for the median of observed discharge, the dashed (dotted) lines the forecast reliability for the 10th (90th) percentiles of observed discharge.

To put these increments into context we also look at the improvement in skill of the two systems (SYS4 and SEAM) against the CLIM benchmark forecast (Fig. 3c–d). The gain from having improved initial conditions in SEAM is similar in comparison with CLIM (Fig. 3c) as with SYS4 (Fig. 3a) in the first week, but the skill deteriorates quicker and the median CRPSS is negative after 5 weeks. Without the increase in skill due to the advantage of the better initial conditions, SEAM still shows a gain against the CLIM forecast with a CRPSS of 0.4 for the first week, although the spread is quite wide (Fig. 3d). Also, SYS4 shows an increase in skill against the CLIM forecast. Both forecasts are less skillful than CLIM for most river points after week 4. It can also be noted that SEAM has a higher spread than SYS4 on longer lead times even though they are forced with the same data from day 47 and onwards. An explanation can be that the ensembles from the two meteorological forecasts are not matched member by member in terms of their relative deviation from the mean, for example matching members from each distribution according to their wetness. If two extreme driving forecasts from the two meteorological forecasts are combined it can lead to members that are further away from the ensemble mean than when only one driving forecast is used.

3.2 Geographical variation in forecast skill

The geographical distribution of skill gain provided by the SEAM and SYS4 prediction reveals a coherent picture with

good scores against the CLIM run over most of Europe (Fig. 4a–b). The gain in the figure is expressed as a difference in the number of weeks into the forecast needed for the CRPSS to drop below zero (i.e., there is no skill in the forecast in comparison with CLIM), which gives an indication of the expected time gain in terms of information provided by the forecast against the reference forecast. Both SYS4 and SEAM are better than CLIM, and SEAM has higher skill than SYS4 for most of Europe. There is a small negative effect over the Alps, southeastern Europe and northern Finland (Fig. 4d). The performance of the operational EFAS in these regions is generally poor, which is caused by the difficulty of having good observations of precipitation in high-altitude stations and the atmospheric models difficulty in resolving steep orography (Alfieri et al., 2014). The snow accumulation and snowmelt are further divided into three elevation zones within a grid in LISFLOOD to better account for orographic effects in mountainous regions. However, this increase in sub-grid resolution is not likely to be high enough to capture the snow variability during the snow accumulation and snowmelt in mountainous regions. Further, precipitation forecasts have documented biases in steep orography (Haiden et al., 2014).

Another interesting aspect to showcase is the relevance of more frequent model version updates for the overall improvement of river discharge for all stations in proximity to the western coasts. This can be attributed to recent developments in the precipitation model scheme, for example a new diagnostic closure introduced in the convection scheme (Bechtold et al., 2014) and a new parameterization of precipitation formation (Haiden et al., 2014).

3.3 Bias and reliability

The relatively sharp decline in CRPSS can, to some extent, be explained by the negative bias (too wet forecast) for both SEAM and SYS4 forecast (Fig. 5). SEAM has lower bias than SYS4, also when the analysis is confined to the first few weeks (Fig. 5b). The slightly better bias in SEAM disappears quickly after the merge (week 7). The bias of the forecast is not spatially consistent, it is generally larger in western and central Europe (Fig. 6). The figure shows the bias for SEAM (a–c) but the pattern is similar for SYS4. SEAM has generally a smaller bias than SYS4 (Fig. 6d). SYS4 has lower bias south of the alps, where it also performs better than SEAM.

The reliability of a forecast in terms of its usefulness for decision making is important. A reliable forecast can be trusted to predict the correct probability of certain events, regardless of the accuracy. An unreliable forecast is in practice of no use and can lead to poor decisions (Weisheimer and Palmer, 2014). Both forecast systems are overconfident when it comes to predicting the median flow, which can be attributed to an underestimation of the ensemble spread (Fig. 7). The results are comparable to a previous study of 2 m temperature and precipitation over Europe with SYS4

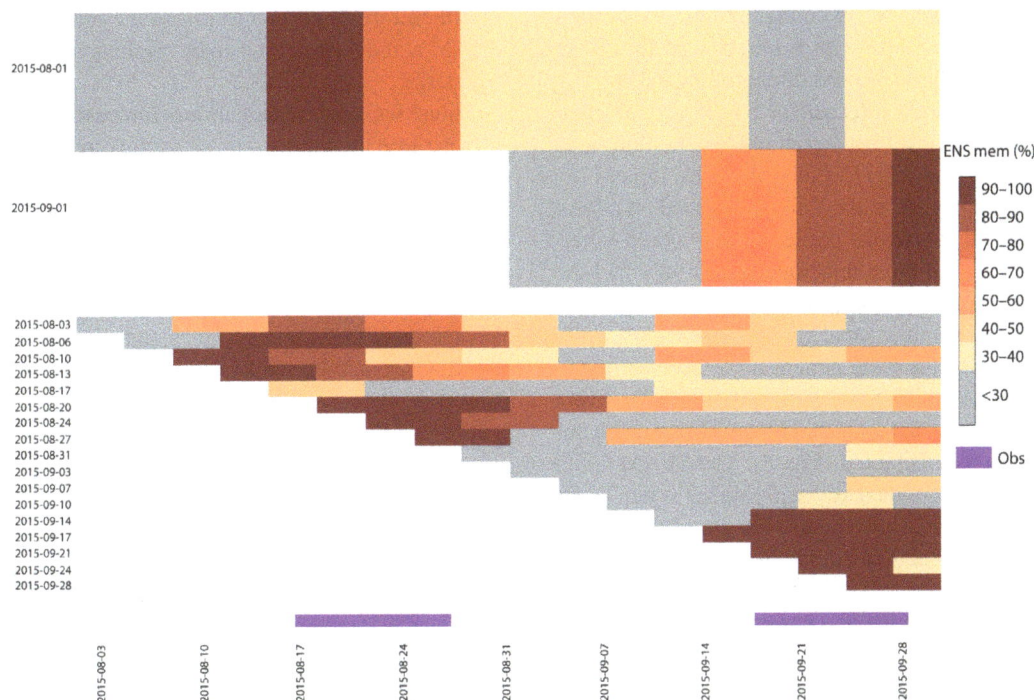

Figure 8. Percentage of ensemble members predicting low-flow anomaly ($< 97\%$) on the river Rhine north of Cologne for summer 2003. The two starting dates in August and September from SYS4 are compared to the 17 starting dates of the seamless forecasting system. In two separate events the discharge was recorded below the 97th percentile, event 1 on 17–27 August and event 2 on 18–28 September 2003.

(Weisheimer and Palmer, 2014). The reliability with regards to low flows (dashed line, Fig. 7) indicates an overprediction of low flows, which can be explained by the wet bias of both systems causing an overestimation of the low flows. SYS4 is performing better than SEAM in this regard. The high flows are generally underestimated by both systems, but SEAM performs slightly better than SYS4 (dotted line, Fig. 7). The skill of the forecasts from both systems could be potentially higher by performing a bias correction, either on the atmospheric input and/or on the discharge. However, in this paper we concentrate on the differences in skills provided by the various configurations and no bias correction has been applied.

3.4 Added value of the seamless forecast

Even though the increase in the overall skill provided by SEAM in comparison with SYS4 is noticeable, the justification for its use in an operational context also depends on the actionable time gain in a response situation. More frequent forecast updates could potentially be useful in decision making. As an example, we analyze the predicted stream flow for the river Rhine at a station just upstream of Cologne, Germany, during the European heat wave in the summer of 2003. It was an exceptional meteorological event which combined significant precipitation deficits with record-setting high temperatures (García-Herrera et al., 2010). At its peak in August, extremely low discharge levels

of rivers were reported in large parts of Europe. Economic losses were huge in many primary economic sectors including transportation (Ciais et al., 2005). For several months, inland navigation was heavily impaired and the major European transport routes in the Danube and Rhine basins ceased completely (Jonkeren et al., 2008).

Despite the fact that 2003 conditions were extreme from the meteorological point of view, the upcoming deficit in precipitation and the high temperatures were well predicted by the ECMWF seasonal systems operational at that time (System-3; Weisheimer et al., 2011). The good predictability of the event is confirmed by the low discharge prediction provided by SYS4 at the Rhine upstream of Cologne (Fig. 8). More then 30 % of the ensemble members forecast extreme low-flow conditions. In fact the observed discharge confirms that the river flow on two separate occasions, event 1 on August 17–27 and event 2 on 18–28 September 2003, went below the third percentile of its climatological value for the season (Fig. 8). While most of SYS4 ensemble members mark the extreme condition 3 to 4 weeks ahead, there is no information of the recovery period observed between event 1 and 2 in the forecast starting the first of August. SYS4 predicts a swift recovery back to normal conditions on the forecast issued on 1 September. A more detailed picture of this intermediate recovery is instead conveyed by the seamless system. Thanks to the more frequent updates, the temporary increase in river flow is correctly picked up giving a potential advantage of 2 to 3 weeks for planning actions. SYS4 does indicate the second low flow with a longer lead time than SEAM. However, SYS4 misses the timing of the event.

Even if this was a good forecast for SYS4, the information it provides is more informative (anomaly condition) than "actionable" (White et al., 2017). In the above example, a decision maker would have to make a decision based on a forecast that was issued 2.5 weeks earlier, which would inherently make the decision more uncertain if you only had the seasonal forecast. With the seamless system available, a decision maker would gain the same early indication of a hazardous event and also have the benefit of frequent updates. In this particular case, the SEAM forecast for the first event was more unstable for some ensemble members, but in general the event was well captured (Fig. 8). The SEAM could also correctly capture the recovery with higher water levels between the extreme low-flow events. The onset of the second low period was correctly modeled by the SEAM system, whereas the timing of the low flow was missed by SYS4. It should be said that using other less extreme thresholds (< 10 and < 5 percentiles) even further strengthened the case for using SEAM.

4   Conclusions

This study compared a set of hydrological hindcast experiments over the European domain with two meteorological forcings: ECMWF's seasonal forecasting system (SYS4) and a merged system of ECMWF extended-range forecast and seasonal forecast system (SEAM). The latter showed a better overall skill and lower bias over most areas in Europe with lead times up to 7 weeks. This increase in skill could be attributed to better initial conditions of the hydrological and meteorological model as well as a better atmospheric model version in SEAM. In some areas, particularly in the Alps and northern Finland, the seasonal forecast outperformed the merged forecast. However, in these areas the predictability of the hydrological model is generally poor, which makes these results quite uncertain. Given that the skill in the sub-seasonal range over Europe is in the range of the extended-range ensemble forecast, this motivates us to use the ENS-ER instead of SYS4 for hydrometeorological predictions.

Still, there is an added benefit of using a seamless forecast over the extended range due to the extension of the forecast horizon for the early detection of upcoming anomalous conditions. Indeed, as an example, this study also highlighted the potential for the use of a sub-seasonal to seasonal forecast in the case of an extreme low-flow situation in the river Rhine. The higher frequency and skill of SEAM has the advantage of being a more "actionable" forecast than seasonal forecasts, given that a decision maker would be able to make use of the extra information. Care should be taken when using the forecasts in decision making since the reliability over Europe is "marginally useful" (Weisheimer and Palmer, 2014).

It is therefore important to assess the reliability and skill of SEAM at the location it is to be implemented over the season of interest.

Future work with the seamless forecasting system is to further explore the limits of predictability, reliability and bias to assess the strengths and limitations of the current setup. The assumption that the forecasts can be randomly concatenated would also need to be tested against a system where the forecasts are matched according to their respective climatology. Bias correction of the forecasts might be a necessity, and the advantage of the extended-range and seasonal forecasts from ECMWF is that the availability of hindcasts enables just that.

Competing interests.   The authors declare that they have no conflict of interest.

Special issue statement.   This article is part of the special issue "Sub-seasonal to seasonal hydrological forecasting". It is a result of the HEPEX workshop on seasonal hydrological forecasting, Norrköping, Sweden, 21–23 September 2015.

Acknowledgements.   This paper was financed through the framework service contract for operating the EFAS computational center in support of to the Copernicus Emergency Management Service (EMS) and Early Warning Systems (EWS) 198702. The authors would like to acknowledge Blazej Krzeminski for setting up the computational framework and Florian Pappenberger for the discussions regarding the seamless forecast system. We would also like to thank Kean Foster, Bastian Klein and Mike Hardeker for useful comments on the discussion paper.

Edited by: Quan J. Wang

References

Alfieri, L., Pappenberger, F., Wetterhall, F., Haiden, T., Richardson, D., and Salamon, P.: Evaluation of ensemble streamflow predictions in Europe, J. Hydrol., 517, 913–922, https://doi.org/10.1016/j.jhydrol.2014.06.035, 2014.

Arnal, L., Cloke, H. L., Stephens, E., Wetterhall, F., Prudhomme, C., Neumann, J., Krzeminski, B., and Pappenberger, F.: Skilful seasonal forecasts of streamflow over Europe?, Hydrol. Earth Syst. Sci., 22, 2057–2072, https://doi.org/10.5194/hess-22-2057-2018, 2018.

Balsamo, G., Albergel, C., Beljaars, A., Boussetta, S., Brun, E., Cloke, H., Dee, D., Dutra, E., Muñoz-Sabater, J., Pappenberger, F., de Rosnay, P., Stockdale, T., and Vitart, F.: ERA-Interim/Land: a global land surface reanalysis data set, Hydrol. Earth Syst. Sci., 19, 389–407, https://doi.org/10.5194/hess-19-389-2015, 2015.

Bartholmes, J. C., Thielen, J., Ramos, M. H., and Gentilini, S.: The european flood alert system EFAS – Part 2: Statistical skill assessment of probabilistic and deterministic operational forecasts, Hydrol. Earth Syst. Sci., 13, 141–153, https://doi.org/10.5194/hess-13-141-2009, 2009.

Bazile, R., Boucher, M.-A., Perreault, L., and Leconte, R.: Verification of ECMWF System 4 for seasonal hydrological forecasting in a northern climate, Hydrol. Earth Syst. Sci., 21, 5747–5762, https://doi.org/10.5194/hess-21-5747-2017, 2017.

Bechtold, P., Semane, N., Lopez, P., Chaboureau, J.-P., Beljaars, A., and Bormann, N.: Representing equilibrium and nonequilibrium convection in large-scale models, J. Atmos. Sci., 71, 734–753, 2014.

Brown, A., Milton, S., Cullen, M., Golding, B., Mitchell, J., and Shelly, A.: Unified modeling and prediction of weather and climate: A 25-year journey, B. Am. Meteorol. Soc., 93, 1865–1877, 2012.

Brunet, G., Shapiro, M., Hoskins, B., Moncrieff, M., Dole, R., Kiladis, G. N., Kirtman, B., Lorenc, A., Mills, B., Morss, R., Polavarapu, S., Rogers, D., Schaake, J., and Shukla, J.: Collaboration of the weather and climate communities to advance subseasonal-to-seasonal prediction, B. Am. Meteorol. Soc., 91, 1397–1406, 2010.

Burek, P., Van Der Knijff, J. M., and De Roo, A.: LISFLOOD – Distributed Water Balance and Flood Simulation Model – Revised User Manual 2013, Report 978-92-79-33191-6 (print); 978-92-79-33190-9, Joint Research Centre, European Commission, https://doi.org/10.2788/24982, 2013.

Ciais, P., Reichstein, M., Viovy, N., Granier, A., Ogée, J., Allard, V., Aubinet, M., Buchmann, N., Bernhofer, C., Carrara, A., Chevallier, F., Noblet, N. De, Friend, A. D., Friedlingstein, P., Grünwald, T., Heinesch, B., Keronen, P., Knohl, A., Krinner, G., Loustau, D., Manca, G., Matteucci, G., Miglietta, F., Ourcival, J. M., Papale, D., Pilegaard, K., Rambal, S., Seufert, G., Soussana, J. F., Sanz, M. J., Schulze, E. D., Vesala, T., and Valentini, R.: Europe-wide reduction in primary productivity caused by the heat and drought in 2003, Nature, 437, 529–533, 2005.

Dee, D. P., Uppala, S. M., Simmons, A. J., Berrisford, P., Poli, P., Kobayashi, S., Andrae, U., Balmaseda, M. A., Balsamo, G., Bauer, P., Bechtold, P., Beljaars, A. C. M., van de Berg, L., Bidlot, J., Bormann, N., Delsol, C., Dragani, R., Fuentes, M., Geer, A. J., Haimberger, L., Healy, S. B., Hersbach, H., Hólm, E. V., Isaksen, L., Kållberg, P., Khler, M., Matricardi, M., McNally, A. P., Monge-Sanz, B. M., Morcrette, J. J., Park, B. K., Peubey, C., de Rosnay, P., Tavolato, C., Thépaut, J. N., and Vitart, F.: The ERA-Interim reanalysis: configuration and performance of the data assimilation system, Q. J. Roy. Meteor. Soc., 137, 553–597, https://doi.org/10.1002/qj.828, 2011.

De Roo, A. P. J., Wesseling, C. G., and Van Deursen, W. P. A.: Physically based river basin modelling within a GIS: the LISFLOOD model, Hydrol. Process., 14, 1981–1992, https://doi.org/10.1002/1099-1085(20000815/30)14:11/12<1981::AID-HYP49>3.0.CO;2-F,, 2000.

Di Giuseppe, F., Molteni, F., and Tompkins, A. M.: A rainfall calibration methodology for impacts modelling based on spatial mapping, Q. J. Roy. Meteor. Soc., 139, 1389–1401, 2013.

Dutra, E., Pozzi, W., Wetterhall, F., Di Giuseppe, F., Magnusson, L., Naumann, G., Barbosa, P., Vogt, J., and Pappenberger, F.: Global meteorological drought – Part 2: Seasonal forecasts, Hydrol. Earth Syst. Sci., 18, 2669–2678, https://doi.org/10.5194/hess-18-2669-2014, 2014.

Ferro, C. A. T., Richardson, D. S., and Weigel, A. P.: On the effect of ensemble size on the discrete and continu-

ous ranked probability scores, Meteorol. Appl., 15, 19–24, https://doi.org/10.1002/met.45, 2008.

García-Herrera, R., Díaz, J., Trigo, R., Luterbacher, J., and Fischer, E.: A review of the European summer heat wave of 2003, Crit. Rev. Env. Sci. Tec., 40, 267–306, 2010.

Haiden, T., Magnusson, L., Tsonevsky, I., Wetterhall, F., Alfieri, L., Pappenberger, F., de Rosnay, P., Muñoz-Sabater, J., Balsamo, G., Albergel, C., Forbes, R., Hewson, T., Malardel, S., and Richardson, D.: ECMWF forecast performance during the June 2013 flood in Central Europe, Report, Euopean Centre for Medium-Range Weather Forecasts, 2014.

Hazeleger,W., Severijns, C., Semmler, T., Stefanescu, S., Yang, S.,Wang, X.,Wyser, K., Dutra, E., Baldasano, J. M., Bintanja, R., Bougeault, P., Caballero, R., Ekman, A. M. L., Christensen, J. H., van den Hurk, B., Jimenez, P., Jones, C., Kållberg, P., Koenigk, T., McGrath, R., Miranda, P., van Noije, T., Palmer, T., Parodi, J. A., Schmith, T., Selten, F., Storelvmo, T., Sterl, A., Tapamo, H., Vancoppenolle, M., Viterbo, P., and Willén, U.: EC-Earth: a seamless earth-system prediction approach in action, B. Am. Meteorol. Soc., 91, 1357–1363, 2010.

Hazeleger, W.,Wang, X., Severijns, C., Stefanescu, S., Bintanja, R., Sterl, A., Wyser, K., Semmler, T., Yang, S., van den Hurk, B., van Noije, T., van der Linden, E., and van der Wiel, K.: EC-Earth V2. 2: description and validation of a new seamless earth system prediction model, Clim. Dynam., 39, 2611–2629, 2012.

Hersbach, H.: Decomposition of the Continuous Ranked Probability Score for Ensemble Prediction Systems, Weather Forecast., 15, 559–570, https://doi.org/10.1175/1520-0434(2000)015<0559:DOTCRP>2.0.CO;2, 2000.

Hoskins, B.: The potential for skill across the range of the seamless weather-climate prediction problem: a stimulus for our science, Q. J. Roy. Meteor. Soc., 139, 573–584, 2013.

Hsu, W.-R. and Murphy, A. H.: The attributes diagram A geometrical framework for assessing the quality of probability forecasts, Int. J. Forecast., 2, 285–293, https://doi.org/10.1016/0169-2070(86)90048-8, 1986.

Hurrell, J., Meehl, G. A., Bader, D., Delworth, T. L., Kirtman, B., and Wielicki, B.: A unified modeling approach to climate system prediction, B. Am. Meteorol. Soc., 90, 1819–1832, 2009.

Jonkeren, O., van Ommeren, J., and Rietveld, P.: Effects of low water levels on the river Rhine on the inland waterway transport sector, in: Economics and Management of Climate Change, 53–64, Springer, 2008.

Meiner, D., Klein, B., and Ionita, M.: Development of a monthly to seasonal forecast framework tailored to inland waterway transport in central Europe, Hydrol. Earth Syst. Sci., 21, 6401–6423, https://doi.org/10.5194/hess-21-6401-2017, 2017.

Molteni, F., Stockdale, T., Balmaseda, M., Balsamo, G., Buizza, R., Ferranti, L., Magnusson, L., Mogensen, K., Palmer, T., and Vitart, F.: The new ECMWF seasonal forecast system (System 4), Report, ECMWF, 2011.

Ntegeka, V., Salamon, P., Gomes, G., Sint, H., Lorini, V., Thielen del Pozo, J., and Zambrano, H.: EFAS-Meteo: A European daily high-resolution gridded meteorological data set for 1990–2011, Report, https://doi.org/10.2788/51262, 2013.

Palmer, T. and Webster, P.: Towards a unified approach to climate and weather prediction, in: Proceedings of 1st Demetra Conference on Climate Change, 1993.

Palmer, T. N., Doblas-Reyes, F. J., Weisheimer, A., and Rodwell, M. J.: Toward seamless prediction: Calibration of Climate

Change Projections Using Seasonal Forecasts, B. Am. Meteorol. Soc., 89, 459–470, 2008.

Pappenberger, F., Wetterhall, F., Dutra, E., Di Giuseppe, F., Bogner, K., Alfieri, L., and Cloke, H. L.: Seamless forecasting of extreme events on a global scale, 3–10, Proceedings of H01, IAHS-IAPSO-IASPEI Assembly, Gothenburg, Sweden, 2013.

Sene, K., Tych, W., and Beven, K.: Exploratory studies into seasonal flow forecasting potential for large lakes, Hydrol. Earth Syst. Sci., 22, 127–141, https://doi.org/10.5194/hess-22-127-2018, 2018.

Smith, P., Pappenberger, F., Wetterhall, F., Thielen, J., Krzeminski, B., Salamon, P., Muraro, D., Kalas, M., and Baugh, C.: On the operational implementation of the European Flood Awareness System (EFAS), Report 778, European Centre for Medium-Range Weather Forecasting, available at: http://www.ecmwf.int/en/elibrary/16337-operational-implementation-european-flood-awareness (last access: 1 April 2018), 2016.

Thielen, J., Bartholmes, J., Ramos, M.-H., and de Roo, A.: The European Flood Alert System – Part 1: Concept and development, Hydrol. Earth Syst. Sci., 13, 125–140, https://doi.org/10.5194/hess-13-125-2009, 2009.

Van Der Knijff, J. M., Younis, J., and De Roo, A. P. J.: LISFLOOD: a GIS-based distributed model for river basin scale water balance and flood simulation, Int. J. Geogr. Inf. Sci., 24, 189–212, https://doi.org/10.1080/13658810802549154, 2010.

Vitart, F., Buizza, R., Alonso Balmaseda, M., Balsamo, G., Bidlot, J. R., Bonet, A., Fuentes, M., Hofstadler, A., Molteni, F., and Palmer, T. N.: The new VarEPS-monthly forecasting system: A first step towards seamless prediction, Q. J. Roy. Meteor. Soc., 134, 1789–1799, 2008.

Weisheimer, A. and Palmer, T. N.: On the reliability of seasonal climate forecasts, J. R. Soc. Interface, 11, 96, https://doi.org/10.1098/rsif.2013.1162, 2014.

Weisheimer, A., Doblas-Reyes, F. J., Jung, T., and Palmer, T.: On the predictability of the extreme summer 2003 over Europe, Geophys. Res. Lett., 38, L05704, https://doi.org/10.1029/2010GL046455, 2011.

White, C. J., Carlsen, H., Robertson, A. W., Klein, R. J., Lazo, J. K., Kumar, A., Vitart, F., Coughlan de Perez, E., Ray, A. J., Murray, V., Bharwani, S., MacLeod, D., James, R., Fleming, L., Morse, A. P., Eggen, B., Graham, R., Kjellström, E., Becker, E., Pegion, K. V., Holbrook, N. J., McEvoy, D., Depledge, M., Perkins-Kirkpatrick, S., Brown, T. J., Street, R., Jones, L., Remenyi, T. A., Hodgson-Johnston, I., Buontempo, C., Lamb, R., Meinke, H., Arheimer, B., and Zebiak, S. E.: Potential applications of subseasonal-to-seasonal (S2S) predictions, Meteorol. Appl., 24, 315–325, https://doi.org/10.1002/met.1654, 2017.

Zajac, Z., Zambrano-Bigiarini, M., Salamon, P., Burek, P., Gentile, A., and Bianchi, A.: Calibration of the lisflood hydrological model for europe – calibration round 2013, available at: http://publications.jrc.ec.europa.eu/repository/bitstream/JRC32044/FEYEN2044EUR22125.pdf (last access: 1 June 2018), 2013.

9

Hybridizing Bayesian and variational data assimilation for high-resolution hydrologic forecasting

Felipe Hernández and Xu Liang

Department of Civil and Environmental Engineering, University of Pittsburgh, Pittsburgh, PA, 15261, USA

Correspondence: Xu Liang (xuliang@pitt.edu)

Abstract. The success of real-time estimation and forecasting applications based on geophysical models has been possible thanks to the two main existing frameworks for the determination of the models' initial conditions: Bayesian data assimilation and variational data assimilation. However, while there have been efforts to unify these two paradigms, existing attempts struggle to fully leverage the advantages of both in order to face the challenges posed by modern high-resolution models – mainly related to model indeterminacy and steep computational requirements. In this article we introduce a hybrid algorithm called OPTIMISTS (Optimized PareTo Inverse Modeling through Integrated STochastic Search) which is targeted at non-linear high-resolution problems and that brings together ideas from particle filters (PFs), four-dimensional variational methods (4D-Var), evolutionary Pareto optimization, and kernel density estimation in a unique way. Streamflow forecasting experiments were conducted to test which specific configurations of OPTIMISTS led to higher predictive accuracy. The experiments were conducted on two watersheds: the Blue River (low resolution) using the VIC (Variable Infiltration Capacity) model and the Indiantown Run (high resolution) using the DHSVM (Distributed Hydrology Soil Vegetation Model). By selecting kernel-based non-parametric sampling, non-sequential evaluation of candidate particles, and through the multi-objective minimization of departures from the streamflow observations and from the background states, OPTIMISTS was shown to efficiently produce probabilistic forecasts with comparable accuracy to that obtained from using a particle filter. Moreover, the experiments demonstrated that OPTIMISTS scales well in high-resolution cases without imposing a significant computational overhead. With the combined advantages of allowing for fast, non-Gaussian, non-linear, high-resolution prediction, the algorithm shows the potential to increase the efficiency of operational prediction systems.

1 Introduction

Decision support systems that rely on model-based forecasting of natural phenomena are invaluable to society (Adams et al., 2003; Penning-Rowsell et al., 2000; Ziervogel et al., 2005). However, despite increasing availability of Earth-sensing data, the problem of estimation or prediction in geophysical systems remains as underdetermined as ever because of the growing complexity of such models (Clark et al., 2017). For example, taking advantage of distributed physics and the mounting availability of computational power, modern models have the potential to more accurately represent impacts of heterogeneities on eco-hydrological processes (Koster et al., 2017). This is achieved through the replacement of lumped representations with distributed ones, which entails the inclusion of numerous parameters and state variables. The inclusion of these additional unknowns has the downside of increasing the level of uncertainty in their estimation. Therefore, in order to be able to rely on these high-resolution models for critical real-time and forecast applications, considerable improvements on parameter and initial state estimation techniques must be made with two main goals: first, to allow for an efficient management of the huge number of unknowns; and second, to mitigate the harmful effects of overfitting – i.e. the loss of forecast skill due to an over-reliance on the calibration and training data (Hawkins, 2004). Because of the numerous degrees of freedom associated with these high-resolution distributed models, overfitting is a much bigger threat due to the phenomenon of equi-finality (Beven, 2006).

There exists a plethora of techniques to initialize the state variables of a model through the incorporation of available observations, and they possess overlapping features that make it difficult to develop clear-cut classifications. However, two main schools can be fairly identified: Bayesian data assimilation and variational data assimilation. Bayesian data assimilation creates probabilistic estimates of the state variables in an attempt to also capture their uncertainty. These state probability distributions are adjusted sequentially to better match the observations using Bayes' theorem. While the Kalman filter (KF) is constrained to linear dynamics and Gaussian distributions, ensemble Kalman filters (EnKF) can support non-linear models (Evensen, 2009), and particle filters (PFs) can also manage non-Gaussian estimates for added accuracy (Smith et al., 2013). The stochastic nature of these Bayesian filters is highly valuable because equifinality can rarely be avoided and because of the benefits of quantifying uncertainty in forecasting applications (Verkade and Werner, 2011; Zhu et al., 2002). While superior in accuracy, PFs are usually regarded as impractical for high-dimensional applications (Snyder et al., 2008), and thus recent research has focused on improving their efficiency (van Leeuwen, 2015).

On the other hand, variational data assimilation is more akin to traditional calibration approaches (Efstratiadis and Koutsoyiannis, 2010) because of its use of optimization methods. It seeks to find a single–deterministic initial-state-variable combination that minimizes the departures (or variations) of the modelled values from the observations (Reichle et al., 2001) and, commonly, from their history. One- to three-dimensional variants are also employed sequentially, but the paradigm lends itself easily to evaluating the performance of candidate solutions throughout an extended time window in four-dimensional versions (4D-Var). If the model's dynamics are linearized, the optimum can be very efficiently found in the resulting convex search space through the use of gradient methods. While this feature has made 4D-Var very popular in meteorology and oceanography (Ghil and Malanotte-Rizzoli, 1991), its application in hydrology has been less widespread because of the difficulty of linearizing land-surface physics (Liu and Gupta, 2007). Moreover, variational data assimilation requires the inclusion of computationally expensive adjoint models if one wishes to account for the uncertainty of the state estimates (Errico, 1997).

Traditional implementations from both schools have interesting characteristics and thus the development of hybrid methods has received considerable attention (Bannister, 2016). For example, Bayesian filters have been used as adjoints in 4D-Var to enable probabilistic estimates (Zhang et al., 2009). Moreover, some Bayesian approaches have been coupled with optimization techniques to select ensemble members (Dumedah and Coulibaly, 2013; Park et al., 2009). The fully hybridized algorithm 4DEnVar (Buehner et al., 2010) is gaining increasing attention for weather predic-

tion (Desroziers et al., 2014; Lorenc et al., 2015). It is especially interesting that some algorithms have defied the traditional choice between sequential and extended-time evaluations. Weakly constrained 4D-Var allows state estimates to be determined at several time steps within the assimilation time window and not only at the beginning (Ning et al., 2014; Trémolet, 2006). Conversely, modifications to EnKFs and PFs have been proposed to extend the analysis of candidate members/particles to span multiple time steps (Evensen and van Leeuwen, 2000; Noh et al., 2011). The success of these hybrids demonstrates that there is a balance to be sought between the allowed number of degrees of freedom and the amount of information to be assimilated at once.

Following these promising paths, in this article we introduce OPTIMISTS (Optimized PareTo Inverse Modelling through Integrated STochastic Search), a hybrid data assimilation algorithm whose design was guided by the two stated goals: (i) to allow for practical scalability to high-dimensional models and (ii) to enable balancing the imperfect observations and the imperfect model estimates to minimize overfitting. Table 1 summarizes the main characteristics of typical Bayesian and variational approaches and their contrasts with those of OPTIMISTS. Our algorithm incorporates the features that the literature has found to be the most valuable from both Bayesian and variational methods while mitigating the deficiencies or disadvantages associated with these original approaches (e.g. the linearity and determinism of 4D-Var and the limited scalability of PFs): Non-Gaussian probabilistic estimation and support for non-linear model dynamics have been long held as advantageous over their alternatives (Gordon et al., 1993; van Leeuwen, 2009) and, similarly, meteorologists favour extended-period evaluations over sequential ones (Gauthier et al., 2007; Rawlins et al., 2007; Yang et al., 2009). As shown in the table, OPTIMISTS can readily adopt these proven strategies.

However, there are other aspects of the assimilation problem for which no single combination of features has demonstrated its superiority. For example, is the consistency with previous states better achieved through the minimization of a cost function that includes a background error term (Fisher, 2003), as in variational methods, or through limiting the exploration to samples drawn from that background state distribution, as in Bayesian methods? Table 1 shows that in these cases OPTIMISTS allows for flexible configurations, and it is an additional objective of this study to test which set of feature interactions allows for more accurate forecasts when using highly distributed models. While many of the concepts utilized within the algorithm have been proposed in the literature before, their combination and broad range of available configurations are unlike those of other methods – including existing hybrids which have mostly been developed around ensemble Kalman filters and convex optimization techniques (Bannister, 2016) – and therefore limited to Gaussian distributions and linear dynamics.

Table 1. Comparison between the main features of standard Bayesian data assimilation algorithms (KF: Kalman filter, EnKF: ensemble KF, PF: particle filter), variational data assimilation (one- to four-dimensional), and OPTIMISTS.

	Bayesian	Variational	OPTIMISTS
Resulting state-variable estimate	Probabilistic: Gaussian (KF, EnKF), non-Gaussian (PF)	Deterministic (unless adjoint model is used)	Probabilistic (using kernel density estimation)
Solution quality criteria	High likelihood given observations	Minimum cost value (error, departure from history)	Flexible: e.g. minimum error, maximum consistency with history
Analysis time step	Sequential	Sequential (1-D–3-D) or entire assimilation window (4-D)	Flexible
Search method	Iterative Bayesian belief propagation	Convex optimization	Coupled belief propagation and multi-objective optimization
Model dynamics	Linear (KF), non-linear (EnKF, PF)	Linearized to obtain convex solution space	Non-linear (non-convex solution space)

2 Data assimilation algorithm

In this section we describe OPTIMISTS, our proposed data assimilation algorithm which combines advantageous features from several Bayesian and variational methods. As will be explained in detail for each of the steps of the algorithm, these features were selected with the intent of mitigating the limitations of existing methods. OPTIMISTS allows selecting a flexible data assimilation time step Δt – i.e. the time window in which candidate state configurations are compared to observations. It can be as short as the model time step or as long as the entire assimilation window. For each assimilation time step at time t a new state probability distribution $S^{t+\Delta t}$ is estimated from the current distribution S^t, the model, and one or more observations $o_{\mathrm{obs}}^{t:t+\Delta t}$. For hydrologic applications, as those explored in this article, these states S include land-surface variables within the modelled watershed such as soil moisture, snow cover and water equivalent, and stream water volume; and observations o are typically of streamflow at the outlet (Clark et al., 2008), soil moisture (Houser et al., 1998), and/or snow cover (Andreadis and Lettenmaier, 2006). However, the description of the algorithm will use field-agnostic terminology to not discourage its application in other disciplines.

State probability distributions S in OPTIMISTS are determined from a set of weighted root or base sample states s_i using multivariate weighted kernel density estimation (West, 1993). This form of non-parametric distributions stands in stark contrast with those from KFs and EnKFs in their ability to model non-Gaussian behaviour – an established advantage of PFs. Each of these samples or ensemble members s_i is comprised of a value vector for the state variables. The objective of the algorithm is then to produce a set of n samples $s_i^{t+\Delta t}$ with corresponding weights w_i for the next assimilation time step to determine the target distribution $S^{t+\Delta t}$.

This process is repeated iteratively each assimilation time step Δt until the entire assimilation time frame is covered, at which point the resulting distribution can be used to perform the forecast simulations. In Sect. 2.1 we describe the main ideas and steps involved in the OPTIMISTS data assimilation algorithm; details regarding the state probability distributions, mainly on how to generate random samples and evaluate the likelihood of particles, are explained in Sect. 2.2; and modifications required for high-dimensional problems are described in Sect. 2.3.

2.1 Description of the OPTIMISTS data assimilation algorithm

Let a "particle" P_i be defined by a "source" (or initial) vector of state variables s_i^t (which is a sample of distribution S^t), a corresponding "target" (or final) state vector $s_i^{t+\Delta t}$ (a sample of distribution $S^{t+\Delta t}$), a set of output values $o_i^{t:t+\Delta t}$ (those that have corresponding observations $o_{\mathrm{obs}}^{t:t+\Delta t}$), a set of fitness metrics f_i, a rank r_i, and a weight w_i. Note that the denomination "particle" stems from the PF literature and is analogous to the "member" term in EnKFs. The fitness metrics f_i are used to compare particles with each other in the light of one or more optimization objectives. The algorithm consists of the following steps, whose motivation and details are included in the sub-subsections below and their interactions illustrated in Fig. 1. Table 2 lists the meaning of each of the seven global parameters (Δt, n, w_{root}, p_{samp}, $k_{\mathrm{F\text{-}class}}$, n_{evo}, and g).

1. Drawing: draw root samples s_i^t from S^t in descending weight order until $\sum w_i \geq w_{\mathrm{root}}$.

2. Sampling: randomly sample S^t until the total number of samples in the ensemble is $p_{\mathrm{samp}} \times n$.

Table 2. List of global parameters in OPTIMISTS.

Symbol	Description	Range
Δt	Assimilation time step (particle evaluation time frame)	$\mathbb{R}^+$
n	Total number of root states s_i in the probability distributions	$\mathbb{N} \geq 2$
w_{root}	Total weight of root samples drawn from S^t	$\mathbb{R} \in [0, 1]$
p_{samp}	Percentage of n corresponding to drawn and random samples	$\mathbb{R} \in [0, 1]$
$k_{\text{F-class}}$	Whether or not to use F-class kernels. If not, use D-class kernels.	true or false
n_{evo}	Samples to be generated by the optimizers per iteration	$\mathbb{N} \geq 2$
g	Level of greed for the assignment of particle weights w_i	$\mathbb{R} \in [-1, 1]$

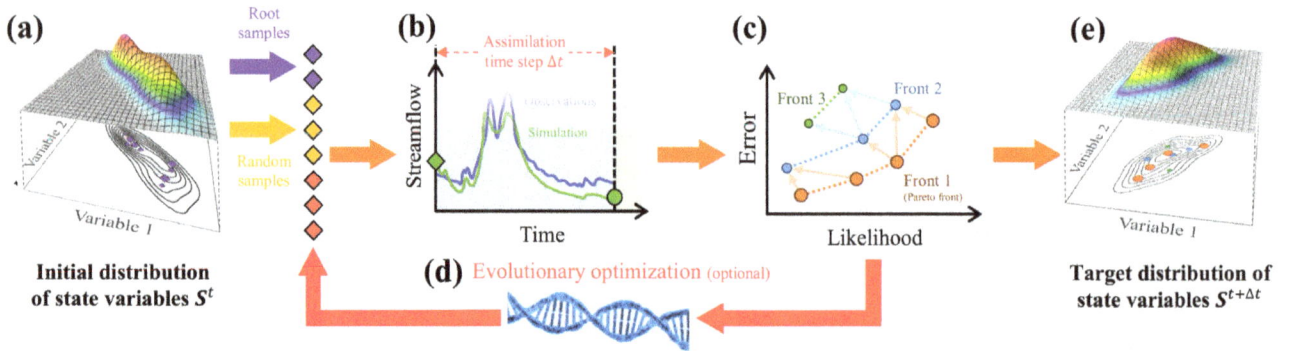

Figure 1. Steps in OPTIMISTS, to be repeated for each assimilation time step Δt. In this example state vectors have two variables, observations are of streamflow, and particles are judged using two user-selected objectives: the likelihood given S^t to be maximized and the error given the observations to be minimized. **(a)** Initial state kernel density distribution S^t from which root samples (purple rhombi) are taken during the drawing step and random samples (yellow rhombi) are taken during the sampling step. **(b)** Execution of the model (simulation step) for each source sample for a time equal to Δt to compute output variables (for comparison with observations) and target samples (circles). **(c)** Evaluation of each particle (evaluation step) based on the objectives and organization into non-domination fronts (ranking step). The dashed lines represent the fronts while the arrows denote domination relationships between particles in adjacent fronts. **(d)** Optional optimization step which can be executed several times and that uses a population-based evolutionary optimization algorithm to generate additional samples (red rhombi). **(e)** Target state kernel density distribution S^{t+t} constructed from the particles' final samples (circles) after being weighted according to the rank of their front (weighting step): kernels centred on samples with higher weight (shown larger) have a higher probability density contribution.

3. Simulation: compute $s_i^{t+\Delta t}$ and $o_i^{t:t+\Delta t}$ from each non-evaluated sample s_i^t using the model.

4. Evaluation: compute the fitness values f_i for each particle P_i.

5. Optimization: create additional samples using evolutionary algorithms and return to 3 (if number of samples is below n).

6. Ranking: assign ranks r_i to all particles P_i using non-dominated sorting.

7. Weighting: compute the weight w_i for each particle P_i based on its rank r_i.

2.1.1 Drawing step

While traditional PFs draw all the root (or base) samples from S^t (Gordon et al., 1993), OPTIMISTS can limit this selection to a subset of them. The root samples with the highest

weight – those that are the best performers – are drawn first, followed by the next ones in descending weight order, until the total weight of the drawn samples $\sum w_i$ reaches w_{root}. w_{root} thus controls what percentage of the root samples to draw, and, if set to one, all of them are selected.

2.1.2 Sampling step

In this step the set of root samples drawn is complemented with random samples. The distinction between root samples and random samples is that the former are those that define the probability distribution S^t (that serve as centroids for the kernels), while the latter are generated stochastically from the kernels. Random samples are generated until the size of the combined set reaches $p_{\text{samp}} \times n$ by following the equations introduced in Sect. 2.2. This second step contributes to the diversity of the ensemble in order to avoid sample impoverishment as seen on PFs (Carpenter et al., 1999) and serves as a replacement for traditional resampling strategies

(Liu and Chen, 1998). The parameter w_{root} therefore controls the intensity with which this feature is applied to offer users some level of flexibility. Generating random samples at the beginning, instead of resampling those that have been already evaluated, could lead to discarding degenerate particles (those with high errors) early on and contribute to improved efficiency, given that the ones discarded are mainly those with the lowest weight as determined in the previous assimilation time step.

2.1.3 Simulation step

In this step, the algorithm uses the model to compute the resulting state vector $s_i^{t+\Delta t}$ and an additional set of output variables $o_i^{t:t+\Delta t}$ for each of the samples (it is possible that state variables double as output variables). The simulation is executed starting at time t for the duration of the assimilation time step Δt (not to be confused with the model time step which is usually shorter). Depending on the complexity of the model, the simulation step can be the one with the highest computational requirements. In those cases, parallelization of the simulations would greatly help in reducing the total footprint of the assimilation process. The construction of each particle P_i is started by assembling the corresponding values computed so far: s_i^t (drawing, sampling, and optimization steps), and $s_i^{t+\Delta t}$ and $o_i^{t:t+\Delta t}$ (simulation step).

2.1.4 Evaluation step

In order to determine which initial state s_i^t is the most desirable, a two-term cost function J is typically used in variational methods that simultaneously measures the resulting deviations of modelled values $o_i^{t:t+\Delta t}$ from observed values $o_{obs}^{t:t+\Delta t}$ and the departures from the background state distribution S^t (Fisher, 2003). The function usually has the form shown in Eq. (1):

$$J_i = c_1 \cdot J_{background}\left(s_i^t, S^t\right) + c_2$$
$$\cdot J_{observations}\left(o_i^{t:t+\Delta t}, o_{obs}^{t:t+\Delta t}\right), \tag{1}$$

where c_1 and c_2 are balancing constants usually set so that $c_1 = c_2$. Such a multi-criteria evaluation is crucial both to guarantee a good level of fit with the observations (second term) and to avoid the optimization algorithm to produce an initial state that is inconsistent with previous states (first term) – which could potentially result in overfitting problems rooted in disproportionate violations of mass and energy conservation laws (e.g. in hydrologic applications a sharp, unrealistic rise in the initial soil moisture could reduce $J_{observations}$ but would increase $J_{background}$). In Bayesian methods, since the consistency with the state history is maintained by sampling only from the prior or background distribution S^t, single-term functions are used instead – which typically measure the probability density or likelihood of the modelled values given a distribution of the observations.

In OPTIMISTS any such fitness metric could be used and, most importantly, the algorithm allows defining several of them. Moreover, users can determine whether each function is to be minimized (e.g. costs or errors) or maximized (e.g. likelihoods). We expect these features to be helpful if one wishes to separate errors when multiple types of observations are available (Montzka et al., 2012) and as a more natural way to consider different fitness criteria (lumping them together in a single function as in Eq. (1) can lead to balancing and "apples and oranges" complications). Moreover, it might prove beneficial to take into account the consistency with the state history both by explicitly defining such an objective here and by allowing states to be sampled from the previous distribution (and thus compounding the individual mechanisms of Bayesian and variational methods). Functions to measure this consistency are proposed in Sect. 2.2. With the set of objective functions defined by the user, the algorithm computes the vector of fitness metrics f_i for each particle during the evaluation step.

2.1.5 Optimization step

The optimization step is optional and is used to generate additional particles by exploiting the knowledge encoded in the fitness values of the current particle ensemble. In a twist to the signature characteristic of variational data assimilation, OPTIMISTS incorporates evolutionary multi-objective optimization algorithms (Deb, 2014) instead of the established gradient-based, single-objective methods. Evolutionary optimizers compensate for their slower convergence speed with the capability of efficiently navigating non-convex solution spaces (i.e. the models and the fitness functions do not need to be linear with respect to the observations and the states). This feature effectively opens the door for variational methods to be used in disciplines where the linearization of the driving dynamics is either impractical, inconvenient, or undesirable. Whereas any traditional multi-objective global optimization method would work, our implementation of OPTIMISTS features a state-of-the-art adaptive ensemble algorithm similar to the algorithm of Vrugt and Robinson (2007), AMALGAM, that allows model simulations to be run in parallel (Crainic and Toulouse, 2010). The optimizer ensemble includes a genetic algorithm (Deb et al., 2002) and a hybrid approach that combines ant colony optimization (Socha and Dorigo, 2008) and Metropolis–Hastings sampling (Haario et al., 2001).

During the optimization step, the group of optimizers is used to generate n_{evo} new sample states s_i^t based on those in the current ensemble. For example, the genetic algorithm selects pairs of base samples with high performance scores f_i and then proceeds to combine their individual values using standard crossover and mutation operators. The simulation and evaluation steps are repeated for these new samples, and then this iterative process is repeated until the particle ensemble has a size of n. Note that w_{root} and p_{samp} thus determine what percentage of the particles is generated in which way. For example, for relatively small values of w_{root} and a p_{samp} of 0.2, 80 % of the particles will be generated by the optimization algorithms. In this way, OPTIMISTS offers its users the flexibility to behave anywhere in the range between fully Bayesian ($p_{samp} = 1$) and fully variational ($p_{samp} = 0$) in terms of particle generation. In the latter case, in which no root and random samples are available, the initial population or ensemble of states s_i^t is sampled uniformly from the viable range of each state variable.

2.1.6 Ranking step

A fundamental aspect of OPTIMISTS is the way in which it provides a probabilistic interpretation to the results of the multi-objective evaluation, thus bridging the gap between Bayesian and variational assimilation. Such method has been used before (Dumedah et al., 2011) and is based on the employment of non-dominated sorting (Deb, 2014), another technique from the multi-objective optimization literature, which is used to balance the potential tensions between various objectives. This sorting approach is centred on the concept of dominance, instead of organizing all particles from the best to the worst. A particle dominates another if it outperforms it according to at least one of the criteria/objectives while simultaneously is not being outperformed according to any of the others. Following this principle, in the ranking step particles are grouped in fronts comprised of members which are mutually non-dominated; that is, none of them is dominated by any of the rest. Particles in a front, therefore, represent the effective trade-offs between the competing criteria.

Figure 1c illustrates the result of non-dominated sorting applied to nine particles being analysed under two objectives: minimum deviation from observations and maximum likelihood given the background state distribution S^t. Note that, if a single-objective function is used, the sorting method assigns ranks from best to worst according to that function, and two particles would only share ranks if their fitness values coincide. In our implementation we use the fast non-dominated sorting algorithm to define the fronts and assign the corresponding ranks r_i (Deb et al., 2002). More efficient non-dominated sorting alternatives are available if performance becomes an issue (Zhang et al., 2015).

2.1.7 Weighting step

In this final step, OPTIMISTS assigns weights w_i to each particle according to its rank r_i as shown in Eqs. (2) and (3). This Gaussian weighting depends on the ensemble size n and the greed parameter g and is similar to the one proposed by Socha and Dorigo (2008). When g is equal to zero, particles in all fronts are weighted uniformly; when g is equal to one, only particles in the Pareto or first front are assigned non-zero weights. With this, the final estimated probability distribution

of state variables for the next time step $S^{t+\Delta t}$ can be established using multivariate weighted kernel density estimation (details in the next sub-section), as demonstrated in Fig. 1e., by taking all target states $s_i^{t+\Delta t}$ (circles) as the centroids of the kernels. The obtained distribution $S^{t+\Delta t}$ can then be used as the initial distribution for a new assimilation time step or, if the end of the assimilation window has been reached, it can be used to perform (ensemble) forecast simulations.

$$w_i = \frac{1}{\sigma\sqrt{2\pi}} e^{-\frac{(r_i-1)^2}{2\sigma^2}} \tag{2}$$

$$\sigma = n \cdot \left[0.1 + 9.9 \cdot (1-g)^5 \right] \tag{3}$$

2.2 Model state probability distributions

As mentioned before, OPTIMISTS uses kernel density probability distributions (West, 1993) to model the stochastic estimates of the state-variable vectors. The algorithm requires two computations related to the state-variable probability distribution S^t: obtaining the probability density p or likelihood $\mathcal{L}$ of a sample and generating random samples. The first computation can be used in the evaluation step as an objective function to preserve the consistency of particles with the state history (e.g. to penalize aggressive departures from the prior conditions). It should be noted that several metrics that try to approximate this consistency exist, from very simple (Dumedah et al., 2011) to quite complex (Ning et al., 2014). For example, it is common in variational data assimilation to utilize the background error term

$$J_{background} = (s - s_b)^T \mathbf{C}^{-1} (s - s_b), \tag{4}$$

where s_b and $\mathbf{C}$ are the mean and the covariance of the background state distribution (S^t in our case), which is assumed to be Gaussian (Fisher, 2003). The term $J_{background}$ is plugged into the cost function shown in Eq. (1). For OPTIMISTS, we propose that the probability density of the weighted state kernel density distribution S^t at a given point (p) be used as a stand-alone objective. The density is given by Eq. (5) (Wand and Jones, 1994). If Gaussian kernels are selected, the kernel function K, parameterized by the bandwidth matrix $\mathbf{B}$, is evaluated using Eq. (6).

$$p(s|s) = \frac{1}{\sum w_i} \sum_{i=1}^{n} \left[w_i \cdot K_B (s - s_i) \right] \tag{5}$$

$$K_B^{Gauss}(\mathbf{z}) = \frac{1}{\sqrt{(2\pi)^n \cdot |\mathbf{B}|}} \exp\left(-\frac{1}{2}\mathbf{z}^T \mathbf{B}^{-1}\mathbf{z}\right) \tag{6}$$

Matrix $\mathbf{B}$ is the covariance matrix of the kernels and thus determines their spread and orientation in the state space. $\mathbf{B}$ is of size $d \times d$, where d is the dimensionality of the state distribution (i.e. the number of variables), and can be thought of as a scaled-down version of the background error covariance matrix $\mathbf{C}$ from the variational literature. In this sense, matrix $\mathbf{B}$, together with the spread of the ensemble of samples s_i, effectively encodes the uncertainty of the state variables. Several optimization-based methods exist to compute **B** by attempting to minimize the asymptotic mean integrated squared error (AMISE) (Duong and Hazelton, 2005; Sheather and Jones, 1991). However, here we opt to use a simplified approach for the sake of computational efficiency: we determine **B** by scaling down the sample covariance matrix **C** using Silverman's rule of thumb, which takes into account the number of samples n and the dimensionality of the distribution d, as shown in Eq. (7) (Silverman, 1986). Figure 1 shows the density of two two-dimensional example distributions using this method (Fig. 1a and e). If computational constraints are not a concern, using AMISE-based methods or kernels with variable bandwidth (Hazelton, 2003; Terrell and Scott, 1992) could result in higher accuracy.

$$\mathbf{B}^{\text{Silverman}} = \left(\frac{4}{d+2}\right)^{\frac{2}{d+4}} \cdot n^{-\frac{2}{d+4}} \cdot \mathbf{C} \qquad (7)$$

Secondly, OPTIMISTS' sampling step requires generating random samples from a multivariate weighted kernel density distribution. This is achieved by dividing the problem into two: we first select the root sample and then generate a random sample from the kernel associated with that base sample. The first step corresponds to randomly sampling a multinomial distribution with n categories and assigning the normalized weights of the particles as the probability of each category. Once a root sample s_{root} is selected, a random sample s_{random} can be generated from a vector v of independent standard normal random values of size d and a matrix **A** as shown in Eq. (8). **A** can be computed from a Cholesky decomposition (Krishnamoorthy and Menon, 2011) such that $\mathbf{AA}^T = \mathbf{B}$. Alternatively, an eigendecomposition can be used to obtain $\mathbf{Q\Lambda Q}^T = \mathbf{B}$ to then set $\mathbf{A} = \mathbf{Q\Lambda}^{\frac{1}{2}}$.

$$s_{\text{random}} = s_{\text{root}} + \mathbf{A}v \qquad (8)$$

Both computations (density or likelihood and sampling) require **B** to be invertible and, therefore, that none of the variables have zero variance or are perfectly linearly dependent on each other. Zero-variance variables must therefore be isolated and **B** marginalized before attempting to use Eq. (6) or to compute **A**. Similarly, linear dependencies must also be identified beforehand. If we include variables one by one in the construction of **C**, we can determine if a newly added one is linearly dependent if the determinant of the extended sample covariance matrix **C** is zero. Once identified, the regression coefficients for the dependent variable can be efficiently computed from **C** following the method described by Friedman et al. (2008). The constant coefficient of the regression must also be calculated for future reference. What this process effectively does is to determine a linear model for each dependent variable that is represented by a set of regression coefficients. Dependent variables are not included in **C**, but they need to be taken into account afterwards (e.g. by deter-

mining their values for the random samples by solving the linear model with the values obtained for the variables in **C**).

2.3 High-dimensional state vectors

When the state vector of the model becomes large (i.e. d increases), as is the case for distributed high-resolution numerical models, difficulties start to arise when dealing with the computations involving the probability distribution. At first, the probability density, as computed with Eqs. (5) and (6), tends to diverge either towards zero or towards infinity. This phenomenon is related to the normalization of the density – so that it can integrate to one – and to its fast exponential decay as a function of the sample's distance from the kernel's centres. In these cases we propose replacing the density computation with an approximated likelihood formulation that is proportional to the inverse square Mahalanobis distance (Mahalanobis, 1936) to the root samples, thus skipping the exponentiation and normalization operations of the Gaussian density. This simplification, which corresponds to the inverse square difference between the sample value and the kernel's mean in the univariate case, is shown in Eq. (9). The resulting distortion of the Gaussian bell-curve shape does not affect the results significantly, given that OPTIMISTS uses the fitness functions only to check for domination between particles – so only the signs of the differences between likelihood values are important and not their actual magnitudes.

$$\mathcal{L}^{\text{Mahalanobis}}(s|S) = \frac{1}{\sum w_i} \sum_{i=1}^{n} \frac{w_i}{\left|(s - s_i)^T \mathbf{B}^{-1}(s - s_i)\right|} \qquad (9)$$

However, computational constraints might also make this simplified approach unfeasible both due to the $O(d^2)$ space requirements for storing the bandwidth matrix **B** and the $O(d^3)$ time complexity of the decomposition algorithms, which rapidly become huge burdens for the memory and the processors. Therefore, we can chose to sacrifice some accuracy by using a diagonal bandwidth matrix **B** which does not include any covariance term – only the variance terms in the diagonal are computed and stored. This implies that, even though the multiplicity of root samples would help in maintaining a large portion of the covariance, another portion is lost by preventing the kernels from reflecting the existing correlations. In other words, variables would not be rendered completely independent, but rather conditionally independent because the kernels are still centred on the set of root samples. Kernels using diagonal bandwidth matrices are referred to as "D-class" kernels while those using the full covariance matrix are referred to as "F-class" kernels. The $k_{\text{F-class}}$ parameter controls which version is used.

With only the diagonal terms of matrix **B** available (b_{jj}), we opt to roughly approximate the likelihood by computing the average of the standardized marginal likelihood value for each variable j, as shown in Eq. (10):

$$\mathcal{L}^{\text{independent}}(s \mid S) = \frac{1}{d\sqrt{2\pi}\sum w_i} \sum_{j=1}^{d} \sum_{i=1}^{n}$$

$$\left\{ w_i \cdot \exp\left[-\frac{\left(s_j - s_{i,j}\right)^2}{2b_{jj}} \right] \right\}, \qquad (10)$$

where s_j represents the jth element of state vector s and $s_{i,j}$ represents the jth element of the ith sample of probability distribution S. Independent and marginal random sampling of each variable can also be applied to replace Eq. (8) by adding random Gaussian residuals to the elements of the selected root sample s_{root}. Sparse bandwidth matrices (Friedman et al., 2008; Ghil and Malanotte-Rizzoli, 1991) or low-rank approximations (Bannister, 2008; Ghorbanidehno et al., 2015; Li et al., 2015) could be worthwhile intermediate alternatives to our proposed quasi-independent approach to be explored in the future.

3 Experimental set-up

In this section we prepare the elements to investigate whether OPTIMISTS can help improve the forecasting skill of hydrologic models. More specifically, the experiments seek to answer the following questions. Which characteristics of Bayesian and variational methods are the most advantageous? How can OPTIMISTS be configured to take advantage of these characteristics? How does the algorithm compare to established data assimilation methods? And how does it perform with high-dimensional applications? To help answer these questions, this section first introduces two case studies and then it describes a traditional PF that was used for comparison purposes.

3.1 Case studies

We coupled a Java implementation of OPTIMISTS with two popular open-source distributed hydrologic modelling engines: Variable Infiltration Capacity (VIC) (Liang et al., 1994, 1996a b; Liang and Xie, 2001, 2003) and the Distributed Hydrology Soil Vegetation Model (DHSVM) (Wigmosta et al., 1994, 2002). VIC is targeted at large watersheds by focusing on vertical subsurface dynamics and also enabling intra-cell precipitation, soil, and vegetation heterogeneity. The DHSVM, on the other hand, was conceived for high-resolution representations of the Earth's surface, allowing for saturated and unsaturated subsurface flow routing and 1-D or 2-D surface routing (Zhang et al., 2018). Both engines needed several modifications so that they could be executed in a non-continuous fashion as required for sequential assimilation. Given the non-Markovian nature of surface routing schemes coupled with VIC that are based either on multiscale approaches (Guo et al., 2004; Wen et al., 2012) or on

the unit hydrograph concept (Lohmann et al., 1998), a simplified routing routine was developed that treats the model cells as channels – albeit with longer retention times. In the simplified method, direct run-off and baseflow produced by each model cell is partly routed through an assumed equivalent channel (slow component) and partly poured directly to the channel network (fast component). Both the channel network and the equivalent channels representing overland flow hydraulics are modelled using the Muskingum method. On the other hand, several important bugs in version 3.2.1 of the DHSVM, mostly related to the initialization of state variables but also pertaining to routing data and physics, were fixed.

We selected two watersheds to perform streamflow forecasting tests using OPTIMISTS: one with the VIC model running at a $1/8°$ resolution for the Blue River in Oklahoma and the other with the DHSVM running at a 100 m resolution for the Indiantown Run in Pennsylvania. Table 3 lists the main characteristics of the two test watersheds and the information of their associated model configurations. Figure 2 shows the land cover map together with the layout of the modelling cells for the two watersheds. The multiobjective ensemble optimization algorithm included in OPTIMISTS was employed to calibrate the parameters of the two models with the streamflow measurements from the corresponding USGS stations. For the Blue River, the traditional ℓ_2-norm Nash–Sutcliffe efficiency (NSE_{ℓ_2}) (which focuses mostly on the peaks of hydrographs), an ℓ_1-norm version of the Nash–Sutcliffe efficiency coefficient (NSE_{ℓ_1}) (Krause et al., 2005), and the mean absolute relative error (MARE) (which focuses mostly on the inter-peak periods) were used as optimization criteria. From 85 600 candidate parameterizations tried, one was chosen from the resulting Pareto front with $\text{NSE}_{\ell_2} = 0.69$, $\text{NSE}_{\ell_1} = 0.56$, and $\text{MARE} = 44.71\%$. For the Indiantown Run, the NSE_{ℓ_2}, MARE, and absolute bias were optimized, resulting in a parameterization, out of 2575, with $\text{NSE}_{\ell_2} = 0.81$, $\text{MARE} = 37.85\%$, and an absolute bias of $11.83\,\text{L}\,\text{s}^{-1}$.

These optimal parameter sets, together with additional sets produced in the optimization process, were used to run the models and determine a set of time-lagged state-variable vectors s to construct the state probability distribution S^0 at the beginning of each of a set of data assimilation scenarios. The state variables include liquid and solid interception; ponding, water equivalent, and temperature of the snow packs; and moisture and temperature of each of the soil layers. While we do not expect all of these variables to be identifiable and sensitive within the assimilation problem, we decided to be thorough in their inclusion – a decision that also increases the challenge for the algorithm in terms of the potential for overfitting. The Blue River model application has 20 cells, with a maximum of seven intra-cell soil–vegetation partitions. After adding the stream network variables, the model has a total of $d = 812$ state variables. The Indiantown Run model application has a total of 1472 cells and $d = 33\,455$ state variables.

Table 3. Characteristics of the two test watersheds: Blue River and Indiantown Run. US hydrologic units are defined in Seaber et al. (1987). Elevation information was obtained from the Shuttle Radar Topography Mission (Rodríguez et al., 2006); land cover and impervious percentage from the National Land Cover Database, NLCD (Homer et al., 2012); soil type from CONUS-SOIL (Miller and White, 1998); and precipitation, evapotranspiration, and temperature from NLDAS-2 (Cosgrove et al., 2003). The streamflow and temperature include their range of variation of 90 % of the time (5 % tails at the high and low end are excluded).

Model characteristic	Blue River	Indiantown Run
USGS station; US hydrologic unit	07332500; 11140102	01572950; 02050305
Area (km^2); impervious	3031; 8.05 %	14.78; 0.83 %
Elevation range; average slope	158–403 m; 3.5 %	153–412 m; 14.5 %
Land cover	43 % grassland, 28 % forest, 21 % pasture/hay	74.6 % deciduous forest
Soil type	Clay loam (26.4 %), clay (24.8 %), sandy loam (20.26 %)	Silt loam (51 %), sandy loam (49 %)
Average streamflow (90 % range)	$9.06\,m^3\,s^{-1}$ (0.59–$44.71\,m^3\,s^{-1}$)	$0.3\,m^3\,s^{-1}$ (0.035–$0.793\,m^3\,s^{-1}$)
Average precipitation; average ET	1086; $748\,mm\,yr^{-1}$	1176; $528\,mm\,yr^{-1}$
Average temperature (90 % range)	17.26 °C (2.5–31° C)	10.9 °C (−3.5–24 °C)
Model cells; stream segments; d	20; 14; 812	1472; 21; 33 455
Resolution	0.125°; daily	100 m; hourly
Calibration	167 parameters; 85 months; objectives: NSE_{ℓ_2}, NSE_{ℓ_1}, MARE	18 parameters; 20 months; objectives: NSE_{ℓ_2}, MARE, absolute bias

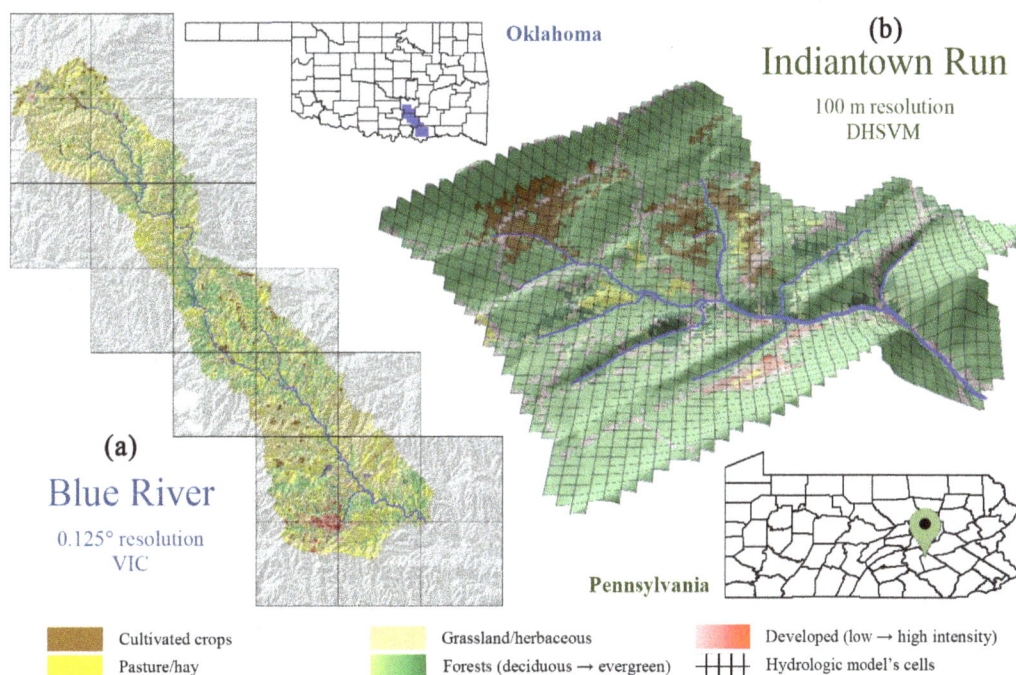

Figure 2. Maps of the two test watersheds in the United States displaying the 30 m resolution land cover distribution from the NLCD (Homer et al., 2012). (a) Oklahoma's Blue River watershed 0.125° resolution VIC model application (20 cells). (b) Pennsylvania's Indiantown Run watershed 100 m resolution DHSVM model application (1472 cells).

Three diverse scenarios were selected for the Blue River, each of them comprised of a 2-week assimilation period (when streamflow observations are assimilated) and a 2-week forecasting period (when the model is run in an open loop using the states obtained at the end of the assimilation period): Scenario 1, starting on 15 October 1996, is rainy through the entire 4 weeks. Scenario 2, which starts on 15 January 1997, has a dry assimilation period and a mildly rainy forecast period. Scenario 3, starting on 1 June 1997, has a relatively rainy assimilation period and a mostly dry forecast period. Two scenarios, also spanning 4 weeks, were selected for the

Table 4. Set-up of the three factorial experiments, including the watershed, the total number of configurations (conf.), the values assigned to OPTIMISTS' parameters, and which objectives (objs.) were used (one objective: minimize MAE given the streamflow observations; two objectives: minimize MAE and maximize likelihood given the source or background state distribution S^t). n_{evo} was set to 25 in all cases. The total number of configurations results from combining all the possible parameter assignments listed for each experiment. Note that for Experiment 3 there are configurations that require a 4-week assimilation period (all others have a length of 2 weeks).

No.	Watershed	Conf.	Δt	n	w_{root}	p_{samp}	$k_{F\text{-class}}$	g	Objs.
1	Blue River	48	1 d, 5 d, 2 w	100, 500	0.95	0.25, 1	false, true	0.75	1, 2
2	Indiantown Run	32	1 h, 2 w	100, 200	0.6, 0.95	0.25, 1	false	0.75	1, 2
3	Indiantown Run	24	1 h, 6 h, 1 d, 3.5 d, 2 w, 4 w	100	0.95	0.4, 1	false	0.5, 1	2

Indiantown Run, one starting on 26 July 2009 and the other on 26 August 2009.

We used factorial experiments (Montgomery, 2012) to test different configurations of OPTIMISTS on each of these scenarios, by first assimilating the streamflow and then measuring the forecasting skill. In this type of experimental designs a set of assignments is established for each parameter and then all possible assignment combinations are tried. The design allows us to establish the statistical significance of altering several parameters simultaneously, providing an adequate framework for determining, for example, whether using a short or a long assimilation time step Δt is preferable, or if utilizing the optional optimization step within the algorithm is worthwhile. Table 4 shows the set-up of each of the three full factorial experiments we conducted, together with the selected set of assignments for OPTIMISTS' parameters. The forecasts were produced in an ensemble fashion, by running the models using each of the samples s_i from the state distribution S at the end of the assimilation time period, and then using the samples' weights w_i to produce an average forecast. Deterministic model parameters (those from the calibrated models) and forcings were used in all simulations.

Observation errors are usually taken into account in traditional assimilation algorithms by assuming a probability distribution for the observations at each time step and then performing a probabilistic evaluation of the predicted value of each particle/member against that distribution. As mentioned in Sect. 2, such a fitness metric, like the likelihood utilized in PFs to weight candidate particles, is perfectly compatible with OPTIMISTS. However, since it is difficult to estimate the magnitude of the observation error in general, and fitness metrics f_i here are only used to determine (non-)dominance between particles, we opted to use the mean absolute error (MAE) with respect to the streamflow observations in all cases.

For the Blue River scenarios, a secondary likelihood objective or metric was used in some cases to select for particles with higher consistency with the state history. It was computed using either Eq. (10) if $k_{F\text{-class}}$ was set to false or Eq. (9) if it was set to true. Equation (10) was used for all Indiantown Run scenarios given the large number of dimensions. The assimilation period was of 2 weeks for most

configurations, except for those in Experiment 3, which have $\Delta t = 4$ weeks. During both the assimilation and the forecasting periods we used unaltered streamflow data from the USGS and forcing data from the North American Land Data Assimilation System (NLDAS-2) (Cosgrove et al., 2003) – even though a forecasted forcing would be used instead in an operational setting (e.g. from systems like NAM, Rogers et al., 2009; or ECMWF, Molteni et al., 1996). While adopting perfect forcings for the forecast period leads to an overestimation of their accuracy, any comparisons with control runs or between methods are still valid as they all share the same benefit. Also, removing the uncertainty in the meteorological forcings allows the analysis to focus on the uncertainty corresponding to the land surface.

3.2 Data assimilation method comparison

Comparing the performance of different configurations of OPTIMISTS can shed light into the adequacy of individual strategies utilized by traditional Bayesian and variational methods. For example, producing all particles with the optimization algorithms ($p_{samp} = 0$), setting long values for Δt, and utilizing a traditional two-term cost function as that in Eq. (1) makes the method behave somewhat as a strongly constrained 4D-Var approach, while sampling all particles from the source state distribution ($p_{samp} = 1$), setting Δt equal to the model time step, and using a single likelihood objective involving the observation error would resemble a PF. Herein we also compare OPTIMISTS with a traditional PF on both model applications. Since the forcing is assumed to be deterministic, the implemented PF uses Gaussian perturbation of resampled particles to avoid degeneration (Pham, 2001). Resampling is executed such that the probability of duplicating a particle is proportional to their weight (Moradkhani et al., 2012).

Additionally, the comparison is performed using a continuous forecasting experiment set-up instead of a scenario-based one. In this continuous test, forecasts are performed every time step and compiled in series for different forecast lead times that span several months. Forecast lead times are of 1, 3, 6, and 12 days for the Blue River and of 6 h, and 1, 4, and 16 days for the Indiantown Run. Before each forecast, both OPTIMISTS and the PF assimilate streamflow observations for the assimilation time step of each algorithm (daily for the PF). The assimilation is performed cumulatively, meaning that the initial state distribution S^t was produced by assimilating all the records available since the beginning of the experiment until time t. The forecasted streamflow series are then compared to the actual measurements to evaluate their quality using deterministic metrics (NSE_{ℓ_2}, NSE_{ℓ_1}, and MARE) and two probabilistic ones: the ensemble-based continuous ranked probability score (CRPS) (Bröcker, 2012), which is computed for each time step and then averaged for the entire duration of the forecast; and the average normalized probability density p of the observed streamflow q_{obs} given the distribution of the forecasted ensemble q_{forecast},

$$
p\left(q_{\mathrm{obs}} | q_{\mathrm{forecast}}\right)
$$
$$
= \frac{\sum_{i=1}^{n} w_i \cdot \left(2\pi b^2\right)^{-2} \cdot \exp\left[-(q_{\mathrm{obs}} - q_i)^2 / (2b^2)\right]}{\sum_{i=1}^{n} w_i}, \quad (11)
$$

where the forecasted streamflow q_{forecast} is composed of values q_i for each particle i and accompanying weight w_i, and b is the bandwidth of the univariate kernel density estimate. The bandwidth b can be obtained by utilizing Silverman's rule of thumb (Silverman, 1986). The probability p is computed every time step, then normalized by multiplying by the standard deviation of the estimate, and then averaged for all time steps. As opposed to the CRPS, which can only give an idea of the bias of the estimate, the density p can detect both bias and under- or overconfidence: high values for the density indicate that the ensemble is producing narrow estimates around the true value, while low values indicate either that the stochastic estimate is spread too thin or is centred far away from the true value.

4 Results and discussion

This section summarizes the forecasting results obtained from the three scenario-based experiments and the continuous forecasting experiments on the Blue River and the Indiantown Run model applications. The scenario-based experiments were performed to explore the effects of multiple parameterizations of OPTIMISTS, and the performance was analysed as follows. The model was run for the duration of the forecast period (2 weeks) using the state configuration encoded in each root state s_i of the distribution S obtained at the end of the assimilation period for each configuration of OPTIMISTS and each scenario. We then computed the mean streamflow time series for each case by averaging the model results for each particle P_i (the average was weighted based on the corresponding weights w_i). With this averaged streamflow series, we compute the three performance metrics – the NSE_{ℓ_2}, the NSE_{ℓ_1}, and the MARE – based on the observations from the corresponding stream gauge. The values

for each experiment, scenario, and configuration are listed in tables in the Supplement. With these, we compute the change in the forecast performance between each configuration and a control open-loop model run (one without the benefit of assimilating the observations).

4.1 Blue River – low-resolution application

The Supplement includes the performance metrics for all the tested configurations on all scenarios and for all scenario-based experiments. Figure 3 summarizes the results for Experiment 1 with the VIC model application for the Blue River watershed, in which the distributions of the changes in MARE after marginalizing the results for each scenario and each of the parameter assignments are shown. That is, each box (and pair of whiskers) represents the distribution of change in MARE of all cases in the specified scenario or for which the specified parameter assignment was used. Negative values in the vertical axis indicate that OPTIMISTS decreased the error, while positive values indicate it increased the error. It can be seen that, on average, OPTIMISTS improves the precision of the forecast in most cases, except for several of the configurations in Scenario 1 (for this scenario the control already produces a good forecast) and when using an assimilation step Δt of 1 day. We performed an analysis of variance (ANOVA) to determine the statistical significance of the difference found for each of the factors indicated in the horizontal axis. While Fig. 3 shows the p values for the main effects, the full ANOVA table for all experiments can be found in the Supplement. From the values in Fig. 3, we can conclude that the assimilation time step, the number of objectives, and the use of optimization algorithms are all statistically significant. On the other hand, the number of particles and the use of F-class kernels are not.

A Δt of 5 days produced the best results overall for the tested case, suggesting that there exists a sweet spot that balances the amount of information being assimilated (larger for a long Δt) and the number of state variables to be modified (larger for a small Δt). Based on such results, it is reasonable to assume that the sweet spot may depend on the time series of precipitation, the characteristics of the watershed, and the temporal and spatial resolutions of its model application. From this perspective, the poor results for a step of 1 day could be explained in terms of overfitting, where there are many degrees of freedom and only one value being assimilated per step. Evaluating particles in the light of two objectives, one minimizing departures from the observations and the other maximizing the likelihood of the source state, resulted in statistically significant improvements compared to using the first objective alone. Additionally, the data suggest that not executing the optional optimization step of the algorithm (optimization = false), but instead relying only on particles sampled from the prior or source distribution, is also beneficial. These two results reinforce the idea that maintaining consistency with the state history to some ex-

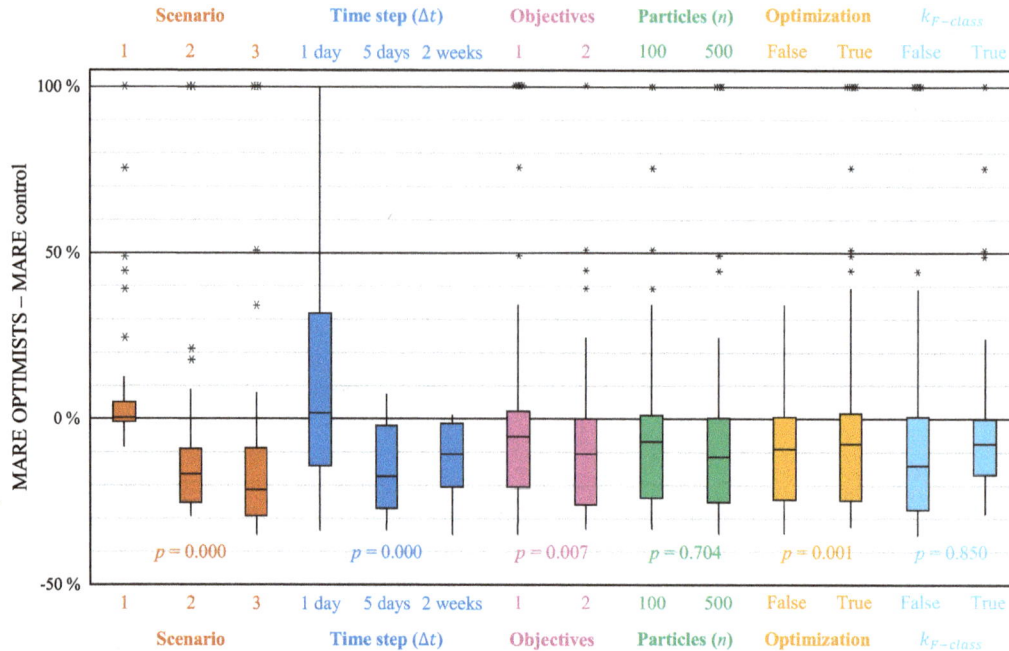

Figure 3. Box plots of the changes in forecasting error (MARE) achieved while using OPTIMISTS on Experiment 1 (Blue River). Changes are relative to an open-loop control run where no assimilation was performed. Each column corresponds to the distribution of the error changes on the specified scenario or assignment to the indicated parameter. Positive values indicate that OPTIMISTS increased the error, while negative values indicate it decreased the error. Outliers are noted as asterisks and values were limited to 100 %. For the one-objective case, the particles' MAE was to be minimized; for the two-objective case, the likelihood given the background was to be maximized in addition. No optimization ("false") corresponds to $p_{samp} = 1.0$ (i.e. all samples are obtained from the prior distribution); "true" corresponds to $p_{samp} = 0.25$. The p values were determined using ANOVA (Montgomery, 2012) and indicate the probability that the differences in means corresponding to boxes of the same colour are produced by chance (e.g. values close to zero indicate certainty that the parameter effectively affects the forecast error).

tent is of paramount importance, perhaps to the point where the strategies used in Bayesian filters and variational methods are insufficient in isolation. Indeed, the best performance was observed only when both sampling was limited to generate particles from the prior state distribution and the particles were evaluated for their consistency with that distribution.

On the other hand, we found it counterintuitive that neither using a larger particle ensemble nor taking into account state-variable dependencies through the use of F-class kernels leads to improved results. In the first case it could be hypothesized that using too many particles could lead to overfitting, since there would be more chances of particles being generated that happen to match the observations better but for the wrong reasons. In the second case, the non-parametric nature of kernel density estimation could be sufficient for encoding the raw dependencies between variables, especially in low-resolution cases like this one, in which significant correlations between variables in adjacent cells are not expected to be too high. Both results deserve further investigation, especially concerning the impact of D- vs. F-class kernels in high-dimensional models.

Interestingly, the ANOVA also yielded small p values for several high-order interactions (see the ANOVA table in

the Supplement). This means that, unlike the general case for factorial experiments as characterized by the sparsity-of-effects principle (Montgomery et al., 2009), specific combinations of multiple parameters have a large effect on the forecasting skill of the model. There are significant interactions (with p smaller than 0.05) between the following groups of factors: objectives and Δt ($p = 0.001$); n and $k_{F\text{-class}}$ ($p = 0.039$); Δt and the use of optimization ($p = 0.000$); the use of optimization and $k_{F\text{-class}}$ ($p = 0.029$); the objectives, Δt, and the use of optimization ($p = 0.043$); n, Δt, and $k_{F\text{-class}}$ ($p = 0.020$); n, the use of optimization, and $k_{F\text{-class}}$ ($p = 0.013$); and n, Δt, the use of optimizers, and $k_{F\text{-class}}$ ($p = 0.006$). These interactions show that, for example, (i) using a single objective is especially inadequate when the time step is 1 day or when optimization is used; (ii) employing optimization is only significantly detrimental when Δt is 1 day – probably because of intensified overfitting; and (iii) choosing F-class kernels leads to higher errors when Δt is small, when n is large, and when the optimizers are being used.

Based on these results, we recommend the use of both objectives and no optimization as the preferred configuration of OPTIMISTS for the Blue River application. A time step

Table 5. Continuous daily streamflow forecast performance metrics for the Blue River application using OPTIMISTS ($\Delta t = 7$ days; three objectives: NSE_{ℓ_2}, MARE, and likelihood; $n = 30$; no optimization; and D-class kernels) and a traditional PF ($n = 30$). The continuous forecast extends from January to June 1997. The NSE_{ℓ_2}, NSE_{ℓ_1}, and MARE (deterministic) are computed using the mean streamflow of the forecast ensembles and contrasting it with the daily observations, while the CRPS and the density (probabilistic) are computed taking into account all the members of the forecasted ensemble.

Algorithm	Lead time	NSE_{ℓ_2}	NSE_{ℓ_1}	MARE	CRPS ($m^3\ s^{-1}$)	Density
OPTIMISTS	1 day	0.497	0.293	51.40 %	7.173	0.061
	3 days	0.527	0.312	50.16 %	6.959	0.065
	6 days	0.534	0.315	50.18 %	6.945	0.073
	12 days	0.516	0.297	51.26 %	7.124	0.078
Particle filter	1 day	0.675	0.522	30.06 %	4.480	0.098
	3 days	0.623	0.493	33.20 %	4.744	0.113
	6 days	0.602	0.473	35.79 %	5.000	0.109
	12 days	0.515	0.432	38.36 %	5.593	0.105

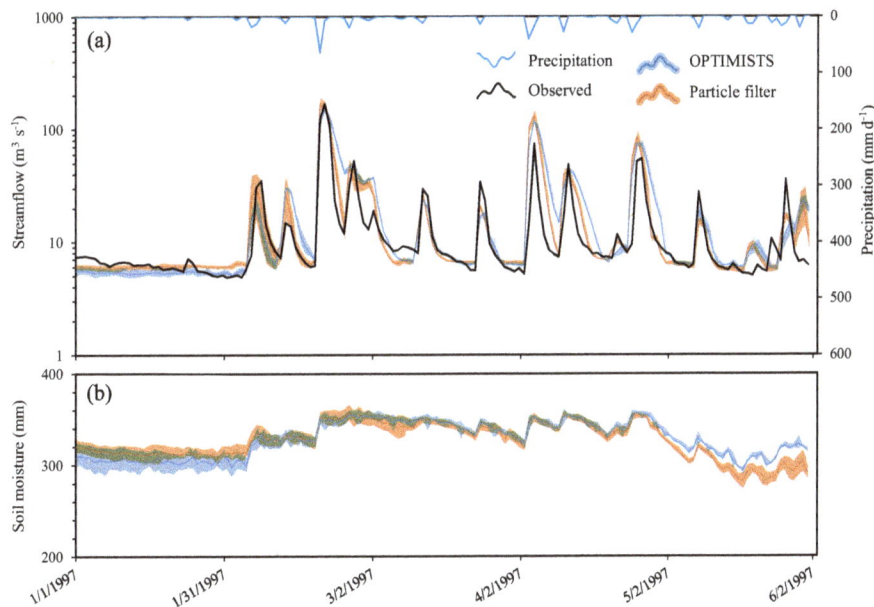

Figure 4. Comparison of 6-day lead time probabilistic streamflow (**a**) and area-averaged soil moisture (**b**) forecasts between OPTIMISTS ($\Delta t = 7$ days; three objectives: NSE_{ℓ_2}, MARE, and likelihood; $n = 30$; no optimization; and D-class kernels) and a traditional PF ($n = 30$) for the Blue River. The dark blue and orange lines indicate the mean of OPTIMISTS' and the PF's ensembles, respectively, while the light blue and light orange bands illustrate the spread of the forecast by highlighting the areas where the probability density of the estimate is at least 50 % of the density at the mode (the maximum) at that time step. The green bands indicate areas where the light blue and light orange bands intersect.

of around 5 days appears to be adequate for this specific model application. Also, without strong evidence for their advantages, we recommend using more particles or kernels of class F only if there is no pressure for computational frugality. However, the number of particles should not be too small to ensure an appropriate sample size.

Table 5 shows the results of the 5-month-long continuous forecasting experiment on the Blue River using a 30-particle PF and a configuration of OPTIMISTS with a 7-day assimilation time step Δt, three objectives (NSE_{ℓ_2}, MARE, and the

likelihood), 30 particles, no optimization, and D-class kernels. This specific configuration of OPTIMISTS was chosen from a few that were tested with the recommendations above applied. The selected configuration was the one that best balanced the spread and the accuracy of the ensemble as some configurations had slightly better deterministic performance but larger ensemble spread for dry weather – which lead to worse probabilistic performance.

Figure 4 shows the probabilistic streamflow forecasts for both algorithms for a lead time of 6 days. The portrayed

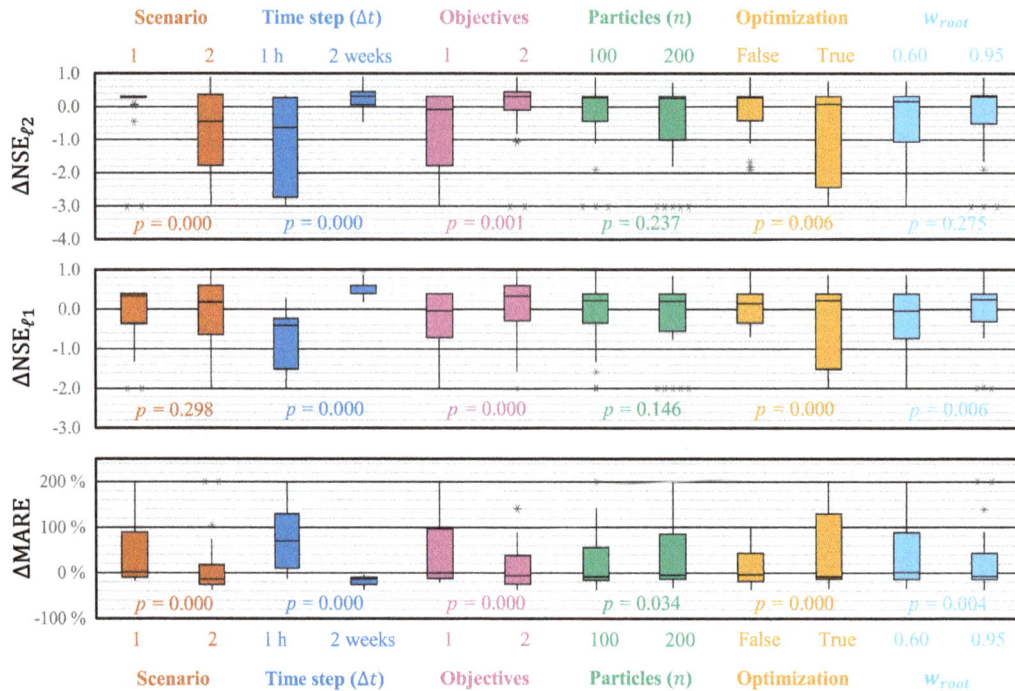

Figure 5. Box plots of the changes in forecasting performance (NSE_{ℓ_2}, NSE_{ℓ_1}, and MARE) achieved while using OPTIMISTS on Experiment 2 (Indiantown Run). Changes are relative to an open-loop control run where no assimilation was performed. Each column corresponds to the distribution of the error metric changes on the specified scenario or assignment to the indicated parameter. Outliers are noted as stars and values were constrained to $NSE_{\ell_2} \geq -3$, $NSE_{\ell_1} \geq -3$, and MARE $\leq 200\,\%$. Positive values indicate improvements for the NSE_{ℓ_2} and the NSE_{ℓ_1}. The meaning for the MARE and for other symbols is the same as those defined in Fig. 3.

evolution of the density, in which the mean does not necessarily correspond to the centre of the ensemble spread, evidences the non-Gaussian nature of both estimates. Both the selected configuration of OPTIMISTS and the PF methods show relatively good performance for all lead times (1, 3, 6, and 12 days) based on the performance metrics. However, the PF generally outperforms OPTIMISTS.

We offer three possible explanations for this result. First, the relatively low dimensionality of this test case does not allow OPTIMISTS to showcase its real strength, perhaps especially since the large scale of the watershed does not allow for tight spatial interactions between state variables. Second, OPTIMISTS can find solutions based on multiple objectives rather than a single one, which could be advantageous when multiple types of observations are available (e.g. of streamflow, evapotranspiration, and soil moisture). Thus, the solutions are likely not the best for each individual objective, but the algorithm balances their overall behaviour across the multiple objectives. Due to the lack of observations on multiple variables, only streamflow observations are used in these experiments even though more than one objective is used. Since it is the case that these objectives are consistent with each other, to a large extent, for the studied watershed, the strengths of using multiple objectives within the Pareto approach in OPTIMISTS cannot be fully evidenced. Third, additional efforts might be needed to find a configuration of the algorithm, together with a set of objectives, that best suits the specific conditions of the tested watershed.

While PFs remain easier to use out of the box because of their ease of configuration, the fact that adjusting the parameters of OPTIMISTS allowed us to trade off deterministic and probabilistic accuracy points to the adaptability potential of the algorithm. This allows for probing the spectrum between exploration and exploitation of candidate particles – which usually leads to higher and lower diversity of the ensemble, respectively.

4.2 Indiantown Run – high-resolution application

Figure 5 summarizes the changes in performance when using OPTIMISTS in Experiment 2. In this case, the more uniform forcing and streamflow conditions of the two scenarios allowed us to statistically analyse all three performance metrics. For Scenario 1 we can see that OPTIMISTS produces a general increase in the Nash–Sutcliffe coefficients, but a decline in the MARE, evidencing tension between fitting the peaks and the inter-peak periods simultaneously. For both scenarios there are configurations that performed very poorly, and we can look at the marginalized results in the box plots for clues into which parameters might have caused this. Similar to the Blue River case, the use of a 1 h time step sig-

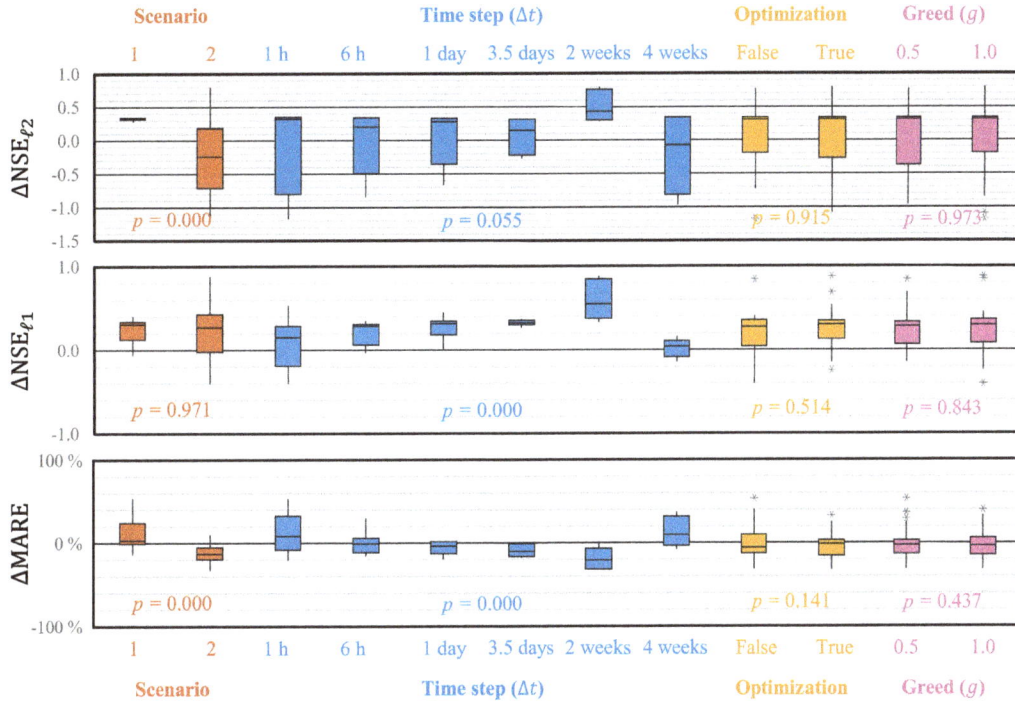

Figure 6. Box plots of the changes in forecasting performance (NSE_{ℓ_2}, NSE_{ℓ_1}, and MARE) achieved while using OPTIMISTS on Experiment 3 (Indiantown Run). Changes are relative to an open-loop control run where no assimilation was performed. Each column corresponds to the distribution of the error metric changes on the specified scenario or assignment to the indicated parameter. Positive values indicate improvements for the NSE_{ℓ_2} and the NSE_{ℓ_1}. See the caption of Fig. 3 for more information.

nificantly reduced the forecast skill, while the longer step almost always improved it; and the inclusion of the secondary history-consistent objective (two objectives) also resulted in improved performance. Not only does it seem that for this watershed the secondary objective mitigated the effects of overfitting, but it was interesting to note some configurations in which using it actually helped to achieve a better fit during the assimilation period.

While the ANOVA also provided evidence against the use of optimization algorithms, we are reluctant to instantly rule them out on the grounds that there were statistically significant interactions with other parameters (see the ANOVA table in the Supplement). The optimizers led to poor results in cases with 1 h time steps or when only the first objective was used. Other statistically significant results point to the benefits of using the root samples more intensively (in opposition to using random samples) and, to a lesser extent, to the benefits of maintaining an ensemble of moderate size.

Figure 6 shows the summarized changes in Experiment 3, where the effect of the time step Δt is explored in greater detail. Once again, there appears to be evidence favouring the hypothesis that there exists a sweet spot, and in this case it appears to be close to the 2-week mark: both shorter and longer time steps led to considerably poorer performance. In this experiment, with all configurations using both optimization objectives, we can see that there are no clear disadvantages

of using optimization algorithms (but also no advantages). Experiment 3 also shows that the effect of the greed parameter g is not very significant. That is, selecting some particles from dominated fronts to construct the target state distribution, and not only from the Pareto front, does not seem to affect the results.

Table 6 and Fig. 7 show the results from comparing continuous forecasts from the PF and from a configuration of OPTIMISTS with a time step of 1 week, two objectives, 50 particles, and no optimization. Both algorithms display overconfidence in their estimations, which is evidenced in Fig. 7 by the bias and narrowness of the ensembles' spread. It is possible that a more realistic incorporation of uncertainties pertaining to model parameters and forcings (which, as mentioned, are trivialized in these tests) would help to compensate for overconfidence. For the time being, these experiments help characterize the performance of OPTIMISTS in contrast with the PF, as both algorithms are deployed under the same circumstances. In this sense, while the forecasts obtained using the PF show slightly better results for lead times of 6 h and 1 day, OPTIMISTS shows a better characterization of the ensemble's uncertainty for the longer lead times.

OPTIMISTS' improved results in the high-resolution test case over those in the low-resolution one suggest that the strengths of the hybrid method might become more apparent as the dimensionality, and therefore the difficulty, of the as-

Table 6. Continuous hourly streamflow forecast performance metrics for the Indiantown Run application using OPTIMISTS ($\Delta t = 7$ days, two objectives, $n = 50$, no optimization, and D-class kernels) and a traditional PF ($n = 50$). The continuous forecast extends from September to December 2009. The NSE_{ℓ_2}, NSE_{ℓ_1}, and MARE (deterministic) are computed using the mean streamflow of the forecast ensembles and contrasting it with the daily observations, while the CRPS and the density (probabilistic) are computed taking into account all the members of the forecasted ensemble.

Algorithm	Lead time	NSE_{ℓ_2}	NSE_{ℓ_1}	MARE	CRPS $(L\,s^{-1})$	Density
OPTIMISTS	6 h	0.574	0.316	32.25 %	97.27	0.016
	1 day	0.609	0.340	31.42 %	93.92	0.013
	4 days	0.573	0.316	32.20 %	97.19	0.025
	16 days	0.521	0.272	33.90 %	103.51	0.013
Particle filter	6 h	0.660	0.480	26.87 %	79.61	0.061
	1 day	0.639	0.464	26.68 %	82.75	0.051
	4 days	0.558	0.401	27.42 %	93.20	0.021
	16 days	0.520	0.346	28.75 %	102.37	0.010

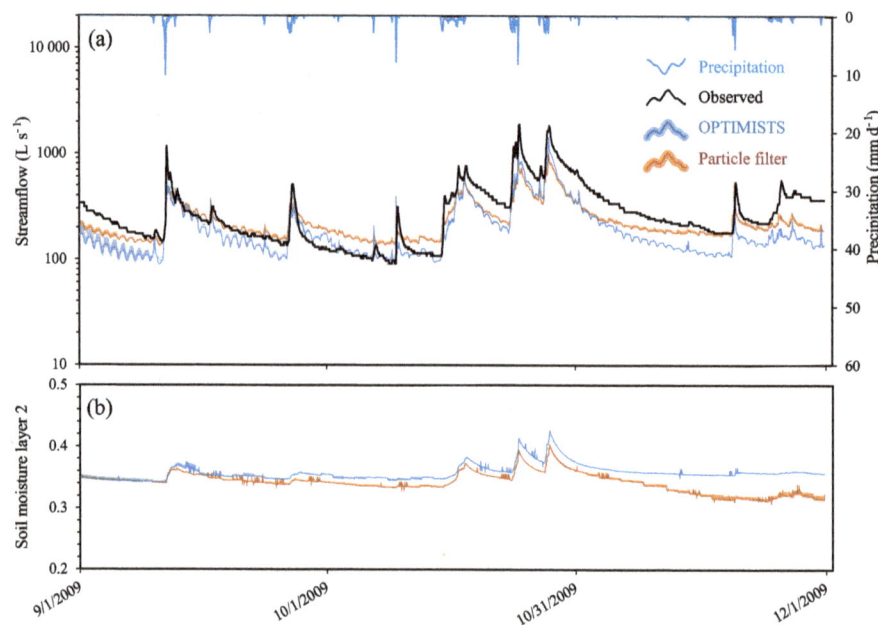

Figure 7. Comparison of 4-day lead time probabilistic streamflow **(a)** and area-averaged soil moisture **(b)** forecasts between OPTIMISTS ($\Delta t = 7$ days, two objectives, $n = 50$, no optimization, and D-class kernels) and a traditional PF ($n = 50$) for the Indiantown Run. The dark blue and orange lines indicate the mean of OPTIMISTS' and the PF's ensembles, respectively, while the light blue and light orange bands illustrate the spread of the forecast by highlighting the areas where the probability density of the estimate is at least 50 % of the density at the mode (the maximum) at that time step. The green bands indicate areas where the light blue and light orange bands intersect. Layer 2 of the soil corresponds to 100 to 250 mm depths.

simulation problem increases. However, while OPTIMISTS was able to produce comparable results to those of the PF, it was not able to provide definite advantages in terms of accuracy. As suggested before, additional efforts might be needed to find the configurations of OPTIMISTS that better match the characteristics of the individual case studies and, as with the Blue River, the limitation related to the lack of observations of multiple variables also applies here. Moreover, the implemented version of the PF did not present the parti-

cle degeneracy or impoverishment problems usually associated with these filters when dealing with high dimensionality, which also prompts further investigation.

4.3 Computational performance

It is worth noting that the longer the assimilation time step, the faster the entire process is. This occurs because, even though the number of hydrological calculations is the same in the end, for every assimilation time step the model files need to be generated accordingly, then accessed, and finally the result files written and accessed. This whole process takes a considerable amount of time. Therefore, everything else being constant, sequential assimilation (like with PFs) automatically imposes additional computational requirements. In our tests we used RAM drive software to accelerate the process of running the models sequentially and, even then, the overhead imposed by OPTIMISTS was consistently below 10 % of the total computation time. Most of the computational effort remained with running the model, both for VIC and the DHSVM. In this sense, model developers may consider allowing their engines to be able to receive input data from main memory, if possible, to facilitate data assimilation and other similar processes.

4.4 Recommendations for configuring OPTIMISTS

Finally, here we summarize the recommended choices for the parameters in OPTIMISTS based on the results of the experiments. In the first place, given their low observed effect, default values can be used for g (around 0.5). A w_{root} higher than 90 % was found to be advantageous. The execution of the optimization step ($p_{samp} < 1$) was, on the other hand, not found to be advantageous and, therefore, we consider it a cleaner approach to simply generate all samples from the initial distribution. Similarly, while not found to be disadvantageous, using diagonal bandwidth (D-class) kernels provide a significant improvement in computational efficiency and are thus recommended for the time being. Future work will be conducted to further explore the effect of the bandwidth configuration in OPTIMISTS.

Even though only two objective functions were tested, one measuring the departures from the observations being assimilated and another measuring the compatibility of initial samples with the initial distribution, the results clearly show that it is beneficial to simultaneously evaluate candidate particles using both criteria. While traditional cost functions like the one in Eq. (1) do indeed consider both aspects, we argue that using multiple objectives has the added benefit of enriching the diversity of the particle ensemble and, ultimately, the resulting probabilistic estimate of the target states.

Our results demonstrated that the assimilation time step is the most sensitive parameter and, therefore, its selection must be done with the greatest involvement. Taking the results together, we recommend that multiple choices be tried for any new case study looking to strike a balance between the amount of information being assimilated and the number of degrees of freedom. This empirical selection should also be performed with a rough sense of what is the range of forecasting lead times that is considered the most important. Lastly, more work is required to provide guidelines to select the number of particles n to be used. While the literature suggests that more should increase forecast accuracy, our tests did not back this conclusion. We tentatively recommend trying different ensemble sizes based on the computational resources available and selecting the one that offers the best observed trade-off between accuracy and efficiency.

5 Conclusions and future work

In this article we introduced OPTIMISTS, a flexible, model-independent data assimilation algorithm that effectively combines the signature elements from Bayesian and variational methods: by employing essential features from particle filters, it allows performing probabilistic non-Gaussian estimates of state variables through the filtering of a set of particles drawn from a prior distribution to better match the available observations. Adding critical features from variational methods, OPTIMISTS grants its users the option of exploring the state space using optimization techniques and evaluating candidate states through a time window of arbitrary length. The algorithm fuses a multi-objective or Pareto analysis of candidate particles with kernel density probability distributions to effectively bridge the gap between the probabilistic and the variational perspectives. Moreover, the use of evolutionary optimization algorithms enables its efficient application on highly non-linear models as those usually found in most geosciences. This unique combination of features represents a clear differentiation from the existing hybrid assimilation methods in the literature (Bannister, 2016), which are limited to Gaussian distributions and linear dynamics.

We conducted a set of hydrologic forecasting factorial experiments on two watersheds, the Blue River with 812 state variables and the Indiantown Run with 33 455, at two distinct modelling resolutions using two different modelling engines: VIC and the DHSVM, respectively. Capitalizing on the flexible configurations available for OPTIMISTS, these tests allowed us to determine which individual characteristics of traditional algorithms prove to be the most advantageous for forecasting applications. For example, while there is a general consensus in the literature favouring extended time steps (4-D) over sequential ones (1-D–3-D), the results from assimilating streamflow data in our experiments suggest that there is an ideal duration of the assimilation time step that is dependent on the case study under consideration, on the spatiotemporal resolution of the corresponding model application, and on the desired forecast length. Sequential time steps not only required considerably longer computational times but also produced the worst results – perhaps given the overwhelming number of degrees of freedom in contrast with the scarce observations available. Similarly, there was a drop in the performance of the forecast ensemble when the algorithm was set to use overly long time steps.

Procuring the consistency of candidate particles, not only with the observations but also with the state history, led to significant gains in predictive skill. OPTIMISTS can be configured to both perform Bayesian sampling and find Pareto-optimal particles that trade off deviations from the observations and from the prior conditions. This Bayesian and multi-objective formulation of the optimization problem was especially beneficial for the high-resolution watershed application, as it allows the model to overcome the risk of overfitting generated by the enlarged effect of equifinality.

On the other hand, our experiments did not produce enough evidence to recommend either exploring the state space with optimization algorithms instead of doing so with simple probabilistic sampling, the use of a larger number of particles above the established baseline of 100, or the computationally intensive utilization of full covariance matrices to encode the dependencies between variables in the kernel-based state distributions. Nevertheless, strong interactions between several of these parameters suggest that some specific combinations could potentially yield strong outcomes. Together with OPTIMISTS' observed high level of sensitivity to the parameters, these results indicate that there could be promise in the implementation of self-adaptive strategies (Karafotias et al., 2014) to assist in their selection in the future. With these experiments, we were able to configure the algorithm to consistently improve the forecasting skill of the models compared to control open-loop runs. Additionally, comparative tests showed that OPTIMISTS was able to reliably produce adequate forecasts that were comparable to those resulting from assimilating the observations with a particle filter in the high-resolution application. While not being able to provide consistent accuracy advantages over the implemented particle filter, OPTIMISTS does offer considerable gains in computational efficiency given its ability to analyse multiple model time steps each time.

Moreover, in this article we offered several alternatives in the implementation of the components of OPTIMISTS whenever there were tensions between prediction accuracy and computational efficiency. In the future, we will focus on incorporating additional successful ideas from diverse assimilation algorithms and on improving components in such a way that both of these goals are attained with ever-smaller compromises. For instance, the estimation of initial states should not be overburdened with the responsibility of compensating for structural and calibration deficiencies in the model. In this sense, we embrace the vision of a unified framework for the joint probabilistic estimation of structures, parameters, and state variables (Liu and Gupta, 2007), where it is important to address challenges associated with approaches that would increase the indeterminacy of the problem by adding unknowns without providing additional information or additional means of relating existing variables. We expect that with continued efforts OPTIMISTS will be a worthy candidate framework to be deployed in operational settings for hydrologic prediction and beyond.

Code and data availability. All the data utilized to construct the models are publicly available through the internet from their corresponding US government agencies' websites. The Java implementation of OPTIMISTS and of the particle filter are available through GitHub (2018) (https://github.com/felherc/). These sources include all the information needed to replicate the experiments in this article.

Author contributions. FH designed and implemented OPTIMISTS, implemented the particle filter, performed the experiments, and drafted the manuscript. XL identified problems for study; provided guidance; supervised the investigation, including experiment design; and finalized the manuscript.

Competing interests. The authors declare that they have no conflict of interest.

Acknowledgements. The authors are thankful to the two referees and the editor for their valuable comments and suggestions. This work was supported in part by the United States Department of Transportation through award no. OASRTRS-14-H-PIT to the University of Pittsburgh and by the William Kepler Whiteford Professorship from the University of Pittsburgh.

Edited by: Dimitri Solomatine

References

Adams, R. M., Houston, L. L., McCarl, B. A., Tiscareño, M. L., Matus, J. G., and Weiher, R. F.: The benefits to Mexican agriculture of an El Niño-southern oscillation (ENSO) early warning system, Agr. Forest Meteorol., 115, 183–194, https://doi.org/10.1016/S0168-1923(02)00201-0, 2003.

Andreadis, K. M. and Lettenmaier, D. P.: Assimilating remotely sensed snow observations into a macroscale hydrology model, Adv. Water Resour., 29, 872–886, https://doi.org/10.1016/j.advwatres.2005.08.004, 2006.

Bannister, R. N.: A review of forecast error covariance statistics in atmospheric variational data assimilation. II: Modelling the forecast error covariance statistics, Q. J. Roy. Meteorol. Soc., 134, 1971–1996, https://doi.org/10.1002/qj.340, 2008.

Bannister, R. N.: A review of operational methods of variational and ensemble-variational data assimilation, Q. J. Roy. Meteorol. Soc., 29, 1–29, https://doi.org/10.1002/QJ.2982, 2016.

Beven, K.: A manifesto for the equifinality thesis, J. Hydrol., 320, 18–36, https://doi.org/10.1016/j.jhydrol.2005.07.007, 2006.

Bröcker, J.: Evaluating raw ensembles with the continuous ranked probability score, Q. J. R. Meteorol. Soc., 138, 1611–1617, https://doi.org/10.1002/qj.1891, 2012.

Buehner, M., Houtekamer, P. L., Charette, C., Mitchell, H. L. and He, B.: Intercomparison of Variational Data Assimilation and the Ensemble Kalman Filter for Global Deterministic NWP. Part II: One-Month Experiments with Real Observations, Mon. Weather Rev., 138, 1567–1586, https://doi.org/10.1175/2009MWR3158.1, 2010.

Carpenter, J., Clifford, P., and Fearnhead, P.: Improved particle filter for nonlinear problems, IEEE Proc. - Radar, Sonar Navig., 146, 2–7, https://doi.org/10.1049/ip-rsn:19990255, 1999.

Clark, M. P., Rupp, D. E., Woods, R. a., Zheng, X., Ibbitt, R. P., Slater, A. G., Schmidt, J., and Uddstrom, M. J.: Hydrological data assimilation with the ensemble Kalman filter: Use of streamflow observations to update states in a distributed hydrological model, Adv. Water Resour., 31, 1309–1324, https://doi.org/10.1016/j.advwatres.2008.06.005, 2008.

Clark, M. P., Bierkens, M. F. P. P., Samaniego, L., Woods, R. A., Uijlenhoet, R., Bennett, K. E., Pauwels, V. R. N. N., Cai, X., Wood, A. W., Peters-Lidard, C. D., Uijlenhoet, R., Bennet, K. E., Pauwels, V. R. N. N., Cai, X., Wood, A. W., and Peters-Lidard, C. D.: The evolution of process-based hydrologic models: Historical challenges and the collective quest for physical realism, Hydrol. Earth Syst. Sci., 21, 3427–3440, https://doi.org/10.5194/hess-21-3427-2017, 2017.

Cosgrove, B. A., Lohmann, D., Mitchell, K. E., Houser, P. R., Wood, E. F., Schaake, J. C., Robock, A., Marshall, C., Sheffield, J., Duan, Q., Luo, L., Higgins, R. W., Pinker, R. T., Tarpley, J. D., and Meng, J.: Real-time and retrospective forcing in the North American Land Data Assimilation System (NLDAS) project, J. Geophys. Res.-Atmos., 108, 1–12, https://doi.org/10.1029/2002JD003118, 2003.

Crainic, T. G. and Toulouse, M.: Parallel Meta-heuristics, in: Handbook of Metaheuristics, vol. 146, edited by: Gendreau, M. and Potvin, J.-Y., Springer US, 497–541, 2010.

Deb, K.: Multi-objective Optimization, in: Search Methodologies: Introductory Tutorials in Optimization and Decision Support Techniques, edited by: Burke, E. K. and Kendall, G., Springer US, 403–449, 2014.

Deb, K., Pratap, A., Agarwal, S., and Meyarivan, T.: A fast and elitist multiobjective genetic algorithm: NSGA-II, IEEE Trans. Evol. Comput., 6, 182–197, https://doi.org/10.1109/4235.996017, 2002.

Desroziers, G., Camino, J. T., and Berre, L.: 4DEnVar: Link with 4D state formulation of variational assimilation and different possible implementations, Q. J. Roy. Meteorol. Soc., 140, 2097–2110, https://doi.org/10.1002/qj.2325, 2014.

Dumedah, G. and Coulibaly, P.: Evolutionary assimilation of streamflow in distributed hydrologic modeling using insitu soil moisture data, Adv. Water Resour., 53, 231–241, https://doi.org/10.1016/j.advwatres.2012.07.012, 2013.

Dumedah, G., Berg, A. A., and Wineberg, M.: An Integrated Framework for a Joint Assimilation of Brightness Temperature and Soil Moisture Using the Nondominated Sorting Genetic Algorithm II, J. Hydrometeorol., 12, 1596–1609, https://doi.org/10.1175/JHM-D-10-05029.1, 2011.

Duong, T. and Hazelton, M. L.: Cross-validation bandwidth matrices for multivariate kernel density estimation, Scand. J. Stat., 32, 485–506, https://doi.org/10.1111/j.1467-9469.2005.00445.x, 2005.

Efstratiadis, A. and Koutsoyiannis, D.: One decade of multi-objective calibration approaches in hydrological modelling: a review, Hydrolog. Sci. J., 55, 58–78, https://doi.org/10.1080/02626660903526292, 2010.

Errico, R. M.: What Is an Adjoint Model?, B. Am. Meteorol. Soc., 78, 2577–2591, https://doi.org/10.1175/1520-0477(1997)078<2577:WIAAM>2.0.CO;2, 1997.

Evensen, G.: Data assimilation: the ensemble Kalman filter, Springer Science & Business Media, 2009.

Evensen, G. and van Leeuwen, P. J.: An ensemble Kalman smoother for nonlinear dynamics, Mon. Weather

Rev., 128, 1852–1867, https://doi.org/10.1175/1520-0493(2000)128<1852:AEKSFN>2.0.CO;2, 2000.

Fisher, M.: Background error covariance modelling, Semin. Recent Dev. Data Assim., 45–63, available at: https://www.ecmwf.int/sites/default/files/elibrary/2003/9404-background-error-covariance-modelling.pdf (last access: 29 October 2018), 2003.

Friedman, J., Hastie, T., and Tibshirani, R.: Sparse inverse covariance estimation with the graphical lasso, Biostatistics, 9, 432–441, https://doi.org/10.1093/biostatistics/kxm045, 2008.

Gauthier, P., Tanguay, M., Laroche, S., Pellerin, S., and Morneau, J.: Extension of 3DVAR to 4DVAR: Implementation of 4DVAR at the Meteorological Service of Canada, Mon. Weather Rev., 135, 2339–2354, https://doi.org/10.1175/MWR3394.1, 2007.

Ghil, M. and Malanotte-Rizzoli, P.: Data assimilation in meteorology and oceanography, Adv. Geophys., 33, 141–266, https://doi.org/10.1016/S0065-2687(08)60442-2, 1991.

Ghorbanidehno, H., Kokkinaki, A., Li, J. Y., Darve, E. and Kitanidis, P. K.: Real-time data assimilation for large-scale systems: The spectral Kalman filter, Adv. Water Resour., 86, 260–272, https://doi.org/10.1016/j.advwatres.2015.07.017, 2015.

GitHub: felherc, available at: https://github.com/felherc/, last access: 1 October 2018.

Gordon, N. J., Salmond, D. J., and Smith, A. F. M.: Novel approach to nonlinear/non-Gaussian Bayesian state estimation, IEEE Proc. F Radar Signal Process., 140, 107–113, https://doi.org/10.1049/ip-f-2.1993.0015, 1993.

Guo, J., Liang, X., and Leung, L. R.: A new multiscale flow network generation scheme for land surface models, Geophys. Res. Lett., 31, 1–4, https://doi.org/10.1029/2004GL021381, 2004.

Haario, H., Saksman, E., and Tamminen, J.: An Adaptive Metropolis Algorithm, Bernoulli, 7, 223–242, https://doi.org/10.2307/3318737, 2001.

Hawkins, D. M.: The Problem of Overfitting, J. Chem. Inf. Comput. Sci., 44, 1–12, https://doi.org/10.1021/ci0342472, 2004.

Hazelton, M. L.: Variable kernel density estimation, Aust. N. Z. J. Stat., 45, 271–284, https://doi.org/10.1111/1467-842X.00283, 2003.

Homer, C., Fry, J., and Barnes, C.: The National Land Cover Database, US Geol. Surv. Fact Sheet, 3020, 1–4, available at: http://pubs.usgs.gov/fs/2012/3020/ (last access: 22 October 2018), 2012.

Houser, P. R., Shuttleworth, W. J., Famiglietti, J. S., Gupta, H. V., Syed, K. H., and Goodrich, D. C.: Integration of soil moisture remote sensing and hydrologic modeling using data assimilation, Water Resour. Res., 34, 3405–3420, 1998.

Karafotias, G., Hoogendoorn, M., and Eiben, A. E.: Parameter Control in Evolutionary Algorithms: Trends and Challenges, IEEE Trans. Evol. Comput., 2, 167–187, https://doi.org/10.1109/TEVC.2014.2308294, 2014.

Koster, R. D., Betts, A. K., Dirmeyer, P. A., Bierkens, M., Bennett, K. E., Déry, S. J., Evans, J. P., Fu, R., Hernández, F., Leung, L. R., Liang, X., Masood, M., Savenije, H., Wang, G., and Yuan, X.: Hydroclimatic variability and predictability: a survey of recent research, Hydrol. Earth Syst. Sci., 21, 3777–3798, https://doi.org/10.5194/hess-21-3777-2017, 2017.

Krause, P., Boyle, D. P., and Bäse, F.: Comparison of different efficiency criteria for hydrological model assessment, Adv. Geosci., 5, 89–97, https://doi.org/10.5194/adgeo-5-89-2005, 2005.

Krishnamoorthy, A. and Menon, D.: Matrix Inversion Using Cholesky Decomposition, CoRR, 10–12, available at: http://arxiv.org/abs/1111.4144 (last access: 1 October 2018), 2011.

Li, J. Y., Kokkinaki, A., Ghorbanidehno, H., Darve, E. F., and Kitanidis, P. K.: The compressed state Kalman filter for nonlinear state estimation: Application to large-scale reservoir monitoring, Water Resour. Res., 51, 9942–9963, https://doi.org/10.1002/2015WR017203, 2015.

Liang, X. and Xie, Z.: A new surface runoff parameterization with subgrid-scale soil heterogeneity for land surface models, Adv. Water Resour., 24, 1173–1193, 2001.

Liang, X. and Xie, Z.: Important factors in land–atmosphere interactions: surface runoff generations and interactions between surface and groundwater, Global Planet. Change, 38, 101–114, 2003.

Liang, X., Lettenmaier, D. P., Wood, E. F., and Burges, S. J.: A simple hydrologically based model of land surface water and energy fluxcs for general circulation models, J. Geophys. Res., 99, 14415, https://doi.org/10.1029/94JD00483, 1994.

Liang, X., Lettenmaier, D. P., and Wood, E. F.: One-dimensional statistical dynamic representation of subgrid spatial variability of precipitation in the two-layer variable infiltration capacity model, J. Geophys. Res.-Atmos., 101, 21403–21422, 1996a.

Liang, X., Wood, E. F., and Lettenmaier, D. P.: Surface soil moisture parameterization of the VIC-2L model: Evaluation and modification, Global Planet. Change, 13, 195–206, 1996b.

Liu, J. S. and Chen, R.: Sequential Monte Carlo Methods for Dynamic Systems, J. Am. Stat. Assoc., 93, 1032–1044, https://doi.org/10.2307/2669847, 1998.

Liu, Y. and Gupta, H. V.: Uncertainty in hydrologic modeling: Toward an integrated data assimilation framework, Water Resour. Res., 43, 1–18, https://doi.org/10.1029/2006WR005756, 2007.

Lohmann, D., Rashke, E., Nijssen, B., and Lettenmaier, D. P.: Regional scale hydrology: I. Formulation of the VIC-2L model coupled to a routing model, Hydrolog. Sci. J., 43, 131–141, https://doi.org/10.1080/02626669809492107, 1998.

Lorenc, A. C., Bowler, N. E., Clayton, A. M., Pring, S. R., and Fairbairn, D.: Comparison of Hybrid-4DEnVar and Hybrid-4DVar Data Assimilation Methods for Global NWP, Mon. Weather Rev., 143, 212–229, https://doi.org/10.1175/MWR-D-14-00195.1, 2015.

Mahalanobis, P. C.: On the generalized distance in statistics, Proc. Natl. Inst. Sci., 2, 49–55, 1936.

Miller, D. A. and White, R. A.: A Conterminous United States Multilayer Soil Characteristics Dataset for Regional Climate and Hydrology Modeling, Earth Interact., 2, 1–26, https://doi.org/10.1175/1087-3562(1998)002<0002:CUSMS>2.0.CO;2, 1998.

Molteni, F., Buizza, R., Palmer, T. N., and Petroliagis, T.: The ECMWF ensemble prediction system: Methodology and validation, Q. J. Roy. Meteorol. Soc., 122, 73–119, https://doi.org/10.1002/qj.49712252905, 1996.

Montgomery, D. C.: Design and analysis of experiments, 8th Edn., John Wiley & Sons, 2012.

Montgomery, D. C., Runger, G. C., and Hubele, N. F.: Engineering statistics, John Wiley & Sons, USA, 2009.

Montzka, C., Pauwels, V. R. N., Franssen, H.-J. H., Han, X., and Vereecken, H.: Multivariate and Multiscale Data Assimilation in Terrestrial Systems: A Review, Sensors, 12, 16291–16333, https://doi.org/10.3390/s121216291, 2012.

Moradkhani, H., DeChant, C. M., and Sorooshian, S.: Evolution of ensemble data assimilation for uncertainty quantification using the particle filter – Markov chain Monte Carlo method, Water Resour. Res., 48, 1–13, https://doi.org/10.1029/2012WR012144, 2012.

Ning, L., Carli, F. P., Ebtehaj, A. M., Foufoula-Georgiou, E., and Georgiou, T. T.: Coping with model error in variational data assimilation using optimal mass transport, Water Resour. Res., 50, 5817–5830, https://doi.org/10.1002/2013WR014966, 2014.

Noh, S. J., Tachikawa, Y., Shiiba, M., and Kim, S.: Applying sequential Monte Carlo methods into a distributed hydrologic model: Lagged particle filtering approach with regularization, Hydrol. Earth Syst. Sci., 15, 3237–3251, https://doi.org/10.5194/hess-15-3237-2011, 2011.

Park, S., Hwang, J. P., Kim, E., and Kang, H. J.: A new evolutionary particle filter for the prevention of sample impoverishment, IEEE Trans. Evol. Comput., 13, 801–809, https://doi.org/10.1109/TEVC.2008.2011729, 2009.

Penning-Rowsell, E. C., Tunstall, S. M., Tapsell, S. M. and Parker, D. J.: The benefits of flood warnings: Real but elusive, and politically significant, J. Chart. Inst. Water Environ. Manage., 14, 7–14, https://doi.org/10.1111/j.1747-6593.2000.tb00219.x, 2000.

Pham, D. T.: Stochastic methods for sequential data assimilation in strongly nonlinear systems, Mon. Weather Rev., 129, 1194–1207, https://doi.org/10.1175/1520-0493(2001)129<1194:SMFSDA>2.0.CO;2, 2001.

Rawlins, F., Ballard, S. P., Bovis, K. J., Clayton, A. M., Li, D., Inverarity, G. W., Lorenc, A. C., and Payne, T. J.: The Met Office global four-dimensional variational data assimilation scheme, Q. J. Roy. Meteorol. Soc., 133, 347–362, https://doi.org/10.1002/qj.32, 2007.

Reichle, R. H., McLaughlin, D. B., and Entekhabi, D.: Variational data assimilation of microwave radiobrightness observations for land surface hydrology applications, IEEE T. Geosci. Remote, 39, 1708–1718, https://doi.org/10.1109/36.942549, 2001.

Rodríguez, E., Morris, C. S., and Belz, J. E.: A global assessment of the SRTM performance, Photogramm. Eng. Remote Sens., 72, 249–260, 2006.

Rogers, E., DiMego, G., Black, T., Ek, M., Ferrier, B., Gayno, G., Janic, Z., Lin, Y., Pyle, M., Wong, V., and Wu, W.-S.: The NCEP North American Mesoscale Modeling System: Recent Changes and Future Plans, in: 23rd Conf. Weather Anal. Forecast. Conf. Numer. Weather Predict., available at: http://ams.confex.com/ams/23WAF19NWP/techprogram/paper_154114.htm (last access: 1 October 2018), 2009.

Seaber, P. R., Kapinos, F. P., and Knapp, G. L.: Hydrologic unit maps, US Government Printing Office Washington, D.C., USA, 1987.

Sheather, S. J. and Jones, M. C.: A Reliable Data-Based Bandwidth Selection Method for Kernel Density Estimation, J. R. Stat. Soc. Ser. B, 53, 683–690, available at: http://www.jstor.org/stable/2345597 (last access: 1 October 2018), 1991.

Silverman, B. B. W.: Density estimation for statistics and data analysis, CRC Press, USA, 1986.

Smith, A., Doucet, A., de Freitas, N., and Gordon, N.: Sequential Monte Carlo methods in practice, Springer Science & Business Media, USA, 2013.

Snyder, C., Bengtsson, T., Bickel, P., and Anderson, J.: Obstacles to High-Dimensional Particle Filtering, Mon. Weather Rev., 136, 4629–4640, https://doi.org/10.1175/2008MWR2529.1, 2008.

Socha, K. and Dorigo, M.: Ant colony optimization for continuous domains, Eur. J. Oper. Res., 185, 1155–1173, https://doi.org/10.1016/j.ejor.2006.06.046, 2008.

Terrell, G. R. and Scott, D. W.: Variable kernel density estimation, Ann. Stat., 20, 1236–1265, https://doi.org/10.1214/aos/1176348768, 1992.

Trémolet, Y.: Accounting for an imperfect model in 4D-Var, Q. J. Roy. Meteorol. Soc., 132, 2483–2504, https://doi.org/10.1256/qj.05.224, 2006.

van Leeuwen, P. J.: Particle Filtering in Geophysical Systems, Mon. Weather Rev., 137, 4089–4114, https://doi.org/10.1175/2009MWR2835.1, 2009.

van Leeuwen, P. J.: Nonlinear Data Assimilation for high-dimensional systems, in Nonlinear Data Assimilation, edited by: Van Leeuwen, J. P., Cheng, Y., and Reich, S., Springer International Publishing, 1–73, 2015.

Verkade, J. S. and Werner, M. G. F.: Estimating the benefits of single value and probability forecasting for flood warning, Hydrol. Earth Syst. Sci., 15, 3751–3765, https://doi.org/10.5194/hess-15-3751-2011, 2011.

Vrugt, J. A. and Robinson, B. A.: Improved evolutionary optimization from genetically adaptive multimethod search, P. Natl. Acad. Sci. USA, 104, 708–711, https://doi.org/10.1073/pnas.0610471104, 2007.

Wand, M. P. and Jones, M. C.: Kernel smoothing, CRC Press, New York, 1994.

Wen, Z., Liang, X., and Yang, S.: A new multiscale routing framework and its evaluation for land surface modeling applications, Water Resour. Res., 48, 1–16, https://doi.org/10.1029/2011WR011337, 2012.

West, M.: Mixture models, Monte Carlo, Bayesian updating, and dynamic models, Comput. Sci. Stat., 1–11, available at: http://www.stat.duke.edu/~mw/MWextrapubs/West1993a.pdf (last access: 1 October 2018), 1993.

Wigmosta, M. S., Vail, L. W., and Lettenmaier, D. P.: A distributed hydrology-vegetation model for complex terrain, Water Resour. Res., 30, 1665–1679, https://doi.org/10.1029/94WR00436, 1994.

Wigmosta, M. S., Nijssen, B., and Storck, P.: The distributed hydrology soil vegetation model, Math. Model. Small Watershed Hydrol. Appl., 7–42, available at: http://ftp.hydro.washington.edu/pub/dhsvm/The-distributed-hydrology-soil-vegetation-model.pdf (last access: 1 October 2018), 2002.

Yang, S.-C., Corazza, M., Carrassi, A., Kalnay, E., and Miyoshi, T.: Comparison of Local Ensemble Transform Kalman Filter, 3DVAR, and 4DVAR in a Quasi-geostrophic Model, Mon. Weather Rev., 137, 693–709, https://doi.org/10.1175/2008MWR2396.1, 2009.

Zhang, F., Zhang, M., and Hansen, J.: Coupling ensemble Kalman filter with four-dimensional variational data assimilation, Adv. Atmos. Sci., 26, 1–8, https://doi.org/10.1007/s00376-009-0001-8, 2009.

Zhang, L., Nan, Z., Liang, X., Xu, Y., Hernández, F., and Li, L.: Application of the MacCormack scheme to overland flow routing for high-spatial resolution distributed hydrological model, J. Hydrol., 558, 421–431, https://doi.org/10.1016/j.jhydrol.2018.01.048, 2018.

Zhang, X., Tian, Y., Cheng, R., and Jin, Y.: An Efficient Approach to Nondominated Sorting for Evolutionary Multiobjective Optimization, IEEE Trans. Evol. Comput., 19, 201–213, https://doi.org/10.1109/TEVC.2014.2308305, 2015.

Zhu, Y., Toth, Z., Wobus, R., Richardson, D., and Mylne, K.: The economic value of ensemble-based weather forecasts, B. Am. Meteorol. Soc., 83, 73–83, https://doi.org/10.1175/1520-0477(2002)083<0073:TEVOEB>2.3.CO;2, 2002.

Ziervogel, G., Bithell, M., Washington, R., and Downing, T.: Agent-based social simulation: A method for assessing the impact of seasonal climate forecast applications among smallholder farmers, Agric. Syst., 83, 1–26, https://doi.org/10.1016/j.agsy.2004.02.009, 2005.

Informing a hydrological model of the Ogooué with multi-mission remote sensing data

Cecile M. M. Kittel[1], Karina Nielsen[2], Christian Tøttrup[3], and Peter Bauer-Gottwein[1]

[1]Department of Environmental Engineering, Technical University of Denmark, Technical University of Denmark, 2800 Kgs. Lyngby, Denmark
[2]National Space Institute, Technical University of Denmark, 2800 Kgs. Lyngby, Denmark
[3]DHI-GRAS, 2970 Hørsholm, Denmark

Correspondence: Cecile M. M. Kittel (ceki@env.dtu.dk)

Abstract. Remote sensing provides a unique opportunity to inform and constrain a hydrological model and to increase its value as a decision-support tool. In this study, we applied a multi-mission approach to force, calibrate and validate a hydrological model of the ungauged Ogooué river basin in Africa with publicly available and free remote sensing observations. We used a rainfall–runoff model based on the Budyko framework coupled with a Muskingum routing approach. We parametrized the model using the Shuttle Radar Topography Mission digital elevation model (SRTM DEM) and forced it using precipitation from two satellite-based rainfall estimates, FEWS-RFE (Famine Early Warning System rainfall estimate) and the Tropical Rainfall Measuring Mission (TRMM) 3B42 v.7, and temperature from ECMWF ERA-Interim. We combined three different datasets to calibrate the model using an aggregated objective function with contributions from (1) historical in situ discharge observations from the period 1953–1984 at six locations in the basin, (2) radar altimetry measurements of river stages by Envisat and Jason-2 at 12 locations in the basin and (3) GRACE (Gravity Recovery and Climate Experiment) total water storage change (TWSC). Additionally, we extracted CryoSat-2 observations throughout the basin using a Sentinel-1 SAR (synthetic aperture radar) imagery water mask and used the observations for validation of the model. The use of new satellite missions, including Sentinel-1 and CryoSat-2, increased the spatial characterization of river stage. Throughout the basin, we achieved good agreement between observed and simulated discharge and the river stage, with an RMSD between simulated and observed water amplitudes at virtual stations of 0.74 m for the TRMM-forced model and 0.87 m for the FEWS-RFE-forced model. The hydrological model also captures overall total water storage change patterns, although the amplitude of storage change is generally underestimated. By combining hydrological modeling with multi-mission remote sensing from 10 different satellite missions, we obtain new information on an otherwise unstudied basin. The proposed model is the best current baseline characterization of hydrological conditions in the Ogooué in light of the available observations.

1 Introduction

River basin hydrology, ecosystem health and human livelihood are intrinsically linked, emphasizing the need for knowledge about hydrological processes on the river basin scale. While hydrological models can increase the understanding of the hydrological regime and its vulnerability to changes (Tanner and Hughes, 2015), physical observations of hydrological states are required to force and calibrate hydrological models and are crucial to produce useful simulations. Paradoxically, in situ gauging networks have thinned out over recent decades (Vörösmarty et al., 2001; Hannah et al., 2011). Satellite remote sensing provides a unique opportunity to acquire information on important components of the land-surface water balance and bridge this gap (Tang et al., 2009; Sneeuw et al., 2014). Remote sensing estimates can supplement and, to some extent, replace in situ observations, where these are insufficient or impossible to acquire (Getirana, 2010; Xie et al., 2012; Knoche et al., 2014). As more remote-sensing-based estimates of hydrological variables become globally and publicly available the need for sound and scientifically founded methods to integrate remote sensing observations with hydrological models increases (van Griensven et al., 2012).

Several studies have benefited from using remote-sensing-based estimates to provide hydrological models with necessary basin-scale information and forcing inputs (e.g., Bauer-Gottwein et al., 2015; Stisen et al., 2008; Awange et al., 2016). A large number of satellite-based products are publicly available, offering gridded, large-scale information on the global scale. Furthermore, hydrological models often contain conceptual parameters, which are either impossible or impractical to measure directly (Xu et al., 2014). In order to estimate the best-fitting parameter values – and subsequently evaluate model performance – model simulations are compared to observations of hydrological variables. Traditionally, hydrological models use discharge measurements to calibrate and validate hydrological models (Bauer-Gottwein et al., 2015; Knoche et al., 2014). However, in many river basins, in situ data are limited or insufficient (Hannah et al., 2011). Instead, remote sensing observations of hydrological state variables such as river level (Schneider et al., 2017), total water storage or soil moisture (Milzow et al., 2011; Xie et al., 2012; Abelen and Seitz, 2013) can be used to calibrate and validate the hydrological model performance and improve parameter estimation. Alvarez-Garreton et al. (2014) improved model parametrization by exploring multiple hydrological state variables in a multi-objective calibration.

A commonly used supporting dataset is total water storage change (TWSC) inferred from gravimetric remote sensing. Since 2002, the Gravity Recovery and Climate Experiment (GRACE) mission has recorded and mapped temporal anomalies in the Earth's gravity field. Changes in terrestrial water storage can be inferred from these anomalies. The dataset has been successfully used to evaluate catchment-scale total water storage and as part of hydrological model calibration (Xie et al., 2012; Awange et al., 2014; Eicker et al., 2014; Mulder et al., 2015).

The use of radar altimetry to infer river levels is a relatively new field of research as the utility of the observations over narrow water bodies is limited by the footprint of the altimeter and consequent risk of contamination from surrounding land, large ground track spacing and low overpass frequency (Schumann and Domeneghetti, 2016). Since the 1990s, technological advances and the improvement of retracking algorithms have enabled the extraction of radar altimetry observations of water heights over inland water bodies (Berry and Benveniste, 2013), with accuracies of between 30 and 70 cm – even for rivers less than several hundred meters wide (Villadsen et al., 2015; Schumann and Domeneghetti, 2016). Michailovsky et al. (2012) used in situ observations from field campaigns and historical records to obtain discharge estimates from radar altimetry observa-

tions and obtained RMSE values ranging from 4.5 to 7.2 % of the mean annual discharge amplitude, corresponding to 19.9 and 69.4 m^3 s^{-1} respectively for a water level RMSE between 30 and 70 cm relative to the in situ levels. Radar altimetry has been used in several studies to inform hydrological models both for calibration and in data assimilation schemes (Michailovsky et al., 2013; Getirana and Peters-Lidard, 2013; Getirana, 2010). Repeat ground track missions, such as Envisat or Jason-2, are typically favored in hydrological studies, as time series can be obtained at fixed locations over the river (virtual stations), similarly to traditional gauging stations. Increasing the spatiotemporal resolution of river level observations by applying a multi-mission approach can improve model calibration (Domeneghetti et al., 2014) and the representation of river hydraulics (Tourian et al., 2016).

Therefore, recent studies have focused on densifying the altimetry dataset by incorporating observations from drifting ground track missions as well (Schneider et al., 2017). The long repeat period results in a higher spatial resolution, as more points are sampled along the river. Water masks with sufficiently high resolution are required to extract the observations properly. Schneider et al. (2017) used a water mask based on the Landsat normalized difference vegetation index (NDVI) observations to extract CryoSat-2 observations over the Brahmaputra. However, optical data are not suitable in tropical regions with frequent cloud cover. New, publicly available synthetic aperture radar (SAR) observations from Sentinel-1 enable the extraction of water masks with high spatial and temporal resolution, facilitating the extraction of CryoSat-2 observations over rivers globally.

The biggest obstacle to using remote sensing data products in hydrological modeling is the difficulty in defining uncertainties in the data (Tang et al., 2009; van Griensven et al., 2012). The latter is still poorly described on a global scale, and current sensors and extraction algorithms are not precise enough to close the water balance based on remote sensing data (Tang et al., 2009). Furthermore, Gebregiorgis et al. (2012) showed a very high correlation between runoff error and precipitation misses (85 %), highlighting the importance of accurate precipitation estimates. Because of the crucial role of precipitation in driving the land-surface water balance, several precipitation datasets are often compared, if possible to in situ observations, and evaluated through their performance as model input prior to selecting a specific product (e.g., Awange et al., 2016; Milzow et al., 2011; Cohen Liechti et al., 2012; Stisen and Sandholt, 2010).

However, if only one hydrological variable is considered, calibration of the hydrological model can compensate for data errors and in turn conceal deficiencies in the model structure. Knoche et al. (2014) and van Griensven et al. (2012), among others, stipulate that while remote sensing input data have allowed for new possibilities in terms of catchment-scale modeling, calibration focused on discharge observations tends to compensate for input-data errors by compromising the representation of other hydrological processes. Awange et al. (2016) recommend evaluating the sensitivity of multiple outputs (e.g., groundwater recharge or actual evapotranspiration) to assess the effect of different data sets and uncover the interdependence between model evaluation and data. Furthermore, Knoche et al. (2014) identified a correlation between sensitivity to input-data errors and model complexity, showing that lumped conceptual models can provide good results in spite of the reduced complexity (Xu et al., 2014). While several studies have investigated the benefits of using a single type of remote sensing data to supplement in situ data, few studies have combined several remote sensing data types with available in situ data to inform hydrological models (Milzow et al., 2011).

The choice of model determines the input requirements as well as the level of parametrization, both of which increase with model complexity. Previous studies have used models with varying complexity ranging from fully distributed physically based hydrologic and hydrodynamic models (Stisen and Sandholt, 2010; Paiva et al., 2011) to semi-distributed models (Xie et al., 2012; Han et al., 2012) and simpler, lumped conceptual rainfall–runoff models (Knoche et al., 2014; Brocca et al., 2010). Whilst gridded remote sensing data offer the possibility to parametrize and drive fully distributed models with high spatiotemporal resolution, the choice of model must reflect the user requirements and capacities as well as the availability and uncertainty of the observations used to define the model (Johnston and Smakhtin, 2014). Furthermore, Paturel et al. (2003) and the follow-up study by Dezetter et al. (2008) highlighted the importance of reliable potential evapotranspiration (PET) estimates and of a robust, suitable numerical model, particularly in arid regions. Here, we select a model structure, which can accommodate the integration of different types of remote sensing observations and is suitable in data-scarce regions and for a wide range of user requirements.

In this study, we investigate how multi-mission remote sensing observations can be used to inform a hydrological model of a large ungauged basin, the Ogooué, Gabon. We show how combining multiple, publicly available datasets can increase the spatiotemporal characterization of river hydrology and improve model parameter definition. Remote sensing observations of precipitation and temperature are used to force the model, and observations of water height and total water storage from satellite altimetry and gravimetric observations respectively are used to supplement historical in situ discharge observations in the model calibration and validation.

2 The Ogooué

The Ogooué is the fourth largest river in Africa by volume of discharge with a mean annual rate of 4700 m^3 s. It is 1200 km long and drains approximately 224 000 km^2, 90 % of which lies within Gabon (Fig. 1). The river originates in the Ntalé

mountains on the Batéké Plateau in the Congo and runs northwest into Gabon. The basin is characterized by plateaus and hills bordering a narrow coastal plain. Although the hills are not very high (mean elevation in the catchment is 450 m), steep slopes and cliffs several hundred meters above the plain below create characteristic chutes and rapids, and between Lastoursville and Ndjolé the river is unnavigable. After Ndjolé, the river runs west and reaches the 100 km wide and 100 km long Ogooué delta. The lower part of the Ogooué is navigable and gentler than the rest of the river, with relatively low bed slopes, between 0.07 and 0.13 m km^{-1}. The river has numerous tributaries. The largest are the Ivindo, which flows from northeast to southwest Gabon before draining into the Ogooué just below the Chutes and Rapids of the Ivindo, and the Ngounié, which flows from the Chaillu Mountains along the southern border of Gabon before joining the Ogooué just upstream of Lambaréné.

The climate is equatorial with two rain seasons: February to May and October to December. Mean annual precipitation is 1831 mm and temperatures vary between 21 and 28 °C. The dense vegetation cover across the basin attenuates the potential flooding from the heavy rain in the two rainy seasons, and the basin is not particularly prone to flooding. Large portions of the river are fed by baseflow during the drier austral winter months, when cooler temperatures greatly reduce evapotranspiration (Mengue Medou et al., 2008).

The main challenge for water resources management in the region is the reconciliation of conservation and development plans. The Ogooué is home to several important ecosystems including several Ramsar sites (i.e., wetlands of international importance) such as the Chutes and Rapids of the Ivindo, or the Mboungou Baduma and the Doumé Rapids. Conservation of these wetlands is intrinsically linked to the hydrological regime in the basin. The Ogooué also plays a significant role in development plans in Gabon, both as part of the energy infrastructure and as a transport waterway (World Bank, 2012). Thousands of endemic species have been identified in the region surrounding the Grand Poubara hydropower station and in potential mining sites, and the risk of pollution from mineral industries and transport combined with changes to the flow regime are not negligible for the riparian ecosystems (Mezui and Boumono Moukoumi, 2013).

Hydrological monitoring efforts by ORSTOM (Office de la Recherche Scientifique et Technique Outre-Mer) hydrologists from the 1950s until the 1980s have produced decade-long time series of discharge measurements at several locations in the basin; however, the most recent publicly available observations are from 1984 at Lambaréné and earlier for all other in situ stations. Published studies focusing on the hydrological regime of the Ogooué have focused on large-scale investigations of West African rivers and on the reconstruction of historical discharge observations (Mahé et al., 1990; Mahé and Olivry, 1999; Mahé et al., 2013). Mahé et al. (1990) highlighted a reduction and temporal shift in

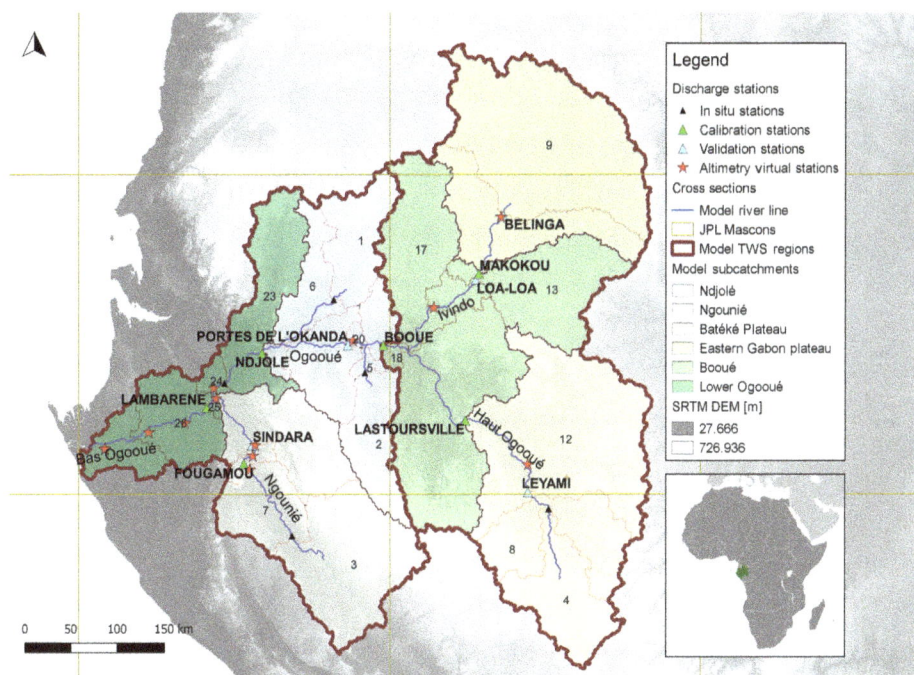

Figure 1. Basemap of the hydrological model of the Ogooué basin along with in situ discharge stations and altimetry virtual stations.

the spring flood at Lambaréné since the 1980s, which was later confirmed in the 2010s in Mahé et al. (2013) and attributed to changes in the regional climate pattern. To the authors' knowledge, there are no previous hydrological modeling studies of the basin.

3 Data and methods

3.1 Climate forcing

Daily temperature and precipitation observations are required to force the hydrological model. We use the ERA-Interim reanalysis from the European Centre for Medium-Range Weather Forecasts (ECMWF) as temperature input. Global, 6-hourly 2 m temperature estimates at 0.75° spatial resolution can be accessed from 1979 to present with 2 months' delay. We select two widely used and well documented satellite-based rainfall estimates to force the model based on results from previous studies comparing satellite rainfall estimates (SRFE) products over the African continent (Thiemig et al., 2013; Stisen and Sandholt, 2010; Awange et al., 2016). The Famine Early Warning System rainfall estimate (FEWS-RFE) has been operational since 2001 and is specifically designed for the African continent. The Tropical Rainfall Measuring Mission (TRMM) is a global mission launched by NASA in late 1997 and operational until 2015. The TRMM 3B42 v.7 product is a reanalysis product produced from observations from the Global Precipitation Measurement (GPM) mission since 2015. The dataset has a temporal resolution of 3 h and a spatial resolution of 0.25° and is provided between 50° S and 50° N. All climate data are aggregated to daily observations. We place virtual climate stations at the centroid coordinates of each model subbasin and transform the gridded precipitation and temperature data to point data using zonal statistics over the subbasins of the hydrological model.

3.2 Intercomparison of precipitation data

We compare the two precipitation products in order to identify any significant differences in precipitation trends. The spatial (Fig. 2a and b) and temporal (Fig. 2c) distribution of rainfall is relatively similar; however, TRMM predicts significantly more rain than FEWS-RFE (1600–2400 mm per year versus 1200–2200 mm). The annual average precipitation and double mass plot (Fig. 2d and e) reveal that, while the overall inter-annual variations are similar, the magnitude varies strongly: ranging from nearly similar annual magnitude in 2010 and 2011 to 500 mm more rain in 2006 and 100 mm less rain in 2014 predicted by TRMM compared to FEWS-RFE.

We compare the satellite observations to historical precipitation observations at four locations in the basin: Booué (1948–1980), Fougamou (1950–1980), Lebamba (1954–1974) and Petit Okano (1954–1976). The comparison reveals that, while both products record more days with rain, TRMM is closest to the observed mean monthly precipitation with an RMSD of between 11 and 19 % of the observed precipitation compared to RMSD values of 18 to 33 % for FEWS-RFE.

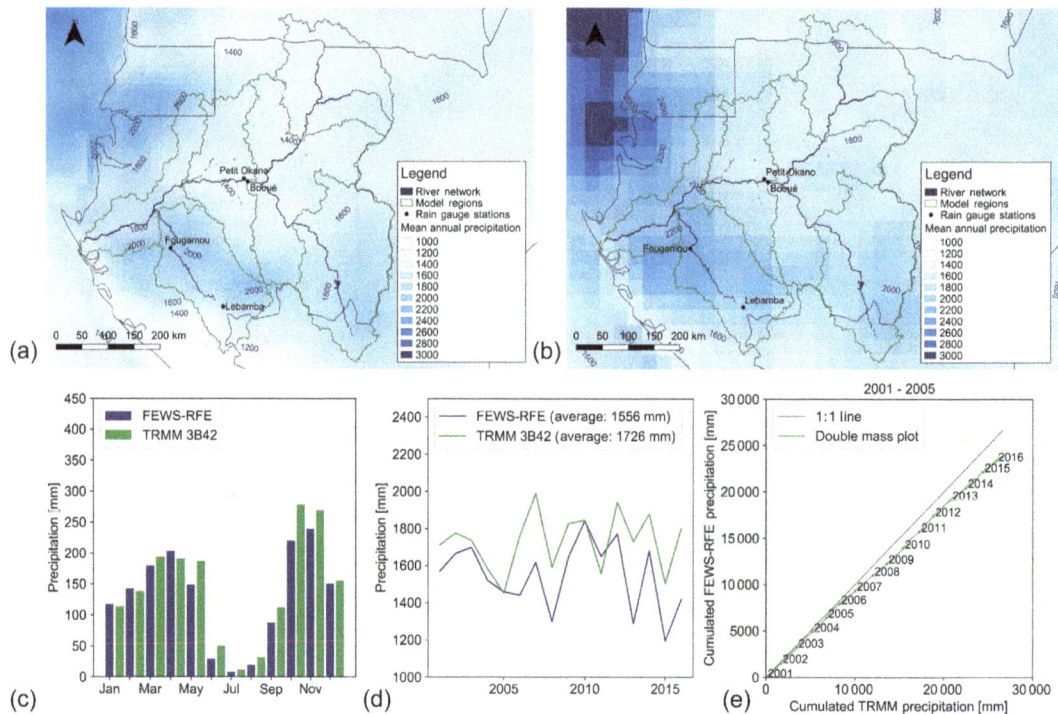

Figure 2. Average annual precipitation in the Ogooué basin based on FEWS-RFE (**a**) and TRMM 3B42 v7 (**b**), long-term monthly average (**c**) and annual average (**d**) precipitation from TRMM 3B42 v7 and FEWS-RFE v2, and double mass plot (**e**).

The satellite-based estimates are gridded data, observing rain events over larger areas than the gauge-stations, thus increasing the probability of recording at least one smaller event every day. Secondly, the period of record differs between the in situ data and the SRFE observations by over two decades, leaving room for changes in the long-term trends. The analysis indicates the products are relatively similar and we find no large discrepancies in terms of trends between the in situ and remotely sensed observations. Without up-to-date in situ precipitation records covering the entire basin it is impossible to conclude which product best reflects the present precipitation patterns over Gabon. Therefore, we estimate the model parameters using both products as model forcing.

3.3 GRACE total water storage

We obtain total water storage observations over the Ogooué from the JPL mascon surface mass change solution applied to Gravity Recovery and Climate Experiment (GRACE) gravimetric observations (Longuevergne et al., 2010; Watkins et al., 2015). Data from April 2002 to present can be derived at monthly intervals. A mascon set of multiplicative gain factors is provided with the dataset and can be applied to compensate for the attenuation of small-scale mass variations due to the sampling and processing of the GRACE observations – for instance in hydrological studies were these may be significant – by reducing the difference between the smoothed and unfiltered total water storage variations (Long et al., 2015).

The gain factors have a spatial resolution of 0.5°; however, at this resolution the correlation between neighboring cells is much higher. We aggregate the scaled solution to the native resolution of the mascons to produce time series for the two regions within the Ogooué using zonal statistics, splitting the basin along the frontier of two mascons (Fig. 1).

3.4 SAR imagery

Sentinel-1 is a two-satellite constellation launched by the European Space Agency (ESA) in 2014 for land and sea monitoring. The two satellites orbit 180° apart, at a 700 km altitude, ensuring optimal coverage and a short revisit time of 6 days on average. Both Sentinel-1 satellites carry a SAR instrument working in C band, which penetrates cloud cover. Over land, the satellite operates in Interferometric Wide (IW) swath mode by default, with a swath width of 250 km and a 5×20 m ground resolution. Sentinel-1 satellites carry dual-polarization SAR instruments, which can transmit and receive signals in vertical (V) and horizontal (H) polarization. In IW mode, dual polarizations VV and VH are available over land.

Level-1 Ground Range Detected (GRD) IW Sentinel-1 images acquired in May and June 2016 over the study area are preprocessed in the ESA Sentinel Application Platform (SNAP) toolbox. The images are (1) calibrated, (2) speckle filtered using the refined Lee filter and (3) geocoded using range-Doppler terrain correction with the 3 arcsec Shuttle

Radar Topography Mission digital elevation model (SRTM DEM) as topographic reference. Due to the lower reflectance of water compared to land, the histogram of the filtered backscatter coefficient is expected to contain two peaks of different magnitudes: very low values of backscatter corresponding to water pixels and higher values representing the land pixels. The threshold separating water from non-water points is the minimum between the two peaks. We define a threshold value for each individual scene and adjust it manually to ensure the best balance between false positives (where soil moisture enhances absorption, thus decreasing backscattering) and false negatives (waves on the water surface enhance reflection and increase backscattering).

3.5 Altimetry

We obtain remotely sensed river stages from Envisat and Jason-2 from the river and lake and Hydroweb project databases (Berry et al., 2005; Santos da Silva et al., 2010) at 12 locations in the basin within the Sentinel-1 water mask, at temporal resolution corresponding to the satellites' return periods: 35 days and 10 days respectively for the periods 2002–2009 and 2008–2012. Additionally we obtain CryoSat-2 Level 2 data from the National Space Institute, Technical University of Denmark (DTU Space) for the period 17 July 2010 to 21 February 2015. The data provided by DTU Space are based on the 20 Hz L1b dataset provided by ESA and have been retracked using an empirical retracker. Details concerning data processing are described in Villadsen et al. (2015). Finally, ICESat laser altimetry observations are obtained from the inland water surface spot heights (IWSH) database for the period 2003–2009. Details on the processing of the ICESat observations can be found in O'Loughlin et al. (2016). The 48 ICESat observations within the Ogooué basin provided on the IWSH database have been filtered using a using a global water mask and transect-averaged (O'Loughlin et al., 2016). All the obtained river stages are transect-averaged. We project all altimetry observations onto the EGM2008 geoid.

We filter the CryoSat-2 observations over the Sentinel-1 river mask using a point-in-polygon approach and reproject the points onto the model river line. A total of 762 CryoSat-2 ground tracks cross the Ogooué basin during the period of record, resulting in 1521 single observations within the river mask. Obvious outliers in the CryoSat-2 dataset are removed using the SRTM DEM. Over the Ogooué, the CryoSat-2 altimeter operates in low resolution mode (LRM). The CryoSat-2 waveform may include topographical noise due to its large footprint in LRM, particularly in the middle part of the river. CryoSat-2 heights are almost consistently smaller than SRTM-derived heights. The difference can be attributed to topographical noise and the density of vegetation in the basin, as SRTM contains averaged topography within 90 m pixels and may be recording the top of the canopy. Furthermore, we identify discrepancies in the longi-

tudinal cross section of the SRTM DEM along the river line. We reduce the risk of removing potentially valid CryoSat-2 observations based on erroneous SRTM heights by correcting the SRTM heights to the immediate downstream value if they exceed the upstream elevation by more than 1 m. We define CryoSat-2 outliers as observations more than 20 m lower than the SRTM height or more than 3 m higher. Most outliers are from single observation transects.

In cases where CryoSat-2 overpasses are parallel to the river line, important spatial variations may be lost in a single transect average. However, as most of the Ogooué runs perpendicular to CryoSat-2 satellite tracks, we transect-average the observations to obtain a time series. For tracks crossing subbasin borders, two separate means are calculated. We obtain 524 transect-averaged observations from the 1342 outlier-filtered single observations. Most observations are concentrated in the lower Ogooué (downstream of Ndjolé), which is the furthest downstream of the river network. Figure 3 shows the longitudinal profile of the SRTM elevation with the ICESat and CryoSat-2 single observations for the entire river (CryoSat-2 outliers are shown in grey). The river network includes confluent branches, resulting in three possible routes in the basin: the Ogooué (from the Batéké Plateau to the delta), the Ivindo (from the eastern Gabon plateau through the confluence to the Ogooué to the delta) and the Ngounié (from the upstream Ngounié to the Ogooué delta). To ensure each point is associated to a single "chainage", we define the latter as the distance of each point on the river to the main outlet in kilometers. Outliers are concentrated between around chainage 150 and downstream of the Batéké Plateau (chainage 780 on the Ogooué, lower branch in Fig. 3). In the upstream regions, the river drops off plateaus and runs through narrow valleys surrounded by steep slopes, increasing the risk of reflections from the surrounding land surface. Visual inspection of the longitudinal profile does not suggest clear bias (see inset in Fig. 3); however, the ICESat observations are generally larger than the CryoSat-2 observations, which is explained by time of observation: 62.5 % of the ICESat observations are sampled during the wet seasons (February–April and September–December) against only 39.0 % of the CryoSat-2 observations.

While the altimetry water heights are referenced to EGM2008, the model simulates water depth. To circumvent this discrepancy, we compare water height anomalies. For repeat ground track missions, the average water height recorded at each virtual station is subtracted from each observations to obtain the water height anomalies. Due to the coarse resolution of CryoSat-2 observations, the observations are interpolated over space and time, considering only the day of year (DOY) of the observation. The mean water height at a given chainage is subtracted from the interpolated water heights and from the individual observations to obtain relative water heights or anomalies. The amplitudes of Envisat and Jason-2 observations are compared to the CryoSat-2 amplitudes in order to evaluate potential inter-satellite bias

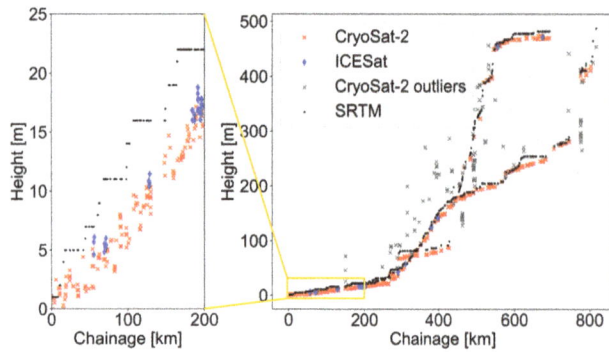

Figure 3. Longitudinal profile of the single CryoSat-2 observations, transect-averaged ICESat observations and corrected SRTM reference heights on the Ogooué and its tributaries.

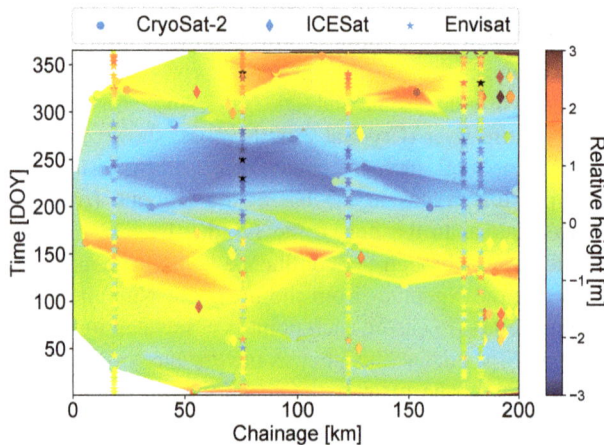

Figure 4. Spatiotemporal characterization of the annual water elevation changes of the lower Ogooué.

throughout the basin (Table 1). Concurrence between the missions strengthens the model evaluation and justifies the multi-mission approach. The interpolated mean annual water elevation at a given chainage is subtracted from the observations and only the day of year of the observation is considered. Figure 4 shows the spatiotemporal distribution of the Envisat observations against the CryoSat-2 observations. The two rain seasons are clearly visible with all missions, with the annual minimum in June–September (DOY 153-244).

3.6 Hydrological model

The hydrologic–hydrodynamic modeling framework used in this study consists of a lumped conceptual rainfall–runoff model based on the Budyko framework and developed by Zhang et al. (2008), coupled to a cascade of linear reservoirs and a Muskingum routing compartment (Chow et al., 1988). Figure 5 shows the model flow chart.

Zhang et al. (2008) simulate catchment water balance down to a daily timescale using a holistic approach based on

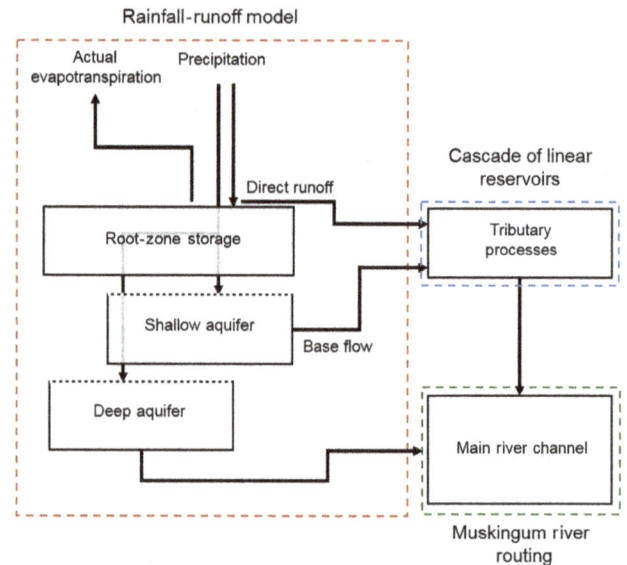

Figure 5. Flow chart of the hydrologic–hydrodynamic model along with the two main modifications to the Zhang et al. (2008) rainfall–runoff model: the deep aquifer and the tributary processes.

the Budyko framework, which assumes that two parameters control the equilibrium water balance: water availability and atmospheric demand. The former is approximated by precipitation, while the latter is represented through potential evapotranspiration. In the Zhang et al. (2008) approach, catchment storage is conceptualized as two compartments: root-zone storage and groundwater storage. In this study, we add a deep aquifer, splitting groundwater recharge using a simple, time-constant partitioning coefficient. The two aquifers have different storage constants used to calculate baseflow in the model. The storage constants are spatially and temporally uniform within each calibration zone but are adjusted in the calibration. At each time step, Budyko's limits concept is used to partition precipitation into direct runoff and catchment rainfall retention, to compute groundwater recharge from the catchment retention and soil storage, and to partition soil water availability into actual evapotranspiration (ET) and the updated soil storage. In natural systems, several processes delay direct runoff before it reaches the main channel (overland flow, transmission losses, evaporation losses, bank storage, etc.; Neitsch et al., 2009) and the basin contains a number of lakes and wetlands, which are not directly resolved by the model. Therefore, we implement conceptual tributary reaches in the form of a Nash cascade of linear reservoirs to route the direct runoff and baseflow from the shallow aquifer to the main channel.

We use Muskingum routing to route discharge from one subbasin outlet node to the next (Chow et al., 1988). The approach has two parameters: a proportionality coefficient, K, between the cross-sectional area of the flood flow and the discharge at a given section and a dimensionless weighting fac-

Table 1. Comparison of CryoSat-2 and Envisat or Jason-2 water height amplitudes for three branches of the Ogooué with sufficiently dense CryoSat-2 observations. The dispersion of the amplitudes predicted by CryoSat-2 are given by the standard deviation for the given river section.

	CryoSat-2 observations	CryoSat-2 amplitude (m)	Virtual stations	Envisat or Jason-2 amplitude (m)
Upstream of Makokou (Ivindo)	32	3.3 ± 1.5	1	2.22
Upstream of Sindara (Ngounié)	41	2.8 ± 0.9	3	2.4–3.2
Downstream of Ndjolé (Ogooué)	156	3.4 ± 0.7	4	2.4–3.7

tor, X. Traditionally, K and X are calibrated using inflow and outflow observations; however, in poorly gauged catchments, the parameters can be fitted through calibration and assumptions about channel properties. We estimate K based on segment lengths and average river flow velocity calculated from Manning's equation using trapezoidal cross sections and a calibrated roughness coefficient (Todini, 2007). In this study, we selected a 1 : 2 run to rise ratio, resulting in relatively limited changes in widths. X and Manning's roughness coefficient, n, are calibrated.

3.7 Watershed delineation

We use the SRTM digital elevation model and TauDEM watershed delineation hydroprocessing routine (Tarboton, 2015) to derive the drainage network and subbasins. The DEM resolution is reduced to approximately 1 km in order to comply with memory and CPU constraints. We place model outlets at points of interest including in situ gauging stations and upstream of key wetlands. The latter are included for reference in future scenario development studies in the catchment. Reach geometry including bed slope, reach lengths and widths are estimated by the hydroprocessing tool and refined based on the Sentinel-1 water mask and a high-resolution SRTM DEM. We further subdivide the main channel into reach segments in order to ensure numerical stability of the routing model. We place cross sections every 5–25 km.

3.8 Calibration

In order to include multiple observations of varying spatiotemporal scale, a holistic calibration approach is used. A warm-up period of 1 year allows the model to stabilize. Based on the basin geography, we divide the basin into six calibration zones with common parameter values (Fig. 1):

- The Batéké Plateau: the Haut-Ogooué province until Lastoursville station (subbasins 4, 8 and 12).

- The eastern Gabon plateau: the upstream Ivindo basin until the Makokou station (subbasins 9 and 10).

- The Ogooué and the Ivindo catchments until the Booué station (subbasins 13, 14, 16, 17 and 18).

- The Ogooué until the Ndjolé station (subbasins 1, 2, 5, 6, 19, 20, 21 and 22).

- The Ngounié (subbasins 3, 7, 11 and 15).

- The lower Ogooué and delta until Port-Gentil, using the Lambaréné station at the outflow of subbasin 25 for calibration (subbasins 11, 15, 23, 24, 25, 26 and 27).

The calibration parameters are shown in Table 2. In total 60 parameters are calibrated.

The following sections describe the individual objective functions combined for the calibration as well as the validation of the model.

The hydrological model is calibrated using a global search algorithm, the shuffled complex evolution method of the University of Arizona (SCEUA) algorithm developed by Duan et al. (1992) and implemented in Python by Houska et al. (2015) in the SPOTPY plugin. The algorithm has been widely used in hydrological studies. The parameters are calibrated by evolving 10 complexes and with convergence criteria of 0.1 % change in objective function and parameter value over 100 model runs. We use an aggregated objective function in order to exploit all available and suitable observations in the basin. The objective function contributions minimize the difference between the observed and simulated

- flow regimes at Lastoursville, Makokou, Booué, Ndjolé, Fougamou and Lambaréné, using historical observations from the 1930s to the 1980s – the flow regime is characterized by the

  - flow duration curves (FDCs) and
  - the daily or monthly climatology benchmark depending on available observations;

- stages at 12 virtual stations throughout the basin;

- catchment total water storage – due to the coarse resolution of GRACE, the calibration regions are aggregated into two calibration zones upstream and downstream of Booué.

When several objective functions are optimized at once, the optimal solutions representing the trade-offs between the

Table 2. Calibrated model parameters – one set of 10 parameters is defined for each calibration region. Calibration ranges are based on early trials, manual calibration and parameter definitions.

Parameter symbol	Description (unit)	Calibration range
α_1	Budyko parameter governing the partition between catchment retention and runoff (–)	[0.1–0.7]
α_2	Budyko parameter governing the partition between catchment retention and runoff (–)	[0.1–0.7]
d	Baseflow recession coefficient (day^{-1})	[0.003–0.7]
S_{max}	Maximum soil water storage (mm)	[100–1500]
n_{Nash}	Number of identical reservoirs in series in the Nash cascade (–)	[1–10]
k_{Nash}	Reservoir storage constant in the Nash cascade (day)	[1–10]
X_{GW}	Partitioning coefficient of recharge to shallow and deep aquifer (–)	[0–0.95]
d_{deep}	Deep aquifer baseflow recession constant (–)	[0.001–0.2]
X	Muskingum weighting factor (–)	[0–0.5]
n	Manning's roughness coefficient ($\text{s m}^{-1/3}$)	[0.015–0.05]

different objectives lie on the so-called Pareto front. However, it is computationally expensive to compute the full Pareto front for a meaningful number of parameter sets and for high-dimensional problems (Madsen, 2000). Instead, priorities can be given to the individual solutions prior to calibration based on the applications of the model to achieve a compromise between the individual contributions. The aggregated objective function ϕ, and calibration objective, was defined as the weighted root mean square deviation (WRMSD) between the objective function value resulting from the simulation and the objective function value $\phi_{\text{ref},i}$ for a perfect fit.

$$\phi = \sqrt{\frac{1}{N}\sum_{i=1}^{N}\left(\phi_{\text{ref},i} - \phi_{\text{sim},i}\right)^2 \times w_i} \qquad (1)$$

Here, w_i is the weight assigned to each individual objective function contribution. We weight the observations within the objective functions prior to aggregation in order to account for input-data error and uncertainty. Because all the objective functions are functions of scaled or weighted residuals, weights of 1 are deemed reasonable for most contributions, except the contributions from GRACE, which are given a weight of 2 to balance the low number of available GRACE observations.

The goodness-of-fit measures used for each partial objective function are different for the different contributions. We calibrate the FDC based on the method described in Westerberg et al. (2011). Selected percentiles are chosen based on a discharge volume interval approach. The area under the FDC is divided into 20 equal discharge volume bins with 5 % volume increments, resulting in 19 equally spaced evaluation points. The performance measure is based on a scaled score approach. At a given evaluation point, i, a perfect fit gives a score, S, of 0, while values differing by more than 10 % are given scores of 1 and −1 respectively. The performance

measure is defined as

$$R_{\text{FDC}} = 1 - \frac{\sum_{i=1}^{N-1}|S_i|}{N-1}. \qquad (2)$$

The remaining contributions are evaluated based on the WRMSD, using data uncertainty or variability as weights, yielding the performance measure

$$\text{WRMSD} = \sqrt{\frac{1}{N}\sum_{i=1}^{N}\left(\frac{y_{\text{sim},t} - y_{\text{obs},t}}{\sigma_t^2}\right)^2}, \qquad (3)$$

where σ_t^2 is the standard deviation of the observations for the climatology of day t, and the observation uncertainty for the TWSC and water height contributions. For the water stage comparison, we select a measurement uncertainty of 0.5 m based on previous studies (e.g., Santos da Silva et al., 2010; Birkinshaw et al., 2010). Villadsen et al. (2015) provide a summary of RMSDs obtained in literature. GRACE measurement uncertainties are provided with the dataset (Longuevergne et al., 2010; Watkins et al., 2015).

No bathymetry observations are available for the Ogooué; therefore, we compare altimetry water heights to simulated relative water depths. Water depth in the middle of a given reach can be estimated directly from the reach storage and combined with the water depth of the prism storage to linearly interpolate the water depth along the river line at any distance from the cross sections.

3.9 Sensitivity analysis

A global sensitivity analysis is carried out based on a latin hypercube sampling (LHS) of the parameter space. We used the extended Fourier amplitude sensitivity test (FAST) (Saltelli et al., 1999) implemented in SPOTPY by Houska et al. (2015). FAST provides two sensitivity measures: the first sensitivity index and a total sensitivity index, which includes contributions from parameter interaction. Over 200 000 model iterations are performed. We use

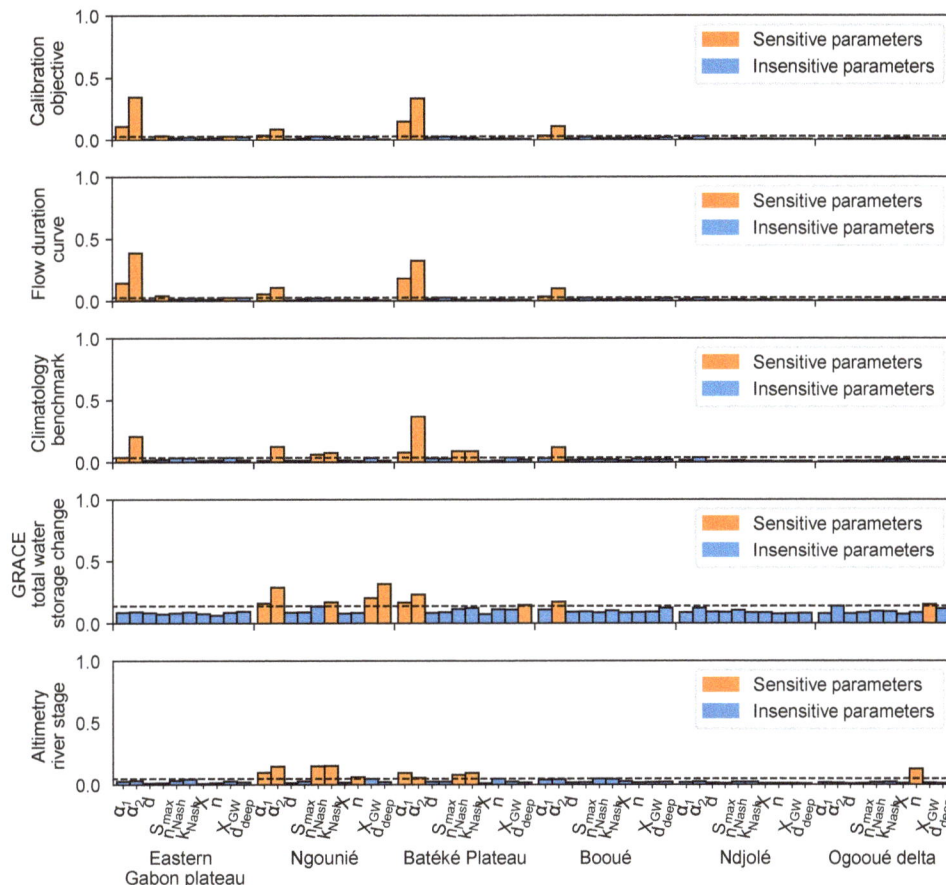

Figure 6. Sensitivity analysis of the model parameters on contributions to aggregated objective function (top) on (from second row down) FDC, climatology, GRACE and altimetry water height. The 10 most sensitive parameters are highlighted for each objective.

a multi-objective approach in order to evaluate the sensitivity of the individual contribution groups to different parameters and identify how including different observation groups constrains different parameters.

4 Results and discussion

4.1 Sensitivity analysis and parameter calibration

The sensitivity analysis provides useful information on how the different contributions to the global objective function constrain different parameters. The sensitivity indices are shown in Fig. 6. We find that the contributions to the calibration objective function are sensitive to different model parameters. For instance, the climatology constrains the Nash cascade parameters, while the FDC performance statistic is not very sensitive to changes in those parameters. The parameter sensitivity indices relative to the GRACE objective are more evenly distributed. The altimetry objective is most sensitive to the routing parameters, in particular channel roughness. Comparison between the calibration objective and the contributions shows a clear dominance of the FDC in the ag-

gregated objective function. Simulating the full Pareto front allows the user to assess trade-offs between individual contributions but is computationally expensive.

The aggregated objective function values are 0.81 and 0.86 for TRMM and FEWS respectively. The models perform very similarly regardless of the climate forcing, although the statistics of the TRMM model are slightly better overall. Evaluation of the parameter space post-calibration in Fig. 7 shows a clear convergence of all parameters to their optimal value. Only X appears to be less constrained in the shown region.

All calibrated parameter values of both models are provided in Table 3. Very few parameters converged to the upper boundary of the a priori parameter interval: k_{Nash} is close but not equal to the lower boundary in the TRMM model in the Ogooué delta and Ndjolé region, and X_{GW} is equal to 0.95 in the FEWS-RFE model in the eastern Gabon plateau region. The a priori parameter interval could be extended to allow larger values of X_{GW} and consequently close to no recharge to the deep aquifer. Parameter correlation between the parameters governing the partitioning of water between different reservoirs and delaying runoff is inevitable. Both

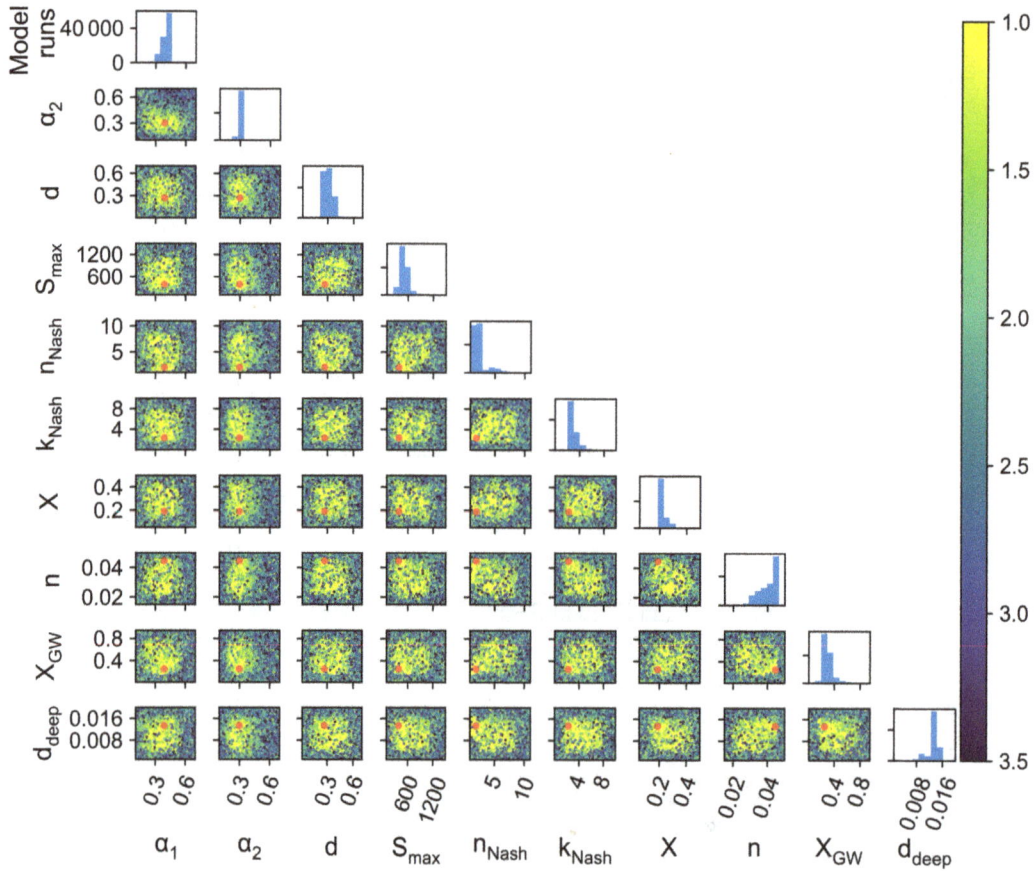

Figure 7. Parameter space post-calibration for the Ngounié calibration zone and the TRMM model. The yellow dots represent the best model runs and the red dots indicate the best parameter values. Example for TRMM model, Ngounié parameters.

parameter sets are physically reasonable and the basin median is very similar between the two models; however, some of the most sensitive parameters are quite different, suggesting a propagation of the difference in precipitation through the model. In particular, the TRMM parameters are more heterogeneous throughout the basin. Furthermore, the TRMM model has a higher retention efficiency in four out of six regions and a higher ET efficiency in all basins (larger α values). The TRMM model also has more recharge to the deep aquifer (smaller X_{GW} in all regions). This reflects that TRMM predicts larger volumes of precipitation throughout the basin.

4.2   Spatial characterization of discharge

Figure 8 shows the observed and simulated flow duration curves and climatology at the downstream calibration station, Lambaréné. The flow regime in the Ogooué consists of precipitation-driven direct runoff peaks as seen from the steep slope of the FDC for low exceedance probabilities and a sizeable baseflow, characterized by a nonzero minimum flow value and a flattening curve at higher exceedance probabilities. Generally, the FDCs simulated by both models are

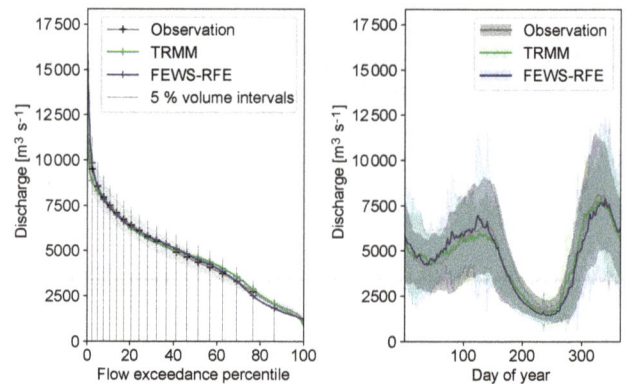

Figure 8. Flow duration curves and daily discharge climatology benchmark at the Lambaréné calibration stations; the surfaces in the climatology plot represent the 90 % confidence interval.

within 10 % of the observed FDC at all six calibration stations ($R_{FDC} \geq 0$). Furthermore, the calibration is deemed reasonable if the simulated climatology falls within 1 standard deviation of the observation (WRMSD ≤ 1). For both models, this is the case at all calibration stations.

Table 3. Calibrated parameters from the two models forced by TRMM and FEWS-RFE precipitation.

	α_1	α_2	d	S_{max}	n_{Nash}	k_{Nash}	X_{GW}	d_{deep}	X	n
Batéké Plateau										
TRMM	0.41	0.25	0.18	466	5	1.17	0.134	0.010	0.22	0.019
FEWS-RFE	0.38	0.23	0.26	633	5	3.25	0.250	0.006	0.22	0.017
Eastern Gabon plateau										
TRMM	0.53	0.30	0.31	559	6	4.43	0.876	0.016	0.29	0.026
FEWS-RFE	0.64	0.27	0.19	795	7	3.34	0.950	0.014	0.13	0.037
Booué										
TRMM	0.57	0.28	0.42	844	5	5.67	0.034	0.018	0.36	0.041
FEWS-RFE	0.43	0.26	0.47	934	4	3.88	0.194	0.015	0.42	0.050
Ndjolé										
TRMM	0.26	0.61	0.30	1142	6	0.20	0.364	0.008	0.22	0.049
FEWS-RFE	0.24	0.53	0.43	737	4	4.84	0.383	0.015	0.35	0.036
Ngounié										
TRMM	0.39	0.30	0.27	380	2	2.30	0.249	0.013	0.19	0.044
FEWS-RFE	0.44	0.27	0.21	152	1	5.69	0.308	0.013	0.26	0.040
Ogooué delta										
TRMM	0.42	0.20	0.47	856	5	0.44	0.095	0.016	0.31	0.036
FEWS-RFE	0.24	0.20	0.64	797	6	5.05	0.520	0.018	0.24	0.034
Basin median										
TRMM	0.42	0.29	0.30	701	5	1.73	0.192	0.014	0.26	0.037
FEWS-RFE	0.41	0.26	0.23	766	5	3.61	0.346	0.015	0.25	0.036

Table 4 shows the performance statistics for the FDC and climatology contributions to the calibration objective at the calibration and validation stations. Both models are within the validation criteria at all calibration stations and two out of five validation stations. Overall, the performances of the two models are similar in terms of simulating flow regime in the basin: the TRMM model performs better based on 10 out of 19 validated performance measures.

The calibration objective incorporates two important evaluation criteria: the model's ability to capture the seasonality and probability distributions of discharge throughout the basin. The results indicate the model is capable of simulating both, regardless of precipitation forcing. Day-to-day comparison with up-to-date discharge is necessary to evaluate the success of the calibration strategy compared to traditional approaches but, in cases where no current observations are available, the approach used in this study is a good compromise.

4.3 Simulated total water storage change

Figure 9 shows the total water storage change in the two basin halves observed by GRACE and simulated by the TRMM- and FEWS-RFE-forced models. The monthly total water storage simulated by the model consists of the sum of water stored in the root zone, the shallow aquifer and deep aquifer, the tributary processes and the main channel. The tributary processes represent 9.2 and 10.1 % of the total storage change throughout the basin in the TRMM and FEWS-RFE model respectively, indicating a significant contribution from water retention processes. This is consistent with the large number of wetlands and lakes in the basin. The deep aquifer holds the lion's share of total water storage change: respectively 70.6 and 83.0 % of total water storage change in the TRMM and FEWS-RFE models. The soil storage contributes 12.2 and 1.9 and around 2.5 % of the change originates from the shallow aquifer in both models. Storage changes in the main channel contribute 5.5 and 2.4 % to the total water storage change respectively. This pattern is due to the monthly aggregation of simulations. Most of the low frequency variations are observed in the deep aquifers, which have smaller storage constants, while high frequency variations are averaged out in the other stores. The largest difference between the models relates to the changes in soil water storage. The TRMM model generally has larger α parameters: larger retention efficiency leads to larger positive

Table 4. Performance measures for the TRMM and FEWS-RFE models based on the discharge observations. Values in bold highlight the best validated performance.

Station (reach)		R_{FDC}		Climatology, WRMSD	
		TRMM	FEWS-RFE	TRMM	FEWS-RFE
Batéké Plateau					
Calibration	Lastoursville	0.39	**0.63**	0.56	**0.33**
Validation	Leyami	−0.08	−0.14	1.43	1.14
Eastern Gabon plateau					
Calibration	Makokou	**0.43**	0.36	0.65	**0.59**
Validation	Belinga	−0.64	−0.76	**0.68**	0.92
Booué					
Calibration	Booué	0.60	**0.79**	**0.41**	0.51
Validation	Loa-Loa	−0.05	−0.04	**0.62**	0.75
Ndjolé					
Calibration	Ndjolé	**0.67**	0.60	**0.52**	0.82
Validation	Portes de l'Okanda	0.58	**0.71**	**0.38**	0.58
Ngounié					
Calibration	Fougamou	**0.76**	0.67	0.37	**0.36**
Validation	Sindara	**0.68**	0.67	0.57	**0.53**
Ogooué delta					
Calibration	Lambaréné	0.67	**0.71**	**0.31**	0.42

soil storage changes and higher evapotranspiration efficiency leads to larger negative soil storage changes.

Table 5 shows the performance statistics for the TWSC contribution. The TRMM model generally performs better although the performance statistics are higher than the validation criteria (WRMSD ≤ 1), suggesting the residuals exceed the observation uncertainty. However, the models both capture the TWSC in the basin quite well, albeit storage change is generally underestimated. The best performance is achieved in the western basin (bottom plots), with WRMSD values below 1.4 for the calibration and validation period. We compute the precipitation anomalies relative to the mean monthly precipitation. On average, the TRMM estimates fluctuate more, as seen in the larger anomalies (5.8 cm per month, compared to 5.0 cm for FEWS-RFE). Due to the delay between precipitation signal and storage response, we obtain better fits in years where the precipitation anomalies match the observed storage change: e.g., in late 2006–early 2007, FEWS-RFE estimated more rain during the rainy season, resulting in an overestimation of the relative total water storage in the subsequent year. Similarly, both products predict little to no positive water storage change in 2009 and have a larger number of negative than positive precipitation anomalies. Thus, the discrepancies between the GRACE observations and the simulated total water storage changes can

Table 5. GRACE objective functions (WRMSD [–]) for the two models for the calibration and validation periods.

	Calibration		Validation	
	TRMM	FEWS-RFE	TRMM	FEWS-RFE
East	2.11	2.19	2.55	2.68
West	1.21	1.33	1.16	1.33

be attributed to three factors: the trade-off between fitting the water storage in the basin versus other calibration objectives; uncertainties in the GRACE observations, particularly considering the size of the study region and the spatial resolution of the observations; and, finally, differences in trends in water storage and precipitation anomalies. The latter can be due to water retention or diversion in the basin not accounted for by the model or to uncertainties in the precipitation estimates.

4.4 River stage

For comparative purposes, we reference the observed and simulated water heights to the long-term mean. At virtual stations, we calculate the long-term mean based only on dates where satellite observations were acquired. The results are shown in Table 6. Simulated water depths depend

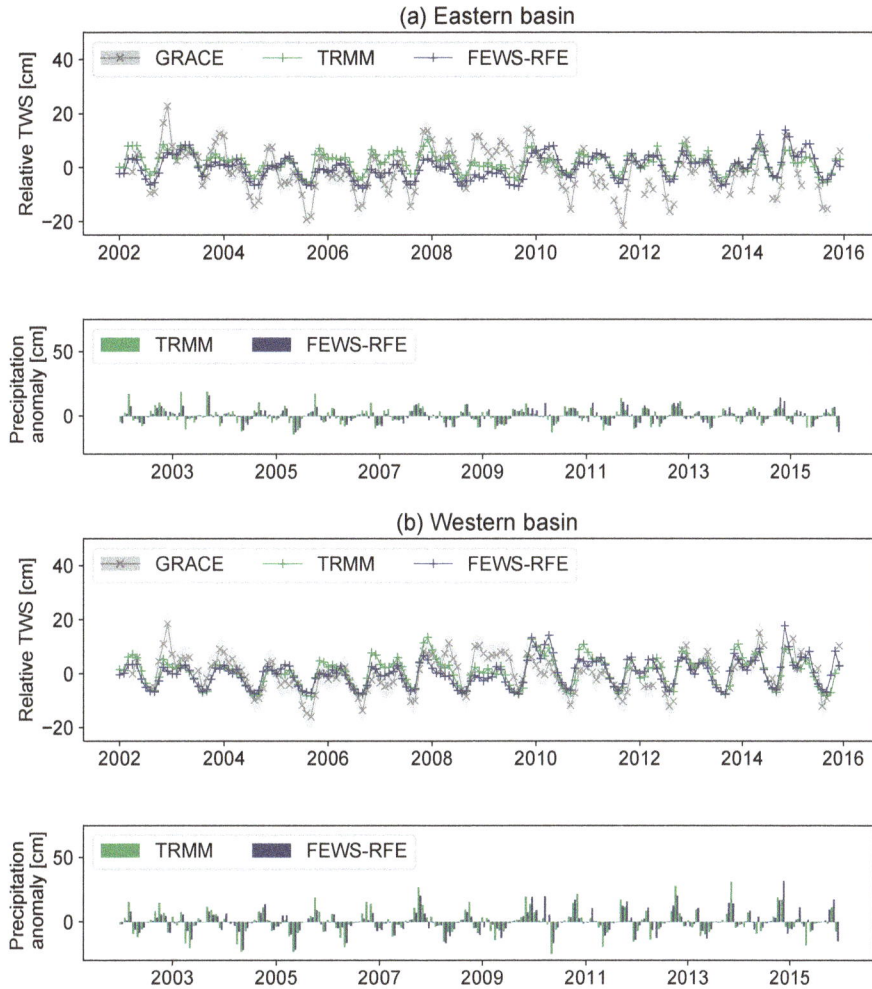

Figure 9. Total water storage change and precipitation anomaly referenced to the monthly climatology over the period of simulation for the eastern (**a**) and western (**b**) basins.

on the river cross-sectional geometry. We do not calibrate river cross-sectional geometry in order to limit the number of fitting parameters. Nevertheless, the simulated depth amplitudes are realistic. The simulated amplitudes are within the 90 % confidence intervals of the observation at all but one virtual station. The Nash–Sutcliffe efficiency (NSE) is above 0.5 during the calibration period in 9 out of 12 virtual stations for the TRMM-forced model and in 8 out of 12 for the FEWS-RFE model. Performance slightly decreases in the validation period, in particular for the Ngounié virtual stations and the FEWS-RFE model. When comparing the simulated water depth amplitudes to those observed at each station, the RMSD is 0.74 m for the TRMM model and 0.87 m for FEWS-RFE, corresponding to 0.85 and 0.94 times the standard deviation of annual water height amplitude (Table 7).This is comparable to the study by Schneider et al. (2017), in which they obtained an average RMSE of 0.83 m for the Brahmaputra after calibrating the river cross sections in a hydrodynamic model against Envisat virtual stations.

Figure 10 shows the water height fluctuations at two of the virtual stations.

Figure 11 shows the simulated water height anomaly climatology from the Batéké Plateau to the delta and all available altimetry observations. Sharp changes in amplitude reflect the confluence of river branches briefly increasing width (e.g., chainage 450–420 at the confluence of the Ivindo and the Ogooué) and the nature of the topography: the river is narrow between Booué and Ndjolé (chainage 420–260), before reaching the plain and eventually the delta, where the river width reaches up to 1300 m. At chainage 180, the Ngounié joins the Ogooué and the river width increases by 500 m. The temporal pattern agrees well and the spatial patterns are comparable. The RMSD between CryoSat-2 anomalies and model simulations is between 1 and 2 m in most regions in the basin (Table 8), part of which can be attributed to the approximated mean water level and to the time of observation. Due to its long repeat period, CryoSat-2 samples more often during certain seasons over different parts

Table 6. Performance statistics for altimetry at virtual stations. The values in parentheses indicate the model subbasin (Fig. 1). The first line shows the statistics for the calibration period and the second for the validation period. Values in bold are within the validation criteria (NSE > 0.5, WRMSD ≤ 1).

Virtual station	Amplitude (m)			NSE		WRMSD	
Mission (subbasin ID) coordinates, chainage	Altimetry mission [90 % CI]	TRMM	FEWS	TRMM	FEWS	TRMM	FEWS
Ogooué							
Envisat (12)	2.35 [1.52–3.87]	1.21	1.08	0.43	0.20	**0.81**	**0.96**
1.224° S, 13.334° E, 695 km				**0.61**	**0.51**	**0.70**	**0.79**
Envisat (20)	4.22 [1.17–5.39]	2.83	2.02	**0.60**	0.41	1.54	1.86
0.061° S, 11.642° E, 385 km				0.21	0.10	1.70	1.81
Envisat (24)	2.87 [1.25–3.65]	1.80	1.82	**0.74**	**0.63**	**0.72**	**0.86**
0.506° S, 10.302° E, 187 km				0.46	0.25	**1.00**	1.18
Envisat (26)	2.87 [1.84–4.71]	2.70	2.72	**0.78**	**0.77**	**0.68**	**0.70**
0.835° S, 10.027° E, 133 km				**0.53**	−0.08	**0.95**	1.44
Envisat (26)	3.74 [2.06–5.80]	3.40	3.71	**0.67**	**0.73**	1.14	1.04
0.921° S, 9.675° E, 83 km				**0.52**	−0.33	1.30	2.15
Envisat (27)	2.42 [1.54–3.96]	2.22	2.27	**0.78**	**0.75**	**0.57**	**0.61**
1.073° S, 9.256° E, 30 km				**0.55**	0.15	**0.90**	1.30
Ivindo							
Jason-2 (10)	4.72 [1.13–5.85]	3.80	4.15	0.34	0.33	2.00	2.02
1.1° N, 13.076° E, 677 km				0.06	−0.12	2.13	2.32
Envisat (14)	2.22 [1.11–3.33]	1.56	1.74	**0.62**	**0.57**	**0.75**	**0.80**
0.251° N, 12.422° E, 533 km				0.39	0.11	**0.83**	**1.00**
Ngounié							
Envisat (7)	2.69 [1.44–4.13]	3.74	2.79	**0.66**	**0.64**	**0.87**	**0.89**
1.272° S, 10.650° E, 305 km				−0.48	−1.72	1.94	2.63
Envisat (11)	2.42 [1.43–3.86]	2.65	2.67	**0.82**	**0.73**	**0.54**	**0.67**
1.142° S, 10.678° E, 273 km				0.05	−0.59	1.38	1.78
Envisat (11)	3.18 [1.19–4.37]	2.86	2.77	0.41	0.39	1.43	1.46
1.042° S, 10.701° E, 263 km				−0.24	−0.68	1.57	1.83
Envisat (15)	2.99 [2.04–5.03]	2.84	2.63	**0.75**	**0.55**	**0.73**	**0.97**
0.601° S, 10.323° E, 183 km				0.39	−0.39	1.22	1.83

Table 7. Basin amplitude statistics at all virtual stations: bias and root mean square deviation (RMSD). The percentages are relative to the mean observed amplitude.

WRMSD		RMSD (m) (%)		Bias (m) (%)	
TRMM	FEWS	TRMM	FEWS	TRMM	FEWS
0.85	0.94	0.74 (24.8 %)	0.87 (28.8 %)	0.41 (13.8 %)	0.42 (14.0 %)

of the river. Schneider et al. (2017) obtained an RMSD of 2.5 m between simulated water heights and CryoSat-2 observations over the Brahmaputra – thus, we deem the obtained results satisfactory in light of the available information.

While altimetry observations from drifting ground track missions increase the spatial resolution, observations from the virtual stations give a temporal characterization of water height fluctuations at specific locations in the basin. The obtained accuracy is on the order of magnitude of values reported in the literature – better results could be obtained with knowledge about the bathymetry or by calibrating the river cross sections. In this study, increasing the number of calibration parameters would not be suitable because only a limited number of CryoSat-2 observations are available and no contemporary discharge observations to validate timing.

Similarly, to the water storage amplitudes, the water level amplitudes are slightly underestimated, particularly in the eastern basin. The model parameters are most sensitive to improving the FDC and climatology benchmark contributions, which are based on historical discharge observations. Changes in precipitation patterns since the time of observation are likely to have affected discharge patterns. The comparison to contemporary satellite altimetry observations

Table 8. CryoSat-2 versus simulated relative water depths.

River stretch	Number of CryoSat-2 observations	RMSD (m)		Bias (m)	
		TRMM	FEWS-RFE	TRMM	FEWS-RFE
Upstream of Makokou (Ivindo)	32	1.76	1.69	0.01	0.07
Upstream of Sindara (Ngounié)	47	2.37	2.06	−0.19	0.10
Between Ndjolé and Lambaréné (Ogooué)	46	0.94	0.98	0.02	−0.02
Downstream of Lambaréné (Ogooué)	110	1.03	1.15	−0.08	−0.14
Ogooué river	353	1.85	1.85	−0.09	−0.21

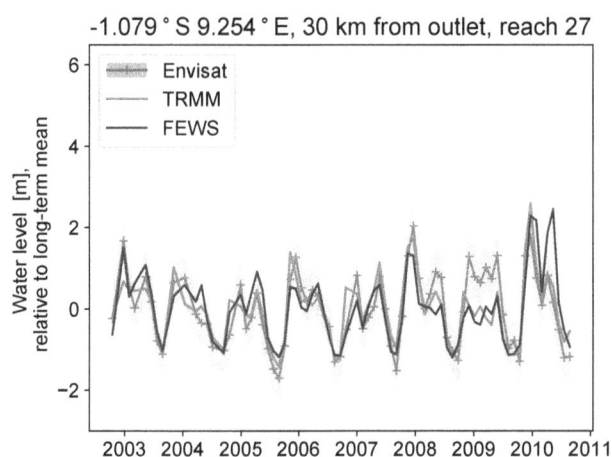

Figure 10. Results for water height simulation at a selected virtual station.

strengthens the validation of the model; however, the underestimation of water height amplitude and total water storage change in the basin may indicate that the model compensates for changes in precipitation patterns and uncertainties in the precipitation products in order to fit the historical discharge dataset.

4.5 Discussion

This study uses free, publicly available remote sensing observations relevant to the proposed model structure to characterize the basin. Several more types of remote sensing products are available but not included. For instance, no reliable soil moisture estimates can be produced for the Ogooué basin because the dense vegetation masks the microwave returns from the underlying soil (Tang et al., 2009). We select the most relevant products and explore how new data sources may supplement existing datasets and extend their applicability. To the authors' knowledge, this study is the first to use SAR imagery from Sentinel-1 to extract CryoSat-2 observations over an inland water body and the first study to evaluate CryoSat-2 observations over the Ogooué. The size of the Ogooué (approximately 1.3 km at its widest and 390 m on average) makes it as an interesting study area for altimetry observations. However, without cloud-penetrating tech-

nologies, it would be very difficult to produce a satisfactory water mask of the river. The possibility to develop detailed water masks for virtually any inland water body from SAR imagery greatly expands the applicability of altimetry observations from drifting ground track missions over rivers.

In poorly gauged basins, the paucity of observations limits the estimation of the model parameters and consequently model complexity (Johnston and Smakhtin, 2014). Remote sensing data have been used in several studies to compensate for gaps in in situ observations and have enabled the definition of distributed or semi-distributed models even in poorly gauged basins (van Griensven et al., 2012). Furthermore, the accessibility of remote sensing observations creates modeling opportunities in basins, where in situ data are insufficient on their own (Johnston and Smakhtin, 2014). This is the case for the Ogooué, which to the authors' knowledge only has discharge observations prior to 1984 at best, and precipitation records at a dozen stations. The model used in this study has a fairly simple and flexible structure with few parameters and limited input-data requirements, which can accommodate several basin and river network configurations. Furthermore, although the model currently does not support reservoir characterization or abstraction losses, these can be implemented within the model structure. By starting with a simple structure and gradually adding complexity (deep aquifer, tributary processes), the principle of parsimony is respected.

The remote sensing observations used in this study help characterize the otherwise ungauged basin and the model can produce valuable information for water managers. Several studies have benefitted from including altimetry observations (Schneider et al., 2017; Michailovsky et al., 2013; Domeneghetti et al., 2014) and total water storage observations (Xie et al., 2012; Milzow et al., 2011) in river basin models. In this study, the altimetry missions generally agree very well and the observations provide valuable information on water heights throughout the river. Although it would be useful to confirm the remotely sensed observations with ground observations, the availability of contemporary observations strengthened the evaluation of the hydrological model of the Ogooué. Without additional observations or on-ground information on the basin, the presented model is the best available representation of the Ogooué basin. However, model simulations can never replace observations and remote

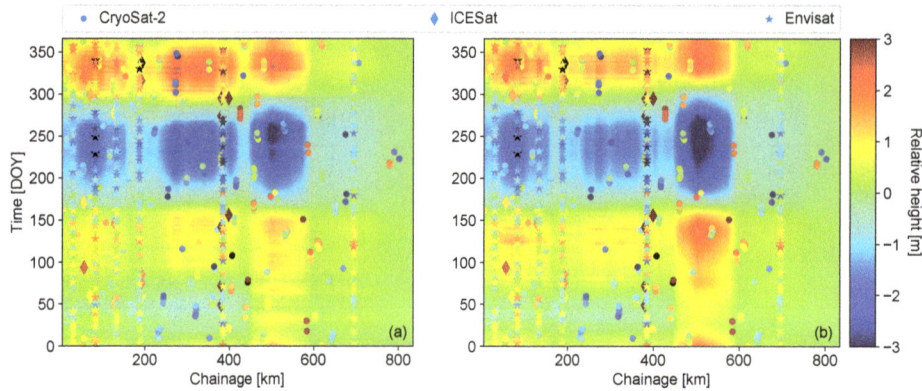

Figure 11. Interpolated relative water heights (m) based on the simulated daily water depth climatology in the Ogooué by the model forced with TRMM 3B42 v7 **(a)** and the FEWS-RFE **(b)** precipitation and altimetry observations from CryoSat-2, ICESat and Envisat. The virtual stations visible in the figure correspond to those in reach 27, 26, 24, 20 and 12.

sensing observations have never been evaluated in the basin before. Therefore, ground truthing efforts and in situ gauging campaigns would greatly strengthen the conclusions of this study.

A model should always be evaluated in light of its intended application (Johnston and Smakhtin, 2014). The model developed in this study is the first model of the Ogooué river basin and provides otherwise unavailable information regarding the baseline river flow regime. It can be used in a broad range of applications, including flood forecasting, climate change evaluation and as an impact assessment tool for planned water infrastructure investments. For instance, the hydrologic impact of hydraulic infrastructures at the inlet to downstream key wetlands resolved by the model can be assessed and compared to the baseline developed in this study.

5   Conclusions

In this study, we explore the use of multi-mission remote sensing to inform a hydrological model of the fourth largest African river by discharge, the Ogooué in Gabon. We set up a lumped conceptual rainfall–runoff model based on the Budyko framework coupled to a Muskingum routing scheme. We force the model using remote sensing precipitation and calibrated using a combination of historical in situ discharge observations from the 1960s and 1970s, as well as total water storage observations from the GRACE mission. Remote sensing enables the evaluation of the model against contemporary observations and helps constrain model parameters by including information other than discharge measurements.

In addition, this study shows the potential of the new ESA Sentinel missions by deriving a detailed river mask from Sentinel-1 radar imagery, which is used to extract altimetry observations from CryoSat-2. The multi-mission approach increases spatial and temporal coverage and acts as a useful supplement to the observed in situ discharge in terms of validation in regions were the missions agree. We validate the water height simulations against the altimetry observations at multiple points in the basin. With the methods applied in this study, a dynamic river mask can be defined and used to extract relevant observations over inland water bodies of interest from existing and new satellite altimetry missions. New radar altimetry missions such as Sentinel-3 carrying state-of-the-art equipment are expected to provide higher accuracy observations. Combined with the water masking method proposed in this study, relevant time series of river water heights can be extracted and used in hydrological modeling studies.

Progress in remote sensing technologies, instruments and extraction algorithms now allows for the observation of most hydrological states and fluxes from space. This offers a unique possibility to obtain observations in poorly gauged or remote areas and to supplement hydrological modeling applications with the necessary input-data and useful observations for parameter estimation. The model used in this study can be applied in scenario evaluations and provides an otherwise unavailable insight into the hydrological regime of the Ogooué on the catchment scale. By combining hydrological modeling with multi-mission remote sensing from 10 different satellite missions, we obtain new information on an otherwise unstudied basin. The proposed model is the best current baseline characterization of hydrological conditions in the Ogooué in light of the available observations.

Code and data availability. The python code used in this study will be publicly available in upcoming versions of the GlobWetlands Africa toolbox. All data sets used in this study are derived from publicly available resources. The model climate input files and river delineation as well as the GRACE and CryoSat-2 observations used in the study.

Competing interests. The authors declare that they have no conflict of interest.

Acknowledgements. We acknowledge funding from the European Space Agency (ESA) through the GlobWetlands Africa project. Daily historical in situ observations of discharge in the river basin recorded by the Office de la Recherche Scientifique et Technique Outre-Mer (ORSTOM) and the Direction Générale des Ressources Hydrauliques (DGRH) in Gabon were recovered from the Institut de Recherche pour le Développement (IRD). The observations are accessible on their Système d'Informations Environnementales sur les Ressources en Eaux et leur Modélisation (SIEREM) website, http://www.hydrosciences.fr/sierem (Boyer et al., 2006).

Edited by: Bettina Schaefli

References

Abelen, S. and Seitz, F.: Relating satellite gravimetry data to global soil moisture products via data harmonization and correlation analysis, Remote Sens. Environ., 136, 89–98, https://doi.org/10.1016/j.rse.2013.04.012, 2013.

Alvarez-Garreton, C., Ryu, D., Western, A. W., Crow, W. T., and Robertson, D. E.: The impacts of assimilating satellite soil moisture into a rainfall-runoff model in a semi-arid catchment, J. Hydrol., 519, 2763–2774, https://doi.org/10.1016/j.jhydrol.2014.07.041, 2014.

Awange, J. L., Gebremichael, M., Forootan, E., Wakbulcho, G., Anyah, R., Ferreira, V. G., Alemayehu, T., Awange, J. L., Gebremichael, M., Forootan, E., Wakbulcho, G., Anyah, R., Ferreira, V. G., and Alemayehu, T.: Characterization of Ethiopian mega hydrogeological regimes using GRACE, TRMM and GLDAS datasets, Adv. Water Resour., 74, 64–78, https://doi.org/10.1016/j.advwatres.2014.07.012, 2014.

Awange, J. L., Ferreira, V. G., Forootan, E., Khandu, Andam-Akorful, S. A., Agutu, N. O., and He, X. F.: Uncertainties in remotely sensed precipitation data over Africa, Int. J. Climatol., 36, 303–323, https://doi.org/10.1002/joc.4346, 2016.

Bauer-Gottwein, P., Jensen, I. H., Guzinski, R., Bredtoft, G. K. T., Hansen, S., and Michailovsky, C. I.: Operational river discharge forecasting in poorly gauged basins: the Kavango River basin case study, Hydrol. Earth Syst. Sci., 19, 1469–1485, https://doi.org/10.5194/hess-19-1469-2015, 2015.

Berry, P. A. M. and Benveniste, J.: Global Inland Water Monitoring from Satellite Radar Altimetry – What Can We Really Do?, in: ESA Living Planet Symposium 2013, Edinburgh, UK, 2013.

Berry, P. A. M., Garlick, J. D., Freeman, J. A., and Mathers, E. L.: Global inland water monitoring from multi-mission altimetry, Geophys. Res. Lett., 32, 1–4, https://doi.org/10.1029/2005GL022814, 2005.

Birkinshaw, S. J., O'Donnell, G. M., Moore, P., Kilsby, C. G., Fowler, H. J., and Berry, P. A. M.: Using satellite altimetry data to augment flow estimation techniques on the Mekong River, Hydrol. Process., 24, 3811–3825, https://doi.org/10.1002/hyp.7811, 2010.

Boyer, J. F., Dieulin, C., Rouche, N., Cres, A., Servat, E., Paturel, J. E., and Mahé: SIEREM: An environmental information system for water resources, Iahs-aish P., 308, 19–25, 2006.

Brocca, L., Melone, F., Moramarco, T., Wagner, W., Naeimi, V., Bartalis, Z., and Hasenauer, S.: Improving runoff prediction through the assimilation of the ASCAT soil moisture product, Hydrol. Earth Syst. Sci., 14, 1881–1893, https://doi.org/10.5194/hess-14-1881-2010, 2010.

Chow, V. T., Maidment, D. R., and Mays, L. W.: Lumped flow routing, in: Applied Hydrology, chap. 8, 242–271, McGraw-Hill, New York, 1988.

Cohen Liechti, T., Matos, J. P., Boillat, J.-L., and Schleiss, A. J.: Comparison and evaluation of satellite derived precipitation products for hydrological modeling of the Zambezi River Basin, Hydrol. Earth Syst. Sci., 16, 489–500, https://doi.org/10.5194/hess-16-489-2012, 2012.

Dezetter, A., Girard, S., Paturel, J. E., Mahé, G., Ardoin-Bardin, S., and Servat, E.: Simulation of runoff in West Africa: Is there a single data-model combination that produces the best simulation results?, J. Hydrol., 354, 203–212, https://doi.org/10.1016/j.jhydrol.2008.03.014, 2008.

Domeneghetti, A., Tarpanelli, A., Brocca, L., Barbetta, S., Moramarco, T., Castellarin, A., and Brath, A.: The use of remote sensing-derived water surface data for hydraulic model calibration, Remote Sens. Environ., 149, 130–141, https://doi.org/10.1016/j.rse.2014.04.007, 2014.

Duan, Q., Sorooshian, S., and Gupta, V.: Effective and Efficient Global Optimization for Conceptual Rainfall-Runoff Models, Water Resour. Res., 28, 1015–1031, 1992.

Eicker, A., Schumacher, M., Kusche, J., Döll, P., and Schmied, H. M.: Calibration/Data Assimilation Approach for Integrating GRACE Data into the WaterGAP Global Hydrology Model (WGHM) Using an Ensemble Kalman Filter: First Results, Surv. Geophys., 35, 1285–1309, https://doi.org/10.1007/s10712-014-9309-8, 2014.

Gebregiorgis, A. S., Tian, Y., Peters-Lidard, C. D., and Hossain, F.: Tracing hydrologic model simulation error as a function of satellite rainfall estimation bias components and land use and land cover conditions, Water Resour. Res., 48, W11509, https://doi.org/10.1029/2011WR011643, 2012.

Getirana, A. C. V.: Integrating spatial altimetry data into the automatic calibration of hydrological models, J. Hydrol., 387, 244–255, https://doi.org/10.1016/j.jhydrol.2010.04.013, 2010.

Getirana, A. C. V. and Peters-Lidard, C.: Estimating water discharge from large radar altimetry datasets, Hydrol. Earth Syst. Sci., 17, 923–933, https://doi.org/10.5194/hess-17-923-2013, 2013.

Han, E., Merwade, V., and Heathman, G. C.: Implementation of surface soil moisture data assimilation with watershed scale distributed hydrological model, J. Hydrol., 416–417, 98–117, https://doi.org/10.1016/j.jhydrol.2011.11.039, 2012.

Hannah, D. M., Demuth, S., van Lanen, H. A., Looser, U., Prudhomme, C., Rees, G., Stahl, K., and Tallaksen, L. M.: Large-scale river flow archives: Importance, current status and future needs, Hydrol. Process., 25, 1191–1200, https://doi.org/10.1002/hyp.7794, 2011.

Houska, T., Kraft, P., Chamorro-Chavez, A., and Breuer, L.: SPOTting model parameters using a ready-made python package, PLoS ONE, 10, 1–22, https://doi.org/10.1371/journal.pone.0145180, 2015.

Johnston, R. and Smakhtin, V.: Hydrological Modeling of Large river Basins: How Much is Enough?, Water Resour. Manag., 28, 2695–2730, https://doi.org/10.1007/s11269-014-0637-8, 2014.

Kittel, C. M. M., Nielsen, K., Tøttrup, C., and Bauer-Gottwein, P.: Dataset used in Kittel et al., 2017 (https://doi.org/10.5194/hess-2017-549), including model files and processed remote sensing observations (CryoSat-2 radar altimetry and GRACE total water storage) [Data set], Zenodo, available at: https://doi.org/10.5281/zenodo.1157344, last access: 21 February 2018.

Knoche, M., Fischer, C., Pohl, E., Krause, P., and Merz, R.: Combined uncertainty of hydrological model complexity and satellite-based forcing data evaluated in two data-scarce semi-arid catchments in Ethiopia, J. Hydrol., 519, 2049–2066, https://doi.org/10.1016/j.jhydrol.2014.10.003, 2014.

Long, D., Longuevergne, L., and Scanlon, B. R.: Global analysis of approaches for deriving total water storage changes from GRACE satellites, Water Resour. Res., 51, 2574–2594, https://doi.org/10.1002/2015WR017096, 2015.

Longuevergne, L., Scanlon, B. R., and Wilson, C. R.: GRACE hydrological estimates for small basins: Evaluating processing approaches on the High Plains aquifer, USA, Water Resour. Res., 46, 1–15, https://doi.org/10.1029/2009WR008564, 2010.

Madsen, H.: Automatic calibration of a conceptual rainfall–runoff model using multiple objectives, J. Hydrol., 235, 276–288, https://doi.org/10.1016/S0022-1694(00)00279-1, 2000.

Mahé, G. and Olivry, J. C.: Assessment of freshwater yields to the ocean along the intertropical Atlantic coast of Africa (1951–1989), CR ACAD SCI II A, 328, 621–626, https://doi.org/10.1016/S1251-8050(99)80159-1, 1999.

Mahé, G., Lerique, J., and Olivry, J.-C.: Le fleuve Ogooué au Gabon: Reconstitution des débits manquants et mise en évidence de variations climatiques à l'équateur, Hydrologie continentale, 5, 105–124, 1990.

Mahé, G., Lienou, G., Descroix, L., Bamba, F., Paturel, J. E., Laraque, A., Meddi, M., Habaieb, H., Adeaga, O., Dieulin, C., Chahnez Kotti, F., and Khomsi, K.: The rivers of Africa: Witness of climate change and human impact on the environment, Hydrol. Process., 27, 2105–2114, https://doi.org/10.1002/hyp.9813, 2013.

Mengue Medou, C., Ondamba Ombanda, F., Mounganga, M.-D., Bayani, E., Ndjokounda, C., and Mikala, R.: Site Ramsar Mboungou-Badouma et de Doume, Tech. Rep. 2006–2008, Ramsar, Libreville, Gabon, available at: https://rsis.ramsar.org/RISapp/files/RISrep/GA1853RIS_1703_fr.pdf (last access: 21 February 2018), 2008.

Mezui, E. and Boumono Moukoumi, V.: Le Bassin de l'Ogooué: caractéristiques et importance, in: Annual International Conference on Geological and Earth Sciences (GEOS), Ministère du Pétrole, de l'Énergie et des Ressources Hydrauliques, available at: http://slideplayer.fr/slide/177651/ (last access: 21 February 2018), 2013.

Michailovsky, C. I., Milzow, C., and Bauer-Gottwein, P.: Assimilation of radar altimetry to a routing model of the Brahmaputra River, Water Resour. Res., 49, 4807–4816, https://doi.org/10.1002/wrcr.20345, 2013.

Milzow, C., Krogh, P. E., and Bauer-Gottwein, P.: Combining satellite radar altimetry, SAR surface soil moisture and GRACE total storage changes for hydrological model calibration in a large poorly gauged catchment, Hydrol. Earth Syst. Sci., 15, 1729–1743, https://doi.org/10.5194/hess-15-1729-2011, 2011.

Mulder, G., Olsthoorn, T. N., Al-Manmi, D. A. M. A., Schrama, E. J. O., and Smidt, E. H.: Identifying water mass depletion in

northern Iraq observed by GRACE, Hydrol. Earth Syst. Sci., 19, 1487–1500, https://doi.org/10.5194/hess-19-1487-2015, 2015.

Neitsch, S. L., Arnold, J. G., Kiniry, J. R., and Williams, J. R.: Soil & Water Assessment Tool – Theoretical Documentation Version 2009, Tech. rep., Texas A&M University System, Temple, Texas, available at: http://swat.tamu.edu/documentation/ (last access: 21 February 2018), 2009.

O'Loughlin, F. E., Neal, J., Yamazaki, D., and Bates, P. D.: ICESat-derived inland water surface spot heights, Water Resour. Res., 52, 3276–3284, https://doi.org/10.1002/2015WR018237, 2016.

Paiva, R. C. D., Collischonn, W., and Tucci, C. E. M.: Large scale hydrologic and hydrodynamic modeling using limited data and a GIS based approach, J. Hydrol., 406, 170–181, https://doi.org/10.1016/j.jhydrol.2011.06.007, 2011.

Paturel, J. E., Ouedraogo, M., Mahé, G., Servat, E., Dezetter, A., and Ardoin, S.: The influence of distributed input data on the hydrological modelling of monthly river flow regimes in West Africa, Hydrolog. Sci. J., 48, 881–890, https://doi.org/10.1623/hysj.48.6.881.51422, 2003.

Saltelli, A., Tarantola, S., and Chan, K. P.: A Quantitative Model-Independent Method for Global Sensitivity Analysis of Model Output A Quantitative Model-Independent Method for Global Sensitivity Analysis of Model Output, Technometrics, 41, 39–56, 1999.

Santos da Silva, J., Calmant, S., Seyler, F., Rotunno Filho, O. C., Cochonneau, G., and Mansur, W. J.: Water levels in the Amazon basin derived from the ERS 2 and ENVISAT radar altimetry missions, Remote Sens. Environ., 114, 2160–2181, https://doi.org/10.1016/j.rse.2010.04.020, 2010.

Schneider, R., Godiksen, P. N., Villadsen, H., Madsen, H., and Bauer-Gottwein, P.: Application of CryoSat-2 altimetry data for river analysis and modelling, Hydrol. Earth Syst. Sci., 21, 751–764, https://doi.org/10.5194/hess-21-751-2017, 2017.

Schumann, G. J. P. and Domeneghetti, A.: Exploiting the proliferation of current and future satellite observations of rivers, Hydrol. Process., 30, 2891–2896, https://doi.org/10.1002/hyp.10825, 2016.

Sneeuw, N., Lorenz, C., Devaraju, B., Tourian, M. J., Riegger, J., Kunstmann, H., and Bárdossy, A.: Estimating Runoff Using Hydro-Geodetic Approaches, Surv. Geophys., 35, 1333–1359, https://doi.org/10.1007/s10712-014-9300-4, 2014.

Stisen, S. and Sandholt, I.: Evaluation of remote-sensing-based rainfall products through predictive capability in hydrological runoff modelling, Hydrol. Process., 24, 879–891, https://doi.org/10.1002/hyp.7529, 2010.

Stisen, S., Jensen, K. H., Sandholt, I., and Grimes, D. I. F.: A remote sensing driven distributed hydrological model of the Senegal River basin, J. Hydrol., 354, 131–148, https://doi.org/10.1016/j.jhydrol.2008.03.006, 2008.

Tang, Q., Gao, H., Lu, H., Dennis, P., and Lettenmaier, D. P.: Remote sensing: hydrology, Prog. Phys. Geog., 33, 490–509, https://doi.org/10.1177/0309133309346650, 2009.

Tanner, J. and Hughes, D.: Surface water–groundwater interactions in catchment scale water resources assessments–understanding and hypothesis testing with a hydrological model, Hydrolog. Sci. J., 6667, 1–16, https://doi.org/10.1080/02626667.2015.1052453, 2015.

Tarboton, D.: TauDEM Version 5, available at: http://hydrology.usu.edu/taudem/taudem5/documentation.html (last access: 21 February 2018), 2015.

Thiemig, V., Rojas, R., Zambrano-Bigiarini, M., and De Roo, A.: Hydrological evaluation of satellite-based rainfall estimates over the Volta and Baro-Akobo Basin, J. Hydrol., 499, 324–338, https://doi.org/10.1016/j.jhydrol.2013.07.012, 2013.

Todini, E.: A mass conservative and water storage consistent variable parameter Muskingum-Cunge approach, Hydrol. Earth Syst. Sci., 11, 1645–1659, https://doi.org/10.5194/hess-11-1645-2007, 2007.

Tourian, M., Tarpanelli, A., Elmi, O., Qin, T., Brocca, L., Moramarco, T., and Sneeuw, N.: Spatiotemporal densification of water level time series by multimission satellite altimetry, Water Resour. Res., 52, 613–615, https://doi.org/10.1002/2015WR017654, 2016.

van Griensven, A., Ndomba, P., Yalew, S., and Kilonzo, F.: Critical review of SWAT applications in the upper Nile basin countries, Hydrol. Earth Syst. Sci., 16, 3371–3381, https://doi.org/10.5194/hess-16-3371-2012, 2012.

Villadsen, H., Andersen, O. B., Stenseng, L., Nielsen, K., and Knudsen, P.: CryoSat-2 altimetry for river level monitoring – Evaluation in the Ganges–Brahmaputra River basin, Remote Sens. Environ., 168, 80–89, https://doi.org/10.1016/j.rse.2015.05.025, 2015.

Vörösmarty, C., Askew, A., Grabs, W., Barry, R. G., Birkett, C., Döll, P., Goodison, B., Hall, A., Jenne, R., Kitaev, L., Landwehr, J., Keeler, M., Leavesley, G., Schaake, J., Strzepek, K., Sundarvel, S. S., Takeuchi, K., and Webster, F.: Global water data: A newly endangered species, Eos, 82, 54–58, https://doi.org/10.1029/01EO00031, 2001.

Watkins, M. M., Wiese, D. N., Yuan, D.-N., Boening, C., and Landerer, F. W.: Improved methods for observing Earth's time variable mass distribution with GRACE using spherical cap mascons, J. Geophys. Res.-Sol. Ea., 120, 2648–2671, https://doi.org/10.1002/2014JB011547, 2015.

Westerberg, I. K., Guerrero, J.-L., Younger, P. M., Beven, K. J., Seibert, J., Halldin, S., Freer, J. E., and Xu, C.-Y.: Calibration of hydrological models using flow-duration curves, Hydrol. Earth Syst. Sci., 15, 2205–2227, https://doi.org/10.5194/hess-15-2205-2011, 2011.

World Bank: Gabon – Country partnership strategy for the period FY2012-FY2016., Tech. rep., World Bank, Washington DC, available at: http://documents.worldbank.org/curated/en/4962314680303171 (last access: 21 February 2018), 2012.

Xie, H., Longuevergne, L., Ringler, C., and Scanlon, B. R.: Calibration and evaluation of a semi-distributed watershed model of Sub-Saharan Africa using GRACE data, Hydrol. Earth Syst. Sci., 16, 3083–3099, https://doi.org/10.5194/hess-16-3083-2012, 2012.

Xu, X., Li, J., and Tolson, B. A.: Progress in integrating remote sensing data and hydrologic modeling, Prog. Phys. Geog., 38, 464–498, https://doi.org/10.1177/0309133314536583, 2014.

Zhang, L., Potter, N., Hickel, K., Zhang, Y., and Shao, Q.: Water balance modeling over variable time scales based on the Budyko framework – Model development and testing, J. Hydrol., 360, 117–131, https://doi.org/10.1016/j.jhydrol.2008.07.021, 2008.

A hydrological routing scheme for the Ecosystem Demography model (ED2+R) tested in the Tapajós River basin in the Brazilian Amazon

Fabio F. Pereira[1,a], Fabio Farinosi[1,2,d], Mauricio E. Arias[1,3], Eunjee Lee[1,b,c], John Briscoe[1,†], and Paul R. Moorcroft[1]

[1]Sustainability Science Program, Kennedy School of Government, Harvard University, Cambridge, MA 02138, USA
[2]Ca' Foscari University of Venice, Venice, Italy
[3]Department of Civil and Environmental Engineering, University of South Florida, Tampa, FL 33620, USA
[a]now at: Department of Renewable Energy Engineering, Federal University of Alagoas, Maceió, AL, Brazil
[b]now at: Goddard Earth Sciences Technology and Research, Universities Space Research Association, Columbia, MD 21046, USA
[c]current address: Global Modeling and Assimilation Office, NASA Goddard Space Flight Center, Greenbelt, MD 22071, USA
[d]now at: European Commission, DG Joint Research Centre, Ispra, Italy
[†]deceased, 12 November 2014

Correspondence to: Fabio Farinosi (fabio.farinosi@gmail.com)

Abstract. Land surface models are excellent tools for studying how climate change and land use affect surface hydrology. However, in order to assess the impacts of Earth processes on river flows, simulated changes in runoff need to be routed through the landscape. In this technical note, we describe the integration of the Ecosystem Demography (ED2) model with a hydrological routing scheme. The purpose of the study was to create a tool capable of incorporating to hydrological predictions the terrestrial ecosystem responses to climate, carbon dioxide, and land-use change, as simulated with terrestrial biosphere models. The resulting ED2+R model calculates the lateral routing of surface and subsurface runoff resulting from the terrestrial biosphere models' vertical water balance in order to determine spatiotemporal patterns of river flows within the simulated region. We evaluated the ED2+R model in the Tapajós, a 476 674 km^2 river basin in the southeastern Amazon, Brazil. The results showed that the integration of ED2 with the lateral routing scheme results in an adequate representation (Nash–Sutcliffe efficiency up to 0.76, Kling–Gupta efficiency up to 0.86, Pearson's R up to 0.88, and volume ratio up to 1.06) of daily to decadal river flow dynamics in the Tapajós. These results are a consistent step forward with respect to the "no river representa-tion" common among terrestrial biosphere models, such as the initial version of ED2.

1 Introduction

Understanding the impacts of deforestation (e.g., Lejeune et al., 2015; Medvigy et al., 2011; Andréassian, 2004) and climate change (e.g., Jiménez-Cisneros et al., 2014) on the Earth's water cycle has been a topic of substantial interest in recent years given its potential implications for ecosystems and society (e.g., Wohl et al., 2012; Brown et al., 2005). Analyses of climate change impacts on the Earth's water cycle increasingly use terrestrial biosphere models, which are capable of estimating changes in the vertical water balance as a function of climate forcing and/or land-use-induced changes in canopy structure and composition (Zulkafli et al., 2013). Terrestrial biosphere models actively used for hydrological and Earth system sciences include the Joint UK Land Environment Simulator (JULES) (Best et al., 2011; Clark et al., 2011), the Community Land Model (CLM) (Lawrence et al., 2011; Oleson et al., 2010), the Lund–Potsdam–Jena (LPJ) land model (Gerten et al., 2004; Sitch et al., 2003), the Max

Planck Institute MPI-JSBACH model (Vamborg et al., 2011; Raddatz et al., 2007), and the Integrated Biosphere Simulator (IBIS) (Kucharik et al., 2000).

Initial formulations of the hydrological processes within terrestrial biosphere models were based on simple "bucket" model formulations (Cox et al., 1999 after Carson, 1982). Moisture within each climatological grid cell of the domain was simulated in a single below-ground pool in which surface temperature and specific soil moisture factors determined evaporation, while runoff was equal to the bucket overflow (Cox et al., 1999; Carson, 1982). Recently, the hydrologic schemes within terrestrial biosphere models have become increasingly sophisticated. In the most recent generation of land surface models, water fluxes in and out of the soil column are vertically resolved and take into account feedback from the different components, for instance, through an explicit formulation of the soil–plant–atmosphere continuum. This enables the models to provide a detailed representation of the interactions between evapotranspiration, soil moisture, and runoff (Clark et al., 2015).

To couple the calculation of the one-dimensional water balance with the estimation of daily river flows, it is necessary to simulate multiple hydrological dynamics involved in the lateral flow propagation through the landscape, ideally including the most complex hydraulic features of floodplains, lakes, and wetlands (Yamazaki et al., 2011). The first step towards representing the finer-scale hydrodynamic processes responsible for patterns in river gauge observations is to consider the topographic and geomorphological features that control water flow (Arora et al., 1999). The coarse spatial resolution of regional land surface models, imposed by computational constraints, does not allow for proper simulation of the complex hydrological dynamics determined by fine-scale topography in river channels and floodplains (Yamazaki et al., 2011; Kauffeldt et al., 2016). However, the combination of the terrestrial models with routing schemes can be used to simulate the implications of global and regional environmental changes for flood/drought forecasting, water resources planning and management, and infrastructure development (Andersson et al., 2015). Consequently, several terrestrial biosphere models have been integrated with routing schemes. For example, JULES has been integrated with the Total Runoff Integrating Pathways (TRIP) to evaluate the accuracy of its estimates of annual streamflow (Oki et al., 1999). This integrated model was used to investigate the status of the global water budget (Oki et al., 2001). Rost et al. (2008) also used a modeling framework composed of the global dynamic vegetation model, LPJ, and a simple water balance model to quantify the global consumption of water for rain-fed and irrigated agriculture. An offline coupling of the dynamic vegetation model, ISIS, and HYDRA – which simulates the lateral transport of water through rivers, lakes, and wetlands – was proposed in Coe et al. (2008) with the purpose of reproducing linkages between land use, hydrology, and climate. Moreover, Liang et al. (1994) developed

and tested the coupling of the well-known VIC model with a general circulation model (GCM) to improve the GCM's ability to capture the interactions between surface hydrology and atmosphere. For the same purpose, the MPI hydrological discharge model was validated with NCEP reanalysis and parameterized for simulating the river routing for climate analysis at global scale (Hagemann and Gates, 2001; Hagemann and Dumenil, 1997). Several routing schemes have been designed to date, including normal depth, modified pulse, simple Muskingum, and Muskingum–Cunge (USACE, 1991). Most notably, the semi-distributed kinematic wave-routing Muskingum–Cunge method has been recognized for its stability over different spatial and temporal modeling resolutions (USACE, 1991; Miller and Cunge, 1975; Cunge, 1969), and has been adopted by the most widely used regional-scale hydrological models, such as VIC, SWAT, and MGB-IPH.

Recent studies have investigated the influence of land use on regional patterns of rainfall and biosphere temperature (Ostberg et al., 2015; Bahn et al., 2014; Pearson et al., 2013). These studies tracked how the occurrence of conversion of land from its natural state over the same time frame as observed fluctuations of rainfall and air temperature occurred – aspects fully analyzed by terrestrial biosphere models (Hurtt et al., 2006; Goldewijk, 2001; Ramankutty and Foley, 1999). However, these models assumed that global and regional changes in the biosphere were a result of dynamics of vegetation in a collection of landscapes given by forests, deserts, and farmland only. Inland surface waters (e.g., rivers, lakes, and wetlands) were not considered as an interactive component of the biosphere and hence the climate system (Cole et al., 2007).

The Ecosystem Demography (ED2) is a terrestrial biosphere model that simulates the coupled water, carbon, and energy dynamics of terrestrial land surfaces (Longo, 2014; Medvigy et al., 2009; Moorcroft et al., 2001) to describe the coupled water, carbon, and energy dynamics of heterogeneous landscapes (Hurtt et al., 2013; Medvigy et al., 2009; Moorcroft et al., 2001). ED2's ability to incorporate sub-grid-scale ecosystem heterogeneity arising from land-use change makes the model suited for investigating how the combined impacts of changes in climate, atmospheric carbon dioxide concentrations, and land cover affect terrestrial ecosystems. For example, ED2 was successfully used to simulate the carbon flux dynamics in the North American continent (Hurtt et al., 2002; Albani et al., 2006) and to assess the impacts on Amazonian ecosystems of changes in climate, atmospheric carbon dioxide, and land use (Zhang et al., 2015). Moreover, ED2, coupled with a regional atmospheric circulation component, has also been successfully applied to assess the impacts of deforestation on the Amazonian climate (Knox et al., 2015; Swann et al., 2015). The aforementioned studies were not aimed at assessing hydrological implications of changes in land use and climate. These works demonstrated the validity of ED2 for assessing impacts of global and regional changes on ecosystem function and built

Short-term dynamics – patch *m*

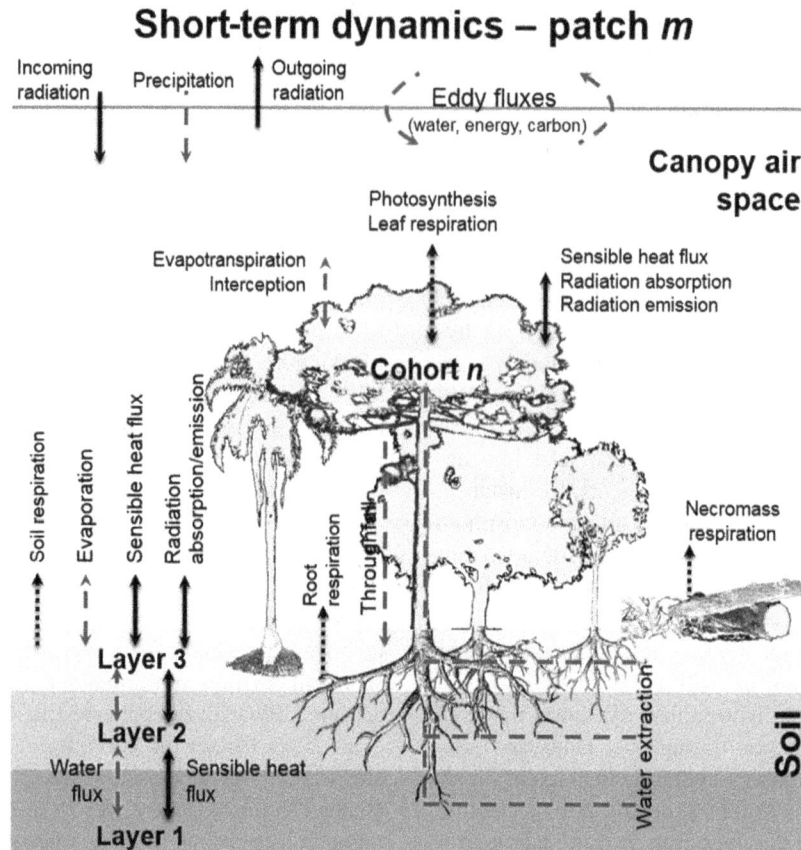

Figure 1. Schematic of the enthalpy fluxes (all arrows) and water fluxes (all but solid black arrows) that are solved in ED2. The schematic is based on Walko et al. (2000) and Medvigy et al. (2009). (Figure courtesy of Marcos Longo.)

the foundations for an integrated tool aimed at analyzing hydrological implications.

In this technical note, we describe the integration of ED2 with a hydrological routing scheme. The hydrological routing scheme chosen was adapted from the MGB-IPH (Collischonn et al., 2007). This exercise aims to calculate the lateral propagation and attenuation of the surface and subsurface runoff resulting from the vertical balance calculations in order to simulate daily river flows through a large river basin. The advantage of the proposed model is its ability to predict the sensitivity of river flows to global and regional environmental changes such as climate and land-use changes. The new product combines the advantages of biosphere and hydrological models, bringing together global-, regional-, and local-scale hydrological dynamics in a single modeling framework. The resulting model is intended to be used in future studies as a computational tool to explore a variety of research questions. In particular, it could be used to analyze how current and future climate and land cover affect water availability in river systems; how land-use-driven changes can influence the water availability for human activities (hydropower, food production, urban supply); and what

the implications of those changes are for water and land resources management.

The identified research areas are in line with key problems raised in the literature, focusing on the importance of large-scale modeling and remote sensing to fill knowledge gaps in water resources and hydrological dynamics (Alsdorf et al., 2007; Prigent et al., 2007). The product obtained from this exercise was tested in the Tapajós Basin, a large river system in the southeastern Amazon, Brazil.

2   Ecosystem Demography (ED2) model

ED2 is a terrestrial biosphere simulation model capable of representing biological and physical processes driving the dynamics of ecosystems as a function of climate and soil properties. Rather than using a conventional "ecosystem as big-leaf" assumption, ED2 is formulated at the scale of functional and age groups of plants. Ecosystem-scale dynamics and fluxes are calculated through a scaling procedure which reproduces the macroscopic behavior of the ecosystem within each climatological grid cell. It simulates ecosystem structure and dynamics as well as the corresponding carbon, energy, and water fluxes (Fig. 1; Hurtt et al., 2013; Med-

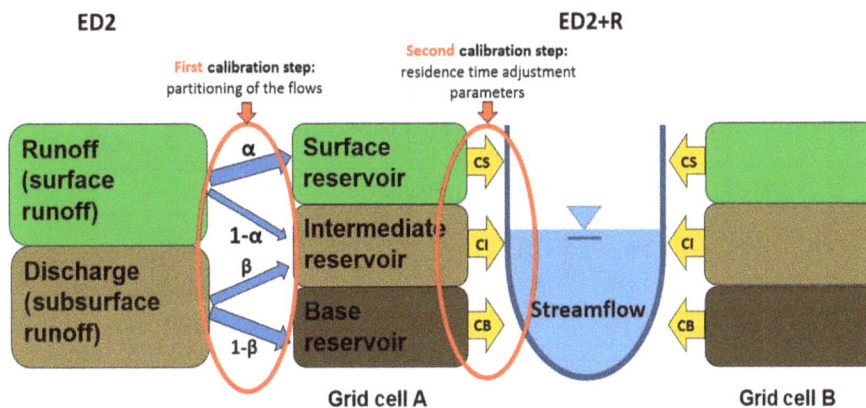

Figure 2. Schematic representation of the connection between the terrestrial biosphere model and the hydrological routing scheme. Calibrating parameters are circled in red. The reservoirs are used to determine the contribution of streamflow that comes from overland flow, interflow, and groundwater flow. The daily sum of these three reservoirs is then moved from each grid cell into the drainage network.

vigy et al., 2009; Moorcroft et al., 2001). ED2 simulates the dynamics of different plant functional types subdivided into tiles with a homogeneous canopy (Swann et al., 2015; Medvigy et al., 2009). The dynamic tiles represent the sub-grid-scale heterogeneity in ecosystem composition within each cell. Grid cell size is determined by the resolution of meteorological forcing and soil characteristics data, typically from 1 to 0.001° (~ 110 to 1 km). ED2 simulates biosphere dynamics by taking into consideration natural disturbances, such as forest fires and plant mortality due to changing environmental conditions, as well as human-caused disturbances, such as deforestation and forest harvesting (Medvigy et al., 2009; Albani et al., 2006). Disturbances are expressed in the model as annual transitions between primary vegetation, secondary vegetation, and agriculture (cropland and pasture) (Albani et al., 2006). Natural disturbance, such as wildfire, is represented in the model by the transition from primary vegetation (forest in the case of the Amazon) to grassland–shrubland, and subsequently to secondary vegetation (forest regrowth); the abandonment of an agricultural area is represented with the conversion from grassland to secondary vegetation, while forest logging is represented by the transition from primary or secondary vegetation to grassland. The model is composed of several modules operating at multiple temporal and spatial scales, including plant mortality, plant growth, phenology, biodiversity, soil biogeochemistry, disturbance, and hydrology (Longo, 2014; Medvigy et al., 2009). A selection of the main parameters and the input used for this study are presented in Table 1, and for a more complete description of the model, we refer the reader to the literature available (Zhang et al., 2015; Longo 2014; Kim et al., 2012; Medvigy et al., 2009; Moorcroft et al., 2001).

2.1　ED2 hydrology module

The hydrological module of the ED2 model is derived from the Land Ecosystem-Atmospheric Feedback model (LEAF-

2) (Walko et al., 2000). The model computes the water cycle through vegetation, air canopy space, and soils, yielding daily estimates of subsurface and surface runoff from each grid cell, isolated from the others in the domain. The number of soil layers and their thickness influence the accuracy with which the model is able to represent the gradients near the surface. Soil composition was derived from Quesada et al. (2010) and from the IGBP-DIS global soil data (Global Soil Data Task, 2014). As described in Zhang et al. (2015), the mean fraction values of sand and clay were assigned to each grid cell at 1 km resolution and then aggregated at 1° resolution. Due to limited data availability, soils were assumed to be homogeneous for a depth of 6 m. Hydraulic conductivity of the soil layers is a function of soil texture and moisture (Longo, 2014). Groundwater exchange is a function of hydraulic conductivity, soil temperature, and terrain topography. Water percolation is limited to the bottom layer by the subsurface drainage, determining the bottom boundary conditions. Vegetation historical records and land-use transitions were derived from the Global Land Use Dataset (Hurtt et al., 2006). A more detailed description of the hydrological subcomponent of the ED2 model is available in Longo (2014).

3　ED2 runoff routing scheme (ED2+R)

River routing schemes are often used to compute the lateral movement of water over land in hydrological models for large river basins. In this way, the prediction performance of models can be evaluated using river discharge measurements. The use of routing schemes was then extended to Earth system models in order to capture the impacts of man-made structures (e.g., dams and reservoirs) and floodplain wetlands on the climate system (Li et al., 2011; Yamazaki et al., 2011).

Daily runoff estimates from ED2 were computed for specific grid cells independently. A hydrological routing scheme

Table 1. ED2+R calibrated parameters (based on Zhang et al., 2015; Longo, 2014; Knox, 2012). Additional information about ED2 parameter calibration for the Amazon Basin are available in Zhang et al. (2015) and Longo (2014).

Input	Source
Meteorological forcing	Sheffield et al. (2006)
Land use	Hurtt et al. (2006)
Topography (DEM)	SRTM, Shuttle Radar Topography Mission 90 m resolution (USGS, 2016)
Soil data	Quesada et al. (2010) – IGBP-DIS global soil data (Global Soil Data Task, 2014)
Geomorphological relations	Coe et al. (2008)
Streamflow observations	HYBAM – ANA (ANA, 2016; Observation Service SO HYBAM, 2016)
Carbon dioxide concentration	378 ppm

Process	Method
Integration scheme	Fourth-order Runge–Kutta method
Energy and water cycles	Knox (2012) and Longo (2014)
Temperature-dependent function for photosynthesis	Q_{10} function
Canopy radiation scheme	Two-stream model
Allometry for height	Based on Poorter et al. (2006)
Allometry for above-ground biomass	Based on Eq. (2) of Baker et al. (2004)
Allometry for leaf biomass	Based on Cole and Ewel (2006) and Calvo-Alvarado et al. (2008)

Parameter	Value	Units
Biophysics time step	600	s
Number of soil layers	16	–
Depth of the deepest soil layer	6	m
Depth of the shallowest soil layer	0.02	m
Cohort water holding capacity	0.11	$kg_w m^{-2}_{leaf+wood}$
Residual stomatal conductance	10 000	$\mu mol\ m^{-2} s^{-1}$
Leaf-level water stress parameter	0.016	$mol_{H_2O}\ mol^{-1}_{Air}$
Oxygenase/carboxylase ratio at 15 °C	4000	–
Power base for oxygenase/carboxylase ratio	0.57	–
Power base for carboxylation rate	2.4	–
Power base for dark respiration rate	2.4	–

Environmentally determined parameters	Value	Units
Weight factor for stress due to light	1.0	–
Maximum environmentally determined mortality rate	5.0	$year^{-1}$
Steepness of logistic curve	10.0	–

Band-dependent radiation parameters *	Value	Units
Dry soil reflectance	(0.20; 0.31; 0.02)	–
Wet soil reflectance	(0.10; 0.20; 0.02)	–
Leaf transmittance	(0.05; 0.20; 0.00)	–
Leaf reflectance (grasses)	(0.10; 0.40; 0.04)	–
Leaf reflectance (trees)	(0.10; 0.40; 0.05)	–
Wood transmittance	(0.05; 0.20; 0.00)	–
Wood reflectance (trees)	(0.05; 0.20; 0.10)	–

Table 1. Continued.

Input	Source	
Plant functional type (PFT)-dependent parameters **	Value	Units
Leaf orientation factor	(0.10; 0.10; 0.10)	–
Leaf clumping factor	(0.80; 0.80; 0.80)	–
Leaf characteristic size	(0.10; 0.10; 0.10)	m
Max. carboxylation rate at 15 °C	(18.75; 12.50; 6.25)	$\mu mol_C\, m_{leaf}^{-2}\, s^{-1}$
Dark respiration rate at 15 °C	(0.272; 0.181; 0.091)	$\mu mol_C\, m_{leaf}^{-2}\, s^{-1}$
Quantum yield	(0.080; 0.080; 0.080)	–
Slope parameter for stomatal conductance	(9.0; 9.0; 9.0)	–
Fine root conductance parameter	(600; 600; 600)	$m^2\, kg_{C_{root}}^{-1}\, year^{-1}$
River routing parameters (Sect. 4)	Value	Units
Grid-cell size (Fig. 4)	0.5 × 0.5	degrees
Flow partitioning parameters (α; β) (Fig. 5)	(0.70; 0.40)	–
Residence time adjustment parameters, respectively, referring to overland (CS), intermediate (CI), and subsurface water flows (CB) (Fig. 7) (CS; CI; CB)	Upper Juruena (2600; 70 000; 90 000) Upper Teles Pires (1600; 1750; 2500) Lower Juruena (1500; 600; 500) Lower Teles Pires (1500; 650; 800) Jamanxim (10; 10; 11) Upper Tapajós (75; 75 000; 75 000) Lower Tapajós (75; 75 000; 75 000)	× 1000***
Initial conditions of the baseflow (Fig. 6)	Upper Juruena (0.0159) Upper Teles Pires (0.009) Lower Juruena (0.0004) Lower Teles Pires (0.011) Jamanxim (0.0001) Upper Tapajós (0.0080) Lower Tapajós (0.0005)	$m^3\, km^2$

* Radiation-dependent parameters are given in the format xPAR, xNIR, and xTIR corresponding to values for photosynthetically active, near infrared, and thermal infrared, respectively. ** PFT-dependent parameters are given in the format xETR, xMTR, and xLTR corresponding to the values for early-, mid-, and late-successional cohorts, respectively. *** The residence time parameters are dimensionless and used to correct the Kirpich formula for time of concentration as explained in Collischonn et al. (2007). Their magnitude is influenced by the size of the grid cell and its topography.

was then linked to this model in order to estimate flow attenuation and accumulation as water moved through the landscape. The hydrological routing scheme chosen was adapted from the original formulation of the MGB-IPH, a rainfall–runoff model that has been used extensively in large river basins in South America (Collischonn et al., 2007). This model was later developed using hydrodynamic solutions and floodplain coupling (Pontes et al., 2015; Paiva et al., 2013b). Although the later development increased the modeling capabilities of the MGB-IPH in representing fine-scale dynamics, given the regional application of our tool, for ED2+R we decided to use the typical application of the MGB-IPH characterized by the Muskingum–Cunge approach. The original MGB-IPH model is composed of four different sub-models: soil water balance, evapotranspiration, intra-cell flow propagation, and inter-cell routing through the river network. In the present study, only the catchment and river routing methods were utilized. The resulting ED2+R model computes the daily total volume of water passing through any given grid cell in the resulting drainage network

in two separate steps: first, ED2 estimates of daily surface and subsurface runoff from each grid cell are divided into three linear reservoirs with different residence times to represent overland flow, interflow, and subsurface water flow (Fig. 2). The reservoirs are used to determine the contribution and attenuation of river flow by different soil layers, characterized by different routing times. The sum of overland flow, interflow, and subsurface water flow is then moved from each grid cell into the drainage network, designed in the preprocessing phase using data from a digital elevation model (DEM) from the Shuttle Radar Topography Mission (SRTM – USGS, 2016) at a 90 m resolution and the Cell Outlet Tracing with an Area Threshold algorithm (COTAT) (Reed, 2003). Each DEM grid cell therefore becomes part of a flow path, which then accumulates water to a final downstream drainage network outlet. A complete description of the technique for defining drainage networks from DEMs employed in this study can be found in Paz et al. (2006). Once water reaches the drainage network, ED2+R adopts the Muskingum–Cunge numerical scheme for the solution of the kinematic wave equation, which also accounts for flow attenuation, using a finite-difference method as a function of river length, width, depth and roughness, as well as terrain elevation slope (Collischonn et al., 2007; Reed, 2003). Statistical relationships for the river morphology were obtained as a function of the drainage area based on geomorphic data collected by Brazil's National Water Agency (ANA) and the Observation Service for the geodynamical, hydrological, and biogeochemical control of erosion/alteration and material transport in the Amazon Basin (HYBAM) at several gauging stations in the Amazon and Tocantins basins as presented by Coe et al. (2008). Further studies successfully derived geomorphological relations in order to estimate river geometric parameters and carry out hydrodynamic simulations of the Amazon River system using a similar approach (Paiva et al., 2011, 2013). Multiple groups of grid cells with common hydrological features, or hydrological response units, can be created in order to parameterize and calibrate ED2+R. In our approach, hydrological traits associated with soil and land cover are primarily computed in ED2; thus, we calibrated ED2+R at the sub-basin level as delineated based on the DEM. Details about the calibration procedure are provided in the next section.

The model's performance was calculated through the adoption of widely used indicators:

- Pearson's R correlation coefficient (Pearson, 1895), calculated as in Eq. (1):

$$R = \frac{\sum \text{sim} \cdot \text{obs} - \frac{(\sum \text{sim})(\sum \text{obs})}{n}}{\sqrt[2]{\left(\sum \text{sim}^2 - \frac{(\sum \text{sim})^2}{n}\right)\left(\sum \text{obs}^2 - \frac{(\sum \text{obs})^2}{n}\right)}}, \quad (1)$$

where sim and obs are the simulated and observed time series, while n is the number of time steps of the simulation period;

- volume ratio, calculated as ratio of the simulated (sim) and observed (obs) total water volume in the simulation period without consideration for the seasonal distribution of flow, as in Eq. (2):

$$VR = \text{Vol}_{\text{sim}}/\text{Vol}_{\text{obs}}; \quad (2)$$

- the Nash–Sutcliffe efficiency (NSE) coefficient (Nash and Sutcliffe, 1970), calculated as in Eq. (3):

$$NSE = 1 - \frac{\sum_1^n |\text{obs}_i - \text{sim}_i|^2}{\sum_1^n |\text{obs}_i - \overline{\text{obs}_i}|^2}, \quad (3)$$

where obs_i and sim_i are the observed and simulated data at time i, $\overline{\text{obs}_i}$ is the mean of the observed data, and n is number of time steps of the simulation period;

- the Kling–Gupta efficiency (KGE) index, both 2009 and 2012 versions, calculated as in Eq. (4):

$$KGE = 1 - \\ \sqrt{(s[1](R-1))^2 + (s[2](\text{vr}_{2009 \text{ or } 2012} - 1))^2 + (s[3](\beta - 1))^2}, \quad (4)$$

where s values are scaling factors (set to 1 in this case), r is Pearson's correlation coefficient, β is the ratio between the mean of the observed values and the mean of the simulated values, and vr is the variability ratio, defined as vr_{2009} (simulated vs. observed standard deviation ratio; Eq. 5) for the 2009 method and vr_{2012} (ratio of coefficient of variation of simulated and coefficient of variation of observed values; Eq. 6) for the 2012 method (Kling et al., 2012; Gupta et al., 2009).

$$\text{vr}_{2009} = \sigma_{\text{sim}}/\sigma_{\text{obs}} \quad (5)$$

$$\text{vr}_{2012} = \frac{\text{CV}_{\text{sim}}}{\text{CV}_{\text{obs}}} = \frac{\sigma_{\text{sim}}/\mu_{\text{sim}}}{\sigma_{\text{obs}}/\mu_{\text{obs}}} \quad (6)$$

The optimal value for the Pearson's R, VR, NSE, and KGE indexes is 1; the closer the indexes are to this value, the more accurately the model reproduces the observed values.

Missing observations in the river flow records (HYBAM and ANA) were filled via linear spatial and temporal interpolation between the series in neighboring gauge stations (Eq. 7):

$$\text{Obs}_y(t) = K + \beta_1 \cdot \text{Obs}_z(t) + \beta_2 \\ \cdot \text{Obs}_q(t) + \beta_3 \cdot \text{Obs}_y(t - 365) + \beta_4 \cdot \text{Obs}_y(t + 365), \quad (7)$$

where z, y, and q are three gauge stations with time series highly correlated (Pearson's $R \geq 0.85$), and t expresses time in days. The estimated β coefficients in Eq. (7) were used for the estimation of the missing observations in site y (Table 2). The interpolation of the gauge historical records was necessary to have continuous time series with a sufficient number of observations to calibrate and validate the ED2+R application in the basin.

For the presentation of the results, in order to compare the simulated and observed values, we also used flow duration curves (FDCs). FDCs are cumulative frequency plots that show the percentage of simulations steps (days in the case presented in this study) in which the discharge is likely to equal or exceed a specific value, without taking into consideration the sequence of the occurrence.

4 Case study: Tapajós River basin

The ED2+R formulation was parameterized and evaluated for the Tapajós River basin, the fifth largest tributary of the Amazon. This basin drains an area of $476\,674\,\text{km}^2$ in the southeastern Amazon, within the Brazilian states of Mato

Table 2. Statistics about the gauge information filling procedure (correlation with the station to be filled, number of original observations, and filled number of observations).

Sub-basin name	Main river gauge station – z in Eq. (7)	Original number of daily gauge records (number of daily observations)	Gap-filling station 1 – q in Eq. (7) (correlation with z)	Gap-filling station 2 – y in Eq. (7) (correlation with z)	Number of daily records after filling procedure (number of daily observations)
Jamanxim	Jamanxim	1928	Jardim do Ouro (0.97)	Novo Progresso (0.96)	5382
Upper Teles Pires	Cachoeirão	10 356	Teles Pires (0.91)	Indeco (0.94)	11 524
Upper Juruena	Fontanilhas	10 469	Foz do Juruena (0.94)	Barra do São Manoel (0.89)	11 688
Lower Teles Pires	Tres Marias	8682	Barra do São Manoel (0.98)	Santa Rosa (0.98)	10 640
Lower Juruena	Foz do Juruena	2074	Barra do São Manoel (0.98)	Jatoba (0.97)	11 447
Upper Tapajós	Jatoba	10 218	Fortaleza (0.99)	Barra do São Manoel (0.98)	11 517
Lower Tapajós	Itaituba	5789	Fortaleza (0.99)	Jatoba (0.98)	11 688

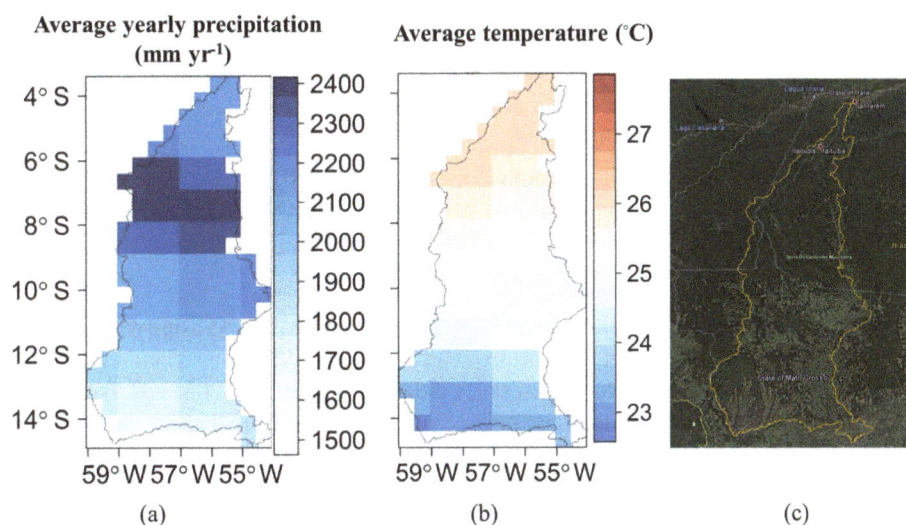

Figure 3. Average precipitation **(a)** and temperature **(b)** in the Tapajós River basin (1986–2005). Re-drafted from Farinosi et al. (2017). **(c)** Aerial imagery the Tapajós River basin illustrating land-cover diversity in the catchment. Source: Google Earth Pro.

Grosso, Pará, and Amazonas. The main rivers in the basin are the Tapajós (with a length greater than 1800 km and average discharge of 11 800 m^3 $^{-1}$), Juruena (length of approximately 1000 km and discharge of 4700 m^3 s^{-1}), and Teles Pires (also known by the name São Manoel, about 1600 km long and average discharge of 3700 m^3 s^{-1}). The river system flows northwards, with terrain elevation ranging from about 800 m a.s.l. (above sea level) in the southern region, to a few meters above sea level at its confluence with the Amazon River (ANA, 2011). The basin ecosystems are mainly represented by tropical evergreen rainforests in the northern region (in the states of Amazonas and Pará), and Cerrado dry vegetation in the south (Mato Grosso). Precipitation ranges from about 1500 mm year^{-1} in the headwaters (southern region) to about 2900 mm year^{-1} towards the basin's outlet (Fig. 3a, b). Rainfall temporal distribution is characterized by a clear seasonal distinction; total precipitation in the wet season (September to May) could be as high as 400 mm month^{-1} in the most tropical areas, whereas in the dry season (June to August), precipitation is close to zero in the Cerrado and as low as 50 mm month^{-1} in the wetter areas (Mohor et al., 2015). As a result of the large rainfall seasonal variability, river flows are also extremely variable: the mean monthly flow of the Tapajós River ranges between about 2300 and 28 600 m^3 s^{-1} according to the historical records used for the calibration of the ED2+R model.

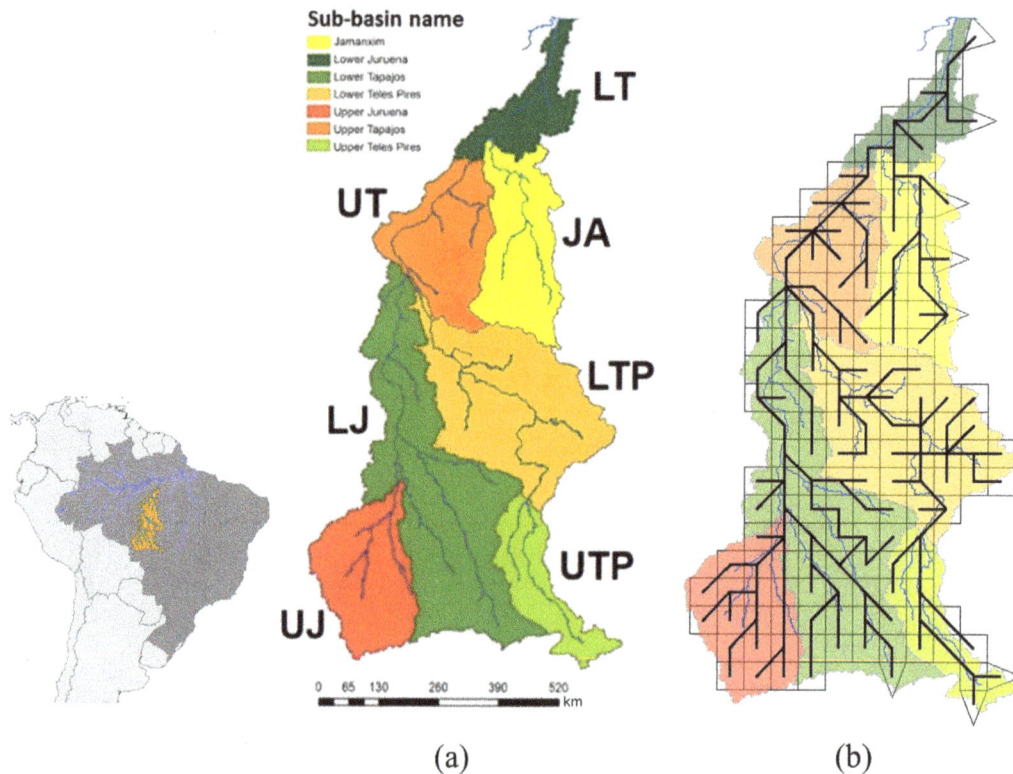

Figure 4. (a) Organization of the Tapajós Basin into seven sub-basins: Upper Juruena (UJ); Lower Juruena (LJ); Upper Teles Pires (UTP); Lower Teles Pires (LTP); Jamanxim (JA); Upper Tapajós (UT); and Lower Tapajós (LT). **(b)** ED2+R represents the domain in grid cells with 0.5° resolution (~ 55 km). The black segments indicate the flow accumulation network.

Soils vary from those typically seen in the Brazilian shield in the south of the basin to alluvial sediments in the north. Land use, almost completely represented by primary forest until the 1970s, was radically changed in recent decades. As estimated from the land-use/land-cover dataset used in this study (Hurtt et al., 2006), in the late 2000s only about 56 % of the basin (270 000 km^2) was covered by the original vegetation cover. Large parts of the basin lying in the territory of Mato Grosso were cleared to make room for agricultural and livestock production, while vast areas around the border between the state of Pará and Mato Grosso were cleared for cattle production. The northern portion of the basin is largely protected by natural parks or indigenous lands, but significant deforestation hotspots could be identified around the cities of Santarém and Itaituba and along the main transportation routes (Fig. 3c). For a more detailed description of the basin's physical characteristics and historical analysis of trends in deforestation, precipitation, and discharge, we refer the reader to Arias et al. (2017) and Farinosi et al. (2017).

For calibration purposes, the basin was divided into seven sub-basins, each of them with a corresponding gauge for which historical daily river flow observations were available (Fig. 4a). The domain was gridded with a spatial resolution of 0.5° by 0.5°, roughly corresponding to 55 km by 55 km. Simulations were carried out for the period 1970–2008. The ED2 model was forced using reconstructed climate (Sheffield et al., 2006) and land-use/land-cover data (Hurtt et al., 2006; Soares-Filho et al., 2006) at 1° spatial resolution. The original meteorological dataset has a 3 h temporal resolution, which was downscaled to an hourly resolution, as described in Zhang et al. (2015). In this technical note, we describe the calibration of the flow-routing component of ED2+R. The parameterization of the ED2 terrestrial biosphere model was developed and evaluated independently using eddy-flux tower observations of carbon, water, and energy fluxes and forest inventory observations of above-ground biomass dynamics. Further details are available in Zhang et al. (2015) and Longo (2014).

ED2+R model calibration

The ED2+R model was manually calibrated using gauge observations (ANA, 2016; Observation Service SO HYBAM, 2016) spanning a period of 17 years from 1976 to 1992 (the period 1970–1975 was not considered in order to avoid simulation initiation effects) through a two-step procedure, as highlighted in Fig. 2. The first step is partitioning the flows from the two reservoirs (surface and subsurface) of the ED2 biosphere model into the three reservoirs (surface, intermediate, base) of the ED2+R routed biosphere model (parameters

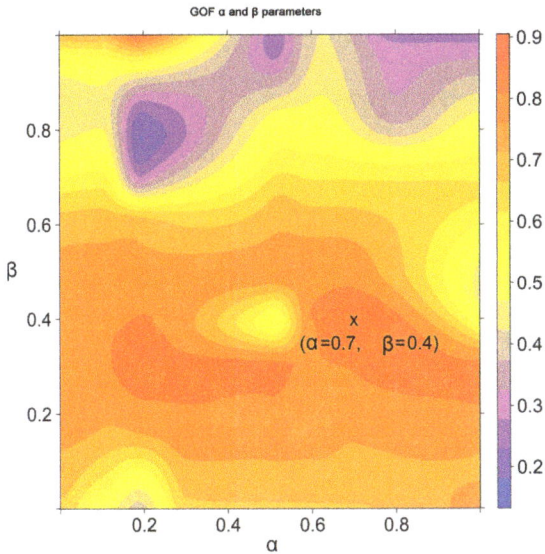

Figure 5. Calibration of flow partitioning (α and β parameters in Fig. 2) between the ED2 and the ED2+R reservoirs. The color bar indicates the NSE values of the simulated vs. the observed river flow values (0: very different; 1: very similar).

α and β in Fig. 2). In particular, α (ranging from 0 to 1, or from 0 to 100 %) represents the share of ED2 surface runoff allocated to the ED2+R surface reservoir. The remaining part $(1 - \alpha)$ is allocated to the ED2+R intermediate reservoir. β represents a similar partitioning coefficient for the ED2 subsurface reservoir to the ED2+R intermediate and base reservoirs. The second step relates to the adjustment of the residence times of the water flows in the three reservoirs for each of the grid cells in each of the sub-basins (overland, intermediate, and subsurface water flows – represented by the adjustment parameters CS, CI, and CB in Fig. 2).

In the first step, following the methodology described by Anderson (2002), the sensitivity of the α and β parameters was tested by running the model multiple times (~ 35). For each run, the NSE (Nash and Sutcliffe, 1970) was quantified by comparing the results of the simulation to historical flow observations. The combinations of the α and β parameters characterized by the largest NSE were selected. Parameters α and β were assumed to be uniform for the whole basin. Figure 5 shows the different combinations of the α and β parameters introduced in Fig. 2. The color bar indicates the NSE resulting from the comparison between the simulated and observed river flow values obtained using different combinations of the parameters α (x axis) and β (y axis). The chosen combination (indicated by an x in Fig. 5) lies in one of the optimal combination areas (NSE ~ 0.8).

In the second step, the residence times (τ) of flow within the ED2+R reservoirs of each grid cell in the domain were calibrated through the adjustment of the non-dimensional parameters (CS, CI, and CB in Fig. 2) used to correct the Kirpich formula for time of concentration (as explained in Col-

Table 3. Calibration and validation results for NSE, Kling–Gupta (2009 and 2012 methods), Pearson's R correlation, and volume ratio. Optimal values are equal to 1 (statistics were calculated using the R package hydroGOF; Zambrano-Bigiarini, 2014). KGE-2009 method values are indicated with square brackets []; KGE-2012 method values are indicated with curly brackets {} (Eqs. 4, 5, and 6).

Sub-basin	|	Calibration period (1976–1992)							|	Validation period (1993–2008)						
	|	Nash–Sutcliffe		Kling–Gupta [2009 method] {2012 method}		Pearson's R correlation		Volume ratio vol sim/vol obs	|	Nash–Sutcliffe		Kling–Gupta [2009 method] {2012 method}		Pearson's R correlation		Volume ratio vol sim/vol obs
	|	ED vs. OBS	ED2+R vs. obs	ED vs. obs	ED2+R vs. obs	ED vs. obs	ED2+R vs. obs	ED vs. obs	ED2+R vs. obs	ED vs. obs	ED2+R vs. obs	ED vs. obs	ED2+R vs. obs	ED vs. obs	ED2+R vs. obs	ED vs. obs	ED2+R vs. obs
Upper Juruena		−26.88	0.45	[−3.60]	[0.50]	0.61	0.68	0.72	0.98	−27.47	0.29	[−3.54]	[0.39]	0.53	0.54	0.68	1.01
Upper Teles Pires		−3.35	0.37	[−5.75] {−0.51}	[0.51] {0.61}	0.53	0.64	0.94	1.01	−3.19	0.28	[−6.10] {−0.51}	[0.38] {0.63}	0.57	0.63	0.96	1.03
Lower Juruena		−1.45	0.65	[−0.23] {−0.64}	[0.64] {0.61}	0.77	0.82	1.02	0.94	−2.17	0.63	[−0.43] {−0.59}	[0.72] {0.63}	0.75	0.81	1.05	1.08
Lower Teles Pires		−0.20	0.71	[−0.18] {0.25}	[0.67] {0.68}	0.80	0.85	1.01	1.02	−0.34	0.67	[0.17] {0.34}	[0.69] {0.67}	0.82	0.85	1.11	1.17
Jamanxim		−0.74	0.67	[0.27] {0.01}	[0.67] {0.79}	0.82	0.85	1.55	1.13	−0.10	0.55	[0.23] {0.75}	[0.75] {0.73}	0.83	0.77	1.43	1.09
Upper Tapajós		−1.01	0.77	[0.01] {−0.13}	[0.78] {0.82}	0.84	0.88	1.20	0.99	−1.23	0.75	[−0.22] {0.52}	[0.84] {0.73}	0.84	0.88	1.21	1.08
Lower Tapajós		−0.40	0.76	[0.39] {−0.09} {0.28}	[0.81] {0.86} {0.83}	0.84	0.88	1.11	1.06	−0.50	0.68	[0.09] {0.16} {0.29}	[0.81] {0.80} {0.76}	0.82	0.86	1.13	1.13

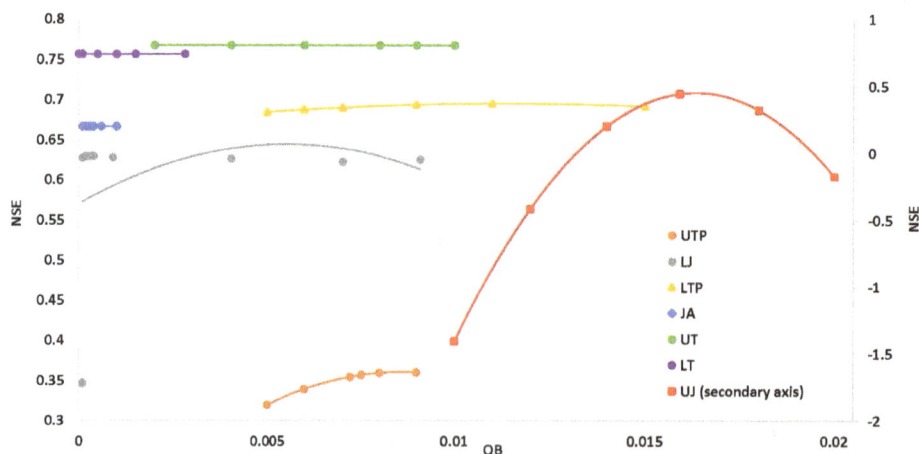

Figure 6. Model sensitivity to the initial conditions of baseflow (QB) for the different ED2+R sub-basins in the domain: Upper Juruena (UJ); Upper Teles Pires (UTP); Lower Juruena (LJ); Lower Teles Pires (LTP); Upper Tapajós (UT); Jamanxim (JA); and Lower Tapajós (LT).

lischonn et al., 2007). The calibration procedure characterizing the second step is similar to the previous one but in this case the calibration is repeated for each sub-basin sequentially. The calibration process was conducted from the furthest upstream sub-basins – headwaters – to the final outlet of the basin (Anderson, 2002). The model was run multiple times (between 30 and 50 per sub-basin) with different combinations of the three parameters (CS, CI, and CB in Fig. 2); for each run, the goodness of fit was quantified. This allowed us to design a sensitivity curve of the model to different combinations of the three parameters for each of the seven sub-basins and to select the combination that best approaches the historical observations. Figure 6 shows how the model is sensitive to marginal variation in initial conditions of baseflow, particularly in the upstream section (i.e., UTP – Upper Teles Pires, UJ – Upper Juruena, and LTP – Lower Teles Pires). Changes in initial subsurface water were controlled by the 5-year initialization period; thus, contributions to the downstream part of the basin had minimal impact (i.e., UT and LT – Upper and Lower Tapajós).

Figure 7 describes the calibration of the residence time adjustment parameters for each of the sub-basins, as well as an approximate calculation of the corresponding time of concentration for each of the reservoirs in the cell. The different combinations of the values assigned to the parameters CS, CI, and CB significantly affect the overall goodness of fit of the river flow simulations (NSE indicator). The calibration process was conducted from the furthest upstream sub-basins – headwaters – (UTP – Upper Teles Pires, UJ – Upper Juruena, and JA – Jamanxim) to the final outlet of the basin (LT – Lower Tapajós). The different combinations are marked with the corresponding NSE value; the optimal combination is marked in red (Fig. 7).

The period 1993–2008 was used for model evaluation. Comparisons between observations and simulated flows (goodness of fit) were carried out using Pearson's R cor-

relation coefficient (Pearson, 1895), volume ratio (VR), the Nash–Sutcliffe efficiency (NSE) coefficient (Nash and Sutcliffe, 1970), and the Kling–Gupta efficiency (KGE) index (Kling et al., 2012; Gupta et al., 2009) (Table 3).

5 Results

The integration of the routing scheme with ED2 increases the ability of the model to reproduce the observed temporal variations in river flows at the basin outlet (Fig. 8). This statement applies to all of the sub-basins, as the application of the routing scheme improved the model's performance between simulated and observed values with respect to all four measures selected (NSE, KGE, Pearson's R correlation, and volume ratio) (Table 3). Both routed (ED2+R) and non-routed (ED2) simulation results manage to reproduce the observed water availability (quantity of water available) in the basin in terms of volume. The volume ratio at the furthest downstream sub-basin (Lower Tapajós), in fact, ranges around the optimal value for both validation and calibration periods (ED2: 1.11–1.13; ED2+R: 1.06–1.13). The routing scheme improves the ability of the model to reproduce the spatiotemporal distribution of water flows across the basin: both the NSE and the KGE indexes reached values ranging between 0.76 and 0.86 in the calibration, and between 0.68 and 0.80 in the validation period (Table 3). Also, the correlation values confirm the results of the other indexes, reaching 0.88 for the calibration and 0.86 for the validation period. The performance of the presented tool is evident also when analyzing FDCs (Fig. 9a–g). The adoption of the river routing scheme allows a more realistic representations of the high discharge values (flow equaled or exceeded 0 to 20–30 % of the time), and low discharge values (flow equaled or exceeded 60 to 100 % of the time) in all the sections of the basin (Fig. 9). The model's performance in simulating river flows is generally

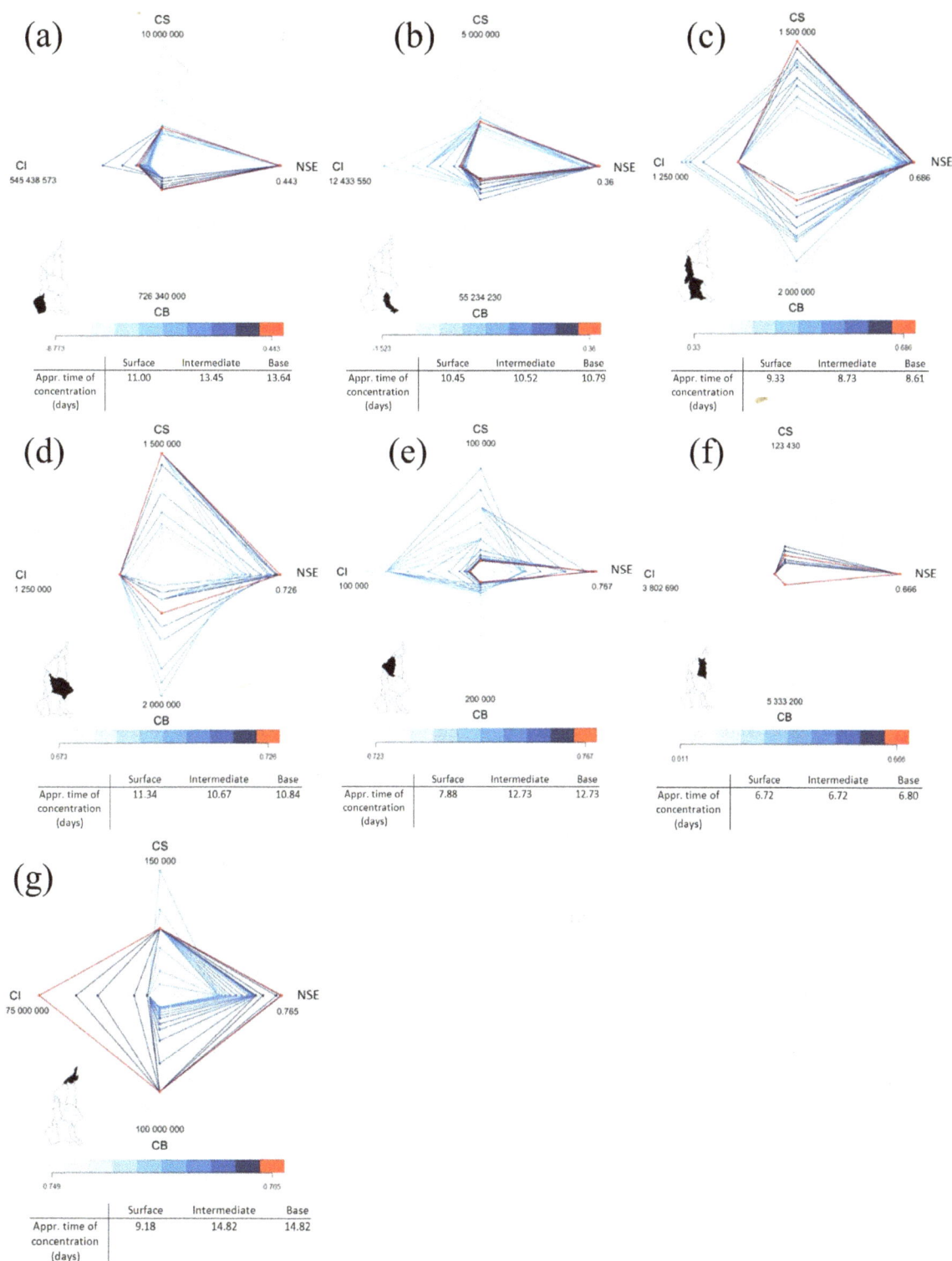

Figure 7. Calibration of the residence times (τ) of the flow within the ED2+R reservoirs of different grid cells in the domain through the adjustment of the non-dimensional C parameters. Overland, intermediate, and subsurface water flows are calibrated, respectively, through the adjustment parameters CS, CI, and CB (Fig. 2). The color bars refer to the model performance (NSE) of the specific parameter combination in the specific sub-basin. In red are the chosen combinations. At the bottom of each graph, there is a table providing the corresponding approximate average time of concentration (in days) for the cells in the sub-basin. **(a)** Upper Juruena (UJ); **(b)** Upper Teles Pires (UTP); **(c)** Lower Juruena (LJ); **(d)** Lower Teles Pires (LTP); **(e)** Upper Tapajós (UT); **(f)** Jamanxim (JA); and **(g)** Lower Tapajós (LT).

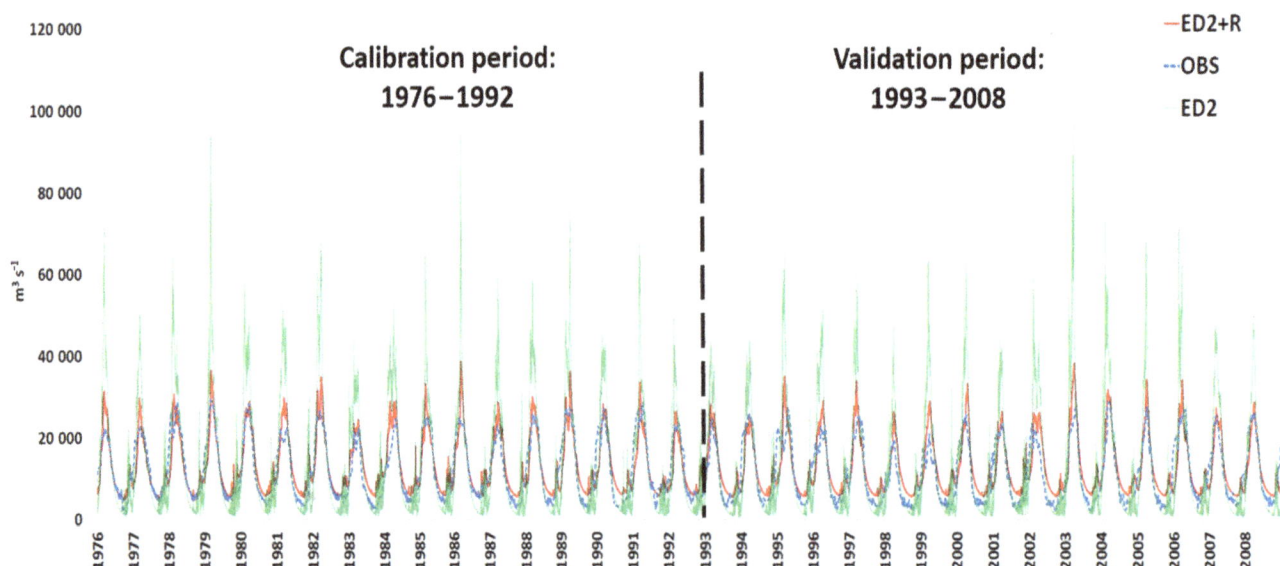

Figure 8. Calibration and validation of the river flow ($m^3\ s^{-1}$) at Itaituba (farthest downstream river gauge – Lower Tapajós sub-basin). ED2 output (green line), ED2+R (red line), and observations (blue dotted line). The dotted black line splits the calibration and validation periods. Similar comparison for each of the seven sub-basins is available in Appendix A.

more robust in the downstream sub-basins (NSE: 0.68–0.77 and KGE: 0.76–0.84 in the Upper and Lower Tapajós) and poorer in the headwaters (NSE: 0.28–0.45 and KGE: 0.38–0.61 in the Upper Juruena and Upper Teles Pires). In the Upper Teles Pires and Upper Juruena, the model achieved the lowest NSE (0.28 and 0.29, respectively, in the calibration, and 0.37 and 0.45 in the validation period), and KGE values (0.61 and 0.50 during calibration, and 0.63 and 0.38 during validation). Although water volumes are correctly reproduced in both the sub-basins (VR between 1.01 and 0.98 in the calibration, and 1.03 and 1.01 in the validation period), the seasonal variability is less accurate (correlation: 0.64–0.68 and 0.63–0.54). The KGE, NSE, and correlation indices are closer to the optimal value in the central and lower parts of the basin, particularly in the Lower Juruena (calibration – NSE: 0.65, KGE: 0.64, and correlation: 0.82; validation – NSE: 0.63, KGE: 0.67, and correlation: 0.81), Lower Teles Pires (calibration – NSE: 0.71, KGE: 0.67, and correlation: 0.85; validation – NSE: 0.67, KGE: 0.60, and correlation: 0.85), Upper Tapajós (calibration – NSE: 0.77, KGE: 0.82, and correlation: 0.88; validation – NSE: 0.75, KGE: 0.81, and correlation: 0.88), and Lower Tapajós (calibration – NSE: 0.76, KGE: 0.83, and correlation: 0.88; validation – NSE: 0.68, KGE: 0.76, and correlation: 0.82) (Table 3).

FDCs, representing the probability of the flow values to exceed a specific discharge, highlight the positive effect of the application of the routing scheme in ED2+R across the entire range of flow variability (Fig. 9). The simulated FDCs follow the same shape of the observed ones in the furthest upstream sub-basins, especially in the cases of the Upper Juruena and Upper Teles Pires, implying that the routing scheme

is effective in maintaining the simulated discharge range (Upper Juruena: 1200–2480 $m^3\ s^{-1}$, Upper Teles Pires: 393–4130 $m^3\ s^{-1}$) in line with the observations (1030–2400 and 302–2767 $m^3\ s^{-1}$, respectively). This is especially true for the lowest flows, where the error between simulated and observed curves is lower than 15 % (Figs. 9a, b, A1). Regarding the intermediate sub-basins, Lower Juruena and Lower Teles Pires, flood duration curves show that the model overestimates the lowest values of the distribution by approximately 30 % of the observed values (flow equaled or exceeded 60 to 100 % of the time in Fig. 9c, d). Similar overestimation of the model could be noticed in the furthest downstream sub-basins, Upper and Lower Tapajós (Fig. 9e–g). The overestimation of the lower discharge values highlighted in Fig. 9g is also evident in the multiyear hydrograph (Fig. 8), which shows that the ED2+R simulation results overestimate (by about 40 % on average in the discharge values included in the range of 60 to 100 % in Fig. 9g) the observations during the dry seasons of the period under consideration.

6 Discussion

As the results in Table 3 and Figs. 8–9 show, the one-way integration of ED2 with a routing scheme improves the performance of simulated daily discharges. Although this could appear obvious from a hydrological modeling perspective, the significance of this study lies in the fact that terrestrial biosphere models, which are widely applied to examine the impacts of climate and land use on the hydrology of the land surface, are typically "no-river representation" models. The incorporation of ecosystem responses to climate, car-

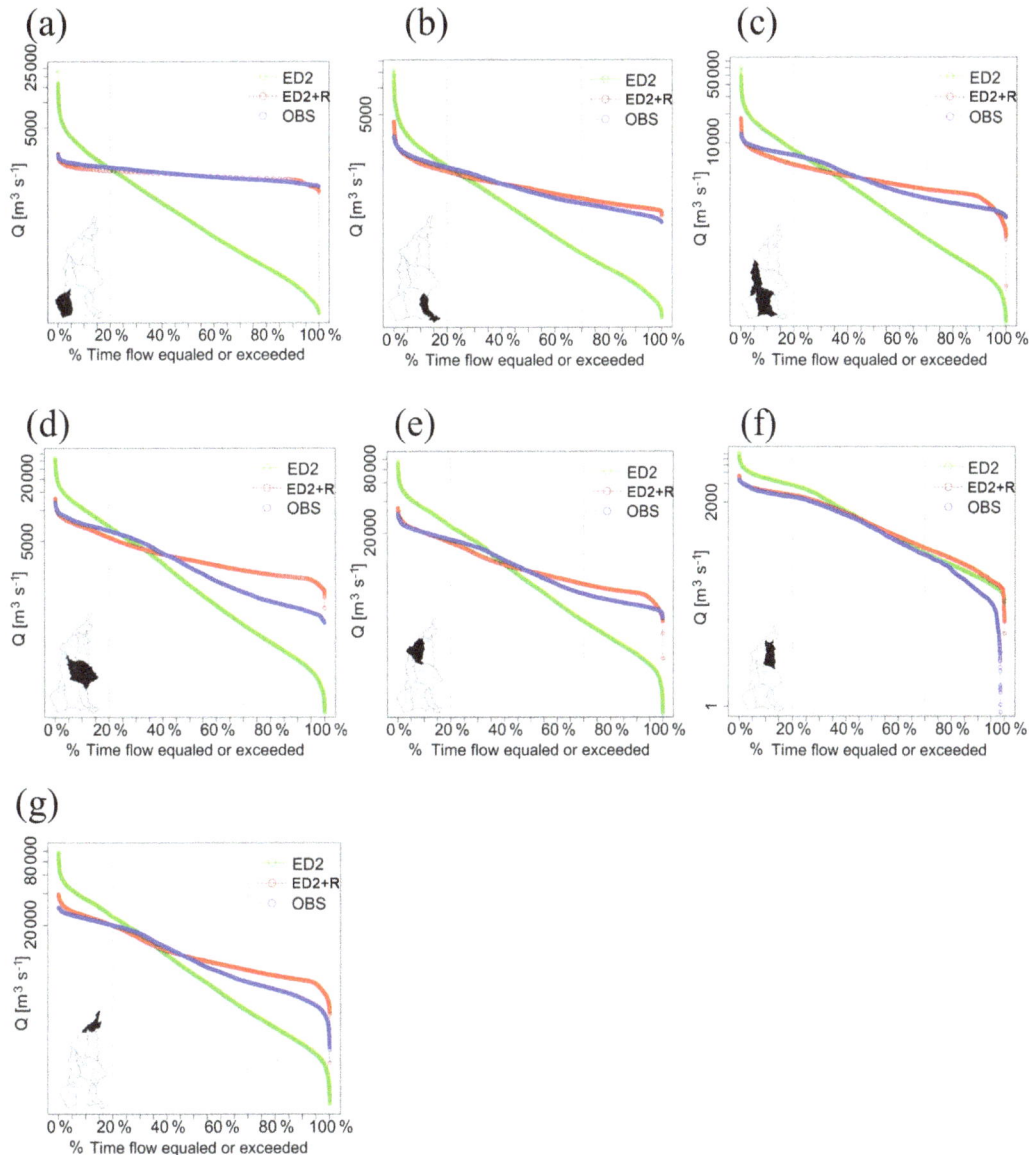

Figure 9. Flow duration curves (percentage of time that flow – m^3 s^{-1} – is likely to equal or exceed determined thresholds) of observed values (blue), ED2 outputs (green), and ED2+R (red) at the outlet of the seven sub-basins. **(a)** Upper Juruena (UJ); **(b)** Upper Teles Pires (UTP); **(c)** Lower Juruena (LJ); **(d)** Lower Teles Pires (LTP); **(e)** Upper Tapajós (UT); **(f)** Jamanxim (JA); and **(g)** Lower Tapajós (LT).

bon dioxide, and land-use changes simulated by terrestrial biosphere models with hydrological modeling improves the representation of the hydrological characteristics of basins characterized by large forest cover and/or large deforestation rates. In applications in the tropics, the one-way integration of the terrestrial biosphere model and the routing scheme (i.e., the two tools are not fully coupled) could lead to a partially inaccurate representation of the seasonally flooded ecosystems, a relevant aspect as documented in the literature (e.g., Cole et al., 2007).

As seen in Fig. 9, the performance of the model in simulating river flows in the basin is generally higher in the downstream sub-basins and poorer in the headwaters. Several factors are likely to cause this issue, both from the simulation of the hydrological dynamics in ED2, the flow partitioning (α and β parameters), and the basin hydraulic characteristics in ED2+R. The accurate calibration of the biosphere model with flux tower observations (Zhang et al., 2015; Longo, 2014) and the optimization of the flow partitioning make us believe that this variation in performance is due to the relatively coarse spatial resolution of the model in combination with the limitations typical of most land surface models in capturing the interactions with deep groundwater (Lobligeois et al., 2014; Zulkafli et al., 2013; Smith et al., 2004). We believe that the error is arising from the complexities associated with deep soils present in the headwaters of the Tapajós Basin. In particular, in the model application developed, soil layers are represented to a depth of 6 m (Table 1), which might be too shallow to realistically represent the conditions in the headwaters of the basin. The importance of groundwater is also evident from the calibration of the residence time parameter of the subsurface water flow: as shown in Fig. 7, in fact, especially in the headwaters, even small variations in the CB parameter greatly affect the model performance (specifically quantified with NSE in Fig. 7). The combined effect of groundwater interactions and spatial resolution is more evident in the upstream sub-basins because of the greater marginal contribution of baseflow in these areas. Surface flow accumulation, in fact, is lower in the headwaters. Therefore, in relative terms, the role of baseflow is more relevant in this portion of any basin. Further downstream, the effect of groundwater interactions and spatial resolution is, at least in part, masked by the larger rainfall–runoff contribution and the overall flow accumulation from the upstream sub-basins. Other recent hydrological simulations of the Tapajós have obtained higher accuracy (e.g., Mohor et al., 2015; Collischonn et al., 2008; Coe et al., 2008); however, these simulations were set up discretizing the basin into a finer spatial resolution grid (9 to 20 km vs. ∼ 55 km grid cells) and using hydrological tools able to reproduce highly detailed hydrodynamic characteristics of complex river systems (i.e., floodplain, lakes, wetlands, backwater effects) that are out of the scope of the tool presented in this study. The advantage of the ED2+R model is the ability to study the sensitivity of the river flows to global and regional changes as computed by traditional terrestrial biosphere models, but adding a more detailed hydrological feature with respect to a very simplistic- or no-river representation. The coarse spatial resolution of the global datasets used as input for ED2+R is, however, a limiting factor. Higher-resolution climatological data, vegetation, and land-use datasets, which would allow a finer resolution of the hydrological grid, are expected to improve the performance of the model by providing more detailed hydrological processes. On the other hand, a finer spatial resolution of the hydrological grid would also require a more detailed representation of the subsurface water in the model. In general, the tool can be used to study how different hydrological systems are being affected by changes in climate forcing and changes in ecosystem composition and structure arising from the combination of changing climate, rising atmospheric carbon dioxide, and land-use transformation. Additionally, ED2+R could potentially bridge one of the missing gaps for diagnosing and assessing feedback between atmosphere and biosphere with inland surface waters being represented as a dynamic system.

7 Conclusion

In this technical note, we present the integration of the terrestrial biosphere model Ecosystem Demography 2 (ED2) with the Muskingum–Cunge routing scheme. We tested the integrated model (ED2+R) in the Tapajós River basin, a large tributary of the Amazon in Brazil, for the period 1970–2008. The results showed that the integration of a biosphere model with a routing scheme improves the ability of the land surface simulation to reproduce the hydrological and river flow dynamics at the basin scale. The main limitations highlighted in this case study were linked to the relatively coarse spatial resolution of the model and the rough representation of subsurface water flow typical of these kinds of models. Moreover, the terrestrial biosphere model ED2 and the routing scheme are presented here in a one-way integration. The full coupling of the routing scheme and ED2 could further improve the tool's ability to reproduce the water balance considering flooded ecosystems, a relevant feature in the simulation of environments like the tropical forest, where local evapotranspiration plays a primary role in the specific ecosystem's dynamics. In this first integration, our goal was to give the terrestrial biosphere model the ability to reproduce river flows through a routing scheme. With a fully coupled (i.e., two-way) integration, the model would be able to determine the grid cells that are likely to be saturated and use this information for the modeling of the ecosystem's dynamics. For instance, this could determine the increase of the mortality rate of plants that are sensitive to inundation. An additional limitation of the model could be identified in its inability to reproduce highly detailed hydrological dynamics of complex river systems (as, for instance, floodplain hydraulic features or backwater effects). However, such a detailed hydrological complexity was out of the scope of this study. Future efforts will address the highlighted limitations, while upcoming studies will use ED2+R to understand historical changes and future projections of the impacts of climate change and deforestation on the Amazon's water resources.

Data availability. Meteorological forcing data are derived from Sheffield et al. (2006) – https://doi.org/10.1175/JCLI3790.1; land use data are derived from Hurtt et al. (2006) – https://doi.org/10.1111/j.1365-2486.2006.01150.x; topographic data are derived from the Shuttle Radar Topography Mission (SRTM) 90 m resolution (USGS, 2016); soil map is derived from Quesada et al. (2010) – https://doi.org/10.5194/bg-7-1515-2010 – and IGBP-DIS global soil data (Global Soil Data Task, 2014) – https://doi.org/10.3334/ORNLDAAC/565; geomorphological relations are obtained from Coe et al. (2008) – https://doi.org/10.1002/hyp.6850; streamflow observations are obtained from Observation Service SO HYBAM (2016) and ANA (2016).

Appendix A

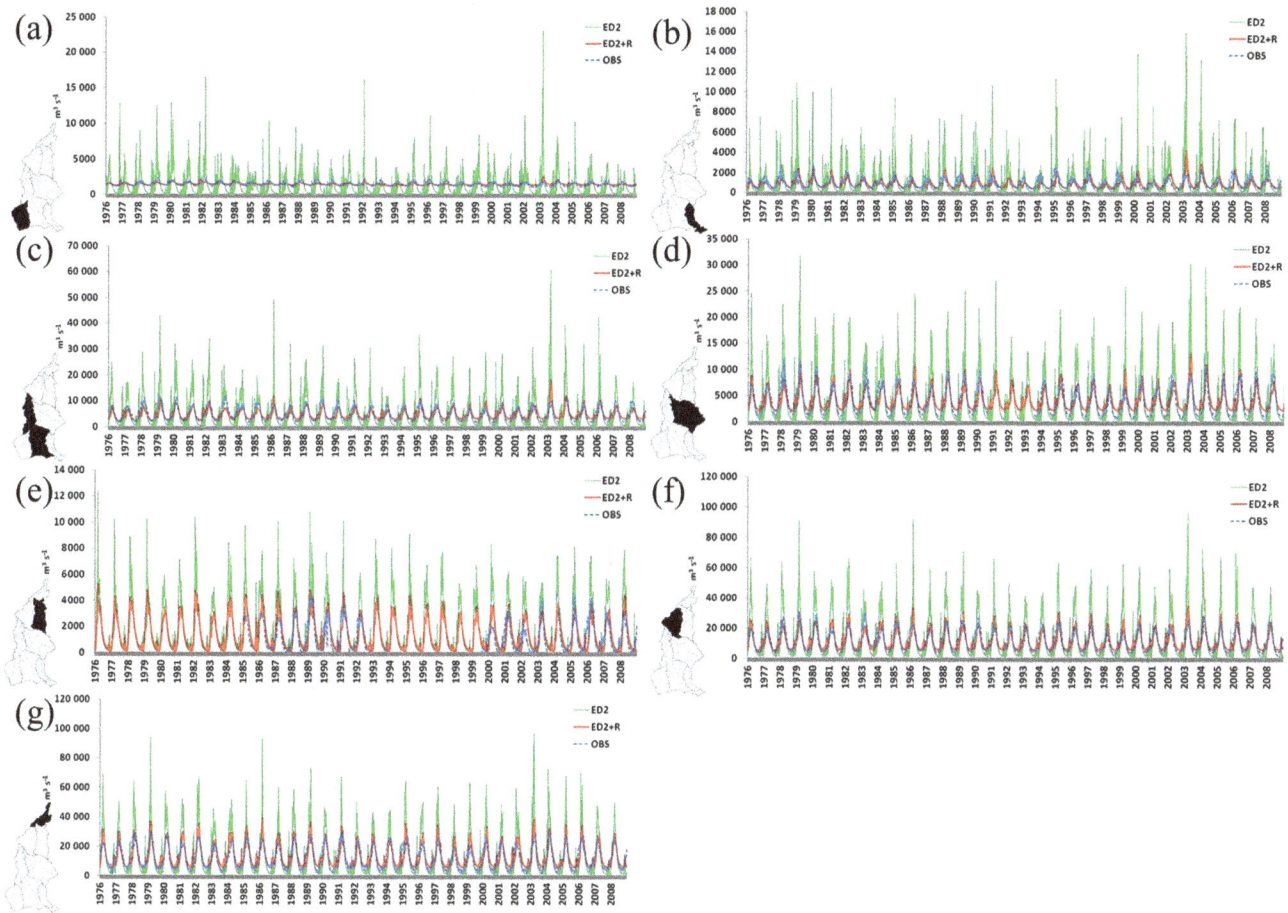

Figure A1. Time series of river flow ($m^3\ s^{-1}$) at the outlet of each sub-basins. ED2 output (green line), ED2+R (red line), and observations (blue dotted line). **(a)** Upper Juruena (UJ); **(b)** Upper Teles Pires (UTP); **(c)** Lower Juruena (LJ); **(d)** Lower Teles Pires (LTP); **(e)** Jamanxim (JA); **(f)** Upper Tapajós (UT); and **(g)** Lower Tapajós (LT).

Author contributions. FP, PM, and JB designed the study; FP developed the ED2+R model code; FF, MA, and EL calibrated the model and carried out the analysis; FF, MA, and PM wrote the paper.

Competing interests. The authors declare that they have no conflict of interest.

Acknowledgements. This work was conducted while Fabio F. Pereira, Fabio Farinosi, Eunjee Lee, and Mauricio E. Arias were Giorgio Ruffolo Fellows in the Sustainability Science Program at Harvard University. Fabio Farinosi was also funded through a doctoral scholarship by Ca' Foscari University of Venice. Support from Italy's Ministry for Environment, Land and Sea is gratefully acknowledged. We would like to thank Marcos Longo for letting us use one of his figures, and Angela Livino for the useful comments. The authors would like to dedicate this study to the late Professor John Briscoe (1948–2014), who envisioned and co-led the Amazon Initiative of Harvard's Sustainability Science Program. We are grateful to the editor, Graham Jewitt, and to Hartley Bulcock and the three other anonymous referees for the valuable comments received during the review process. Finally, the authors would like to thank Erin Ciccone and Madeleine Marino for proofreading the manuscript.

Edited by: Graham Jewitt

References

Albani, M., Medvigy, D., Hurtt, G. C., and Moorcroft, P. R.: The contributions of land-use change, CO_2 fertilization, and climate variability to the Eastern US carbon sink, Glob. Change Biol., 12, 2370–2390, https://doi.org/10.1111/j.1365-2486.2006.01254.x, 2006.

Alsdorf, D. E., Rodríguez, E., and Lettenmaier, D. P.: Measuring surface water from space, Rev. Geophys., 45, RG2002, https://doi.org/10.1029/2006RG000197, 2007.

ANA: Plano Estratégico de Recursos Hídricos da Bacia Amazônica – Afluentes da Margem Direita, Brasilia, Brazil, Brazil, available at: http://margemdireita.ana.gov.br/ (last access: 7 September 2017), 2011 (in Portuguese).

ANA: Hidroweb –Sistema de informações hidrologicas, available from: http://www.snirh.gov.br/hidroweb/, (last access: 7 September 2017), 2016.

Anderson, E. A.: Calibration of Conceptual Models for Use in River Forecasting, available at: http://www.nws.noaa.gov/oh/hrl/calb/calibration1102/main.htm (last access: 7 September 2017), 2002.

Andersson, J. C. M., Pechlivanidis, I. G., Gustafsson, D., Donnelly, C., and Arheimer, B.: Key factors for improving large-scale hydrological model performance, Eur. Water, 49, 77–88, 2015.

Andréassian, V.: Waters and forests: from historical controversy to scientific debate, J. Hydrol., 291, 1–27, https://doi.org/10.1016/j.jhydrol.2003.12.015, 2004.

Arias, M. E., Lee, E., Farinosi, F., Pereira, F. F., Moorcroft, P. R., and Briscoe, J.: Decoupling the effects of deforestation and climate variability in large tropical river basins, J. Hydrol., in review, 2017.

Arora, V. K., Chiew, F. H. S., and Grayson, R. B.: A river flow routing scheme for general circulation models, J. Geophys. Res., 104, 14347, https://doi.org/10.1029/1999JD900200, 1999.

Bahn, M., Reichstein, M., Dukes, J. S., Smith, M. D., and McDowell, N. G.: Climate-biosphere interactions in a more extreme world, New Phytol., 202, 356–359, https://doi.org/10.1111/nph.12662, 2014.

Baker, T. R., Phillips, O. L., Malhi, Y., Almeida, S., Arroyo, L., Di Fiore, A., Erwin, T., Killeen, T. J., Laurance, S. G., Laurance, W. F., Lewis, S. L., Lloyd, J., Monteagudo, A., Neill, D. A., Patino, S., Pitman, N. C. A., Silva, J. N. M., and Vasquez Martinez, R.: Variation in wood density determines spatial patterns in Amazonian forest biomass, Glob. Change Biol., 10, 545–562, https://doi.org/10.1111/j.1365-2486.2004.00751.x, 2004.

Best, M. J., Pryor, M., Clark, D. B., Rooney, G. G., Essery, R. L. H., Ménard, C. B., Edwards, J. M., Hendry, M. A., Porson, A., Gedney, N., Mercado, L. M., Sitch, S., Blyth, E., Boucher, O., Cox, P. M., Grimmond, C. S. B., and Harding, R. J.: The Joint UK Land Environment Simulator (JULES), model description – Part 1: Energy and water fluxes, Geosci. Model Dev., 4, 677–699, https://doi.org/10.5194/gmd-4-677-2011, 2011.

Brown, A. E., Zhang, L., McMahon, T. A., Western, A. W., and Vertessy, R. A.: A review of paired catchment studies for determining changes in water yield resulting from alterations in vegetation, J. Hydrol., 310, 28–61, https://doi.org/10.1016/j.jhydrol.2004.12.010, 2005.

Calvo-Alvarado, J., McDowell, N., and Waring, R.: Allometric relationships predicting foliar biomass and leaf area: sapwood area ratio from tree height in five Costa Rican rain forest species, Tree Physiol., 11, 1601–1608, 2008.

Carson, D.: Current parametrisations of land-surface processes in atmospheric general circulation models, in: Land surface processes in atmospheric general circulation models, edited by: Eagleson, P., Cambridge University Press, Cambridge, UK, 1982.

Clark, D. B., Mercado, L. M., Sitch, S., Jones, C. D., Gedney, N., Best, M. J., Pryor, M., Rooney, G. G., Essery, R. L. H., Blyth, E., Boucher, O., Harding, R. J., Huntingford, C., and Cox, P. M.: The Joint UK Land Environment Simulator (JULES), model description – Part 2: Carbon fluxes and vegetation dynamics, Geosci. Model Dev., 4, 701–722, https://doi.org/10.5194/gmd-4-701-2011, 2011.

Clark, M. P., Fan, Y., Lawrence, D. M., Adam, J. C., Bolster, D., Gochis, D. J., Hooper, R. P., Kumar, M., Leung, L. R., Mackay, D. S., Maxwell, R. M., Shen, C., Swenson, S. C., and Zeng, X.: Improving the representation of hydrologic processes in Earth System Models, Water Resour. Res., 51, 5929–5956, https://doi.org/10.1002/2015WR017096, 2015.

Coe, M. T., Costa, M. H., and Howard, E. A.: Simulating the surface waters of the Amazon River basin: impacts of new river geomorphic and flow parameterizations, Hydrol. Process., 22, 2542–2553, https://doi.org/10.1002/hyp.6850, 2008.

Cole, J. J., Prairie, Y. T., Caraco, N. F., McDowell, W. H., Tranvik, L. J., Striegl, R. G., Duarte, C. M., Kortelainen, P., Downing, J. A., Middelburg, J. J., and Melack, J.: Plumbing the Global Carbon Cycle: Integrating Inland Waters into the Terrestrial Carbon Budget, Ecosystems, 10, 172–185, https://doi.org/10.1007/s10021-006-9013-8, 2007.

Cole, T. G. and Ewel, J. J.: Allometric equations for four valuable tropical tree species, For. Ecol. Manage., 229, 351–360, https://doi.org/10.1016/j.foreco.2006.04.017, 2006.

Collischonn, B., Collischonn, W., and Tucci, C. E. M.: Daily hydrological modeling in the Amazon basin using TRMM rainfall estimates, J. Hydrol., 360, 207–216, https://doi.org/10.1016/j.jhydrol.2008.07.032, 2008.

Collischonn, W., Allasia, D., Da Silva, B. C., and Tucci, C. E. M.: The MGB-IPH model for large-scale rainfall–runoff modelling, Hydrol. Sci. J., 52, 878–895, https://doi.org/10.1623/hysj.52.5.878, 2007.

Cox, P. M., Betts, R. A., Bunton, C. B., Essery, R. L. H., Rowntree, P. R., and Smith, J.: The impact of new land surface physics on the GCM simulation of climate and climate sensitivity, Clim. Dynam., 15, 183–203, https://doi.org/10.1007/s003820050276, 1999.

Cunge, J. A.: On The Subject Of A Flood Propagation Computation Method (Musklngum Method), J. Hydraul. Res., 7, 205–230, https://doi.org/10.1080/00221686909500264, 1969.

Farinosi, F., Arias, M. E., Lee, E., Longo, M., Pereira, F. F., Livino, A., Moorcroft, P. R., and Briscoe, J.: Future climate and land use change impacts on river flows in the Tapajós Basin in the Brazilian Amazon, Earth's Future, in review, 2017.

Gerten, D., Schaphoff, S., Haberlandt, U., Lucht, W., and Sitch, S.: Terrestrial vegetation and water balance – hydrological evaluation of a dynamic global vegetation model, J. Hydrol., 286, 249–270, https://doi.org/10.1016/j.jhydrol.2003.09.029, 2004.

Global Soil Data Task: Global Soil Data Products CD-ROM Contents (IGBP-DIS), Data Set, Oak Ridge National Laboratory Distributed Active Archive Center, Oak Ridge, Tennessee, USA, https://doi.org/10.3334/ORNLDAAC/565, 2014.

Goldewijk, K. K.: Estimating global land use change over the past 300 years: The HYDE Database, Global Biogeochem. Cy., 15, 417–433, https://doi.org/10.1029/1999GB001232, 2001.

Gupta, H. V., Kling, H., Yilmaz, K. K., and Martinez, G. F.: Decomposition of the mean squared error and NSE performance criteria: Implications for improving hydrological modelling, J. Hydrol., 377, 80–91, https://doi.org/10.1016/j.jhydrol.2009.08.003, 2009.

Hagemann, S. and Dumenil, L.: A parametrization of the lateral waterflow for the global scale, Clim. Dynam., 14, 17–31, https://doi.org/10.1007/s003820050205, 1997.

Hagemann, S. and Gates, L. D.: Validation of the hydrological cycle of ECMWF and NCEP reanalyses using the MPI hydrological discharge model, J. Geophys. Res., 106, 1503, https://doi.org/10.1029/2000JD900568, 2001.

Hurtt, G. C., Pacala, S. W., Moorcroft, P. R., Caspersen, J., Shevliakova, E., Houghton, R. A., and Moore, B.: Projecting the future of the U.S. carbon sink, P. Natl. Acad. Sci. USA, 99, 1389–1394, https://doi.org/10.1073/pnas.012249999, 2002.

Hurtt, G. C., Frolking, S., Fearon, M. G., Moore, B., Shevliakova, E., Malyshev, S., Pacala, S. W., and Houghton, R. A.: The underpinnings of land-use history: three centuries of global gridded land-use transitions, wood-harvest activity, and resulting secondary lands, Glob. Chang. Biol., 12, 1208–1229, https://doi.org/10.1111/j.1365-2486.2006.01150.x, 2006.

Hurtt, G. C., Moorcroft, P. R., and Pacala, S. W.: Ecosystem Demography Model: Scaling Vegetation Dynamics Across South America, Ecosyst. Demogr. Model Scaling Veg. Dyn. Across South Am. Model Prod., available at: http://daac.ornl. gov/MODELS/guides/EDM_SA_Vegetation.html (last access: 7 September 2017), 2013.

Jiménez-Cisneros, B. E., Oki, T., Arnell, N. W., Benito, G., Cogley, J. G., Döll, P., Jiang, T., and Mwakalila, S. S.: Freshwater resources, in: Climate Change 2014: Impacts, Adaptation, and Vulnerability. Part A: Global and Sectoral Aspects. Contribution of Working Group II to the Fifth Assessment Report of the Intergovernmental Panel on Climate Change, edited by: Field, C. B., Barros, V. R., Dokken, D. J., Mach, K. J., Mastrandrea, M. D., Bilir, T. E., Chatterjee, M., Ebi, K. L., Estrada, Y. O., Genova, R. C., Girma, B., Kissel, E. S., Levy, A. N., MacCracken, S., Mastrandrea, P. R., and White, L. L., 229–269, Cambridge University Press, Cambridge, United Kingdom and New York, NY, USA, available at: https://www.ipcc.ch/pdf/assessment-report/ar5/wg2/WGIIAR5-Chap3_FINAL.pdf (last access: 7 September 2017), 2014.

Kauffeldt, A., Wetterhall, F., Pappenberger, F., Salamon, P., and Thielen, J.: Technical review of large-scale hydrological models for implementation in operational flood forecasting schemes on continental level, Environ. Model. Softw., 75, 68–76, https://doi.org/10.1016/j.envsoft.2015.09.009, 2016.

Kim, Y., Knox, R. G., Longo, M., Medvigy, D., Hutyra, L. R., Pyle, E. H., Wofsy, S. C., Bras, R. L., and Moorcroft, P. R.: Seasonal carbon dynamics and water fluxes in an Amazon rainforest, Glob. Change Biol., 18, 1322–1334, https://doi.org/10.1111/j.1365-2486.2011.02629.x, 2012.

Kling, H., Fuchs, M., and Paulin, M.: Runoff conditions in the upper Danube basin under an ensemble of climate change scenarios, J. Hydrol., 424–425, 264–277, https://doi.org/10.1016/j.jhydrol.2012.01.011, 2012.

Knox, R. G.: Land Conversion in Amazonia and Northern South America: Influences on Regional Hydrology and Ecosystem Response, PhD Thesis, Massachusetts Institute of Technology, available at: https://dspace.mit.edu/handle/1721.1/79489 (last access: 7 September 2017), 2012.

Knox, R. G., Longo, M., Swann, A. L. S., Zhang, K., Levine, N. M., Moorcroft, P. R., and Bras, R. L.: Hydrometeorological effects of historical land-conversion in an ecosystem-atmosphere model of Northern South America, Hydrol. Earth Syst. Sci., 19, 241–273, https://doi.org/10.5194/hess-19-241-2015, 2015.

Kucharik, C. J., Foley, J. A., Delire, C., Fisher, V. A., Coe, M. T., Lenters, J. D., Young-Molling, C., Ramankutty, N., Norman, J. M., and Gower, S. T.: Testing the performance of a dynamic global ecosystem model: Water balance, carbon balance, and vegetation structure, Global Biogeochem. Cy., 14, 795–825, https://doi.org/10.1029/1999GB001138, 2000.

Lawrence, D. M., Oleson, K. W., Flanner, M. G., Thornton, P. E., Swenson, S. C., Lawrence, P. J., Zeng, X., Yang, Z.-L., Levis, S., Sakaguchi, K., Bonan, G. B., and Slater, A. G.: Parameterization improvements and functional and structural advances in Version 4 of the Community Land Model, J. Adv. Model. Earth Syst., 3, M03001, https://doi.org/10.1029/2011MS000045, 2011.

Lejeune, Q., Davin, E. L., Guillod, B. P., and Seneviratne, S. I.: Influence of Amazonian deforestation on the future evolution of regional surface fluxes, circulation, surface temperature and precipitation, Clim. Dynam., 44, 2769–2786, https://doi.org/10.1007/s00382-014-2203-8, 2015.

Li, R., Chen, Q., and Ye, F.: Modelling the impacts of reservoir operations on the downstream riparian vegetation and fish habitats in the Lijiang River, J. Hydroinformatics, 13, 229, https://doi.org/10.2166/hydro.2010.008, 2011.

Liang, X., Lettenmaier, D. P., Wood, E. F., and Burges, S. J.: A simple hydrologically based model of land surface water and energy fluxes for general circulation model, J. Geophys. Res., 99, 14415–14428, 1994.

Lobligeois, F., Andréassian, V., Perrin, C., Tabary, P., and Loumagne, C.: When does higher spatial resolution rainfall information improve streamflow simulation? An evaluation using 3620 flood events, Hydrol. Earth Syst. Sci., 18, 575–594, https://doi.org/10.5194/hess-18-575-2014, 2014.

Longo, M.: Amazon Forest Response to Changes in Rainfall Regime: Results from an Individual-Based Dynamic Vegetation Model, Harvard University, available at: http://dash.harvard.edu/handle/1/11744438 (last access: 7 September 2017), 2014.

Medvigy, D., Wofsy, S. C., Munger, J. W., Hollinger, D. Y., and Moorcroft, P. R.: Mechanistic scaling of ecosystem function and dynamics in space and time: Ecosystem Demography model version 2, J. Geophys. Res.-Biogeo., 114, G01002, https://doi.org/10.1029/2008JG000812, 2009.

Medvigy, D., Walko, R. L., and Avissar, R.: Effects of Deforestation on Spatiotemporal Distributions of Precipitation in South America, J. Climate, 24, 2147–2163, https://doi.org/10.1175/2010JCLI3882.1, 2011.

Miller, W. A. and Cunge, J. A.: Simplified equations of unsteady flow, in: Unsteady Flow in Open Channels, edited by: Mahmood, K. and Yevjevich, V., Colorado State University, Water Resources Publication, Fort Collins, CO, USA, 1975.

Mohor, G. S., Rodriguez, D. A., Tomasella, J., and Siqueira Júnior, J. L.: Exploratory analyses for the assessment of climate change impacts on the energy production in an Amazon run-of-river hydropower plant, J. Hydrol. Reg. Stud., 4, 41–59, https://doi.org/10.1016/j.ejrh.2015.04.003, 2015.

Moorcroft, P. R., Hurtt, G. C., and Pacala, S. W.: A method for scaling vegetation dynamics: The ecosystem demography model (ED), Ecol. Monogr., 71, 557–586, https://doi.org/10.1890/0012-9615(2001)071[0557:AMFSVD]2.0.CO;2, 2001.

Nash, E. and Sutcliffe, V.: River flow forecasting Through conceptual models PART I- A Discussion of principles, J. Hydrol., 10, 282–290, 1970.

Observation Service SO HYBAM: SO HYBAM – Geodynamical, hydrological and biogeochemical control of erosion/alteration and material transport in the Amazon, Orinoco and Congo basins, available from: http://www.ore-hybam.org/index.php/eng, (last access: 7 September 2017), 2016.

Oki, T., Nishimura, T., and Dirmeyer, P.: Assessment of Annual Runoff from Land Surface Models Using Total Runoff Integrating Pathways (TRIP), J. Meteorol. Soc. Japan, 77, 235–255, 1999.

Oki, T., Agata, Y., Kanae, S., Saruhashi, T., Yang, D., and Musiake, K.: Global assessment of current water resources using total runoff integrating pathways, Hydrol. Sci. J., 46, 983–995, https://doi.org/10.1080/02626660109492890, 2001.

Oleson, K. W., Lawrence, D. M., Bonan, G. B., Flanner, M. G., Kluzek, E., Lawrence, P. J., Levis, S., Swenson, S. C., and Thornton, P. E.: Technical Description of version 4.0 of the Community Land Model (CLM), Boulder, CO, USA, available at: http://www.cesm.ucar.edu/models/cesm1.0/clm/CLM4_Tech_Note.pdf (last access: 7 September 2017), 2010.

Ostberg, S., Schaphoff, S., Lucht, W., and Gerten, D.: Three centuries of dual pressure from land use and climate change on the biosphere, Environ. Res. Lett., 10, 044011, https://doi.org/10.1088/1748-9326/10/4/044011, 2015.

Paiva, R. C. D., Collischonn, W., and Tucci, C. E. M.: Large scale hydrologic and hydrodynamic modeling using limited data and a GIS based approach, J. Hydrol., 406, 170–181, https://doi.org/10.1016/j.jhydrol.2011.06.007, 2011.

Paiva, R. C. D., Buarque, D. C., Collischonn, W., Bonnet, M. P., Frappart, F., Calmant, S., and Bulhões Mendes, C. A.: Large-scale hydrologic and hydrodynamic modeling of the Amazon River basin, Water Resour. Res., 49, 1226–1243, https://doi.org/10.1002/wrcr.20067, 2013a.

Paiva, R. C. D., Collischonn, W., and Buarque, D. C.: Validation of a full hydrodynamic model for large-scale hydrologic modelling in the Amazon, Hydrol. Process., 27, 333–346, https://doi.org/10.1002/hyp.8425, 2013b.

Paz, A. R., Collischonn, W., and Lopes da Silveira, A. L.: Improvements in large-scale drainage networks derived from digital elevation models, Water Resour. Res., 42, W08502, https://doi.org/10.1029/2005WR004544, 2006.

Pearson, K.: Note on regression and inheritance in the case of two parents, Proc. R. Soc. London, 58, 240–242, https://doi.org/10.1098/rspl.1895.0041, 1895.

Pearson, R. G., Phillips, S. J., Loranty, M. M., Beck, P. S. A., Damoulas, T., Knight, S. J., and Goetz, S. J.: Shifts in Arctic vegetation and associated feedbacks under climate change, Nat. Clim. Chang., 3, 673–677, https://doi.org/10.1038/nclimate1858, 2013.

Pontes, P. R. M., Collischonn, W., Fan, F. M., Paiva, R. C. D., and Buarque, D. C.: Modelagem hidrológica e hidráulica de grande escala com propagação inercial de vazões, Rev. Bras. Recur. Hídricos, 20, 888–904, 2015.

Poorter, L., Bongers, L., and Bongers, F.: Architecture of 54 moist-forest tree species: traits, trade-offs, and functional groups, Ecology, 87, 1289–1301, https://doi.org/10.1890/0012-9658(2006)87[1289:AOMTST]2.0.CO;2, 2006.

Prigent, C., Papa, F., Aires, F., Rossow, W. B., and Matthews, E.: Global inundation dynamics inferred from multiple satellite observations, 1993–2000, J. Geophys. Res., 112, D12107, https://doi.org/10.1029/2006JD007847, 2007.

Quesada, C. A., Lloyd, J., Schwarz, M., Patiño, S., Baker, T. R., Czimczik, C., Fyllas, N. M., Martinelli, L., Nardoto, G. B., Schmerler, J., Santos, A. J. B., Hodnett, M. G., Herrera, R., Luizão, F. J., Arneth, A., Lloyd, G., Dezzeo, N., Hilke, I., Kuhlmann, I., Raessler, M., Brand, W. A., Geilmann, H., Moraes Filho, J. O., Carvalho, F. P., Araujo Filho, R. N., Chaves, J. E., Cruz Junior, O. F., Pimentel, T. P., and Paiva, R.: Variations in chemical and physical properties of Amazon forest soils in relation to their genesis, Biogeosciences, 7, 1515–1541, https://doi.org/10.5194/bg-7-1515-2010, 2010.

Raddatz, T. J., Reick, C. H., Knorr, W., Kattge, J., Roeckner, E., Schnur, R., Schnitzler, K.-G., Wetzel, P., and Jungclaus, J.: Will the tropical land biosphere dominate the climate–carbon cycle feedback during the twenty-first century?, Clim. Dynam., 29, 565–574, https://doi.org/10.1007/s00382-007-0247-8, 2007.

Ramankutty, N. and Foley, J. A.: Estimating historical changes in global land cover: Croplands from

1700 to 1992, Global Biogeochem. Cy., 13, 997–1027, https://doi.org/10.1029/1999GB900046, 1999.

Reed, S. M.: Deriving flow directions for coarse-resolution (1–4 km) gridded hydrologic modeling, Water Resour. Res., 39, SWC 4, https://doi.org/10.1029/2003WR001989, 2003.

Rost, S., Gerten, D., Bondeau, A., Lucht, W., Rohwer, J., and Schaphoff, S.: Agricultural green and blue water consumption and its influence on the global water system, Water Resour. Res., 44, W09405, https://doi.org/10.1029/2007WR006331, 2008.

Sheffield, J., Goteti, G., and Wood, E. F.: Development of a 50-Year High-Resolution Global Dataset of Meteorological Forcings for Land Surface Modeling, J. Climate, 19, 3088–3111, https://doi.org/10.1175/JCLI3790.1, 2006.

Sitch, S., Smith, B., Prentice, I. C., Arneth, A., Bondeau, A., Cramer, W., Kaplan, J. O., Levis, S., Lucht, W., Sykes, M. T., Thonicke, K., and Venevsky, S.: Evaluation of ecosystem dynamics, plant geography and terrestrial carbon cycling in the LPJ dynamic global vegetation model, Glob. Change Biol., 9, 161–185, https://doi.org/10.1046/j.1365-2486.2003.00569.x, 2003.

Smith, M. B., Koren, V. I., Zhang, Z., Reed, S. M., Pan, J.-J., and Moreda, F.: Runoff response to spatial variability in precipitation: an analysis of observed data, J. Hydrol., 298, 267–286, https://doi.org/10.1016/j.jhydrol.2004.03.039, 2004.

Soares-Filho, B. S., Nepstad, D. C., Curran, L. M., Cerqueira, G. C., Garcia, R. A., Ramos, C. A., Voll, E., McDonald, A., Lefebvre, P., and Schlesinger, P.: Modelling conservation in the Amazon basin, Nature, 440, 520–523, https://doi.org/10.1038/nature04389, 2006.

Swann, A. L. S., Longo, M., Knox, R. G., Lee, E., and Moorcroft, P. R.: Future deforestation in the Amazon and consequences for South American climate, Agric. For. Meteorol., 214–215, 12–24, https://doi.org/10.1016/j.agrformet.2015.07.006, 2015.

USACE: A Muskingum-Cunge Channel Flow Routing Method for Drainage Networks, available at: http://www.hec.usace.army.mil/publications/TechnicalPapers/TP-135.pdf (last access: 7 September 2017), 1991.

USGS: Shuttle Radar Topography Mission (SRTM), available from: https://lta.cr.usgs.gov/SRTM1Arc, (last access: 7 September 2017), 2016.

Vamborg, F. S. E., Brovkin, V., and Claussen, M.: The effect of a dynamic background albedo scheme on Sahel/Sahara precipitation during the mid-Holocene, Clim. Past, 7, 117–131, https://doi.org/10.5194/cp-7-117-2011, 2011.

Walko, R. L., Band, L. E., Baron, J., Kittel, T. G. F., Lammers, R., Lee, T. J., Ojima, D., Pielke, R. A., Taylor, C., Tague, C., Tremback, C. J., and Vidale, P. L.: Coupled Atmosphere–Biophysics–Hydrology Models for Environmental Modeling, J. Appl. Meteorol., 39, 931–944, https://doi.org/10.1175/1520-0450(2000)039<0931:CABHMF>2.0.CO;2, 2000.

Wohl, E., Barros, A., Brunsell, N., Chappell, N. A., Coe, M., Giambelluca, T., Goldsmith, S., Harmon, R., Hendrickx, J. M. H., Juvik, J., McDonnell, J., and Ogden, F.: The hydrology of the humid tropics, Nat. Clim. Chang., 2, 655–662, https://doi.org/10.1038/nclimate1556, 2012.

Yamazaki, D., Kanae, S., Kim, H., and Oki, T.: A physically based description of floodplain inundation dynamics in a global river routing model, Water Resour. Res., 47, W04501, https://doi.org/10.1029/2010WR009726, 2011.

Zambrano-Bigiarini, M.: hydroGOF: Goodness-of-fit functions for comparison of simulated and observed hydrological time series. R package version 0.3–8, available at: http://cran.r-project.org/package=hydroGOF (last access: 7 September 2017), 2014.

Zhang, K., de Almeida Castanho, A. D., Galbraith, D. R., Moghim, S., Levine, N. M., Bras, R. L., Coe, M. T., Costa, M. H., Malhi, Y., Longo, M., Knox, R. G., McKnight, S., Wang, J., and Moorcroft, P. R.: The fate of Amazonian ecosystems over the coming century arising from changes in climate, atmospheric CO_2, and land use, Glob. Change Biol., 21, 2569–2587, https://doi.org/10.1111/gcb.12903, 2015.

Zulkafli, Z., Buytaert, W., Onof, C., Lavado, W., and Guyot, J. L.: A critical assessment of the JULES land surface model hydrology for humid tropical environments, Hydrol. Earth Syst. Sci., 17, 1113–1132, https://doi.org/10.5194/hess-17-1113-2013, 2013.

Rainfall–runoff modelling using Long Short-Term Memory (LSTM) networks

Frederik Kratzert[1,*]**, Daniel Klotz**[1]**, Claire Brenner**[1]**, Karsten Schulz**[1]**, and Mathew Herrnegger**[1]

[1]Institute of Water Management, Hydrology and Hydraulic Engineering, University of Natural Resources and Life Sciences, Vienna, 1190, Austria
[*] *Invited contribution by Frederik Kratzert, recipient of the EGU Hydrological Sciences Outstanding Student Poster and PICO Award 2016.*

Correspondence: Frederik Kratzert (f.kratzert@gmail.com)

Abstract. Rainfall–runoff modelling is one of the key challenges in the field of hydrology. Various approaches exist, ranging from physically based over conceptual to fully data-driven models. In this paper, we propose a novel data-driven approach, using the Long Short-Term Memory (LSTM) network, a special type of recurrent neural network. The advantage of the LSTM is its ability to learn long-term dependencies between the provided input and output of the network, which are essential for modelling storage effects in e.g. catchments with snow influence. We use 241 catchments of the freely available CAMELS data set to test our approach and also compare the results to the well-known Sacramento Soil Moisture Accounting Model (SAC-SMA) coupled with the Snow-17 snow routine. We also show the potential of the LSTM as a regional hydrological model in which one model predicts the discharge for a variety of catchments. In our last experiment, we show the possibility to transfer process understanding, learned at regional scale, to individual catchments and thereby increasing model performance when compared to a LSTM trained only on the data of single catchments. Using this approach, we were able to achieve better model performance as the SAC-SMA + Snow-17, which underlines the potential of the LSTM for hydrological modelling applications.

1 Introduction

Rainfall–runoff modelling has a long history in hydrological sciences and the first attempts to predict the discharge as a function of precipitation events using regression-type approaches date back 170 years (Beven, 2001; Mulvaney, 1850). Since then, modelling concepts have been further developed by progressively incorporating physically based process understanding and concepts into the (mathematical) model formulations. These include explicitly addressing the spatial variability of processes, boundary conditions and physical properties of the catchments (Freeze and Harlan, 1969; Kirchner, 2006; Schulla, 2007). These developments are largely driven by the advancements in computer technology and the availability of (remote sensing) data at high spatial and temporal resolution (Hengl et al., 2017; Kollet et al., 2010; Mu et al., 2011; Myneni et al., 2002; Rennó et al., 2008).

However, the development towards coupled, physically based and spatially explicit representations of hydrological processes at the catchment scale has come at the price of high computational costs and a high demand for necessary (meteorological) input data (Wood et al., 2011). Therefore, physically based models are still rarely used in operational rainfall–runoff forecasting. In addition, the current data sets for the parameterization of these kind of models, e.g. the 3-D information on the physical characteristics of the sub-surface, are mostly only available for small, experimental watersheds, limiting the model's applicability for larger river basins in an operational context. The high computational costs further limit their application, especially if uncertainty estimations and multiple model runs within an ensemble forecasting framework are required (Clark et al., 2017).

Thus, simplified physically based or conceptual models are still routinely applied for operational purposes (Adams and Pagaon, 2016; Herrnegger et al., 2018; Lindström et al., 2010; Stanzel et al., 2008; Thielen et al., 2009; Wesemann et al., 2018). In addition, data-based mechanistic modelling concepts (Young and Beven, 1994) or fully data-driven approaches such as regression, fuzzy-based or artificial neural networks (ANNs) have been developed and explored in this context (Remesan and Mathew, 2014; Solomatine et al., 2009; Zhu and Fujita, 1993).

ANNs are especially known to mimic highly non-linear and complex systems well. Therefore, the first studies using ANNs for rainfall–runoff prediction date back to the early 1990s (Daniell, 1991; Halff et al., 1993). Since then, many studies applied ANNs for modelling runoff processes (see for example Abrahart et al., 2012; ASCE Task Committee on Application of Artificial Neural Networks, 2000, for a historic overview). However, a drawback of feed-forward ANNs, which have mainly been used in the past, for time series analysis is that any information about the sequential order of the inputs is lost. Recurrent neural networks (RNNs) are a special type of neural network architecture that have been specifically designed to understand temporal dynamics by processing the input in its sequential order (Rumelhart et al., 1986). Carriere et al. (1996) and Hsu et al. (1997) conducted the first studies using RNNs for rainfall–runoff modelling. The former authors tested the use of RNNs within laboratory conditions and demonstrated their potential use for event-based applications. In their study, Hsu et al. (1997) compared a RNN to a traditional ANN. Even though the traditional ANN in general performed equally well, they found that the number of delayed inputs, which are provided as driving inputs to the ANN, is a critical hyperparameter. However, the RNN, due to its architecture, made the search for this number obsolete. Kumar et al. (2004) also used RNNs for monthly streamflow prediction and found them to outperform a traditional feed-forward ANN.

For problems however for which the sequential order of the inputs matters, the current state-of-the-art network architecture is the so-called "Long Short-Term Memory" (LSTM), which in its initial form was introduced by Hochreiter and Schmidhuber (1997). Through a specially designed architecture, the LSTM overcomes the problem of the traditional RNN of learning long-term dependencies representing e.g. storage effects within hydrological catchments, which may play an important role for hydrological processes, for example in snow-driven catchments.

In recent years, neural networks have gained a lot of attention under the name of deep learning (DL). As in hydrological modelling, the success of DL approaches is largely facilitated by the improvements in computer technology (especially through graphic processing units or GPUs; Schmidhuber, 2015) and the availability of huge data sets (Halevy et al.,

2009; Schmidhuber, 2015). While most well-known applications of DL are in the field of computer vision (Farabet et al., 2013; Krizhevsky et al., 2012; Tompson et al., 2014), speech recognition (Hinton et al., 2012) or natural language processing (Sutskever et al., 2014) few attempts have been made to apply recent advances in DL to hydrological problems. Shi et al. (2015) investigated a deep learning approach for precipitation nowcasting. Tao et al. (2016) used a deep neural network for bias correction of satellite precipitation products. Fang et al. (2017) investigated the use of deep learning models to predict soil moisture in the context of NASA's Soil Moisture Active Passive (SMAP) satellite mission. Assem et al. (2017) compared the performance of a deep learning approach for water flow level and flow predictions for the Shannon River in Ireland with multiple baseline models. They reported that the deep learning approach outperforms all baseline models consistently. More recently, D. Zhang et al. (2018) compared the performance of different neural network architectures for simulating and predicting the water levels of a combined sewer structure in Drammen (Norway), based on online data from rain gauges and water-level sensors. They confirmed that LSTM (as well as another recurrent neural network architecture with cell memory) are better suited for for multi-step-ahead predictions than traditional architectures without explicit cell memory. J. Zhang et al. (2018) used an LSTM for predicting water tables in agricultural areas. Among other things, the authors compared the resulting simulation from the LSTM-based approach with that of a traditional neural network and found that the former outperforms the latter. In general, the potential use and benefits of DL approaches in the field of hydrology and water sciences has only recently come into the focus of discussion (Marçais and de Dreuzy, 2017; Shen, 2018; Shen et al., 2018). In this context we would like to mention Shen (2018) more explicitly, since he provides an ambitious argument for the potential of DL in earth sciences/hydrology. In doing so he also provides an overview of various applications of DL in earth sciences. Of special interest for the present case is his point that DL might also provide an avenue for discovering emergent behaviours of hydrological phenomena.

Regardless of the hydrological modelling approach applied, any model will be typically calibrated for specific catchments for which observed time series of meteorological and hydrological data are available. The calibration procedure is required because models are only simplifications of real catchment hydrology and model parameters have to effectively represent non-resolved processes and any effect of subgrid-scale heterogeneity in catchment characteristics (e.g. soil hydraulic properties) (Beven, 1995; Merz et al., 2006). The transferability of model parameters (regionalization) from catchments where meteorological and runoff data are available to ungauged or data-scarce basins is one of the ongoing challenges in hydrology (Buytaert and Beven, 2009; He et al., 2011; Samaniego et al., 2010).

The aim of this study is to explore the potential of the LSTM architecture (in the adapted version proposed by Gers et al., 2000) to describe the rainfall–runoff behaviour of a large number of differently complex catchments at the daily timescale. Additionally, we want to analyse the potential of LSTMs for regionalizing the rainfall–runoff response by training a single model for a multitude of catchments. In order to allow for a more general conclusion about the suitability of our modelling approach, we test this approach on a large number of catchments of the CAMELS data set (Addor et al., 2017b; Newman et al., 2014). This data set is freely available and includes meteorological forcing data and observed discharge for 671 catchments across the contiguous United States. For each basin, the CAMELS data set also includes time series of simulated discharge from the Sacramento Soil Moisture Accounting Model (Burnash et al., 1973) coupled with the Snow-17 snow model (Anderson, 1973). In our study, we use these simulations as a benchmark, to compare our model results with an established modelling approach.

The paper is structured in the following way: in Sect. 2, we will briefly describe the LSTM network architecture and the data set used. This is followed by an introduction into three different experiments: in the first experiment, we test the general ability of the LSTM to model rainfall–runoff processes for a large number of individual catchments. The second experiment investigates the capability of LSTMs for regional modelling, and the last tests whether the regional models can help to enhance the simulation performance for individual catchments. Section 3 presents and discusses the results of our experiments, before we end our paper with a conclusion and outlook for future studies.

2 Methods and database

2.1 Long Short-Term Memory network

In this section, we introduce the LSTM architecture in more detail, using the notation of Graves et al. (2013). Beside a technical description of the network internals, we added a "hydrological interpretation of the LSTM" in Sect. 3.5 in order to bridge differences between the hydrological and deep learning research communities.

The LSTM architecture is a special kind of recurrent neural network (RNN), designed to overcome the weakness of the traditional RNN to learn long-term dependencies. Bengio et al. (1994) have shown that the traditional RNN can hardly remember sequences with a length of over 10. For daily streamflow modelling, this would imply that we could only use the last 10 days of meteorological data as input to predict the streamflow of the next day. This period is too short considering the memory of catchments including groundwater, snow or even glacier storages, with lag times between precipitation and discharge up to several years.

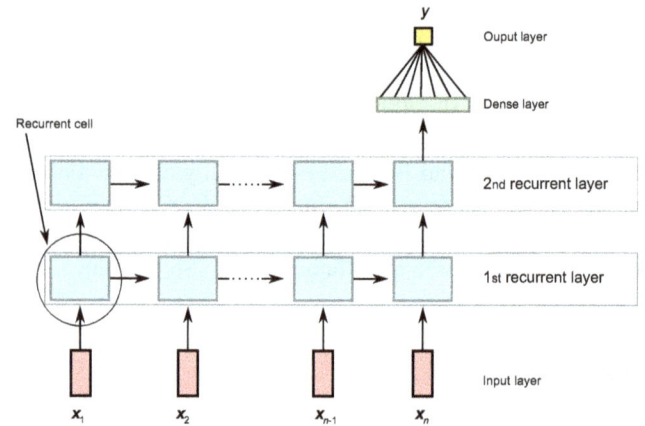

Figure 1. A general example of a two-layer recurrent neural network unrolled over time. The outputs from the last recurrent layer (second layer in this example) and the last time step (x_n) are fed into a dense layer to calculate the final prediction (y).

To explain how the RNN and the LSTM work, we unfold the recurrence of the network into a directed acyclic graph (see Fig. 1). The output (in our case discharge) for a specific time step is predicted from the input $x = [x_1, ..., x_n]$ consisting of the last n consecutive time steps of independent variables (in our case daily precipitation, min/max temperature, solar radiation and vapour pressure) and is processed sequentially. In each time step t ($1 \leq t \leq n$), the current input x_t is processed in the recurrent cells of each layer in the network.

The differences of the traditional RNN and the LSTM are the internal operations of the recurrent cell (encircled in Fig. 1) that are depicted in Fig. 2.

In a traditional RNN cell, only one internal state h_t exists (see Fig. 2a), which is recomputed in every time step by the following equation:

$$h_t = g\left(\mathbf{W}x_t + \mathbf{U}h_{t-1} + b\right), \tag{1}$$

where $g(\cdot)$ is the activation function (typically the hyperbolic tangent), $\mathbf{W}$ and $\mathbf{U}$ are the adjustable weight matrices of the hidden state h and the input x, and b is an adjustable bias vector. In the first time step, the hidden state is initialized as a vector of zeros and its length is an user-defined hyperparameter of the network.

In comparison, the LSTM has (i) an additional cell state or cell memory c_t in which information can be stored, and (ii) gates (three encircled letters in Fig. 2b) that control the information flow within the LSTM cell (Hochreiter and Schmidhuber, 1997). The first gate is the forget gate, introduced by Gers et al. (2000). It controls which elements of the cell state vector c_{t-1} will be forgotten (to which degree):

$$f_t = \sigma\left(\mathbf{W}_f x_t + \mathbf{U}_f h_{t-1} + b_f\right), \tag{2}$$

where f_t is a resulting vector with values in the range $(0, 1)$, $\sigma(\cdot)$ represents the logistic sigmoid function and $\mathbf{W}_f$, $\mathbf{U}_f$ and

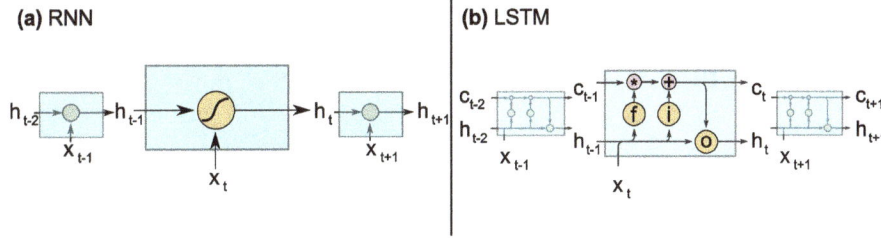

Figure 2. (a) The internal operation of a traditional RNN cell: $\boldsymbol{h}_t$ stands for hidden state and $\boldsymbol{x}_t$ for the input at time step t. **(b)** The internals of a LSTM cell, where $\boldsymbol{f}$ stands for the forget gate (Eq. 2), $\boldsymbol{i}$ for the input gate (Eqs. 3–4), and $\boldsymbol{o}$ for the output gate (Eqs. 6–7). $\boldsymbol{c}_t$ denotes the cell state at time step t and $\boldsymbol{h}_t$ the hidden state.

$\boldsymbol{b}_f$ define the set of learnable parameters for the forget gate, i.e. two adjustable weight matrices and a bias vector. As for the traditional RNN, the hidden state $\boldsymbol{h}$ is initialized in the first time step by a vector of zeros with a user-defined length.

In the next step, a potential update vector for the cell state is computed from the current input ($\boldsymbol{x}_t$) and the last hidden state ($\boldsymbol{h}_{t-1}$) given by the following equation:

$$\widetilde{\boldsymbol{c}}_t = \tanh\left(\mathbf{W}_{\widetilde{c}}\boldsymbol{x}_t + \mathbf{U}_{\widetilde{c}}\boldsymbol{h}_{t-1} + \boldsymbol{b}_{\widetilde{c}}\right), \tag{3}$$

where $\widetilde{\boldsymbol{c}}_t$ is a vector with values in the range $(-1, 1)$, $\tanh(\cdot)$ is the hyperbolic tangent and $\mathbf{W}_{\widetilde{c}}$, $\mathbf{U}_{\widetilde{c}}$ and $\boldsymbol{b}_{\widetilde{c}}$ are another set of learnable parameters.

Additionally, the second gate is compute, the input gate, defining which (and to what degree) information of $\widetilde{\boldsymbol{c}}_t$ is used to update the cell state in the current time step:

$$\boldsymbol{i}_t = \sigma\left(\mathbf{W}_i\boldsymbol{x}_t + \mathbf{U}_i\boldsymbol{h}_{t-1} + \boldsymbol{b}_i\right), \tag{4}$$

where $\boldsymbol{i}_t$ is a vector with values in the range $(0, 1)$, and $\mathbf{W}_i$, $\mathbf{U}_i$ and $\boldsymbol{b}_i$ are a set of learnable parameters, defined for the input gate.

With the results of Eqs. (2)–(4) the cell state $\boldsymbol{c}_t$ is updated by the following equation:

$$\boldsymbol{c}_t = \boldsymbol{f}_t \odot \boldsymbol{c}_{t-1} + \boldsymbol{i}_t \odot \widetilde{\boldsymbol{c}}_t, \tag{5}$$

where $\odot$ denotes element-wise multiplication. Because the vectors $\boldsymbol{f}_t$ and $\boldsymbol{i}_t$ have both entries in the range $(0, 1)$, Eq. (5) can be interpreted in the way that it defines, which information stored in $\boldsymbol{c}_{t-1}$ will be forgotten (values of $\boldsymbol{f}_t$ of approx. 0) and which will be kept (values of $\boldsymbol{f}_t$ of approx. 1). Similarly, $\boldsymbol{i}_t$ decides which new information stored in $\widetilde{\boldsymbol{c}}_t$ will be added to the cell state (values of $\boldsymbol{i}_t$ of approx. 1) and which will be ignored (values of $\boldsymbol{i}_t$ of approx. 0). Like the hidden state vector, the cell state is initialized by a vector of zeros in the first time step. Its length corresponds to the length of the hidden state vector.

The third and last gate is the output gate, which controls the information of the cell state $\boldsymbol{c}_t$ that flows into the new hidden state $\boldsymbol{h}_t$. The output gate is calculated by the following equation:

$$\boldsymbol{o}_t = \sigma\left(\mathbf{W}_o\boldsymbol{x}_t + \mathbf{U}_o\boldsymbol{h}_{t-1} + \boldsymbol{b}_o\right), \tag{6}$$

where $\boldsymbol{o}_t$ is a vector with values in the range $(0, 1)$, and $\mathbf{W}_o$, $\mathbf{U}_o$ and $\boldsymbol{b}_o$ are a set of learnable parameters, defined for the output gate. From this vector, the new hidden state $\boldsymbol{h}_t$ is calculated by combining the results of Eqs. (5) and (6):

$$\boldsymbol{h}_t = \tanh(\boldsymbol{c}_t) \odot \boldsymbol{o}_t. \tag{7}$$

It is in particular the cell state ($\boldsymbol{c}_t$) that allows for an effective learning of long-term dependencies. Due to its very simple linear interactions with the remaining LSTM cell, it can store information unchanged over a long period of time steps. During training, this characteristic helps to prevent the problem of the exploding or vanishing gradients in the backpropagation step (Hochreiter and Schmidhuber, 1997). As with other neural networks, where one layer can consist of multiple units (or neurons), the length of the cell and hidden state vectors in the LSTM can be chosen freely. Additionally, we can stack multiple layers on top of each other. The output from the last LSTM layer at the last time step ($\boldsymbol{h}_n$) is connected through a traditional dense layer to a single output neuron, which computes the final discharge prediction (as shown schematically in Fig. 1). The calculation of the dense layer is given by the following equation:

$$y = \mathbf{W}_d\boldsymbol{h}_n + \boldsymbol{b}_d, \tag{8}$$

where y is the final discharge, $\boldsymbol{h}_n$ is the output of the last LSTM layer at the last time step derive from Eq. (7), $\mathbf{W}_d$ is the weight matrix of the dense layer, and $\boldsymbol{b}_d$ is the bias term.

To conclude, Algorithm 1 shows the pseudocode of the entire LSTM layer. As indicated above and shown in Fig. 1, the inputs for the complete sequence of meteorological observations $x = [\boldsymbol{x}_1, ..., \boldsymbol{x}_n]$, where $\boldsymbol{x}_t$ is a vector containing the meteorological inputs of time step t, is processed time step by time step and in each time step Eqs. (2)–(7) are repeated. In the case of multiple stacked LSTM layers, the next layer takes the output $h = [\boldsymbol{h}_1, ..., \boldsymbol{h}_n]$ of the first layer as input. The final output, the discharge, is then calculated by Eq. (8), where $\boldsymbol{h}_n$ is the last output of the last LSTM layer.

2.2 The calibration procedure

In traditional hydrological models, the calibration involves a defined number of iteration steps of simulating the entire

---

Algorithm 1 Pseudocode of the LSTM layer

1: **Input:** $x = [x_1, ..., x_n], x_t \in \mathbb{R}^m$
2: **Given parameters:** $\mathbf{W}_f, \mathbf{U}_f, b_f, \mathbf{W}_{\widetilde{c}}, \mathbf{U}_{\widetilde{c}}, b_{\widetilde{c}}, \mathbf{W}_i, \mathbf{U}_i, b_i,$
   $\mathbf{W}_o, \mathbf{U}_o, b_o$
3: **Initialize** $h_0, c_0 = \vec{0}$ of length p
4: **for** t=1, ..., n **do**
5:     **Calculate** f_t (Eq. 2), $\widetilde{c}_t$ (Eq. 3), i_t (Eq. 4)
6:     **Update cell state** c_t (Eq. 5)
7:     **Calculate** o_t (Eq. 6), h_t (Eq. 7)
8: **end for**
9: **Output:** $h = [h_1, ..., h_n], h_t \in \mathbb{R}^p$

---

calibration period with a given set of model parameters and evaluating the model performance with some objective criteria. The model parameters are, regardless of the applied optimization technique (global and/or local), perturbed in such a way that the maximum (or minimum) of an objective criteria is found. Regarding the training of a LSTM, the adaptable (or *learnable*) parameters of the network, the weights and biases, are also updated depending on a given loss function of an iteration step. In this study we used the mean-squared error (MSE) as an objective criterion.

In contrast to most hydrological models, the neural network exhibits the property of differentiability of the network equations. Therefore, the gradient of the loss function with respect to any network parameter can always be calculated explicitly. This property is used in the so-called back-propagation step in which the network parameters are adapted to minimize the overall loss. For a detailed description see e.g. Goodfellow et al. (2016).

A schematic illustration of one iteration step in the LSTM training/calibration is is provided in Fig. 3. One iteration step during the training of LSTMs usually works with a subset (called *batch* or *mini-batch*) of the available training data. The number of samples per batch is a hyperparameter, which in our case was defined to be 512. Each of these samples consists of one discharge value of a given day and the meteorological input of the n preceding days. In every iteration step, the loss function is calculated as the average of the MSE of simulated and observed runoff of these 512 samples. Since the discharge of a specific time step is only a function of the meteorological inputs of the last n days, the samples within a batch can consist of random time steps (depicted in Fig. 3 by the different colours), which must not necessarily be ordered chronologically. For faster convergence, it is even advantageous to have random samples in one batch (LeCun et al., 2012). This procedure is different from traditional hydrological model calibration, where usually all the information of the calibration data is processed in each iteration step, since all simulated and observed runoff pairs are used in the model evaluation.

Within traditional hydrological model calibration, the *number of iteration steps* defines the total number of model runs performed during calibration (given an optimization al-

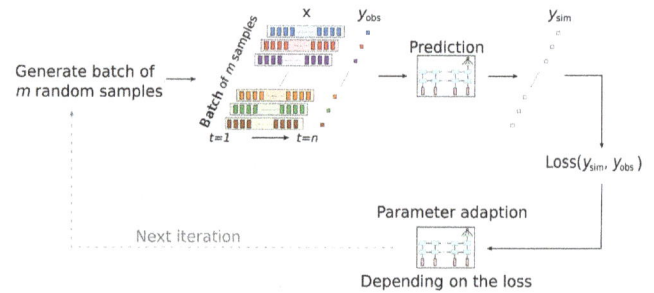

Figure 3. Illustration of one iteration step in the training process of the LSTM. A random batch of input data x consisting of m independent training samples (depicted by the colours) is used in each step. Each training sample consists of n days of look-back data and one target value (y_{obs}) to predict. The loss is computed from the observed discharge and the network's predictions y_{sim} and is used to update the network parameters.

gorithm without a convergence criterion). The corresponding term for neural networks is called *epoch*. One epoch is defined as the period in which each training sample is used once for updating the model parameters. For example, if the data set consists of 1000 training samples and the batch size is 10, one epoch would consist of 100 iteration steps (number of training samples divided by the number of samples per batch). In each iteration step, 10 of the 1000 samples are taken without replacement until all 1000 samples are used once. In our case this means, each time step of the discharge time series in the training data is simulated exactly once. This is somewhat similar to one iteration in the calibration of a classical hydrological model, with the significant difference however that every sample is generated independently of each other. Figure 4 shows the learning process of the LSTM over a number of training epochs. We can see that the network has to learn the entire rainfall–runoff relation from scratch (grey line of random weights) and is able to better represent the discharge dynamics with each epoch.

For efficient learning, all input features (the meteorological variables) as well as the output (the discharge) data are normalized by subtracting the mean and dividing by the standard deviation (LeCun et al., 2012; Minns and Hall, 1996). The mean and standard deviation used for the normalization are calculated from the calibration period only. To receive the final discharge prediction, the output of the network is retransformed using the normalization parameters from the calibration period (Fig. 4 shows the retransformed model outputs).

2.3   Open-source software

Our research heavily relies on open source software. The programming language of choice is Python 3.6 (van Rossum, 1995). The libraries we use for preprocessing our data and for data management in general are Numpy (Van Der Walt et al., 2011), Pandas (McKinney, 2010) and

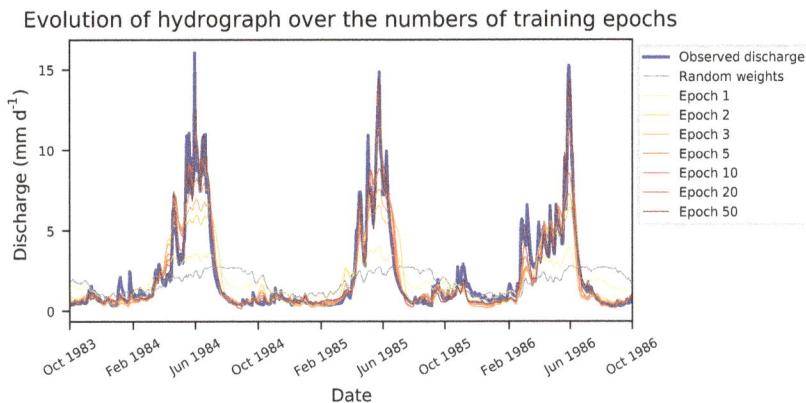

Figure 4. Improvement of the runoff simulation during the learning process of the LSTM. Visualized are the observed discharge and LSTM output after various epochs for the basin 13337000 of the CAMELs data set from 1 October 1983 until 30 September 1986. Random weights represent randomly initialized weights of a LSTM before the first iteration step in the training process.

Scikit-Learn (Pedregosa et al., 2011). The Deep-Learning frameworks we use are TensorFlow (Abadi et al., 2016) and Keras (Chollet, 2015). All figures are made using Matplotlib (Hunter, 2007).

2.4 The CAMELS data set

The underlying data for our study is the CAMELS data set (Addor et al., 2017b; Newman et al., 2014). The acronym stands for "Catchment Attributes for Large-Sample Studies" and it is a freely available data set of 671 catchments with minimal human disturbances across the contiguous United States (CONUS). The data set contains catchment aggregated (lumped) meteorological forcing data and observed discharge at the daily timescale starting (for most catchments) from 1980. The meteorological data are calculated from three different gridded data sources (Daymet, Thornton et al., 2012; Maurer, Maurer et al., 2002; and NLDAS, Xia et al., 2012) and consists of day length, precipitation, shortwave downward radiation, maximum and minimum temperature, snow-water equivalent and humidity. We used the Daymet data, since it has the highest spatial resolution (1 km grid compared to 12 km grid for Maurer and NLDAS) as a basis for calculating the catchment averages and all available meteorological input variables with exception of the snow-water equivalent and the day length.

The 671 catchments in the data set are grouped into 18 hydrological units (HUCs) following the U.S. Geological Survey's HUC map (Seaber et al., 1987). These groups correspond to geographic areas that represent the drainage area of either a major river or the combined drainage area of a series of rivers.

In our study, we used 4 out of the 18 hydrological units with their 241 catchments (see Fig. 5 and Table 1) in order to cover a wide range of different hydrological conditions on one hand and to limit the computational costs on the other hand. The New England region in the northeast contains 27 more or less homogeneous basins (e.g. in terms of snow influence or aridity). The Arkansas-White-Red region in the center of CONUS has a comparable number of basins, namely 32, but is completely different otherwise. Within this region, attributes e.g. aridity and mean annual precipitation have a high variance and strong gradient from east to west (see Fig. 5). Also comparable in size but with disparate hydro-climatic conditions are the South Atlantic-Gulf region (92 basins) and the Pacific Northwest region (91 basins). The latter spans from the Pacific coast till the Rocky Mountains and also exhibits a high variance of attributes across the basins, comparable to the Arkansas-White-Red region. For example, there are very humid catchments with more than $3000\,\mathrm{mm\,yr^{-1}}$ precipitation close to the Pacific coast and very arid (aridity index 2.17, mean annual precipitation $500\,\mathrm{mm\,yr^{-1}}$) basins in the south-east of this region. The relatively flat South Atlantic-Gulf region contains more homogeneous basins, but in contrast to the New England region is not influenced by snow.

Additionally, the CAMELS data set contains time series of simulated discharge from the calibrated Snow-17 models coupled with the Sacramento Soil Moisture Accounting Model. Roughly 35 years of meteorological observations and streamflow records are available for most basins. The first 15 hydrological years with streamflow data (in most cases 1 October 1980 until 30 September 1995) are used for calibrating the model, while the remaining data are used for validation. For each basin, 10 models were calibrated, starting with different random seeds, using the shuffled complex evolution algorithm by Duan et al. (1993) and the root mean squared error (RMSE) as objective function. Of these 10 models, the one with the lowest RMSE in the calibration period is used for validation. For further details see Newman et al. (2015).

Table 1. Overview of the HUCs considered in this study and some region statics averaged over all basins in that region. For each variable mean and standard deviation is reported.

HUC	Region name	No. of basins	Mean precipitation (mm day^{-1})	Mean aridity[1] (–)	Mean altitude (m)	Mean snow frac.[2] (–)	Mean seasonality[3] (–)
01	New England	27	3.61 ± 0.26	0.60 ± 0.03	316 ± 182	0.24 ± 0.06	0.10 ± 0.08
03	South Atlantic-Gulf	92	3.79 ± 0.49	0.87 ± 0.14	189 ± 179	0.02 ± 0.02	0.12 ± 0.26
11	Arkansas-White-Red	31	2.86 ± 0.89	1.18 ± 0.50	613 ± 713	0.08 ± 0.13	0.25 ± 0.29
17	Pacific Northwest	91	5.22 ± 2.03	0.59 ± 0.40	1077 ± 589	0.33 ± 0.23	-0.72 ± 0.17

[1] PET/P; see Addor et al. (2017a). [2] Fraction of precipitation falling on days with temperatures below 0 °C. [3] Positive values indicate that precipitation peaks in summer, negative values that precipitation peaks in the winter month, and values close to 0 that the precipitation is uniform throughout the year (see Addor et al., 2017a).

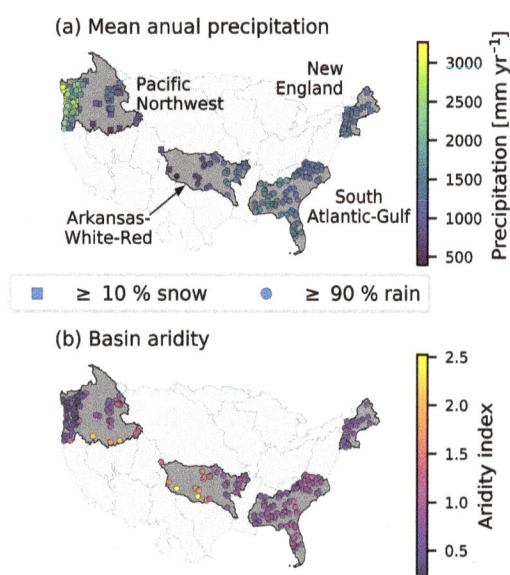

(a) Mean anual precipitation

(b) Basin aridity

Figure 5. Overview of the location of the four hydrological units from the CAMELS data set used in this study, including all their basins. Panel **(a)** shows the mean annual precipitation of each basin, whereas the type of marker symbolizes the snow influence of the basin. Panel **(b)** shows the aridity index of each basin, calculated as PET/P (see Addor et al., 2017a).

2.5 Experimental design

Throughout all of our experiments, we used a two-layer LSTM network, with each layer having a cell/hidden state length of 20. Table 2 shows the resulting shapes of all model parameters from Eqs. (2) to (8). Between the layers, we added dropout, a technique to prevent the model from overfitting (Srivastava et al., 2014). Dropout sets a certain percentage (10 % in our case) of random neurons to zero during training in order to force the network into a more robust feature learning. Another hyperparameter is the length of the input sequence, which corresponds to the number of days of meteorological input data provided to the network for the prediction of the next discharge value. We decided to keep

Table 2. Shapes of learnable parameters of all layers.

Layer	Parameter	Shape
1st LSTM layer	$\mathbf{W}_f, \mathbf{W}_{\widetilde{c}}, \mathbf{W}_i, \mathbf{W}_o$	[20, 5]
	$\mathbf{U}_f, \mathbf{U}_{\widetilde{c}}, \mathbf{U}_i, \mathbf{U}_o$	[20, 20]
	$\boldsymbol{b}_f, \boldsymbol{b}_{\widetilde{c}}, \boldsymbol{b}_i, \boldsymbol{b}_o$	[20]
2nd LSTM layer	$\mathbf{W}_f, \mathbf{W}_{\widetilde{c}}, \mathbf{W}_i, \mathbf{W}_o$	[20, 20]
	$\mathbf{U}_f, \mathbf{U}_{\widetilde{c}}, \mathbf{U}_i, \mathbf{U}_o$	[20, 20]
	$\boldsymbol{b}_f, \boldsymbol{b}_{\widetilde{c}}, \boldsymbol{b}_i, \boldsymbol{b}_o$	[20]
Dense layer	$\mathbf{W}_d$	[20, 1]
	$\boldsymbol{b}_d$	[1]

this value constant at 365 days for this study in order to capture at least the dynamics of a full annual cycle.

The specific design of the network architecture, i.e. the number of layers, cell/hidden state length, dropout rate and input sequence length were found through a number of experiments in several seasonal-influenced catchments in Austria. In these experiments, different architectures (e.g. one or two LSTM layers or 5, 10, 15, or 20 cell/hidden units) were varied manually. The architecture used in this study proved to work well for these catchments (in comparison to a calibrated hydrological model we had available from previous studies; Herrnegger et al., 2018) and was therefore chosen to be applied here without further tuning. A systematic sensitivity analysis of the effects of different hyper-parameters was however not done and is something to do in the future.

We want to mention here that our calibration scheme (see description in the three experiments below) is not the standard way for calibrating and selecting data-driven models, especially neural networks. As of today, a widespread calibration strategy for DL models is to subdivide the data into three parts, referred to as training, validation and test data (see Goodfellow et al., 2016). The first two splits are used to derive the parametrization of the networks and the remainder of the data to diagnose the actual performance. We decided to not implement this splitting strategy, because we are limited to the periods Newman et al. (2015) used so that our models are comparable with their results. Theoretically, it would be possible to split the 15-year calibration period of Newman et al. (2015) further into a training and validation set. However, this would lead to (a) a much shorter period of data that is used for the actual weight updates or (b) a high risk of overfitting to the short validation period, depending on how this 15-year period is divided. In addition to that, LSTMs with a low number of hidden units are quite sensitive to the initialization of their weights. It is thus common practice to repeat the calibration task several times with different random seeds to select the best performing realization of the model (Bengio, 2012). For the present purpose we decided not to implement these strategies, since it would make it more difficult or even impossible to compare the LSTM approach to the SAC-SMA + Smow-17 reference model. The goal of this study is therefore not to find the best per-catchment model, but rather to investigate the general potential of LSTMs for the task of rainfall–runoff modelling. However, we think that the sample size of 241 catchment is large enough to infer some of the (average) properties of the LSTM-based approach.

2.5.1 Experiment 1: one model for each catchment

With the first experiment, we test the general ability of our LSTM network to model rainfall–runoff processes. Here, we train one network separately for each of the 241 catchments. To avoid the effect of overfitting of the network on the training data, we identified the number of epochs (for a definition of an epoch, see Sect. 2.2) in a preliminary step, which yielded, on average, the highest Nash–Sutcliffe efficiency (NSE) across all basins for an independent validation period. For this preliminary experiment, we used the first 14 years of the 15-year calibration period as training data and the last, fifteenth, year as the independent validation period. With the 14 years of data, we trained a model for in total 200 epochs for each catchment and evaluated each model after each epoch with the validation data. Across all catchments, the highest mean NSE was achieved after 50 epochs in this preliminary experiment. Thus, for the final training of the LSTM with the full 15 years of the calibration period as training data, we use the resulting number of 50 epochs for all catchments. Experiment 1 yields 241 separately trained networks, one for each of the 241 catchments.

2.5.2 Experiment 2: one regional model for each hydrological unit

Our second experiment is motivated by two different ideas: (i) deep learning models really excel, when having many training data available (Hestness et al., 2017; Schmidhuber, 2015), and (ii) regional models as potential solution for prediction in ungauged basins.

Regarding the first motivation, having a huge training data set allows the network to learn more general and abstract patterns of the input-to-output relationship. As for all data-driven approaches, the network has to learn the entire "hydrological model" purely from the available data (see Fig. 4). Therefore, having more than just the data of a single catchment available would help to obtain a more general understanding of the rainfall–runoff processes. An illustrative example are two similarly behaving catchments of which one lacks high precipitation events or extended drought periods in the calibration period, while having these events in the validation period. Given that the second catchment experienced these conditions in the calibration set, the LSTM could learn the response behaviour to those extremes and use this knowledge in the first catchment. Classical hydrological models have the process understanding implemented in the model structure itself and therefore – at least in theory – it is not strictly necessary to have these kind of events in the calibration period.

The second motivation is the prediction of runoff in ungauged basins, one of the main challenges in the field of hydrology (Blöschl et al., 2013; Sivapalan, 2003). A regional model that performs reasonably well across all catchments within a region could potentially be a step towards the prediction of runoff for such basins.

Therefore, the aim of the second experiment is to analyse how well the network architecture can generalize (or regionalize) to all catchments within a certain region. We use the HUCs that are used for grouping the catchments in the CAMELS data set for the definition of the regions (four in this case). The training data for these regional models are the combined data of the calibration period of all catchments within the same HUC.

To determine the number of training epochs, we performed the same preliminary experiment as described in Experiment 1. Across all catchments, the highest mean NSE was achieved after 20 epochs in this case. Although the number of epochs is smaller compared to Experiment 1, the number of weight updates is much larger. This is because the number of available training samples has increased and the same batch size as in Experiment 1 is used (see Sect. 2.2 for an explanation of the connection of number of iterations, number of training samples and number of epochs). Thus, for the final training, we train one LSTM for each of the four used HUCs for 20 epochs with the entire 15-year long calibration period.

2.5.3 Experiment 3: fine-tuning the regional model for each catchment

In the third experiment, we want to test whether the more general knowledge of the regional model (Experiment 2) can help to increase the performance of the LSTM in a single catchment. In the field of DL this is a common approach called fine-tuning (Razavian et al., 2014; Yosinski et al., 2014), where a model is first trained on a huge data set to learn general patterns and relationships between (meteorological) input data and (streamflow) output data (this is referred to as *pre-training*). Then, the pre-trained network is further trained for a small number of epochs with the data of a specific catchment alone to adapt the more generally learned processes to a specific catchment. Loosely speaking, the LSTM first learns the general behaviour of the runoff generating processes from a large data set, and is in a second step adapted in order to account for the specific behaviour of a given catchment (e.g. the scaling of the runoff response in a specific catchment).

In this study, the regional models of Experiment 2 serve as pre-trained models. Therefore, depending on the affiliation of a catchment to a certain HUC, the specific regional model for this HUC is taken as a starting point for the fine-tuning. With the initial LSTM weights from the regional model, the training is continued only with the training data of a specific catchment for a few epochs (ranging from 0 to 20, median 10). Thus, similar to Experiment 1, we finally have 241 different models, one for each of the 241 catchments. Different from the two previous experiments, we do not use a global number of epochs for fine-tuning. Instead, we used the 14-year/1-year split to determine the optimal number of epochs for each catchment individually. The reason is that the regional model fits individual catchments within a HUC differently well. Therefore, the number of epochs the LSTM needs to adapt to a certain catchment before it starts to overfit is different for each catchment.

2.6 Evaluation metrics

The metrics for model evaluation are the Nash–Sutcliffe efficiency (Nash and Sutcliffe, 1970) and the three decompositions following Gupta et al. (2009). These are the correlation coefficient of the observed and simulated discharge (r), the variance bias (α) and the total volume bias (β). While all of these measures evaluate the performance over the entire time series, we also use three different signatures of the flow duration curve (FDC) that evaluate the performance of specific ranges of discharge. Following Yilmaz et al. (2008), we calculate the bias of the 2 % flows, the peak flows (FHV), the bias of the slope of the middle section of the FDC (FMS) and the bias of the bottom 30 % low flows (FLV).

Because our modelling approach needs 365 days of meteorological data as input for predicting one time step of discharge, we cannot simulate the first year of the calibration period. To be able to compare our models to the SAC-SMA + Snow-17 benchmark model, we recomputed all metrics for the benchmark model for the same simulation periods.

3   Results and discussion

We start presenting our results by showing an illustrative comparison of the modelling capabilities of traditional RNNs and the LSTM to highlight the problems of RNNs to learn long-term dependencies and its deficits for the task of rainfall–runoff modelling. This is followed by the analysis of the results of Experiment 1, for which we trained one network separately for each basin and compare the results to the SAC-SMA + Snow-17 benchmark model. Then we investigate the potential of LSTMs to learn hydrological behaviour at the regional scale. In this context, we compare the performance of the regional models from Experiment 2 against the models of Experiment 1 and discuss their strengths and weaknesses. Lastly, we examine whether our fine-tuning approach enhances the predictive power of our models in the individual catchments. In all cases, the analysis is based on the data of the 241 catchments of the calibration (the first 15 years) and validation (all remaining years available) periods.

3.1   The effect of (not) learning long-term dependencies

As stated in Sect. 2.1, the traditional RNN can only learn dependencies of 10 or less time steps. The reason for this is the so-called "vanishing or exploding gradients" phenomenon (see Bengio et al., 1994, and Hochreiter and Schmidhuber, 1997), which manifests itself in an error signal during the backward pass of the network training that either diminishes towards zero or grows against infinity, preventing the effective learning of long-term dependencies. However, from the perspective of hydrological modelling, a catchment contains various processes with dependencies well above 10 days (which corresponds to 10 time steps in the case of daily streamflow modelling), e.g. snow accumulation during winter and snowmelt during spring and summer. Traditional hydrological models need to reproduce these processes correctly in order to be able to make accurate streamflow predictions. This is in principle not the case for data-driven approaches.

To empirically test the effect of (not) being able to learn long-term dependencies, we compared the modelling of a snow-influenced catchment (basin 13340600 of the Pacific Northwest region) with a LSTM and a traditional RNN. For this purpose we adapted the number of hidden units of the RNN to be 41 for both layers (so that the number of learnable parameters of the LSTM and RNN is approximately the same). All other modelling boundary conditions, e.g. input data, the number of layers, dropout rate, and number of training epochs, are kept identical.

Figure 6a shows 2 years of the validation period of observed discharge as well as the simulation by LSTM and RNN. We would like to highlight three points. (i) The hydrograph simulated by the RNN has a lot more variance compared to the smooth line of the LSTM. (ii) The RNN underestimates the discharge during the melting season and early summer, which is strongly driven snowmelt and by the precipitation that has fallen through the winter months. (iii) In the winter period, the RNN systematically overestimates observed discharge, since snow accumulation is not accounted

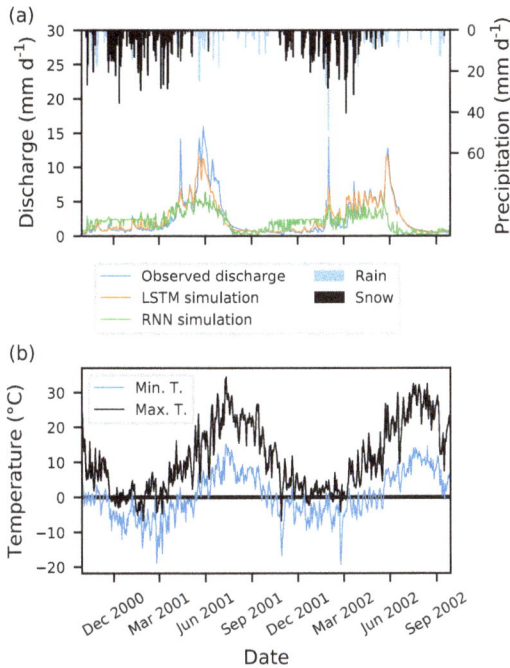

Figure 6. (a) Two years of observed as well as the simulated discharge of the LSTM and RNN from the validation period of basin 13340600. The precipitation is plotted from top to bottom and days with minimum temperature below zero are marked as snow (black bars). **(b)** The corresponding daily maximum and minimum temperature.

for. These simulation deficits can be explained by the lack of the RNN to learn and store long-term dependencies, while especially the last two points are interesting and connected. Recall that the RNN is trained to minimize the average RMSE between observation and simulation. The RNN is not able to store the amount of water which has fallen as snow during the winter and is, in consequence, also not able to generate sufficient discharge during the time of snowmelt. The RNN, minimizing the average RMSE, therefore overestimates the discharge most times of the year by a constant bias and underestimates the peak flows, thus being closer to predicting the mean flow. Only for a short period at the end of the summer is it close to predicting the low flow correctly.

In contrast, the LSTM seems to have (i) no or fewer problems with predicting the correct amount of discharge during the snowmelt season and (ii) the predicted hydrograph is much smoother and fits the general trends of the hydrograph much better. Note that both networks are trained with the exact same data and have the same data available for predicting a single day of discharge.

Here we have only shown a single example for a snow-influenced basin. We also compared the modelling behaviour in one of the arid catchments of the Arkansas-White-Red region, and found that the trends and conclusion were similar. Although only based on a single illustrative example that

shows the problems of RNNs with long-term dependencies, we can conclude that traditional RNNs should not be used if (e.g. daily) discharge is predicted only from meteorological observations.

3.2 Using LSTMs as hydrological models

Figure 7a shows the spatial distribution of the LSTM performances for Experiment 1 in the validation period. In over 50 % of the catchments, an NSE of 0.65 or above is found, with a mean NSE of 0.63 over all catchments. We can see that the LSTM performs better in catchments with snow influence (New England and Pacific Northwest regions) and catchments with higher mean annual precipitation (also the New England and Pacific Northwest regions, but also basins in the western part of the Arkansas-White-Red region; see Fig. 5a for precipitation distribution). The performance deteriorates in the more arid catchments, which are located in the western part of the Arkansas-White-Red region, where no discharge is observed for longer periods of the year (see Fig. 5b). Having a constant value of discharge (zero in this case) for a high percentage of the training samples seems to be difficult information for the LSTM to learn and to reproduce this hydrological behaviour. However, if we compare the results for these basins to the benchmark model (Fig. 7b), we see that for most of these dry catchments the LSTM outperforms the latter, meaning that the benchmark model did not yield satisfactory results for these catchments either. In general, the visualization of the differences in the NSE shows that the LSTM performs slightly better in the northern, more snow-influenced catchments, while the SAC-SMA + Snow-17 performs better in the catchments in the south-east. This clearly shows the benefit of using LSTMs, since the snow accumulation and snowmelt processes are correctly reproduced, despite their inherent complexity. Our results suggest that the model learns these long-term dependencies, i.e. the time lag between precipitation falling as snow during the winter period and runoff generation in spring with warmer temperatures. The median value of the NSE differences is -0.03, which means that the benchmark model slightly outperforms the LSTM. Based on the mean NSE value (0.58 for the benchmark model, compared to 0.63 for the LSTM of this Experiment), the LSTM outperforms the benchmark results.

In Fig. 8, we present the cumulative density functions (CDF) for various metrics for the calibration and validation period. We see that the LSTM and the benchmark model work comparably well for all but the FLV (bias of the bottom 30 % low flows) metric. The underestimation of the peak flow in both models could be expected when using the MSE as the objective function for calibration (Gupta et al., 2009). However, the LSTM underestimates the peaks more strongly compared to the benchmark model (Fig. 8d). In contrast, the middle section of the FDC is better represented in the LSTM (Fig. 8e). Regarding the performance in terms of the NSE, the LSTM shows fewer negative outliers and thus seems to

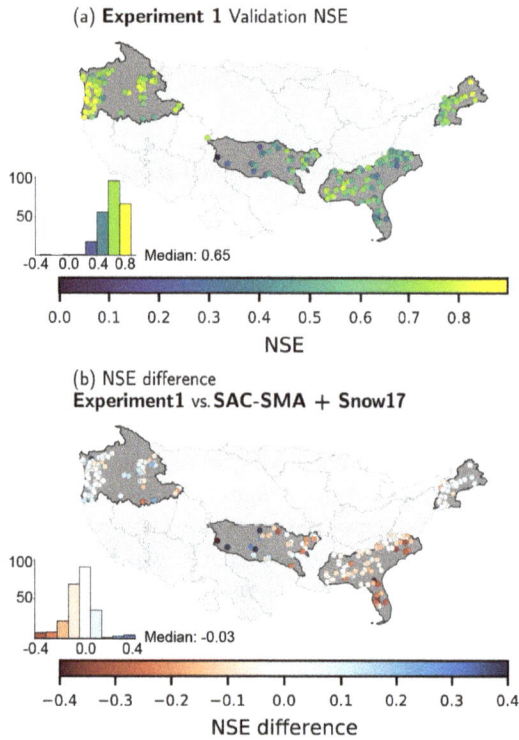

Figure 7. Panel **(a)** shows the NSE of the validation period of the models from Experiment 1 and panel **(b)** the difference of the NSE between the LSTM and the benchmark model (blue colours (> 0) indicate that the LSTM performs better than the benchmark model, red (< 0) the other way around). The colour maps are limited to [0, 1] for the NSE and [−0.4, 0.4] for the NSE differences for better visualization.

be more robust. The poorest model performance in the validation period is an NSE of −0.42 compared to −20.68 of the SAC-SMA + Snow-17. Figure 8f shows large differences between the LSTM and the SAC-SMA + Snow-17 model regarding the FLV metric. The FLV is highly sensitive to the one single minimum flow in the time series, since it compares the area between the FDC and this minimum value in the log space of the observed and simulated discharge. The discharge from the LSTM model, which has no exponential outflow function like traditional hydrological models, can easily drop to diminutive numbers or even zero, to which we limited our model output. A rather simple solution for this issue is to introduce just one additional parameter and to limit the simulated discharge not to zero, but to the minimum observed flow from the calibration period. Figure 9 shows the effect of this approach on the CDF of the FLV. We can see that this simple solution leads to better FLV values compared to the benchmark model. Other metrics, such as the NSE, are almost unaffected by this change, since these low-flow values only marginally influence the resulting NSE values (not shown here).

From the CDF of the NSE in Fig 8a, we can also observe a trend towards higher values in the calibration compared to

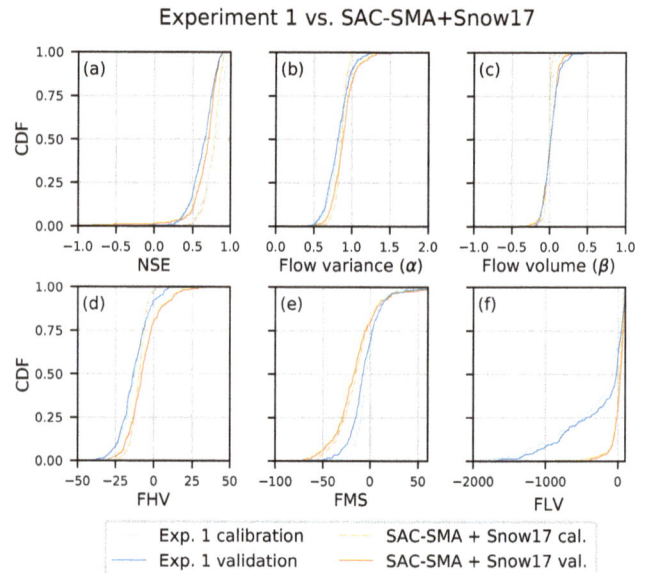

Figure 8. Cumulative density functions for various metrics of the calibration and validation period of Experiment 1 compared to the benchmark model. FHV is the bias of the top 2 % flows, the peak flows, FMS is the slope of the middle section of the flow duration curve and FLV is the bias of the bottom 30 % low flows.

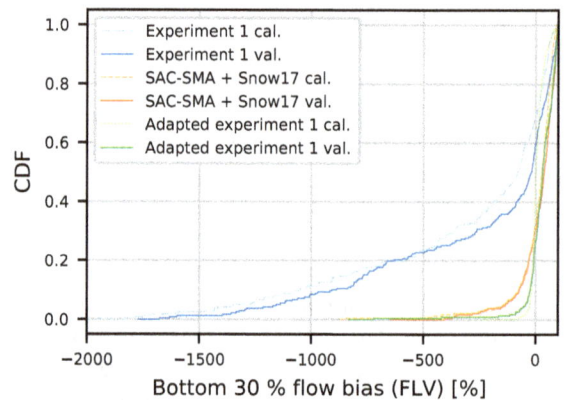

Figure 9. The effect of limiting the discharge prediction of the network not to zero (blue lines) but instead to the minimum observed discharge of the calibration period (green lines) on the FLV. Benchmark model (orange lines) for comparison.

the validation period for both modelling approaches. This is a sign of overfitting, and in the case of the LSTM, could be tackled by a smaller network size, stronger regularization or more data. However, we want to highlight again that achieving the best model performance possible was not the aim of this study, but rather testing the general ability of the LSTM to reproduce runoff processes.

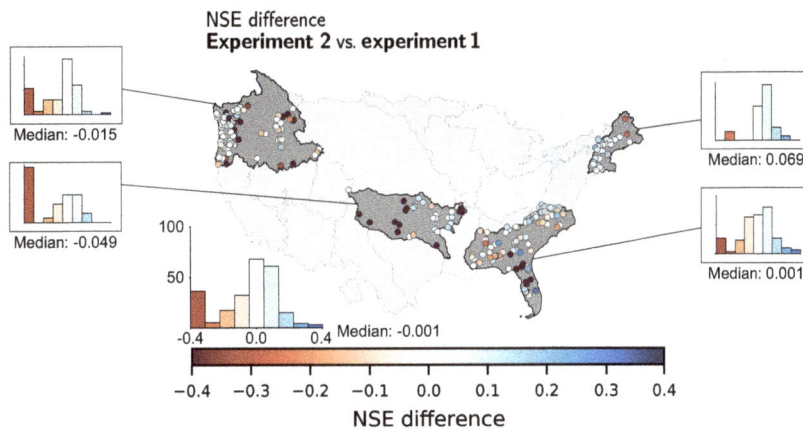

Figure 10. Difference of the regional model compared to the models from Experiment 1 for each basin regarding the NSE of the validation period. Blue colours (> 0) mean the regional model performed better than the models from Experiment 1, red (< 0) the other way around.

3.3 LSTMs as regional hydrological models

We now analyse the results of the four regional models that we trained for the four investigated HUCs in Experiment 2.

Figure 10 shows the difference in the NSE between the model outputs from Experiments 1 and 2. For some basins, the regional models perform significantly worse (dark red) than the individually trained models from Experiment 1. However, from the histograms of the differences we can see that the median is almost zero, meaning that in 50 % of the basins the regional model performs better than the model specifically trained for a single basin. Especially in the New England region the regional model performed better for almost all basins (except for two in the far north-east). In general, for all HUCs and catchments, the median difference is −0.001.

From Fig. 11 it is evident that the increased data size of the regional modelling approach (Experiment 2) helps to attenuate the drop in model performance between the calibration and validation periods, which could be observed in Experiment 1 probably as a result of overfitting. From the CDF of the NSE (Fig. 11a) we can see that Experiment 2 performed worse for approximately 20 % of the basins, while being comparable or even slightly better for the remaining watersheds. We can also observe that the regional models show a more balanced under- and over-estimation, while the models from Experiment 1 as well as the benchmark model tend to underestimate the discharge (see Fig. 11d–f, e.g. the flow variance, the top 2 % flow bias or the bias of the middle flows). This is not too surprising, since we train one model on a range of different basins with different discharge characteristics, where the model minimizes the error between simulated and observed discharge for all basins at the same time. On average, the regional model will therefore equally over- and under-estimate the observed discharge.

The comparison of the performances of Experiment 1 and 2 shows no clear consistent pattern for the investigated

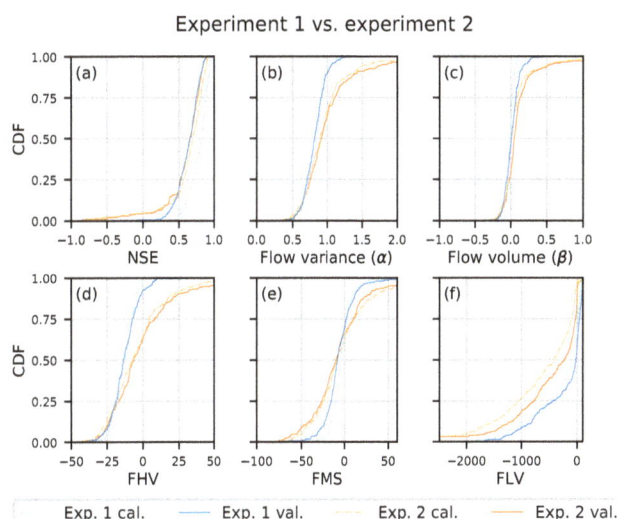

Figure 11. Cumulative density functions for several metrics of the calibration and validation period of the models from Experiment 1 compared to the regional models from Experiment 2. FHV is the bias of the top 2 % flows, the peak flows, FMS is the slope of the middle section of the flow duration curve and FLV is the bias of the bottom 30 % low flows.

HUCs, but reveals a trend toward higher NSE values in the New England region and to lower NSE values in the Arkansas-White-Red region. The reason for these differences might become clearer once we look at the correlation in the observed discharge time series of the basins within both HUCs (see Fig. 12). We can see that in the New England region (where the regional model performed better for most of the catchments compared to the individual models of Experiment 1) many basins have a strong correlation in their discharge time series. Conversely, for Arkansas-White-Red region the overall image of the correlation plot is much different. While some basins exist in the eastern part of the HUC

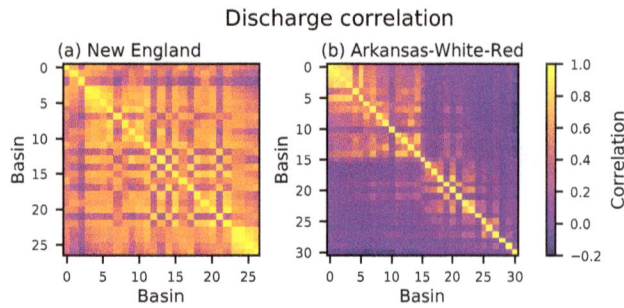

Figure 12. Correlation matrices of the observed discharge of all basins in **(a)** the New England region and **(b)** Arkansas-White-Red region. The basins for both subplots are ordered by longitude from east to west.

with discharge correlation, cspccially the basins in the western, more arid part have no inter-correlation at all. The results suggest that a single, regionally calibrated LSTM could generally be better in predicting the discharge of a group of basins compared to many LSTMs trained separately for each of the basins within the group especially when the group's basins exhibit a strong correlation in their discharge behaviour.

3.4   The effect of fine-tuning

In this section, we analyse the effect of fine-tuning the regional model for a few number of epochs to a specific catchment.

Figure 13 shows two effects of the fine-tuning process. In the comparison with the model performance of Experiment 1, and from the histogram of the differences (Fig. 13a), we see that in general the pre-training and fine-tuning improves the NSE of the runoff prediction. Comparing the results of Experiment 3 to the regional models of Experiment 2 (Fig. 13b), we can see the biggest improvement in those basins in which the regional models performed poorly (see also Fig. 10). It is worth highlighting that, even though the models in Experiment 3 have seen the data of their specific basins for fewer epochs in total than in Experiment 1, they still perform better on average. Therefore, it seems that pre-training with a bigger data set before fine-tuning for a specific catchment helps the model to learn general rainfall–runoff processes and that this knowledge is transferable to single basins. It is also worth noting that the group of catchments we used as one region (the HUC) can be quite inhomogeneous regarding their hydrological catchment properties.

Figure 14 finally shows that the models of Experiment 3 and the benchmark model perform comparably well over all catchments. The median of the NSE for the validation period is almost the same (0.72 and 0.71 for Experiment 3 and the benchmark model), while the mean for the models of Experiment 3 is about 15 % higher (0.68 compared to 0.58). In addition, more basins have an NSE above a threshold of 0.8

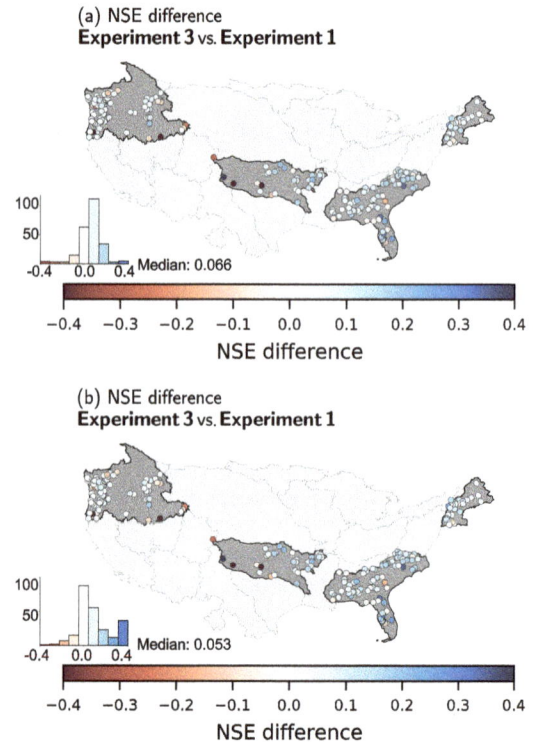

Figure 13. Panel **(a)** shows the difference of the NSE in the validation period of Experiment 3 compared to the models of Experiment 1 and panel **(b)** in comparison to the models of Experiment 2. Blue colours (> 0) indicate in both cases that the fine-tuned models of Experiment 3 perform better and red colours (< 0) the opposite. The NSE differences are capped at $[-0.4, 0.4]$ for better visualization.

(27.4 % of all basins compared to 17.4 % for the benchmark model), which is often taken as a threshold value for reasonably well-performing models (Newman et al., 2015).

3.5   A hydrological interpretation of the LSTM

To round off the discussion of this manuscript, we want to come back to the LSTM and try to explain it again in comparison to the functioning of a classical hydrological model. Similar to continuous hydrological models, the LSTM processes the input data time step after time step. In every time step, the input data (here meteorological forcing data) are used to update a number of values in the LSTM internal cell states. In comparison to traditional hydrological models, the cell states can be interpreted as storages that are often used for e.g. snow accumulation, soil water content, or groundwater storage. Updating the internal cell states (or storages) is regulated through a number of so-called gates: one that regulates the depletion of the storages, a second that regulates the increase in the storages and a third that regulates the outflow of the storages. Each of these gates comes with a set of adjustable parameters that are adapted during a calibration period (referred to as *training*). During the validation period,

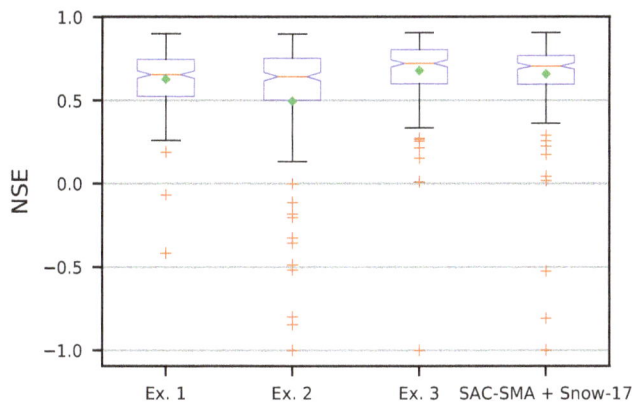

Figure 14. Boxplot of the NSE of the validation period for our three Experiments and the benchmark model. The NSE is capped to -1 for better visualization. The green square diamond marks the mean in addition to the median (red line).

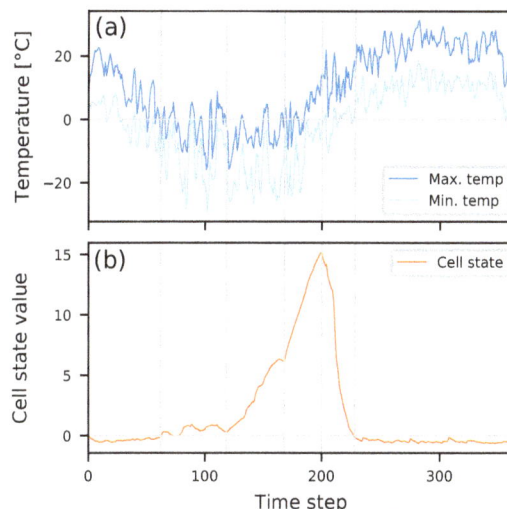

Figure 15. Evolution of a specific cell state in the LSTM (**b**) compared to the daily min and max temperature, with accumulation in winter and depletion in spring (**a**). The vertical grey lines are included for better guidance.

updates of the cell states depend only on the input at a specific time step and the states of the last time step (given the *learned* parameters of the calibration period).

In contrast to hydrological models, however, the LSTM does not "know" the principle of water/mass conservation and the governing process equations describing e.g. infiltration or evapotranspiration processes a priori. Compared to traditional hydrological models, the LSTM is optimized to predict the streamflow as well as possible, and has to learn these physical principles and laws during the calibration process purely from the data.

Finally, we want to show the results of a preliminary analysis in which we inspect the internals of the LSTM. Neural networks (as well as other data-driven approaches) are often criticized for their "black box"-like nature. However, here we want to argue that the internals of the LSTM can be inspected as well as interpreted, thus taking away some of the "black-box-ness".

Figure 15 shows the evolution of a single LSTM cell (c_t; see Sect. 2.1) of a trained LSTM over the period of one input sequence (which equals 365 days in this study) for an arbitrary, snow-influenced catchment. We can see that the cell state matches the dynamics of the temperature curves, as well as our understanding of snow accumulation and snowmelt. As soon as temperatures fall below 0 °C the cell state starts to increase (around time step 60) until the minimum temperature increases above the freezing point (around time step 200) and the cell state depletes quickly. Also, the fluctuations between time steps 60 and 120 match the fluctuations visible in the temperature around the freezing point. Thus, albeit the LSTM was only trained to predict runoff from meteorological observations, it has learned to model snow dynamics without any forcing to do so.

4   Summary and conclusion

This contribution investigated the potential of using Long Short-Term Memory networks (LSTMs) for simulating runoff from meteorological observations. LSTMs are a special type of recurrent neural networks with an internal memory that has the ability to learn and store long-term dependencies of the input–output relationship. Within three experiments, we explored possible applications of LSTMs and demonstrated that they are able to simulate the runoff with competitive performance compared to a baseline hydrological model (here the SAC-SMA + Snow-17 model). In the first experiment we looked at classical single basin modelling, in a second experiment we trained one model for all basins in each of the regions we investigated, and in a third experiment we showed that using a pre-trained model helps to increase the model performance in single basins. Additionally, we showed an illustrative example why traditional RNNs should be avoided in favour of LSTMs if the task is to predict runoff from meteorological observations.

The goal of this study was to explore the potential of the method and not to obtain the best possible realization of the LSTM model per catchment (see Sect. 2.5). It is therefore very likely that better performing LSTMs can be found by an exhaustive (catchment-wise) hyperparameter search. However, with our simple calibration approach, we were already able to obtain comparable (or even slightly higher) model performances compared to the well-established SAC-SMA + Snow-17 model.

In summary, the major findings of the present study are the following.

a. LSTMs are able to predict runoff from meteorological observations with accuracies comparable to the well-established SAC-SMA + Snow-17 model.

b. The 15 years of daily data used for calibration seem to constitute a lower bound of data requirements.

c. Pre-trained knowledge can be transferred into different catchments, which might be a possible approach for reducing the data demand and/or regionalization applications, as well as for prediction in ungauged basins or basins with few observations.

The data intensive nature of the LSTMs (as for any deep learning model) is a potential barrier for applying them in data-scarce problems (e.g. for the usage within a single basin with limited data). We do believe that the use of "pre-trained LSTMs" (as explored in Experiment 3) is a promising way to reduce the large data demand for an individual basin. However, further research is needed to verify this hypothesis. Ultimately, however, LSTMs will always strongly rely on the available data for calibration. Thus, even if less data are needed, it can be seen as a disadvantage in comparison to physically based models, which – at least in theory – are not reliant on calibration and can thus be applied with ease to new situations or catchments. However, more and more large-sample data sets are emerging which will catalyse future applications of LSTMs. In this context, it is also imaginable that adding physical catchment properties as an additional input layer into the LSTM may enhance the predictive power and ability of LSTMs to work as regional models and to make predictions in ungauged basins.

An entirely justifiable barrier of using LSTMs (or any other data-driven model) in real-world applications is their black-box nature. Like every common data-driven tool in hydrology, LSTMs have no explicit internal representation of the water balance. However, for the LSTM at least, it might be possible to analyse the behaviour of the cell states and link them to basic hydrological patterns (such as the snow accumulation melt processes), as we showed briefly in Sect. 3.5. We hypothesize that a systematic interpretation or the interpretability in general of the network internals would increase the trust in data-driven approaches, especially those of LSTMs, leading to their use in more (novel) applications in the environmental sciences in the near future.

Author contributions. FK and DK designed all experiments. FK conducted all experiments and analysed the results. FK prepared the paper with contributions from DK, CB, KS and MH.

Competing interests. The authors declare that they have no conflict of interest.

Acknowledgements. Part of the research was funded by the Austrian Science Fund (FWF) through project P31213-N29. Furthermore, we would like to thank the two anonymous reviewers for their comments that helped to improve this paper.

Edited by: Uwe Ehret

References

Abadi, M., Agarwal, A., Barham, P., Brevdo, E., Chen, Z., Citro, C., Corrado, G. S., Davis, A., Dean, J., Devin, M., Ghemawat, S., Goodfellow, I., Harp, A., Irving, G., Isard, M., Jia, Y., Jozefowicz, R., Kaiser, L., Kudlur, M., Levenberg, J., Mane, D., Monga, R., Moore, S., Murray, D., Olah, C., Schuster, M., Shlens, J., Steiner, B., Sutskever, I., Talwar, K., Tucker, P., Vanhoucke, V., Vasudevan, V., Viegas, F., Vinyals, O., Warden, P., Wattenberg, M., Wicke, M., Yu, Y., and Zheng, X.: TensorFlow: Large-Scale Machine Learning on Heterogeneous Systems, available at: https://www.tensorflow.org/ (last access: 21 November 2018), 2016.

Abrahart, R. J., Anctil, F., Coulibaly, P., Dawson, C. W., Mount, N. J., See, L. M., Shamseldin, A. Y., Solomatine, D. P., Toth, E., and Wilby, R. L.: Two decades of anarchy? Emerging themes and outstanding challenges for neural network river forecasting, Prog. Phys. Geog., 36, 480–513, 2012.

Adams, T. E. and Pagaon, T. C. (Eds.): Flood Forecasting: A Global Perspective, Academic Press, Boston, MA, USA, 2016.

Addor, N., Newman, A. J., Mizukami, N., and Clark, M. P.: The CAMELS data set: catchment attributes and meteorology for large-sample studies, Hydrol. Earth Syst. Sci., 21, 5293–5313, https://doi.org/10.5194/hess-21-5293-2017, 2017a.

Addor, N., Newman, A. J., Mizukami, N., and Clark, M. P.: Catchment attributes for large-sample studies, UCAR/NCAR, Boulder, CO, USA, https://doi.org/10.5065/D6G73C3Q, 2017b.

Anderson, E. A.: National Weather Service River Forecast System - Snow Accumulation and Ablation Model, Tech. Rep. November, US Department of Commerce, Silver Spring, USA, 1973.

ASCE Task Committee on Application of Artificial Neural Networks: Artificial Neural Networks in Hydrology. II: Hydrologic Applications, J. Hydrol. Eng., 52, 124–137, 2000.

Assem, H., Ghariba, S., Makrai, G., Johnston, P., Gill, L., and Pilla, F.: Urban Water Flow and Water Level Prediction Based on Deep Learning, in: ECML PKDD 2017: Machine Learning and Knowledge Discovery in Databases, 317–329, Springer, Cham., 2017.

Bengio, Y.: Practical recommendations for gradient-based training of deep architectures, in: Neural networks: Tricks of the trade, 437–478, Springer, Berlin, Heidelberg, 2012.

Bengio, Y., Simard, P., and Frasconi, P.: Learning long-term dependencies with gradient descent is difficult, IEEE T. Neural Networ., 5, 157–166, 1994.

Beven, K.: Linking parameters across scales: subgrid parameterizations and scale dependent hydrological models, Hydrol. Process., 9, 507–525, 1995.

Beven, K.: Rainfall-Runoff Modelling: The Primer, John Wiley & Sons, Chichester, UK, 2001.

Blöschl, G., Sivapalan, M., Wagener, T., Viglione, A., and Savenije, H. (Eds.): Runoff Prediction in Ungauged Basins: Synthesis Across Processes, Places and Scales, Cambridge University Press, UK, 465 pp., 2013.

Burnash, R. J. C., Ferral, R. L., and McGuire, R. A.: A generalised streamflow simulation system conceptual modelling for digital computers., Tech. rep., US Department of Commerce National Weather Service and State of California Department of Water Resources, Sacramento, CA, USA, 1973.

Buytaert, W. and Beven, K.: Regionalization as a learning process, Water Resour. Res., 45, 1–13, 2009.

Carriere, P., Mohaghegh, S., and Gaskar, R.: Performance of a Virtual Runoff Hydrographic System, Water Resources Planning and Management, 122, 120–125, 1996.

Chollet, F.: Keras, available at: https://github.com/fchollet/keras (last access: 1 April 2018), 2015.

Clark, M. P., Bierkens, M. F. P., Samaniego, L., Woods, R. A., Uijlenhoet, R., Bennett, K. E., Pauwels, V. R. N., Cai, X., Wood, A. W., and Peters-Lidard, C. D.: The evolution of process-based hydrologic models: historical challenges and the collective quest for physical realism, Hydrol. Earth Syst. Sci., 21, 3427–3440, https://doi.org/10.5194/hess-21-3427-2017, 2017.

Daniell, T. M.: Neural networks. Applications in hydrology and water resources engineering, in: Proceedings of the International Hydrology and Water Resource Symposium, vol. 3, 797–802, Institution of Engineers, Perth, Australia, 1991.

Duan, Q., Gupta, V. K., and Sorooshian, S.: Shuffled complex evolution approach for effective and efficient global minimization, J. Optimiz. Theory App., 76, 501–521, 1993.

Fang, K., Shen, C., Kifer, D., and Yang, X.: Prolongation of SMAP to Spatiotemporally Seamless Coverage of Continental U.S. Using a Deep Learning Neural Network, Geophys. Res. Lett., 44, 11030–11039, 2017.

Farabet, C., Couprie, C., Najman, L., and Lecun, Y.: Learning Hierarchical Features for Scence Labeling, IEEE T. Pattern Anal., 35, 1915–1929, 2013.

Freeze, R. A. and Harlan, R. L.: Blueprint for a physically-based, digitally-simulated hydrologic response model, J. Hydrol., 9, 237–258, 1969.

Gers, F. A., Schmidhuber, J., and Cummins, F.: Learning to Forget: Continual Prediction with LSTM, Neural Comput., 12, 2451–2471, 2000.

Goodfellow, I., Bengio, Y., and Courville, A.: Deep Learning, MIT Press, available at: http://www.deeplearningbook.org (last access: 1 April 2018), 2016.

Graves, A., Mohamed, A.-R., and Hinton, G.: Speech recognition with deep recurrent neural networks, in: Acoustics, speech and signal processing (ICASSP), 2013 IEEE International Conference on, 6645–6649, Vancouver, Canada, 2013.

Gupta, H. V., Kling, H., Yilmaz, K. K., and Martinez, G. F.: Decomposition of the mean squared error and NSE performance criteria: Implications for improving hydrological modelling, J. Hydrol., 377, 80–91, 2009.

Halevy, A., Norvig, P., and Pereira, F.: The Unreasonable Effectiveness of Data, IEEE Intell. Syst., 24, 8–12, 2009.

Halff, A. H., Halff, H. M., and Azmoodeh, M.: Predicting runoff from rainfall using neural networks, in: Engineering hydrology, ASCE, 760–765, 1993.

He, Y., Bárdossy, A., and Zehe, E.: A review of regionalisation for continuous streamflow simulation, Hydrol. Earth Syst. Sci., 15, 3539–3553, https://doi.org/10.5194/hess-15-3539-2011, 2011.

Hengl, T., Mendes de Jesus, J., Heuvelink, G. B. M., Ruiperez Gonzalez, M., Kilibarda, M., Blagotic, A., Shangguan, W., Wright, M. N., Geng, X., Bauer-Marschallinger, B., Guevara, M. A., Vargas, R., MacMillan, R. A., Batjes, N. H., Leenaars, J. G. B., Ribeiro, E., Wheeler, I., Mantel, S., and Kempen, B.: SoilGrids250m: Global gridded soil information based on machine learning, PLOS ONE, 12, 1–40, https://doi.org/10.1371/journal.pone.0169748, 2017.

Herrnegger, M., Senoner, T., and Nachtnebel, H. P.: Adjustment of spatio-temporal precipitation patterns in a high Alpine environment, J. Hydrol., 556, 913–921, 2018.

Hestness, J., Narang, S., Ardalani, N., Diamos, G. F., Jun, H., Kianinejad, H., Patwary, M. M. A., Yang, Y., and Zhou, Y.: Deep Learning Scaling is Predictable, Empirically, available at: https://arxiv.org/abs/1712.00409 (last access: 21 November 2018), 2017.

Hinton, G., Deng, L., Yu, D., Dahl, G. E., Mohamed, A.-R., Jaitly, N., Senior, A., Vanhoucke, V., Nguyen, P., Sainath, T. N., and Kingsbury, B.: Deep Neural Networks for Acoustic Modeling in Speech Recognition: The Shared Views of Four Research Groups, IEEE Signal Proc. Mag., 29, 82–97, 2012.

Hochreiter, S. and Schmidhuber, J.: Long Short-Term Memory, Neural Comput., 9, 1735–1780, 1997.

Hsu, K., Gupta, H. V., and Soroochian, S.: Application of a recurrent neural network to rainfall-runoff modeling, Proc., Aesthetics in the Constructed Environment, ASCE, New York, 68–73, 1997.

Hunter, J. D.: Matplotlib: A 2D graphics environment, Comput. Sci. Eng., 9, 90–95, 2007.

Kirchner, J. W.: Getting the right answers for the right reasons: Linking measurements, analyses, and models to advance the science of hydrology, Water Resour. Res., 42, 1–5, 2006.

Kollet, S. J., Maxwell, R. M., Woodward, C. S., Smith, S., Vanderborght, J., Vereecken, H., and Simmer, C.: Proof of concept of regional scale hydrologic simulations at hydrologic resolution utilizing massively parallel computer resources, Water Resour. Res., 46, 1–7, 2010.

Krizhevsky, A., Sutskever, I., and Hinton, G. E.: ImageNet Classification with Deep Convolutional Neural Networks, Adv. Neur. In., 1097–1105, 2012.

Kumar, D. N., Raju, K. S., and Sathish, T.: River Flow Forecasting using Recurrent NeuralNetworks, Water Resour. Manag., 18, 143–161, 2004.

LeCun, Y. A., Bottou, L., Orr, G. B., and Müller, K. R.: Efficient backprop, Springer, Berlin, Heidelberg, Germany, 2012.

Lindström, G., Pers, C., Rosberg, J., Strömqvist, J., and Arheimer, B.: Development and testing of the HYPE (Hydrological Predictions for the Environment) water quality model for different spatial scales, Hydrol. Res., 41, 295–319, 2010.

Marçais, J. and de Dreuzy, J. R.: Prospective Interest of Deep Learning for Hydrological Inference, Groundwater, 55, 688–692, 2017.

Maurer, E. P., Wood, A. W., Adam, J. C., Lettenmaier, D. P., and Nijssen, B.: A long-term hydrologically based dataset of land surface fluxes and states for the conterminous United States, J. Climate, 15, 3237–3251, 2002.

McKinney, W.: Data Structures for Statistical Computing in Python, Proceedings of the 9th Python in Science Conference, Location: Austin, Texas, USA, 1697900, 51–56, available at: http://conference.scipy.org/proceedings/scipy2010/mckinney.html (last access: 1 April 2018), 2010.

Merz, R., Blöschl, G., and Parajka, J.: Regionalisation methods in rainfall-runoff modelling using large samples, Large Sample Basin Experiments for Hydrological Model Parameterization: Results of the Model Parameter Experiment–MOPEX, IAHS Publ., 307, 117–125, 2006.

Minns, A. W. and Hall, M. J.: Artificial neural networks as rainfall-runoff models, Hydrolog. Sci. J., 41, 399–417, 1996.

Mu, Q., Zhao, M., and Running, S. W.: Improvements to a MODIS global terrestrial evapotranspiration algorithm, Remote Sens. Environ., 115, 1781–1800, 2011.

Mulvaney, T. J.: On the use of self-registering rain and flood gauges in making observations of the relations of rainfall and of flood discharges in a given catchment, in: Proceedings Institution of Civil Engineers, Dublin, Vol. 4, 18–31, 1850.

Myneni, R. B., Hoffman, S., Knyazikhin, Y., Privette, J. L., Glassy, J., Tian, Y., Wang, Y., Song, X., Zhang, Y., Smith, G. R., Lotsch, A., Friedl, M., Morisette, J. T., Votava, P., Nemani, R. R., and Running, S. W.: Global products of vegetation leaf area and fraction absorbed PAR from year one of MODIS data, Remote Sens. Environ., 83, 214–231, 2002.

Nash, J. E. and Sutcliffe, J. V.: River Flow Forecasting Through Conceptual Models Part I-a Discussion of Principles, J. Hydrol., 10, 282–290, 1970.

Newman, A., Sampson, K., Clark, M., Bock, A., Viger, R., and Blodgett, D.: A large-sample watershed-scale hydrometeorological dataset for the contiguous USA, UCAR/NCAR, Boulder, CO, USA, https://doi.org/10.5065/D6MW2F4D, 2014.

Newman, A. J., Clark, M. P., Sampson, K., Wood, A., Hay, L. E., Bock, A., Viger, R. J., Blodgett, D., Brekke, L., Arnold, J. R., Hopson, T., and Duan, Q.: Development of a large-sample watershed-scale hydrometeorological data set for the contiguous USA: data set characteristics and assessment of regional variability in hydrologic model performance, Hydrol. Earth Syst. Sci., 19, 209–223, https://doi.org/10.5194/hess-19-209-2015, 2015.

Pedregosa, F., Varoquaux, G., Gramfort, A., Michel, V., Thirion, B., Grisel, O., Blondel, M., Prettenhofer, P., Weiss, R., Dubourg, V., Vanderplas, J., Passos, A., Cournapeau, D., Brucher, M., Perrot, M., and Duchesnay, E.: Scikit-learn: Machine Learning in Python, J. Mach. Learn. Res., 12, 2825–2830, 2011.

Razavian, A. S., Azizpour, H., Sullivan, J., and Carlsson, S.: CNN features off-the-shelf: An astounding baseline for recognition, IEEE Computer Society Conference on Computer Vision and Pattern Recognition Workshops, 24–28 June 2014, Columbus, Ohio, USA, 512–519, 2014.

Remesan, R. and Mathew, J.: Hydrological data driven modelling: a case study approach, vol. 1, Springer International Publishing, 2014.

Rennó, C. D., Nobre, A. D., Cuartas, L. A., Soares, J. V., Hodnett, M. G., Tomasella, J., and Waterloo, M. J.: HAND, a new terrain descriptor using SRTM-DEM: Mapping terra-firme rainforest environments in Amazonia, Remote Sens. Environ., 112, 3469–3481, 2008.

Rumelhart, D. E., Hinton, G. E., and Williams, R. J.: Learning internal representations by error propagation (No. ICS-8506), California Univ San Diego La Jolla Inst for Cognitive Science, 1985.

Samaniego, L., Kumar, R., and Attinger, S.: Multiscale parameter regionalization of a grid-based hydrologic model at the mesoscale, Water Resour. Res., 46, 1–25, 2010.

Schmidhuber, J.: Deep learning in neural networks: An overview, Neural Networks, 61, 85–117, 2015.

Schulla, J.: Model description WaSiM (Water balance Simulation Model), completely revised version 2012, last change: 19 June 2012, available at: http://www.wasim.ch/downloads/doku/wasim/wasim_2012_ed2_en.pdf (last access: 1 April 2018), 2017.

Seaber, P. R., Kapinos, F. P., and Knapp, G. L.: Hydrologic Unit Maps, Tech. rep., U.S. Geological Survey, Water Supply Paper 2294, Reston, Virginia, USA, 1987.

Shen, C.: A transdisciplinary review of deep learning research and its relevance for water resources scientists, Water Resour. Res., 54, https://doi.org/10.1029/2018WR022643, 2018.

Shen, C., Laloy, E., Elshorbagy, A., Albert, A., Bales, J., Chang, F.-J., Ganguly, S., Hsu, K.-L., Kifer, D., Fang, Z., Fang, K., Li, D., Li, X., and Tsai, W.-P.: HESS Opinions: Incubating deep-learning-powered hydrologic science advances as a community, Hydrol. Earth Syst. Sci., 22, 5639–5656, https://doi.org/10.5194/hess-22-5639-2018, 2018.

Shi, X., Chen, Z., Wang, H., Yeung, D.-Y., Wong, W.-K., and Woo, W.-C.: Convolutional LSTM network: A machine learning approach for precipitation nowcasting, Adv. Neur. In., 28, 802–810, 2015.

Sivapalan, M.: Prediction in ungauged basins: a grand challenge for theoretical hydrology, Hydrol. Process., 17, 3163–3170, 2003.

Solomatine, D., See, L. M., and Abrahart, R. J.: Data-driven modelling: concepts, approaches and experiences, in: Practical hydroinformatics, 17–30, Springer, Berlin, Heidelberg, 2009.

Srivastava, N., Hinton, G., Krizhevsky, A., Sutskever, I., and Salakhutdinov, R.: Dropout: A Simple Way to Prevent Neural Networks from Overfitting, J. Mach. Learn. Res., 15, 1929–1958, 2014.

Stanzel, P., Kahl, B., Haberl, U., Herrnegger, M., and Nachtnebel, H. P.: Continuous hydrological modelling in the context of real time flood forecasting in alpine Danube tributary catchments, IOP C. Ser. Earth Env., 4, 012005, https://doi.org/10.1088/1755-1307/4/1/012005, 2008.

Sutskever, I., Vinyals, O., and Le, Q. V.: Sequence to sequence learning with neural networks, in: Advances in neural information processing systems, 3104–3112, 2014.

Tao, Y., Gao, X., Hsu, K., Sorooshian, S., and Ihler, A.: A Deep Neural Network Modeling Framework to Reduce Bias in Satellite Precipitation Products, J. Hydrometeorol., 17, 931–945, 2016.

Thielen, J., Bartholmes, J., Ramos, M.-H., and de Roo, A.: The European Flood Alert System – Part 1: Concept and development, Hydrol. Earth Syst. Sci., 13, 125–140, https://doi.org/10.5194/hess-13-125-2009, 2009.

Thornton, P. E., Thornton, M. M., Mayer, B. W., Wilhelmi, N., Wei, Y., Devarakonda, R., and Cook, R.: Daymet: Daily surface weather on a 1 km grid for North America, 1980–2008, Oak

Ridge National Laboratory (ORNL) Distributed Active Archive Center for Biogeochemical Dynamics (DAAC), Oak Ridge, Tennessee, USA, 2012.

Tompson, J., Jain, A., LeCun, Y., and Bregler, C.: Joint Training of a Convolutional Network and a Graphical Model for Human Pose Estimation, in: Proceedings of Advances in Neural Information Processing Systems, 27, 1799–1807, 2014.

Van Der Walt, S., Colbert, S. C., and Varoquaux, G.: The NumPy array: A structure for efficient numerical computation, Comput. Sci. Eng., 13, 22–30, 2011.

van Rossum, G.: Python tutorial, Technical Report CS-R9526, Tech. rep., Centrum voor Wiskunde en Informatica (CWI), Amsterdam, the Netherlands, 1995.

Wesemann, J., Herrnegger, M., and Schulz, K.: Hydrological modelling in the anthroposphere: predicting local runoff in a heavily modified high-alpine catchment, J. Mt. Sci., 15, 921–938, 2018.

Wood, E. F., Roundy, J. K., Troy, T. J., van Beek, L. P. H., Bierkens, M. F. P., Blyth, E., de Roo, A., Döll, P., Ek, M., Famiglietti, J., Gochis, D., van de Giesen, N., Houser, P., Jaffé, P. R., Kollet, S., Lehner, B., Lettenmaier, D. P., Peters-Lidard, C., Sivapalan, M., Sheffield, J., Wade, A., and Whitehead, P.: Hyperresolution global land surface modeling: Meeting a grand challenge for monitoring Earth's terrestrial water, Water Resour. Res., 47, W05301, https://doi.org/10.1029/2010WR010090, 2011.

Xia, Y., Mitchell, K., Ek, M., Sheffield, J., Cosgrove, B., Wood, E., Luo, L., Alonge, C., Wei, H., Meng, J., and Livneh, B.: Continental-scale water and energy flux analysis and validation for the North American Land Data Assimilation System project phase 2 (NLDAS-2): 1. Intercomparison and application of model products, J. Geophys. Res.-Atmos., 117, D03109, https://doi.org/10.1029/2011JD016048, 2012.

Yilmaz, K. K., Gupta, H. V., and Wagener, T.: A process-based diagnostic approach to model evaluation: Application to the NWS distributed hydrologic model, Water Resour. Res., 44, 1–18, 2008.

Yosinski, J., Clune, J., Bengio, Y., and Lipson, H.: How transferable are features in deep neural networks?, Adv. Neur. In., 27, 1–9, 2014.

Young, P. C. and Beven, K. J.: Data-based mechanistic modelling and the rainfall-flow non-linearity, Environmetrics, 5, 335–363, 1994.

Zhang, D., Lindholm, G., and Ratnaweera, H.: Use long short-term memory to enhance Internet of Things for combined sewer overflow monitoring, J. Hydrol., 556, 409–418, 2018.

Zhang, J., Zhu, Y., Zhang, X., Ye, M., and Yang, J.: Developing a Long Short-Term Memory (LSTM) based model for predicting water table depth in agricultural areas, J. Hydrol., 561, 918–929, 2018.

Zhu, M. and Fujita, M.: Application of neural networks to runoff forecast, vol. 3, Springer, Dodrecht, the Netherlands, 1993.

Assessment of actual evapotranspiration over a semiarid heterogeneous land surface by means of coupled low-resolution remote sensing data with an energy balance model: comparison to extra-large aperture scintillometer measurements

Sameh Saadi[1,2], Gilles Boulet[1], Malik Bahir[1], Aurore Brut[1], Émilie Delogu[1], Pascal Fanise[1], Bernard Mougenot[1], Vincent Simonneaux[1], and Zohra Lili Chabaane[2]

[1]Centre d'Etudes Spatiales de la Biosphère, Université de Toulouse, CNRS, CNES, IRD, UPS, Toulouse, France
[2]Université de Carthage/Institut National Agronomique de Tunisie/ LR17AGR01-GREEN-TEAM, Tunis, Tunisia

Correspondence: Sameh Saadi (saadi_sameh@hotmail.fr)

Abstract. In semiarid areas, agricultural production is restricted by water availability; hence, efficient agricultural water management is a major issue. The design of tools providing regional estimates of evapotranspiration (ET), one of the most relevant water balance fluxes, may help the sustainable management of water resources.

Remote sensing provides periodic data about actual vegetation temporal dynamics (through the normalized difference vegetation index, NDVI) and water availability under water stress (through the surface temperature T_{surf}), which are crucial factors controlling ET.

In this study, spatially distributed estimates of ET (or its energy equivalent, the latent heat flux LE) in the Kairouan plain (central Tunisia) were computed by applying the Soil Plant Atmosphere and Remote Sensing Evapotranspiration (SPARSE) model fed by low-resolution remote sensing data (Terra and Aqua MODIS). The work's goal was to assess the operational use of the SPARSE model and the accuracy of the modeled (i) sensible heat flux (H) and (ii) daily ET over a heterogeneous semiarid landscape with complex land cover (i.e., trees, winter cereals, summer vegetables).

SPARSE was run to compute instantaneous estimates of H and LE fluxes at the satellite overpass times. The good correspondence ($R^2 = 0.60$ and 0.63 and RMSE $= 57.89$ and $53.85\,\mathrm{W\,m^{-2}}$ for Terra and Aqua, respectively) between instantaneous H estimates and large aperture scintillometer (XLAS) H measurements along a path length of 4 km over the study area showed that the SPARSE model presents satisfactory accuracy. Results showed that, despite the fairly large scatter, the instantaneous LE can be suitably estimated at large scales (RMSE $= 47.20$ and $43.20\,\mathrm{W\,m^{-2}}$ for Terra and Aqua, respectively, and $R^2 = 0.55$ for both satellites). Additionally, water stress was investigated by comparing modeled (SPARSE) and observed (XLAS) water stress values; we found that most points were located within a 0.2 confidence interval, thus the general tendencies are well reproduced. Even though extrapolation of instantaneous latent heat flux values to daily totals was less obvious, daily ET estimates are deemed acceptable.

1 Introduction

In water-scarce regions, especially arid and semiarid areas, the sustainable use of water by resource conservation as well as the use of appropriate technologies to do so is a priority for agriculture (Amri et al., 2014; Pereira et al., 2002).

Water use rationalization is especially needed for countries actually suffering from water scarcity, or for countries that probably would suffer from water restrictions according to climate change scenarios. Indeed, the Mediterranean region is one of the most prominent "hot spots" in future climate change projections (Giorgi and Lionello, 2008) due to an expected larger warming than the global average and to a pronounced increase in precipitation interannual variability. The major part of the southern Mediterranean countries, among others Tunisia, already suffer from water scarcity and show a growing water deficit, due to the combined effect of the growth in water needs (soaring demography and irrigated areas extension) and the reduction of resources (temporary drought and/or climate change). This implies that closely monitoring the water budget components is a major issue (Oki and Kanae, 2006).

The estimation of evapotranspiration (ET) is of paramount importance since it represents the preponderant component of the terrestrial water balance; it is the second largest component after precipitation (Glenn et al., 2007); hence ET quantification is a key factor for scarce water resources management. Direct measurement of ET is only possible at local scale (single field) using the eddy covariance method, for example, whereas it is much more difficult at larger scales (irrigated perimeter or watershed) due to the complexity not only of the hydrological processes (Minacapilli et al., 2007) but also of the hydrometeorological processes. Indeed, at landscape scale, surface heterogeneity influences regional and local climate, inducing for example cloudiness, precipitation and temperature pattern differences between areas of higher elevation (hills and mountains surrounding the Kairouan plain) and the plain downstream. Moreover, at these scales, land cover is usually heterogeneous and this affects the land–atmosphere exchanges of heat, water and other constituents (Giorgi and Avissar, 1997). ET estimates for various temporal and spatial scales, from hourly to monthly to seasonal time steps, and from field to global scales, are required for hydrologic applications in water resource management (Anderson et al., 2011). Techniques using remote sensing (RS) information are therefore essential when dealing with processes that cannot be represented by point measurements only (Su, 2002).

In fact, the contribution of RS in vegetation's physical characteristics monitoring in large areas was identified years ago (Tucker, 1978); RS provides periodic data about some major ET drivers, amongst others surface temperature and vegetation properties (e.g., normalized difference vegetation index NDVI and leaf area index LAI) from field to regional scales (Li et al., 2009; Mauser and Schädlich, 1998). Many methods using remotely sensed data to estimate ET are reviewed in Courault et al. (2005). ICARE (Gentine et al., 2007) and SiSPAT (Braud et al., 1995) are examples of complex physically based land surface models (LSMs) using RS data. They include a detailed description of the vegetation water uptake in the root zone, the interactions among groundwater, root zone and surface water. However, the lateral surface and subsurface flows are neglected. This can lead to inaccurate results when applied in areas where such interactions are important (Overgaard et al., 2006).

Moreover, RS can provide estimates of large area fluxes in remote locations, although those estimates are based on the spatial and temporal scales of the measuring systems and thus vary one from another. Hence, one solution is to up-

scale local micrometeorological measurements to larger spatial scales in order to acquire an optimum representation of land–atmosphere interactions (Samain et al., 2012). However, such an upscaling process is not always possible and results might not be reliable in comparison to the RS distributed products.

Water and energy exchange in the soil–plant–atmosphere continuum have been simulated through several land surface models (Bastiaanssen et al., 2007; Feddes et al., 1978). Among them, two different approaches use remote sensing data to estimate spatially distributed ET (Minacapilli et al., 2009): one that is based on the soil water balance (SWB) and one that solves the surface energy budget (SEB). The SWB approach exploits only visible near-infrared (VIS-NIR) observations to perceive the spatial variability of crop parameters. The SEB modeling approach uses visible (VIS), near-infrared (NIR) and thermal infrared (TIR) data to solve the SEB equation by forcing remotely sensed estimates of the SEB components (mainly the surface temperature T_{surf}). In fact, there is a strong link between water availability in the soil and surface temperature under water stress; hence, in order to estimate soil moisture status as well as actual ET at relevant space and timescales, information in the TIR domain (8–14 μm) is frequently used (Boulet et al., 2007). The SWB approach has the advantage of high resolution and high-frequency VIS-NIR remote sensing data availability against limited availability of high-resolution thermal imagery for the SEB approach. Indeed, satellite data such as Landsat or Advanced Spaceborne Thermal Emission and Reflection Radiometer (ASTER) data provide field-scale (30–100 m) estimates of ET (Allen et al., 2011), but they have a low temporal resolution (16 days to monthly) (Anderson et al., 2011).

The RS-based SWB models provide estimates of ET, soil water content and irrigation requirements in a continuous way. For instance, at field scale, estimates of seasonal ET and irrigation can be obtained by SWB modeling using high-resolution remote sensing forcing as done in the study with the SAtellite Monitoring of IRrigation (SAMIR) model by Saadi et al. (2015) over the Kairouan plain. However, for an appropriate estimation of ET, the SWB model requires knowledge of the water inputs (precipitation and irrigation) and an assessment of the extractable water from the soil (mostly derived from the soil moisture characteristics: actual available water content in the root zone, wilting point and field capacity), whereas significant biases are found mainly when dealing with large areas and long periods, due to the spatial variability of the water input uncertainties as well as the inaccuracy in estimating other flux components such as the deep drainage (Calera et al., 2017). Hence, the major limitation of the SWB method is the high number of needed inputs whose estimation is highly uncertain, especially over a heterogeneous land surface due to hydrologic processes complexity. Moreover, spatially distributed SWB models, typically those using the Food and Agriculture Organization-

FAO guidelines (Allen et al., 1998) for crop ET estimation, generally parameterize the vegetation characteristics on the basis of land use maps (Bounoua et al., 2015; Xie et al., 2008), and different parameters are used for different land use classes. Nevertheless, SWB modelers generally do not have the possibility to carry out RS-based land use change mapping due to time, budget or capacity constraints and often use very generic classes, potentially leading to modeling errors (Hunink et al., 2017). In addition, the lack of data about the soil properties (controlling field capacity, wilting point and the water retention) as well as the actual root depths lead to limited practical use of the SWB models (Calera et al., 2017). The same applies to the soil evaporation whose estimation generally relies on the FAO guidelines approach (Allen et al., 1998). However, it was shown that under high evaporation conditions, the FAO-56 (Allen et al., 1998) daily evaporation computed on the basis of the readily evaporable water (REW) is overestimated at the beginning of the dry down phase (i.e., the period after rain or irrigation where the soil moisture is decreasing due to evapotranspiration and drainage, Mutziger et al., 2005; Torres and Calera, 2010). Hence, to improve its estimation a reduction factor proposed by Torres and Calera (2010) was applied to deal with this problem in several studies (e.g., Odi-Lara et al., 2016; Saadi et al., 2015). Furthermore, SWB models such as SWAP (Kroes et al., 2017, Cropsyst (Stöckle et al., 2003), AquaCrop (Steduto et al., 2009) and SAMIR (Simonneaux et al., 2009) are able to take irrigation into account, either as an estimated amount provided by the farmer (as an input if available) or a predicted amount through a module triggering irrigation according to, say, critical soil moisture levels (as an output). However, the limited knowledge of the actual irrigation scheduling is a critical limitation for the validation protocol of irrigation requirement estimates by SWB modeling. Therefore, SWB modelers must deal with the lack of information about real irrigation, which induces unreliable estimations.

Consequently, ET estimation at regional scale is often achieved using SEB approaches, by combining surface temperature from medium- to low-resolution (kilometer scale) remote sensing data with vegetation parameters and meteorological variables (Liou and Kar, 2014). Recently, many efforts have been made to feed remotely sensed surface temperature into ET modeling platforms in combination with other critical variables, e.g., NDVI and albedo (Kalma et al., 2008; Kustas and Anderson, 2009). A wide range of satellite-based ET models were developed, and these methods are reviewed in Liou and Kar (2014). The majority of SEB-based models are single-source models; their algorithms compute a total latent heat flux as the sum of the evaporation and the transpiration components using a remotely sensed surface temperature. However, separate estimates of evaporation and transpiration make the dual-source models more useful for agrohydrological applications (water stress detection, irrigation monitoring, etc.) (Boulet et al., 2015).

Contrarily to SWB models, most SEB models are run in their most standardized version, using observed RS-based parameters such as albedo in conjunction with a set of input parameters taken from literature or in situ data. However, the SEB model validation with enough data in space and time is difficult to achieve, due to the limited availability of high-resolution thermal images (Chirouze et al., 2014). Therefore, it is usually possible to evaluate SEB models results only at similar scale (km) to medium- or low-resolution images. Indeed, the pixel size of thermal remote sensing images, except for the scarce Landsat7 images (60 m), covers a range of 1000 m (Moderate Sensors Resolution Imaging Spectroradiometer MODIS), to 4000 m (Geostationary Operational Environmental Satellite GEOS). However, direct methods measuring sensible heat fluxes (eddy covariance for example) only provide point measurements with a footprint considerably smaller than a satellite pixel. Therefore, scintillometry techniques have emerged as one of the best tools aiming to quantify averaged fluxes over heterogeneous land surfaces (Brunsell et al., 2011). They provide area-averaged sensible heat flux over areas comparable to those observed by satellites (Hemakumara et al., 2003; Lagouarde et al., 2002). Scintillometry can provide sensible heat using different wavelengths (optical wavelength ranges), aperture sizes (15–30 cm) and configurations (long-path and short-path scintillometry) (Meijninger et al., 2002). The upwind area contributing to the flux (i.e., the flux footprint) varies as wind direction and atmospheric stability and must be estimated for the surface measurements in order to compare them to SEB estimates of the flux which are representative of the pixel (Brunsell et al., 2011). Assessing the upwind area contributing to the flux can be done using several footprint models (Schmid, 2002). Although footprint analysis ensures ad hoc spatial intersecting area between ground measurements and satellite-based surface fluxes, the spatial heterogeneity at subpixel scale should be further considered in validating low-resolution satellite data (Bai et al., 2015). The LAS technique has been validated over heterogeneous landscapes against eddy covariance measurements (Bai et al., 2009; Chehbouni et al., 2000; Hoedjes et al., 2009) and also against modeled fluxes (Marx et al., 2008; Samain et al., 2012; Watts et al., 2000). Few studies dealt with extra-large aperture scintillometer (XLAS) data (Kohsiek et al., 2002, 2006; Moene et al., 2006). Historical survey, theoretical background and recent works in applied research concerning scintillometry are reviewed in De Bruin and Wang (2017). Since the scintillometer provides large-scale area-average sensible heat flux (H_XLAS), the corresponding latent heat flux (LE_XLAS) can then be computed as the energy balance residual term (LE_XLAS = $R_n - G -$ H_XLAS); hence, the estimation of a representative value for the available energy (AE = $R_n - G$) is always crucial for the accuracy of the retrieved values of LE_XLAS. This assumption is valid only under the similarity hypothesis of Monin–Obukhov (MO) (Monin and Obukhov, 1954), i.e., surface homogeneity and

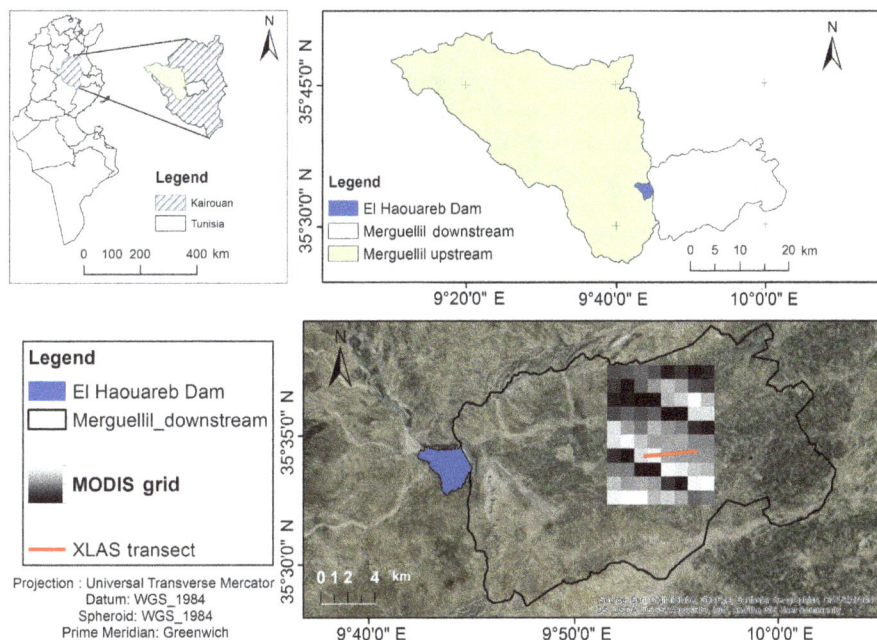

Figure 1. The study area: the downstream Merguellil subbasin is the so-called Kairouan plain; the MODIS grid is the extracted $10\,km \times 8\,km$ MODIS subimage and in red the scintillometer XLAS transect.

stationary flows. These hypotheses are verified in our study area where the topography is flat, and the landscape is heterogeneous only from an agronomic point of view since we find different land uses (cereals, market gardening and fruit trees, mainly olive trees with considerable spacing of bare soil); however, this heterogeneity in landscape features at field scale is randomly distributed and there is no drastic change in height and density of the vegetation at the scale of the XLAS transect (i.e., little heterogeneity at the kilometer scale, most MODIS pixels have similar NDVI values for instance).

In this study, spatially distributed estimates of surface energy fluxes (sensible heat H and latent heat fluxes LE) over an irrigated area located in the Kairouan plain (central Tunisia) were obtained by the SEB method, using the Soil Plant Atmosphere and Remote Sensing Evapotranspiration (SPARSE) model (Boulet et al., 2015) fed by 1 km thermal data and 1 km NDVI data from MODIS sensors on the Terra and Aqua satellites. The main objective of this paper is to compare the modeled H and LE simulated by the SPARSE model with, respectively, the H measured by the XLAS and the LE reconstructed from the XLAS measurements acquired in the course of 2 years over a large, heterogeneous area. We explore the consistency between the instantaneous H and LE estimates at the satellite overpass times, the water stress estimates and also ET derived at daily time steps from both approaches.

2 Experimental site and datasets

2.1 Study area

The study site is a semiarid region located in central Tunisia, the Kairouan plain ($35°1'-35°55'$ N, $9°23'-10°17'$ E; Fig. 1). The landscape is mainly flat, and the vegetation is dominated by agricultural production (cereals, olive groves, fruit trees, market gardening; Zribi et al., 2011). Water management in the study area is typical of semiarid regions with an upstream subcatchment that transfers surface and subsurface flows collected by a dam (the El Haouareb dam) and a downstream plain (Kairouan plain) supporting irrigated agriculture (Fig. 1). Agriculture consumes more than 80% of the total amount of water extracted each year from the Kairouan aquifer (Poussin et al., 2008). Most farmers in the plain use their own wells to extract water for irrigation (Pradeleix et al., 2015), while a few depends on public irrigation schemes based on collective networks of water distribution pipelines all linked to a main borehole. The crop intensification in recent decades, associated with increasing irrigation, has led to growing water demand, and an over-exploitation of the groundwater (Leduc et al., 2004).

2.2 Experimental setup and remote sensing data

An optical Kipp & Zonen extra-large aperture scintillometer (XLAS) was operated continuously for more than 2 years (1 March 2013 to 3 June 2015) over a relatively flat terrain (maximum difference in elevation of about 18 m). The scintillometer consists of a transmitter and a receiver, both

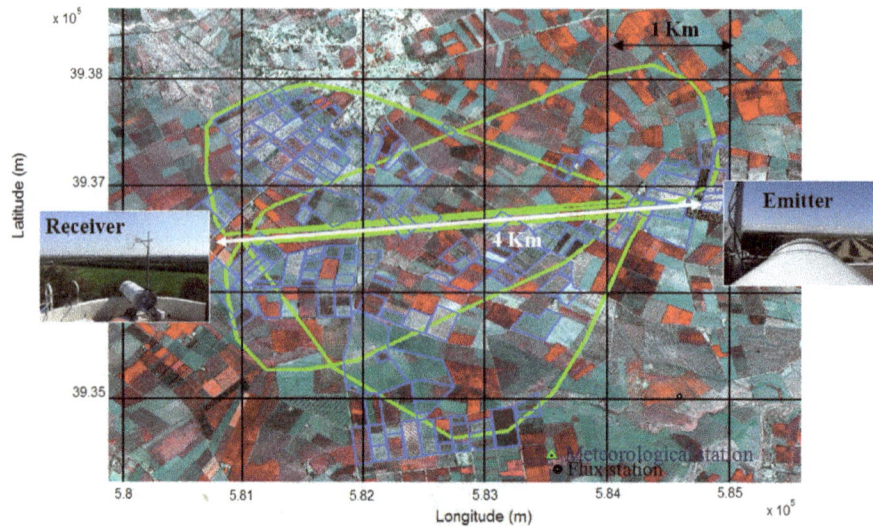

Figure 2. XLAS setup: XLAS transect (white), for which the emitter and the receiver are located at the extremity of each white arrow; half-hourly XLAS footprint for selected typical wind conditions (green); MODIS grid (black); orchards (blue); and the location of the Ben Salem meteorological and flux stations. Background is a three-color (red, green, blue) composite of SPOT5 bands 3 (NIR), 2 (VIS-red) and 1 (VIS-green) acquired on 9 April 2013 and showing in red the cereal plots.

with an aperture diameter of 0.3 m, which allows longer path length. The wavelength of the light beam emitted by the transmitter is 940 nm. The transmitter was located on an eastern water tower (coordinates: 35°34′0.7″ N, 9°53′25.19″ E; 127 m a.s.l. – above sea level) and the receiver on a western water tower (coordinates: 35°34′17.22″ N, 9°56′7.30″ E; 145 m a.s.l.) separated by a path length of 4 km (Fig. 2).

The scintillometer transect was above a mixed-vegetation canopy: trees (mainly olive orchards) with some annual crops (cereals and market gardening) and the mean vegetation height is estimated to be about 1.17 m along the transect. Both instruments were installed at 20 m height as recommended in the Kipp & Zonen instruction manual for LAS and XLAS (Kipp & Zonen, 2017). At this height and for a 4 km path length, the devices are high enough to minimize measurement saturation and assumed to be above or close to the blending height where MO applied.

Furthermore, two automatic Campbell Scientific (Logan, USA) eddy covariance (EC) flux stations were also positioned at the same level on the two water-tower-top platforms. Half-hourly turbulent fluxes in the western and the eastern EC stations were measured using a sonic anemometer CSAT3 (Campbell Scientific, USA) at a rate of 20 Hz and a sonic anemometer RM 81000 (Young, USA) at a rate of 10 Hz, respectively. The western station data were more reliable with less measurement errors and gaps; hence, the western EC setup was used to initialize friction velocity u^* values and the Obukhov length L_o in the scintillometer flux computation (Sect. 3.1).

Half-hourly standard meteorological measurements including incoming long wave radiation, i.e., global incoming radiation (R_{g30}), the incoming longwave radiation (i.e., atmospheric radiation, R_{atm-30}), wind speed (u_{30}), wind direction (u_{d30}), air temperature (T_{30}), relative humidity (RH_{30}) and barometric pressure (P_{30}) were recorded using an automated weather station installed in the study area (Fig. 2), referred to as the Ben Salem meteorological station (35°33′1.44″ N, 9°55′18.11″ E). Meteorological data were used either to force the SPARSE model or as input data in XLAS-derived sensible and latent heat flux. The global incoming radiation was also used in the extrapolation method to scale instantaneous observed (Sect. 3.3.2) and modeled (Sect. 4.2) available energy as well as modeled sensible heat flux (Sect. 4.2) to daily values.

In addition, an EC flux station, referred to as the Ben Salem flux station (a few tens of meters away from the meteorological station) was installed from November 2012 to June 2013 in an irrigated wheat field (Fig. 2) to measure half-hourly convective fluxes exchanged between the surface and the atmosphere (H_{BS-30} and LE_{BS-30}) combined with measurements of the net radiation R_{nBS-30} and the soil heat flux G_{BS-30}. Net radiation and soil heat flux measurements were transferred to the meteorological station from June 2013 till June 2015. Since there are no R_n and G measurements in the two water towers EC stations, R_{nBS} and G_{BS} measurements were among the input data to derive sensible and latent heat fluxes from the XLAS measurements. In addition, measured available energy ($AE_{BS} = R_{nBS} - G_{BS}$) and H_{BS} were used to calibrate the extrapolation relationship of the available energy and the sensible heat flux, respectively (Sects. 3.3.2 and 4.2).

Remotely sensed data were acquired for the study period (1 September 2012 to 30 June 2015) at the resolution of the MODIS sensor at 1 km, embarked on board of the satellites Terra (overpass time around 10:30 local solar time) and Aqua (overpass time around 13:30 local solar time). Downloaded MODIS products were (i) MOD11A1 and MYD11A1 for Terra and Aqua, respectively (surface temperature T_{surf}, surface emissivity ε_{surf} and viewing angle ϕ); (ii) MOD13A2 and MYD13A2 for Terra and Aqua, respectively (NDVI); and (iii) MCD43B1, MCD43B2 and MCD43B3 (albedo α). These MODIS data provided in sinusoidal projection were reprojected in the Universal Transverse Mercator coordinate system using the MODIS Reprojection Tool. Then, subimages of 10 km × 8 km centered on the XLAS transect (Fig. 1) were extracted. The daily MODIS T_{surf} and viewing angle, 8-day MODIS albedo, and 16-day MODIS NDVI contain some missing or unreliable data; hence, days with missing data (35 % of all dates) in MODIS pixels regarding the scintillometer footprint (see later footprint computation in Sect. 3.2) were excluded. Albedo products (MCD43) are available every 8 days; the day of interest is the central date. Both Terra and Aqua data are used in the generation of this product, providing the highest probability for quality input data and designating it as a combined product. Moreover, the 1 km and 16-day NDVI products (MOD13A2/MYD13A2) are available every 16 days and separately for Terra and Aqua. Algorithms generating this product operate on a per-pixel basis and require multiple daily observations to generate a composite NDVI value that will represent the full period (16 days). For both products, data are linearly interpolated over the available dates in order to get daily estimates. For each pixel, the quality index supplied with each product is used to select the best data.

3 Extra-large aperture scintillometer (XLAS): data processing

3.1 Scintillometer-derived fluxes

Scintillometer measurements are based on the scintillation theory; fluxes of sensible heat and momentum cause atmospheric turbulence close to the ground, and create, with surface evaporation, refractive index fluctuations due mainly to air temperature and humidity fluctuations (Hill et al., 1980). The fluctuation intensity of the refractive index is directly linked to sensible and latent heat fluxes. The light beam emitted by the XLAS transmitter towards the receiver is dispersed by the atmospheric turbulence. The scintillations representing the intensity fluctuations are analyzed at the XLAS receiver and are expressed as the structure parameter of the refractive index of air integrated along the optical path C_n^2 ($m^{-2/3}$) (Tatarskii, 1961). The sensitivity of the scintillometer to C_n^2 along the beam is not uniform and follows a bell-shaped curve due to the symmetry of the devices. This means

that the measured flux is more sensitive to sources located towards the transect center and is less affected by those close to the transect extremities.

In order to compute the XLAS sensible heat flux, C_n^2 was converted to the structure parameter of temperature turbulence C_T^2 ($K^2 m^{-2/3}$) by introducing the Bowen ratio (ratio of sensible to latent heat fluxes), hereafter referred to as β, which is a temperature–humidity correlation factor. Moreover, the height of the scintillometer beam above the surface varies along the path. In our study site, the terrain is very flat leading to little beam height variation across the landscape, except for what is induced by the different roughness values of the individual fields. Since the interspaces between trees are large, the effective roughness of the orchards is not significantly different from that of annual crop fields. Consequently, C_n^2 and therefore C_T^2 are not only averaged horizontally but vertically as well.

At visible wavelengths, the refractive index is sensitive to temperature fluctuations. Then, we can relate the C_n^2 to C_T^2 as follows:

$$C_n2 = \left(\frac{-0.78 \times 10^{-6} \times P}{T^2}\right)^2 C_T^2 \left(1 + \frac{0.03}{\beta}\right)^2, \quad (1)$$

with T the air temperature (°K) and P the atmospheric pressure (Pa).

Green and Hayashi (1998) proposed another method to compute XLAS sensible heat flux (H_XLAS) assuming full energy budget closure and using an iterative process without the need for β as an input parameter. This method is called the "β-closure method" (BCM, Twine et al., 2000). In the calculation algorithm, β is estimated iteratively with the BCM method, as described in Solignac et al. (2009) with an initial guess using R_{nBS} and G_{BS} from the Ben Salem flux station and initial u_* coming from the western water tower EC station.

Then, the similarity relationship proposed by Andreas (1988) is used to relate the C_T^2 to the temperature scale T_* in unstable atmospheric conditions as follows:

$$\frac{C_T^2(z_{LAS} - d)^{\frac{2}{3}}}{T_*^2} = 4.9\left(1 - 6.1\left(\frac{z_{LAS} - d}{L_O}\right)^{-\frac{2}{3}}\right), \quad (2)$$

and for stable atmospheric conditions,

$$\frac{C_T^2(z_{LAS} - d)^{\frac{2}{3}}}{T_*^2} = 4.9\left(1 + 2.2\left(\frac{z_{LAS} - d}{L_O}\right)^{\frac{2}{3}}\right), \quad (3)$$

where L_O (m) is the Monin–Obukhov length, Z_{LAS} (m) is the scintillometer height, and d (m) is the displacement height, which corresponds to two-thirds of the averaged vegetation height z_v.

The variable z_v counts for the various heights within the selected footprint using angular zones originating from the center of the transect and supported by high-resolution remote sensing data (see Sect. 4.1).

Furthermore, considering the size of the surface changes in roughness (mean or effective vegetation height ~ 1.5 m), the XLAS measurement height is assumed to be either close to the blending height or higher. Thus, the fluxes measured by XLAS are area-averaged and the MO similarity hypothesis can be applied in the flux algorithm computation.

Moreover, the fluxes measured with the XLAS were calculated for a stability index (Z_{LAS}/L_O) between -2 and 0 ($\sim 73\,\%$ of the cases). Then, according to Gruber and Fochesatto (2013), the sensitivity of the turbulent flux measurements to uncertainties in source measurements is rather similar between iterative algorithms and analytical solution, in this range of atmospheric stability.

From T_* and the friction velocity u_* (computed based on an iteration approach in the BCM method), the sensible heat flux can be derived as follows:

$$H = -\rho c_p T_* u_*, \tag{4}$$

where ρ (kg m^{-3}) the density of air and c_p (J kg^{-1} K^{-1}) the specific heat of air at constant pressure.

H_XLAS was computed at a half-hourly time step. Before flux computation, a strict filtering was applied to the XLAS data to remove outliers depending on weak demodulation signal. Negative nighttime data were set to zero and daytime flux missing data (one to three 30 min data) were gap filled using simple interpolation. Furthermore, half-hourly H_XLAS aberrant values due to measurement errors and values higher than 400 W m^{-2}, arising from measurement saturation, were ruled out (3 % of the total measurement throughout the experiment duration). Finally, daily H_XLAS was computed as the average of the half-hourly H_XLAS.

3.2 XLAS footprint computation

The footprint of a flux measurement defines the spatial context of the measurement and the source area that influences the sensors. In case of inhomogeneous surfaces like patches of various land covers and moisture variability due to irrigation, the measured signal is dependent on the fraction of the surface having the strongest influence on the sensor and thus on the footprint size and location. Footprint models (Horst and Weil, 1992; Leclerc and Thurtell, 1990) have been developed to determine what area is contributing to the heat fluxes as well as the relative weight of each particular cell inside the footprint limits. Contributions of upwind locations to the measured flux depend on the height of the vegetation, height of the instrumentation, wind speed, wind direction and atmospheric stability conditions (Chávez et al., 2005).

According to the model of Horst and Weil (1992), for a one-point measurement system, the footprint function f relates the spatial distribution of surface fluxes, $F_0(x, y)$, to the measured flux at height, z_m, $F(x, y, z_m)$, as follows:

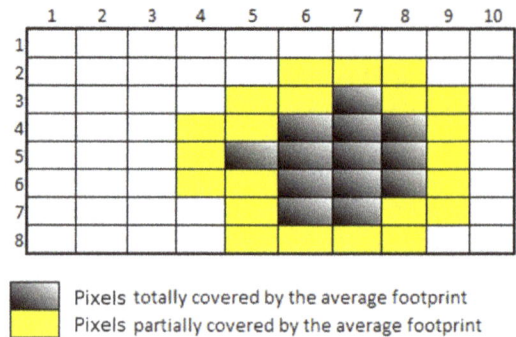

Pixels totally covered by the average footprint
Pixels partially covered by the average footprint

Figure 3. MODIS pixels partially or totally covered by XLAS source area.

$$F(x, y, z_m) = \int_{-\infty}^{\infty} \int_{-\infty}^{x} F_0(x', y') f\left(x - x', y - y', z_m\right) \mathrm{d}x' \mathrm{d}y'. \tag{5}$$

The footprint function f is computed as follows:

$$\overline{f}^y (x, z_m) = \frac{\mathrm{d}\overline{z}}{\mathrm{d}x} \frac{z_m}{\overline{z}^2} \frac{\overline{u}(z_m)}{\overline{u}(c\overline{z})} A e^{-(z_m/b\overline{z})^r}, \tag{6}$$

where $\overline{u}(z)$ is the mean wind speed profile and $\overline{z}$ is the mean plume height for diffusion from a surface source. The variables A, b and c are scale factors and r is a scale factor of the Gamma function. In the case of a scintillometer measurement, the footprint function has to be combined with the spatial weighting function $W(x)$ of the scintillometer to account for the sensor integration along its path. Thus, the sensible heat flux footprint mainly depends on the scintillometer effective height Z_{LAS} (Hartogensis et al., 2003), which includes the topography below the path and the transmitter and receiver heights, the wind direction, and the Obukhov length L_O, which characterizes the atmospheric stability (Solignac et al., 2009). In a subsequent step, daily footprints were computed as a weighted sum of the half-hourly footprints by the XLAS sensible heat flux.

In fact, there is an issue with the MODIS pixel heterogeneity and notably the distribution of the land use classes at the intersection between the square pixel and the XLAS footprint (Bai et al., 2015). Hence, in order to provide a first guess on these relative heterogeneities, land use classes within each MODIS pixel of the 10 km $\times$ 8 km subimage were studied based on the land use map of the 2013–2014 season (Chahbi et al., 2016). The average footprint of all half-hourly footprints for the whole study period was computed and overlaid on the MODIS grid in order to identify the MODIS pixels partially or totally covered by footprint (Fig. 3).

The percentage of land use classes was computed for (i) the part of each pixel that lies within the footprint and (ii) the complementary part of the pixel located outside of the footprint (Fig. 4). Results show that differences in percentages of each land use classes for the pixel fractions located within or outside the footprint are low, with 1.8, 1.7,

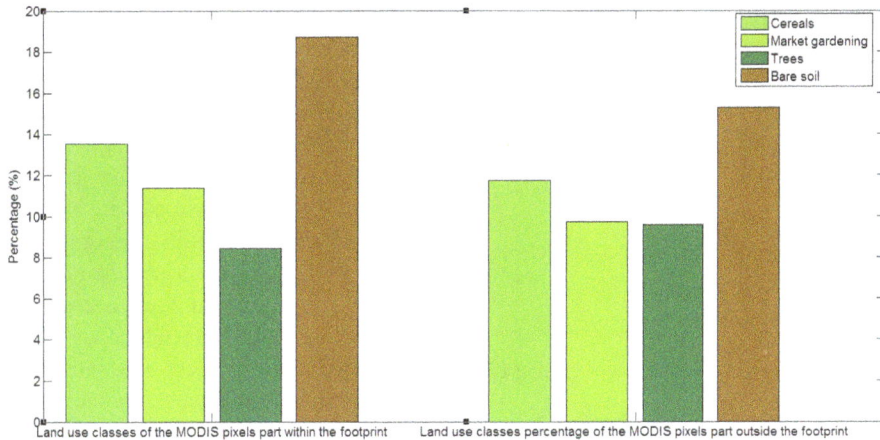

Figure 4. Land use classes' percentage of the MODIS pixels within or outside the footprint.

1.0 and 3.5 % for cereals, market gardening, trees and bare soil, respectively. Moreover, the major part of the area above the transect is covered by fallow and orchards. The land use classes' partition inside the 13 MODIS pixels totally covered by the average footprint is comparable.

3.3 XLAS-derived latent heat flux

Instantaneous (LE_residual_XLAS$_{t\text{-FP}}$) and daily (LE_residual_XLAS$_{\text{day-FP}}$) XLAS-derived latent heat fluxes (i.e., residual latent heat flux) of the XLAS upwind area were computed using the energy budget closure of the XLAS measured sensible heat flux (H_XLAS) with additional estimations of remotely sensed net surface radiation R_n and soil heat flux G, as available energy (AE $= R_n - G$), as follows:

$$\text{LE\_residual\_XLAS}_{t\text{-FP}} = \text{AE}_{t\text{-FP}} - \text{H\_XLAS}_t, \qquad (7)$$

$$\text{LE\_residual\_XLAS}_{\text{day-FP}} = \text{AE}_{\text{day-FP}} - \text{H\_XLAS}_{\text{day}}. \qquad (8)$$

H_XLAS$_t$ and H_XLAS$_{\text{day}}$ are respectively the instantaneous and daily measured H at the times of the satellite overpass interpolated from the half-hourly fluxes measurements. Daily available energy within the footprint (AE$_{\text{day-FP}}$) was computed from instantaneous available energy (AE$_{t\text{-FP}}$) as detailed in Sect. 3.3.1 and Sect. 3.3.2. The subscripts "30", "day" and "t" refer to half-hourly, daily and instantaneous (at the times of Terra and Aqua overpasses) variables, respectively, while the subscript "FP" means that the footprint is taken into account, i.e., instantaneous or the daily (depending on timescale) footprint was multiplied by the variable.

3.3.1 Instantaneous available energy

Net surface radiation is the balance of energy between incoming and outgoing shortwave and longwave radiation fluxes at the land–atmosphere interface. Remotely sensed surface radiative budget components provide unparalleled

spatial and temporal information, and thus several studies have attempted to estimate net radiation by combining remote sensing observations with surface and atmospheric data. Net radiation equation can be written as follows:

$$R_n = (1 - \alpha) R_g + \varepsilon_{\text{surf}} R_{\text{atm}} - \varepsilon_{\text{surf}} \sigma T_{\text{surf}}^4, \qquad (9)$$

where R_g is the incoming shortwave radiation (W m^{-2}), R_{atm} is the incoming longwave radiation (W m^{-2}), α is the albedo, T_{surf} is the surface emissivity, $\varepsilon_{\text{surf}}$ is the surface temperature ($^{\circ}$K) and σ is the Stefan–Boltzmann constant (W^{-2} K^4). The soil heat flux G depends on the soil type and water content as well as the vegetation type (Allen et al., 2005). The direct estimation of G by remote sensing data is not possible (Allen et al., 2011); however, empirical relations can estimate the fraction $\xi = G/R_n$ as a function of soil and vegetation characteristics using satellite image data, such as the LAI, NDVI, α and T_{surf}. Generally, G represents 5–20 % of R_n during daylight hours (Kalma et al., 2008). In order to estimate the G/R_n ratio, several methods have been tested for various types of surfaces at different locations. The most common methods parameterize ξ as a constant for the entire day or at satellite overpass time (Ventura et al., 1999), according to NDVI (Jackson et al., 1987; Kustas and Daughtry, 1990), LAI (Choudhury et al., 1987; Kustas et al., 1993; Tasumi et al., 2005), vegetation fraction (f_c) (Su, 2002), T_{surf} and α (Bastiaanssen, 1995), or only T_{surf} (Santanello Jr. and Friedl, 2003). These empirical methods are suitable for specific conditions; therefore, estimating G, especially in this type of environment where NDVI values are low and thus G/R_n values are large, is a critical issue. The approach adopted here was drawn on Danelichen et al. (2014), who evaluated the parameterization of these different models in three sites in Mato Grosso state in Brazil and found that the model proposed by Bastiaanssen (1995) showed the best performance for all sites, followed by the model from Choudhury et al. (1987) and Jackson et al. (1987):

Bastiaanssen (1995):

$$G = R_n (T_{surf} - 273.16) (0.0038 + 0.0074\alpha)$$
$$\left(1 - 0.98\text{NDVI}^4\right) \tag{10}$$

Choudhury et al. (1987):

$$G = 0.4 Rn(\exp(-0.5\text{LAI})) \tag{11}$$

Jackson et al. (1987)

$$G = 0.583 Rn(\exp(-2.13\text{NDVI})). \tag{12}$$

Hence, these three methods were tested for the Ben Salem flux station measurements, by comparing the measured $G_{BS\text{-}t}$ and the computed G using measured $R_{n_{BS\text{-}t}}$, $T_{surf\text{-}BS\text{-}t}$, α_{BS}, NDVI_{BS} and LAI_{BS} at Terra and Aqua overpass times (results not shown). The best results are issued from the Bastiaanssen (1995) method with a root mean square error (RMSE) of 0.09 (average value of the two satellite overpass times) followed by Jackson et al. (1987) and Choudhury et al. (1987) with RMSE values of 0.15 and 0.2, respectively. Moreover, daily measured $G_{BS\text{-}day}$ was computed and a G accumulation is generally found as has been already mentioned by Clothier et al. (1986), who showed that G is neither constant nor negligible on diurnal timescales and can constitute as much as 50 % of R_n over sparsely vegetated area. Since G estimation was the most uncertain variable, the three above methods were tested to compute the distributed remotely sensed AE. The Ben Salem meteorological station was used to provide R_{g_t} and $R_{atm\text{-}t}$. Remote sensing variables α, T_{surf}, ε_{surf} and NDVI came from MODIS products. Remotely sensed LAI was computed from the MODIS NDVI using a single equation (Clevers, 1989) for all crops in the study area:

$$\text{LAI} = -\frac{1}{k} \ln \left(\frac{\text{NDVI}_\infty - \text{NDVI}}{\text{NDVI}_\infty - \text{NDVI}_{soil}} \right). \tag{13}$$

The calibration of this relationship was done over the Yaqui irrigated perimeter (Mexico) during the 2007–2008 growing season using hemispherical LAI measured in all the studied fields (Chirouze et al., 2014). Calibration results gave the asymptotical values of NDVI, $\text{NDVI}_\infty = 0.97$ and $\text{NDVI}_{soil} = 0.05$, as well as the extinction factor $k = 1.13$. As this relationship was calibrated over a heterogeneous land surface but on herbaceous vegetation only, its relevance for trees was checked. For that purpose, clump-LAI measurements on an olive tree, as well as allometric measurements, i.e., mean distance between trees and mean crown size done using Pleiades satellite data (Mougenot et al., 2014; Touhami, 2013) were obtained. Clump LAI is the value of the LAI of an isolated element of vegetation (tree, shrub, etc.); if this element occupies a fraction cover f and is surrounded by bare soil, then the clump LAI value is equal to the area-average LAI divided by f. Hence, we checked that the pixels

with tree-dominant cover show LAI values close to what was expected (of the order of 0.3 to 0.4 given the interrow distance of 12 m on average).

Remote sensed available energy was computed for the 10 km × 8 km MODIS subimages at Terra MODIS and Aqua MODIS overpass times, using the three methods estimating G. Since the measured heat fluxes H_XLAS_t represent only the weighted contribution of the fluxes from the upwind area to the tower (footprint), then instantaneous footprints at the times of Terra and Aqua overpass were selected among the two half-hours preceding and following the satellite's time of overpass (lowest time interval) and then was multiplied by the instantaneous remotely sensed available energy AE_t to get the available energy of the upwind area $AE_{t\text{-}FP}$.

3.3.2 Daily available energy

Most methods using TIR domain data rely on once-a-day, late morning (such as Terra-MODIS overpass time) or early afternoon (such as Aqua-MODIS overpass time) acquisitions. Thus, they provide a single instantaneous estimate of energy budget components. In order to obtain daily AE from these instantaneous measurements and to reconstruct hourly variations of AE, we considered that its evolution was proportional to another variable whose diurnal evolution can be easily known.

The extrapolation from an instantaneous flux estimate to a daytime flux assumes that the surface energy budget is "self-preserving", i.e., the relative partitioning among components of the budget remains constant throughout the day. However, many studies (Brutsaert and Sugita, 1992; Gurney and Hsu, 1990; Sugita and Brutsaert, 1990) showed that the self-preservation method gives daytime latent heat estimates that are smaller than observed values by 5–10 %. Moreover, Anderson et al. (1997) found that the evaporative fraction computed from instantaneous measured fluxes tends to underestimate the daytime average by about 10 %; hence, a corrected parameterization was used and a coefficient = 1.1 was applied. Similarly, Delogu et al. (2012) found an overestimation of about 10 % between estimated and measured daily components of the available energy, and thus a coefficient = 0.9 was applied. The corrected parameterization proposed by Delogu et al. (2012) was tested, but this coefficient did not give consistent results; therefore, the extrapolation relationship was calibrated in order to get accurate daily results of AE.

Thereby, the applied extrapolation method was tested using in situ Ben Salem flux station measurements. The incoming short-wavelength radiation was used to scale available energy from instantaneous to daily values, but only for clear-sky days for which MODIS images can be acquired and remote sensing data used to compute AE are available. Clear-sky days were selected based on the ratio of daily measured incoming short-wavelength radiation $R_{g_{day}}$ to the theoretical clear-sky radiation Rso as proposed by the FAO-56 method (Allen et al., 1998). A day was defined as clear if the measured $R_{g_{day}}$ is higher than 85 % of the theoretical clear-sky radiation at the satellite overpass times (Delogu et al., 2012).

Daily measured available energy $AE_{BS\text{-}day}$, computed as the average of half-hourly measured $AE_{BS\text{-}30}$, was compared to daily available energy ($AE_{BS\text{-}day\text{-}Terra}$ and $AE_{BS\text{-}day\text{-}Aqua}$) computed using the extrapolation method from instantaneous measured $AE_{BS\text{-}t\text{-}Terra}$ and $AE_{BS\text{-}t\text{-}Aqua}$ at Terra and Aqua overpass times, respectively (Eq. 14).

$$AE_{BS\text{-}day\text{-}Terra} = a_{Terra} R_{g_{day}} \frac{AE_{BS\text{-}t\text{-}Terra}}{R_{g_{t\text{-}Terra}}} + b_{Terra},$$

$$AE_{BS\text{-}day\text{-}Aqua} = a_{Aqua} R_{g_{day}} \frac{AE_{BS\text{-}t\text{-}Aqua}}{R_{g_{t\text{-}Aqua}}} + b_{Aqua}, \qquad (14)$$

where $R_{g_{day}}$ is the daily measured incoming short-wavelength radiation in the Ben Salem meteorological station; $R_{g_{t\text{-}Terra}}$ and $R_{g_{t\text{-}Aqua}}$ are the instantaneous incoming short-wavelength radiation measured at Terra and Aqua overpass time, respectively, and $AE_{BS\text{-}t\text{-}Terra}$ and $AE_{BS\text{-}t\text{-}Aqua}$ are the instantaneous measured available energy in the Ben Salem flux station, at Terra and Aqua overpass times.

Results gave an overestimation of about 15 %. The corrected parameterizations of AE (Table 1), needed to remove the bias between measured ($AE_{BS\text{-}day}$) and computed AE ($AE_{BS\text{-}day\text{-}Terra}$ and $AE_{BS\text{-}day\text{-}Aqua}$), were applied to compute daily remotely sensed AE (AE_{day}) from instantaneous AE (AE_t) following the extrapolation method shown in Eq. (14).

Then AE_{day} was multiplied by the corresponding daily footprint to get the daily available energy of the upwind area $AE_{day\text{-}FP}$. Finally, estimates of Terra and Aqua observed daily LE (LE_residual_XLAS$_{day\text{-}FP}$) were obtained based on the three methods used to compute G.

4 SPARSE model

4.1 Energy fluxes derived from SPARSE model

The SPARSE dual-source model solves the energy budgets of the soil and the vegetation. Here we use the "layer approach", for which the resistance network relating the soil and vegetation heat sources to a main reference level through a common aerodynamic level use a series electrical branching. The main unknowns are the component temperatures, i.e., soil (T_s) and vegetation (T_v) temperatures. Totals at the reference height (the measurement height of the meteorological forcing), as well as the longwave radiation budget, are also solved so that altogether a system of five equations can be built:

Table 1. Corrected parameterizations of available energy for the diurnal reconstitution.

Terra	a_{Terra}	0.85
	b_{Terra}	−19.81
Aqua	a_{Aqua}	0.87
	b_{Aqua}	−18.94

$$\begin{cases} H = H_s + H_v \\ LE = LE_s + LE_v \\ R_{ns} = G + H_s + LE_s \\ R_{nv} = H_v + LE_v \\ \varepsilon_{surf} \sigma T_{surf}^4 = \varepsilon_{surf} R_{atm} - R_{an}, \end{cases} \qquad (15)$$

where R_{atm} is the atmospheric radiation (W m^{-2}), R_{an} is net longwave radiation which depends on T_s and T_v (W m^{-2}), and T_{surf} and ε_{surf} are respectively the surface temperature (°K) and emissivity as observed by the satellite; indexes "s" and "v" designate the soil and the vegetation, respectively.

The first two (Eq. 15) express the continuity of the latent and sensible heat fluxes from the sources to the aerodynamic level through to the reference level, the third and the fourth (Eq. 15) are the soil and vegetation energy budgets, and the fifth (Eq. 15) relates the surface temperature T_{surf} derived from observed MODIS surface temperature to T_s and T_v.

The SPARSE model system of equations is fully described in Boulet et al. (2015). SPARSE is similar to the TSEB model (Kustas and Norman, 1999) but includes the expressions of the aerodynamic resistances of Choudhury and Monteith (1988) and Shuttleworth and Gurney (1990). This system can be solved in a forward mode for which the surface temperature is an output (prescribed conditions), and an inverse mode when the surface temperature is an input derived from satellite observations or in situ measurements in the thermal infrared domain (retrieval conditions). Figure 5 illustrates a diagram showing the flowchart of the model algorithm. System (15) is solved step by step by following similar guidelines as in the TSEB model: the first step assumes that the vegetation transpiration (LE$_v$) is at its maximum, and evaporation (LE$_s$) is computed. If this soil latent heat flux (LE$_s$) is below a minimum positive threshold for vegetation stress detection of 30 W m^{-2}, the hypothesis that the vegetation is unstressed is no longer valid. In that case, the vegetation is assumed to suffer from water stress and the soil surface is assumed to be already long dry. Then, LE$_s$ is set to 30 W m^{-2}. This value accounts for the small but nonnegligible vapor flow reaching the surface (Boulet et al., 1997). The system is then solved for vegetation latent heat flux (LE$_v$). If LE$_v$ is also negative, both LE$_s$ and LE$_v$ values are set to zero, whatever the value of T_{surf}. The system of equation can also be solved for T_s and T_v only if the efficiencies representing stress levels (dependent on surface soil moisture for the evaporation, and root zone soil moisture for the transpiration) are

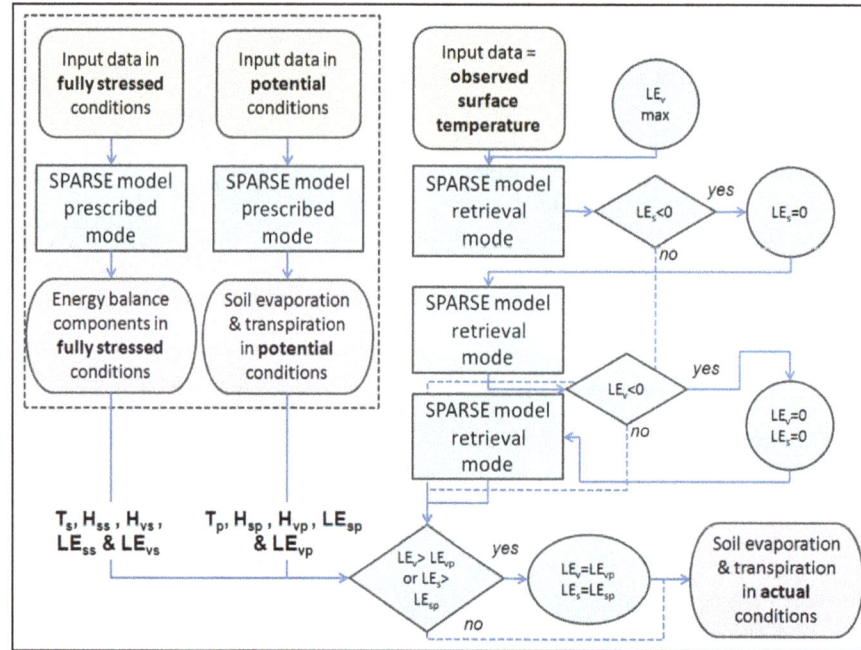

Figure 5. Flowchart of the SPARSE algorithm; T_s, H_{ss}, H_{vs}, LE_{ss} and LE_{vs} are surface temperature, soil sensible heat flux, vegetation sensible heat flux, soil latent heat flux and vegetation latent heat flux in fully stressed conditions, respectively; T_p, H_{sp}, H_{vp}, LE_{sp} and LE_{vp} are surface temperature, soil sensible heat flux, vegetation sensible heat flux, soil latent heat flux and vegetation latent heat flux in potential conditions, respectively.

known. In that case the sole first four equations are solved. This prescribed mode allows all the fluxes in known limiting soil moisture levels to be computed (very dry, e.g., fully stressed, and wet enough, e.g., potential). It limits unrealistically high values of component fluxes, latent heat flux values above the potential rates or sensible heat flux values above that of a nonevaporating surface. The potential evaporation and transpiration rates used later on are computed using this prescribed mode with minimum surface resistance to evaporation and transpiration, respectively.

Some of the model parameters were remotely sensed data while others were taken from the bibliography or measured in situ. Remotely sensed data fed into SPARSE are T_{surf}, ε_{surf}, ϕ, NDVI, LAI and α. A grid of the vegetation height (z_v) was also necessary as input in the SPARSE model; for herbaceous crops, vegetation height was interpolated with the help of NDVI time series between fixed minimum (0.05 m) and maximum (0.8 m) values, while for trees, the roughness length (z_{om}) was linked to the allometric measurements (mentioned before) and computed as a function of canopy area index, drag coefficient and canopy height using the drag partition approach proposed by Raupach (1994) for tall sparse vegetative environments. Then, since SPARSE deals with vegetation height and not roughness length, the same simple rule of the thumb as the one used in SPARSE was used to reconstruct z_v for the tree cover types ($z_v = z_{om}/0.13$). In a final step, to get spatial vegetation height, z_v was averaged

over the MODIS pixels. In situ parameters used in SPARSE were mainly meteorological data: R_g, R_{atm}, T_a, H_a and u. No calibration was performed on the model parameters shown in Table 2.

The retrieval and prescribed modes of the SPARSE model were run for the $10\,\text{km} \times 8\,\text{km}$ subimages at the times of Terra and Aqua overpasses, to get instantaneous modeled fluxes H_SPARSE$_t$, LE_SPARSE$_t$ and AE_SPARSE$_t$ as well as sensible heat flux ($H_{s\text{-}t} = H_{ss\text{-}t} + H_{vs\text{-}t}$) in fully stressed conditions and latent heat ($LE_{p\text{-}t} = LE_{sp\text{-}t} + LE_{vp\text{-}t}$) and sensible heat ($H_{p\text{-}t} = H_{sp\text{-}t} + H_{vp\text{-}t}$) fluxes in potential conditions. Modeled values were then multiplied by the nearest half-hourly footprint to the satellite overpass time, in order to get fluxes corresponding to the upwind area: H_SPARSE$_{t\text{-FP}}$, LE_SPARSE$_{t\text{-FP}}$, AE_SPARSE$_{t\text{-FP}}$, $H_{s\text{-}t\text{-FP}}$, $H_{p\text{-}t\text{-FP}}$ and LE$_{p\text{-}t\text{-FP}}$.

In a subsequent step, the retrieval and prescribed modes of SPARSE model was run at a half-hourly time step using the half-hourly meteorological measurements to get half-hourly latent heat flux at potential conditions LE$_{p\text{-}30}$ and half-hourly modeled available energy AE_SPARSE$_{30}$. The potential LE weighted by the corresponding half-hourly footprint (LE$_{p\text{-}30\text{-FP}}$) is used later when computing daily LE based on the stress factor method, while the half-hourly AE weighted by the corresponding half-hourly footprint (AE_SPARSE$_{30\text{-FP}}$) is used to compute daily LE based on the evaporative fraction method (Sect. 4.2).

Table 2. SPARSE parameters.

	Definition	Value	Data sources
Remote sensing parameters			
NDVI	Normalized difference vegetation index		Satellite imagery
T_{surf} (K)	Surface temperature (K)		Satellite imagery
α	Albedo		Satellite imagery
ε_{surf}	Surface emissivity		Satellite imagery
Φ (rad)	View zenith angle		Satellite imagery
Meteorological parameters			
R_g (W m^{-2})	Incoming solar radiation		In situ data
R_{atm} (W m^{-2})	Incoming atmospheric radiation		In situ data
T_a (K)	Air temperature at reference level		In situ data
RH_a (%)	Air relative humidity		In situ data
u_a (m s^{-1})	Horizontal wind speed at reference level		In situ data
Fixed parameters			
z_a (m)	Atmospheric forcing height	2.32	In situ data
z_v (m)	Vegetation height		Derived from land cover
β_{pot}	Evapotranspiration efficiency in full potential conditions	1.000	
β_{stress}	Evapotranspiration efficiency in fully stressed conditions	0.001	
r_{stmin} (s^{-1})	Minimum stomatal resistance	100	Boulet et al. (2015)
w (m)	Leaf width	0.05	Braud et al. (1995)
ε_v	Vegetation emissivity	0.98	Braud et al. (1995)
α_v	Vegetation albedo	0.25	Estimation
Constants			
ρ_{cp} (J kg^{-1} K^{-1})	Product of air density and specific heat	1170	Braud et al. (1995)
σ (W m^{-2} k^4)	Stefan–Boltzmann constant	5.66×10^{-8}	Braud et al. (1995)
γ (Pa K^{-1})	Psychrometric constant	0.66	Braud et al. (1995)
$z_{om,s}$ (m)	Equivalent roughness length of the underlying bare soil in the absence of vegetation	5×10^{-3}	Braud et al. (1995)
n_{SW}	Coefficient in r_{av} (aerodynamic resistance between the vegetation and the aerodynamic level)	2.5	Boulet et al. (2015)
ξ	Ratio of soil heat flux G to available net radiation on the bare soil R_{ns}	0.4	Braud et al. (1995)

4.2 Reconstruction of daily modeled ET from instantaneous latent heat flux

Daily ET is usually required for applications in hydrology or agronomy for instance, whereas most SEB methods provide a single instantaneous latent heat flux because the energy budget is only computed at the satellite overpass times (Delogu et al., 2012). In order to scale daily ET from one instantaneous estimate, there are various methods relying on the preservation, during the day, of the ratio of the latent heat flux to a scale factor with known diurnal evolution.

4.2.1 Stress factor (SF) method

The stress factor SF (Eq. 16) is assumed to be invariant during the same day, and the diurnal modeled fluxes are accounted for by recovering the diurnal course of potential ET.

$$SF = 1 - \frac{LE\_SPARSE_{t\text{-FP}}}{LE_{p\text{-}t\text{-FP}}} \qquad (16)$$

The daily modeled ET (LE_SPARSE$_{day\text{-FP}}$) can be expressed as the product of the instantaneous estimate of SF at the satellite overpass times and the daily potential evapotranspiration:

$$LE\_SPARSE_{day\text{-FP}} = (1 - SF)LE_{p\text{-}day\text{-FP}}. \qquad (17)$$

LE$_{p\text{-}day\text{-FP}}$ was calculated as the sum of the half-hourly modeled latent heat fluxes at potential conditions LE$_{p\text{-}30\text{-FP}}$.

4.2.2 Evaporative fraction method

The evaporative fraction (EF) self-preservation is a valid assumption under dry conditions but no longer under wet conditions (Hoedjes et al., 2008). For these conditions, assuming a constant EF underestimates actual EF and therefore ET (Lhomme and Elguero, 1999). Indeed, according to Gentine et al. (2007), the diurnal shape of EF depends on both atmospheric forcing and surface conditions. Therefore EF was computed every 30 min using the following empirical parameterization (Delogu et al., 2012):

$$EF_{30} = \left[1.2 - \left(0.4\frac{R_{g30}}{1000} + 0.5\frac{RH_{30}}{100}\right)\right]\left(\frac{EF_{SPARSE\text{-}t}}{EF_{met\text{-}t}}\right), \quad (18)$$

where R_{g30} and RH_{30} are respectively the half-hourly incoming short-wavelength radiation and relative humidity EF$_{SPARSE\text{-}t}$ and EF$_{met\text{-}t}$ are respectively SPARSE EF (Eq. 19) and computed EF using the evaporative fraction method (Eq. 20) at the satellite overpass times.

$$\text{EF}_{\text{SPARSE-}t} = \frac{\text{LE\_SPARSE}_{t\text{-FP}}}{\text{AE\_SPARSE}_{t\text{-FP}}}, \tag{19}$$

$$\text{EF}_{\text{met-}t} = \left[1.2 - \left(0.4\frac{R_{g_t}}{1000} + 0.5\frac{\text{RH}_t}{100}\right)\right], \tag{20}$$

where $\text{LE\_SPARSE}_{t\text{-FP}}$ and $\text{AE\_SPARSE}_{t\text{-FP}}$ are respectively the latent heat flux and the available energy modeled by SPARSE at the satellite overpass times; R_{g_t} and RH_t are respectively the incoming short-wavelength radiation and relative humidity measured at the times of the satellite overpasses.

The half-hourly modeled ET ($\text{LE\_SPARSE}_{30\text{-FP}}$) was computed as the product of the half-hourly EF estimate and the half-hourly modeled available energy $\text{AE\_SPARSE}_{30\text{-FP}}$ (Eq. 21). AE_SPARSE_{30} was computed from instantaneous modeled available energy (AE_SPARSE_t) using the same approach detailed in Sect. 3.3.2 and applying Eq. (14) for a half-hourly time step (instead of a daily time step). AE_SPARSE_{30} was weighted by the corresponding half-hourly footprint to get the modeled AE of the upwind area $\text{AE\_SPARSE}_{30\text{-FP}}$. The daily modeled ET ($\text{LE\_SPARSE}_{\text{day-FP}}$) was computed as the sum of the half-hourly $\text{LE\_SPARSE}_{30\text{-FP}}$ (Eq. 22).

$$\text{LE\_SPARSE}_{30\text{-FP}} = \text{EF}_{30} \times \text{AE\_SPARSE}_{30\text{-FP}} \tag{21}$$

$$\text{LE\_SPARSE}_{\text{day-FP}} = \sum \text{LE\_SPARSE}_{30\text{-FP}} \tag{22}$$

4.2.3 Residual method

Besides, daily modeled ET ($\text{LE\_SPARSE}_{\text{day-FP}}$) was also estimated as a residual term of the surface energy budget using daily modeled sensible heat flux ($\text{H\_SPARSE}_{\text{day-FP}}$) and available energy ($\text{AE\_SPARSE}_{\text{day-FP}}$) as follows:

$$\begin{aligned}\text{LE\_SPARSE}_{\text{day-FP}} = \ &\text{AE\_SPARSE}_{\text{day-FP}} \\ &- \text{H\_SPARSE}_{\text{day-FP}}.\end{aligned} \tag{23}$$

$\text{H\_SPARSE}_{\text{day}}$ was computed from modeled sensible heat flux (H_SPARSE_t) following the same extrapolation method used for the available energy (see Sect. 3.3.2). The corrected parameterizations of H were obtained from the comparison of daily measured sensible heat flux $H_{\text{BS-day}}$ computed as the average of half-hourly measured $H_{\text{BS-30}}$ and daily sensible heat flux ($H_{\text{BS-day-Terra}}$ and $H_{\text{BS-day-Aqua}}$) computed using the extrapolation method from instantaneous measured $H_{\text{BS-}t\text{-Terra}}$ and $H_{\text{BS-}t\text{-Aqua}}$ at Terra and Aqua overpass times, respectively (Eq. 24).

$$H_{\text{BS-day-Terra}} = a'_{\text{Terra}} R_{g_{\text{day}}} \frac{H_{\text{BS-}t\text{-Terra}}}{R_{g_{t}\text{-Terra}}} + b'_{\text{Terra}},$$

$$H_{\text{BS-day-Aqua}} = a'_{\text{Aqua}} R_{g_{\text{day}}} \frac{H_{\text{BS-}t\text{-Aqua}}}{R_{g_{t}\text{-Aqua}}} + b'_{\text{Aqua}}, \tag{24}$$

where $H_{\text{BS-}t\text{-Terra}}$ and $H_{\text{BS-}t\text{-Aqua}}$ are the instantaneous measured sensible heat flux in the Ben Salem flux station.

Table 3. Corrected parameterizations of sensible heat flux for the diurnal reconstitution.

Terra	a'_{Terra}	1.02
	b'_{Terra}	−17.31
Aqua	a'_{Aqua}	1.00
	b'_{Aqua}	−14.83

Therefore, the corrected parameterizations of H (Table 3), needed to remove the bias between measured ($H_{\text{BS-day}}$) and computed H ($H_{\text{BS-day-Terra}}$ and $\text{AE}_{\text{BS-day-Aqua}}$), were applied to compute daily modeled H ($\text{H\_SPARSE}_{\text{day}}$) from instantaneous modeled H (H_SPARSE_t) following the extrapolation method shown in Eq. (21). Finally, $\text{H\_SPARSE}_{\text{day}}$ was weighted by the corresponding daily footprint to get the daily modeled H of the upwind area $\text{H\_SPARSE}_{\text{day-FP}}$.

5 Water stress estimates

Water stress estimation is crucial to deduce the root zone soil moisture level using remote sensing data (Hain et al., 2009). Water stress results in a drop of actual evapotranspiration below the potential rate. Its intensity is usually represented by a stress factor as defined in Sect. 4.2, ranging between 0 (unstressed surface) and 1 (fully stressed surface).

Modeled values of SF at the times of Terra and Aqua overpasses (SF_{mod}) have been computed from modeled potential LE ($\text{LE}_{\text{p-}t\text{-FP}}$) as follows:

$$\text{SF}_{\text{mod}} = 1 - \frac{\text{LE\_SPARSE}_{t\text{-FP}}}{\text{LE}_{\text{p-}t\text{-FP}}} \tag{25}$$

where $\text{LE\_SPARSE}_{t\text{-FP}}$ and $\text{LE}_{\text{p-}t\text{-FP}}$ are the modeled latent heat fluxes in actual and potential conditions, respectively.

Furthermore, surface water stress factor derived from XLAS measurement, named SF_{obs}, at the times of Terra and Aqua overpasses was computed as follows (Su, 2002):

$$\text{SF}_{\text{obs}} = \frac{H\_XLAS_t - H_{\text{p-}t\text{-FP}}}{H_{\text{s-}t\text{-FP}} - H_{\text{p-}t\text{-FP}}}, \tag{26}$$

where $H_{\text{s-}t\text{-FP}}$ and $H_{\text{p-}t\text{-FP}}$ are the modeled sensible heat flux in actual and potential conditions, respectively, and H_XLAS_t is the XLAS sensible heat flux at the satellite overpass times.

6 Results and discussion

6.1 XLAS- and model-derived instantaneous sensible heat fluxes

Our primary focus is the comparison between scintillometer measurements and the modeled sensible heat fluxes computed using the Terra and Aqua remotely sensed data. The scintillometer H at the times of the two satellite overpasses (H_XLAS_t) are interpolated from the half-hourly H measurements. Heat flux determination was possible for typically about 87 % of the daytime measurements during the summer, availability of XLAS heat flux values was lower during the cold season due to poor visibility and/or stable stratification.

H_SPARSE was weighted by the XLAS footprint in order to be able to compare the modeled values ($H\_SPARSE_{t\text{-FP}}$) with the XLAS measurements (H_XLAS_t). Therefore, due to XLAS and remote sensing data availability, we got 175 and 118 values for Terra and Aqua respectively. In order to highlight H interseasonality between the drier 2012–2013 and the wetter 2013–2014 seasons, we present an example of 2 days, each in one season: DOY 2013-082 shows a H value ranging between 163 and 342 W m^{-2} while DOY 2014-208 shows a H value ranging between 97 and 311 W m^{-2} (Fig. 6). The colored area shows the modeled flux and the contours shows the surface source area contributing to the scintillometer measurements. Day 2013-82 (23 March 2013) is chosen in the cold season while day 208-2014 (27 July 2014) is in the warm season to focus on land cover impact on T_{surf} and thus on modeled H (trees and cereals in winter vs. only irrigated trees and market gardening in summer). Moreover, the first day experiences a strong southern wind while there is a light northern wind during the second day. Generally, a small number of MODIS pixels brings a high contribution to the signal; among these, two are hot pixels (pixel with high T_{surf} and low NDVI) in which the land use is mainly arboriculture.

Prediction performance is assessed using RMSE and the coefficient of determination (R^2). Results for the sensible heat flux are illustrated in Fig. 7 and show good agreement between modeled and measured H at the times of satellite overpasses. This is illustrated by linear regressions of $H\_SPARSE_{t\text{-FP}} = 1.065\,H\_XLAS_t - 14.788$ ($R^2 = 0.6$; RMSE $= 57.89$ W m^{-2}) and $H\_SPARSE_{t\text{-FP}} = 1.12\,H\_XLAS_t - 10.57$ ($R^2 = 0.63$; RMSE $= 53.85$ W m^{-2}) for Terra and Aqua, respectively. This result is of great interest considering that the SPARSE model was run with no prior calibration. However, we noted that bias is a function of the flux level and most outliers are recorded for H greater than 200 W m^{-2}. This can be explained by (i) the XLAS measurement saturation (according to the "Kipp & Zonen LAS and XLAS instruction manual" (Kipp & Zonen, 2017), for a path length of 4 km and a scintillometer height of 20 m, the saturation measurement problems start from H values higher than 300 W m^{-2}), (ii) uncertainties on the correction of stability using the universal stability function, and (iii) potential inconsistencies between the area average MODIS surface temperature and the air temperature measured locally at the meteorological station.

Whereas there are several studies dealing with large aperture scintillometer data whose measurements are compared to modeled fluxes, in the few studies dealing with extra large aperture scintillometer data, the comparison is generally

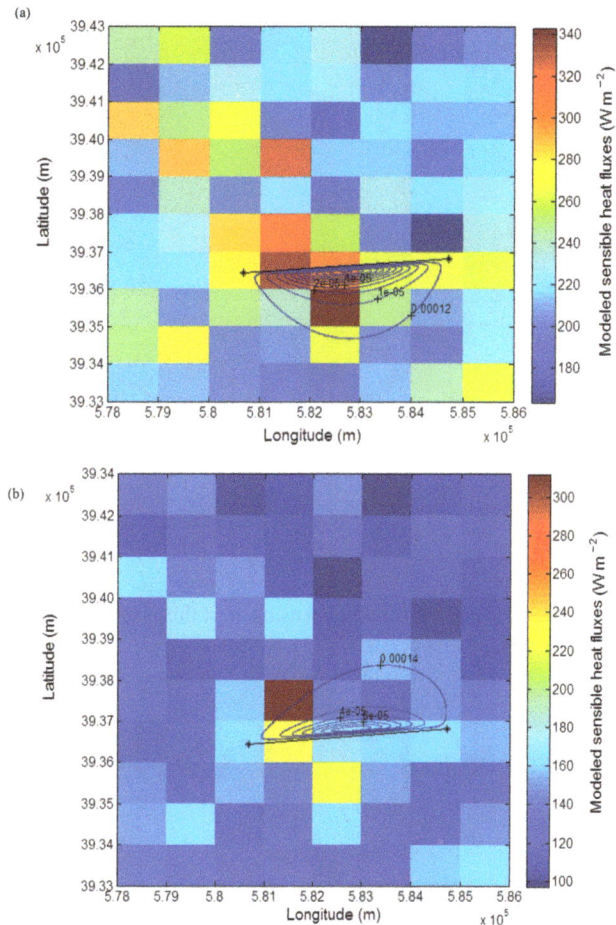

Figure 6. Model-derived sensible heat fluxes and footprints for (a) DOY 2013-082 at Aqua overpass time and (b) DOY 2014-208 at Terra overpass time. The colored area shows the modeled flux and the contours shows the surface source area contributing to the scintillometer measurements.

done with eddy covariance station measurements (Kohsiek et al., 2002; Moene et al., 2006). Indeed, our results are in agreement with those found by Marx et al. (2008), who compared LAS-derived and satellite-derived H (SEBAL was applied with NOAA-AVHRR images providing maps of surface energy fluxes at a 1 km × 1 km spatial resolution) and found that modeled H is underestimated with a RMSE of 39 W m^{-2} for the site Tamale and 104 W m^{-2} for the site Ejura. Moreover, Watts et al. (2000) compared the satellite (AVHRR radiometer) estimates of H to those from LAS over semiarid grassland in north-western Mexico during the summer of 1997. They found RMSE values of 31 and 43 W m^{-2} for LAS path lengths of 300 m and 600 m respectively and showed that LAS measurements are less good than those derived from a 3-D sonic anemometer. They also suggested longer LAS path length (greater than 1.1 km) since the LAS is rather insensitive to the surface near the receiver and the emitter.

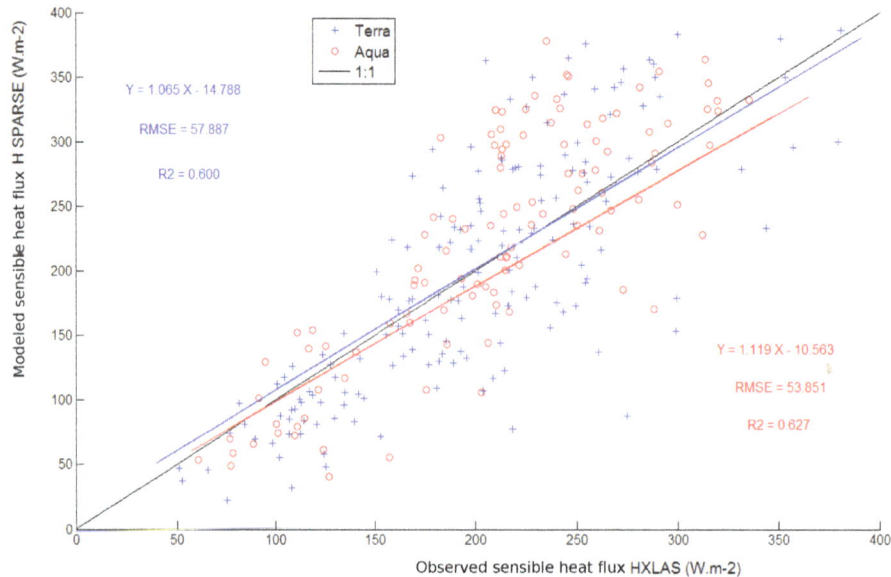

Figure 7. Modeled vs. observed sensible heat fluxes at Terra and Aqua overpass times.

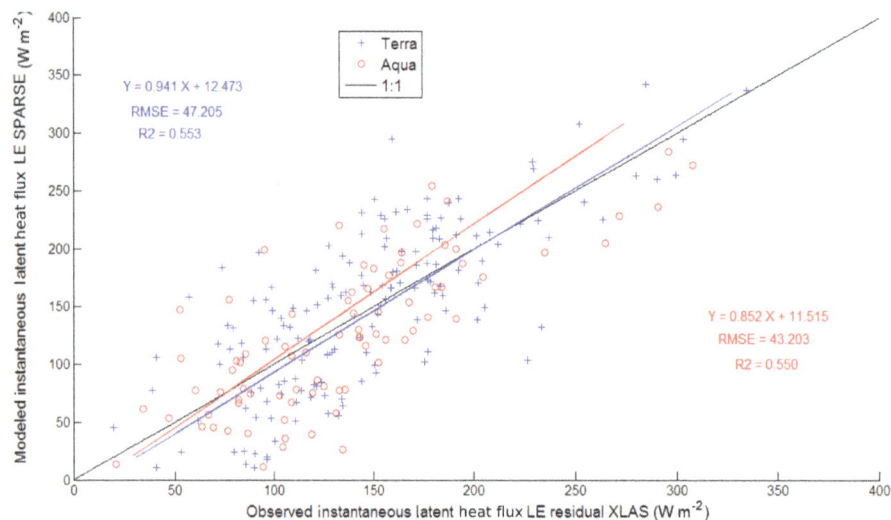

Figure 8. Modeled vs. observed latent heat fluxes at Terra and Aqua overpass times.

6.2 XLAS- and model-derived instantaneous latent heat fluxes

In a subsequent step, SPARSE-derived LE (LE_SPARSE$_{t\text{-FP}}$) was compared to observed LE (LE_residual_XLAS$_{t\text{-FP}}$). Results are illustrated in Fig. 8, showing a good agreement between modeled and observed LE. However, these results are less good than for the H results, as shown by the following linear regressions:
LE_SPARSE$_{t\text{-FP}} = 0.94$ LE_residual_XLAS$_{t\text{-FP}} + 12.47$
(RMSE $= 47.20$ W m^{-2}) and
LE_SPARSE$_{t\text{-FP}} = 0.85$ LE_residual_XLAS$_{t\text{-FP}} + 11.51$
(RMSE $= 43.20$ W m^{-2}) for Terra and Aqua respectively, with an overall R^2 of 0.55 for both satellites. We note

a greater scatter for latent heat flux than for the sensible heat flux (Fig. 7), which can be explained by the fact that LE is here a residual term affected by estimation errors in both AE and H. Despite this moderate discrepancy, the good agreement between both approaches indicates that the methodology adopted in SPARSE for estimating H and AE using MODIS imagery is appropriate for modeling latent heat fluxes.

6.3 Water stress

The scattered values of the stress factor as shown in Fig. 9 are consistent with previous studies such as Boulet et al. (2015). SEB retrieval of stress is limited by the scale mismatch be-

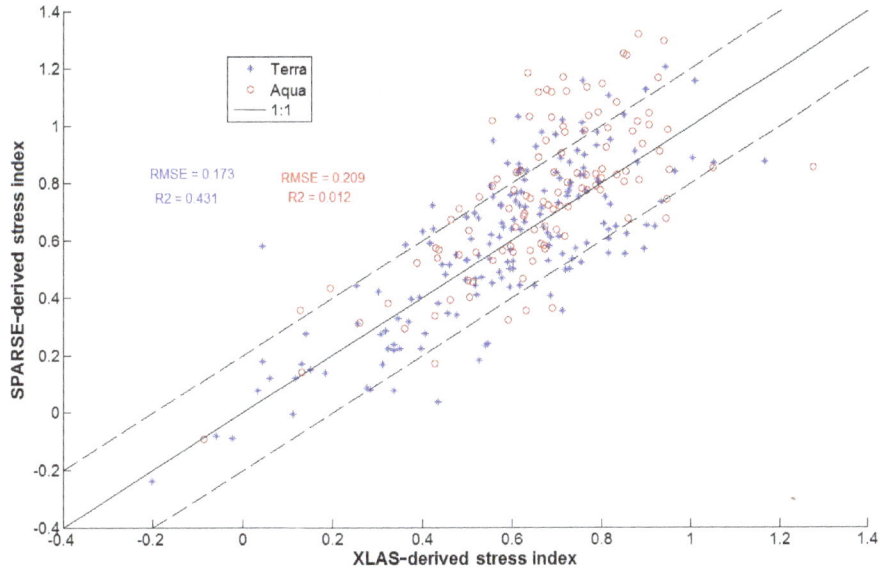

Figure 9. Modeled vs. XLAS-derived stress index SF at Terra and Aqua overpass times.

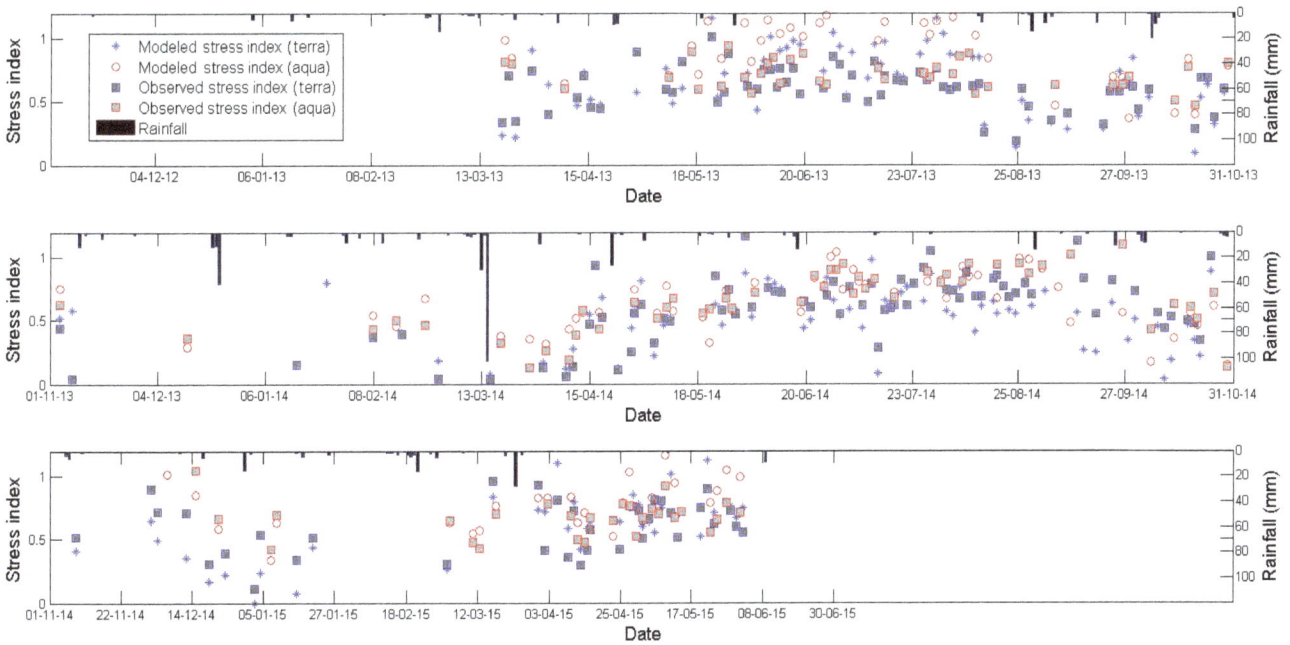

Figure 10. Modeled and observed stress index evolution at Terra and Aqua overpass times compared to daily rainfall.

tween the instantaneous estimate of the surface temperature during the satellite overpass (which can be influenced by high-frequency turbulence) and the aggregated values of other forcing data which are derived from half-hourly averages (Lagouarde et al., 2013, 2015). However, general tendencies are well reproduced, with most points located within a 0.2 confidence interval (illustrated by dotted lines along the 1 : 1 line) as found by Boulet et al. (2015) at field scale, which is encouraging from the perspective of assimilating ET or SF in a water balance model for example. Moreover, it is

noted that results include small LE and LE$_p$ values with the same order of magnitude as the measurement uncertainty itself. Most outliers with greater water stress (~ 1) correspond to high evaporation from bare soil since the dominant land use in the study area is arboriculture, but also, this could be due to saturation of scintillation which led to an underestimation of H XLAS measurements, as pointed out by Frehlich and Ochs (1990) and Kohsiek et al. (2002).

Modeled and observed stress indexes at Terra and Aqua overpass times show a consistent evolution with daily rain-

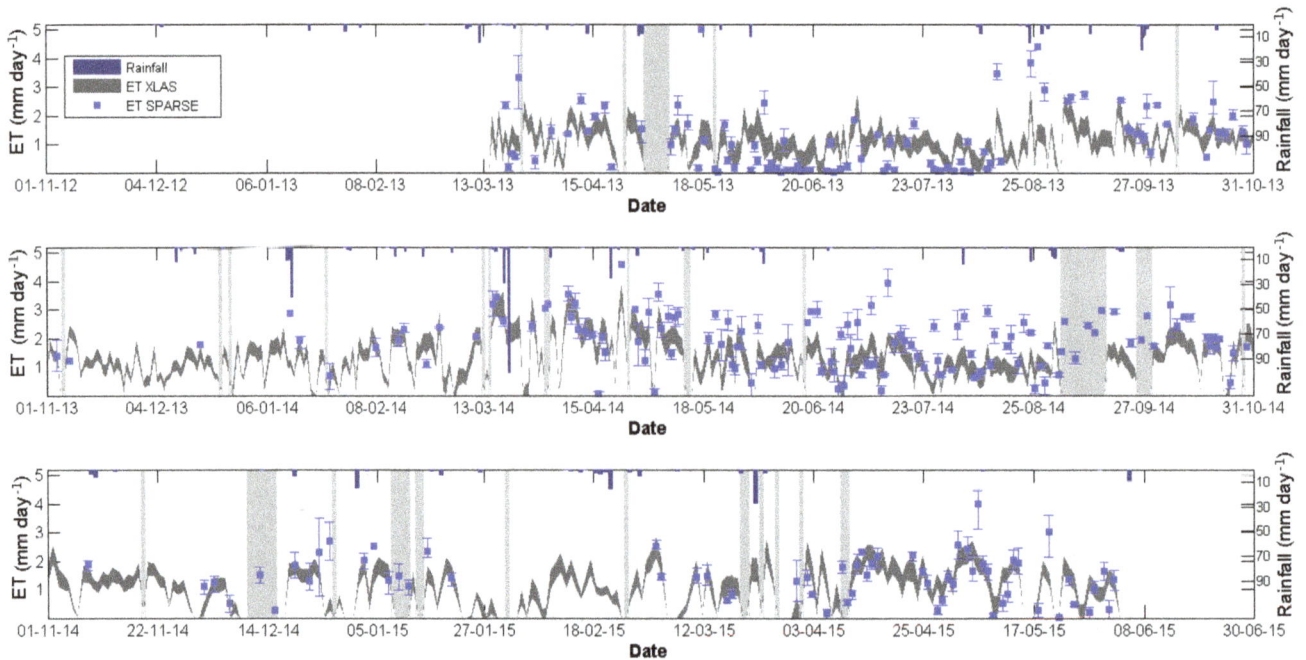

Figure 11. Modeled vs. observed daily latent heat fluxes. Dark grey color shows minimum and maximum daily observed LE. Light grey vertical bars show gaps in XLAS data. Error bars for the modeled ET show the minimum and the maximum daily ET resulting from the three methods used to compute daily ET from instantaneous modeled ET.

fall (Fig. 10), although the modeled stress shows a greater dispersion than the observed one. During a rainy episode (or an eventual irrigation period), the surface temperature decreases towards the unstressed surface temperature, thus marking an unstressed state, and SF tends to 0. Conversely, after a long dry down, the water stress appears and the surface temperature increases towards the equilibrium surface temperature computed by SPARSE under stressed conditions, and SF tends towards 1. Besides, it is noted that modeled stress indexes computed on the basis of Aqua MODIS's T_{surf} are often greater than those computed using Terra MODIS's T_{surf} due to higher T_{surf} (higher global solar radiation) at the time of Terra overpass (around midday).

6.4 XLAS- and model-derived daily latent heat fluxes

Daily observed ET, i.e., LE_residual_XLAS$_{day-FP}$, was computed using the residual method; hence, six estimates of the daily observed ET were obtained by combining the two satellite datasets and three methods to compute G and thus AE (see Sect. 3.3). Only the residual method was used to estimate daily observed ET for two reasons; on the one hand, to reduce the computations approach since, already, three methods to compute AE have been tested and, on the other hand, because the application of the EF method was not possible since we do not have a measured spatially distributed potential evapotranspiration (only point potential evapotranspiration data at the Ben Salem meteorological station are available). From daily observed ET estimates, minimum and max-

imum ET were selected for each day and minimum and maximum daily ET time series were interpolated between successive days based on the self preservation of the ratio of AE to R_g as scale factor (Fig. 11).

In addition, three methods were used to compute SPARSE daily ET for the Terra and Aqua overpasses (see Sect. 4.2), providing six estimates of the daily modeled ET. For each day average ET was plotted (260 days) with error bars showing minimum and maximum values, along with precipitation to understand the rainfall impact on the ET evolution (Fig. 11).

Despite the uncertainty in reconstructing the daily ET from instantaneous ET, overall results show a good agreement between XLAS-derived and SPARSE-derived ET values with similar seasonal dynamics. Daily observed and modeled ET over the whole study period were both in the range of 0–4 mm day^{-1} with an RMSE of 0.7 mm day^{-1}, which is consistent with the land use present in the XLAS path: mainly trees spaced by a considerable fraction of bare soil and fewer herbaceous soil-covering crops (see Sect. 3.2). As expected, ET rates decrease significantly during dry periods (summers) since arid conditions limit the latent heat flux in favor of sensible heat flux and increase immediately after rainfall events due to the high amount of water evaporated from soil. The rainfall peaks that occurred on 3 September 2013 (about 10 mm), 6 October 2013 (about 20 mm), 15 March 2014 (about 100 mm) and 22 April 2014 (about 25 mm) are followed by well-reproduced dry downs.

At seasonal scale, we note a good agreement between modeled and observed daily ET for the 2013–2014 and 2014–2015 seasons, especially when vegetation cover was more developed: from March to July 2014 and from March to May 2015; these periods correspond to cereal vegetation peaks in some plots (March–April) and to market gardening crops (e.g., tomato, water melon, pepper) cultivated generally from spring to the beginning of autumn in the inter-row area of trees plots, which is a common farming practice in the Kairouan plain. However, the 2012–2013 season was dry compared with the two other ones, and less accurate results were obtained. Some points with little to null ET were recorded from May to July 2013 which can be explained by the very dry conditions and scattered vegetation cover with a considerable amount of bare soil. This behavior was not observed in the same period of 2014, because 2014 was a rainy year in comparison to 2013; therefore, even supposing that the farmers have the same attitude and cultivate the same crop types between the two years (which is not true in the context of our study area and farmers always change crop types), precipitation favors the growth of spontaneous vegetation over fallows which contribute to ET rise. However, since this year experiences more rain, farmers cultivate a larger part of the land and diversify the crop types; the vegetation cover is denser and contributes to an overall increase in ET. Overall, lower ET values are recorded in autumn (October and November), which correspond to evapotranspiration from trees only, since the latest summer crops (market gardening crops) have been already harvested and the winter crops (mainly cereals) are not yet sown.

Moreover, it can be seen that occasionally SPARSE overestimated ET. For example, three dates can be selected in August 2013 (15, 25 and 29 August 2013) for which modeled ET were 3.30, 3.80 and 2.80 mm while maximum observed ET were 2.0, 2.40 and 1.20 mm, respectively; broader amplitude between modeled (4.00 mm) and observed ET (1.40 mm) was also recorded on 18 May 2013. SPARSE also overestimates ET throughout 10 days in August 2014 with an average difference of 1.1 mm and a maximum difference of 1.60 mm recorded in 23 August 2014. These discrepancies are always recorded under wet conditions (minimum stress factor) which show the difficulty in accurately representing the conditions close to the potential ET. This might be related to the theoretical limit of the model for low vegetation stress, especially when coupled with low evaporation efficiencies (i.e., dry soil surface), as already reported by Boulet et al. (2015) for senescent vegetation. The average difference between SPARSE and XLAS-derived LE estimates when both are available indicate that SPARSE can predict evapotranspiration with accuracies approaching 5 % of that of the XLAS.

7 Conclusions

This study evaluated the performances of the SPARSE model forced by MODIS remote sensing products in an operational context (no model calibration) to estimate instantaneous and daily evapotranspiration. The validation protocol was based on an unprecedented dataset with an extra large aperture scintillometer. Indeed, to our knowledge, this is the first work based on XLAS measurements acquired during the course of more than 2 years, as compared to 3 months in previous works (Kohsiek et al., 2002; Moene et al., 2006). The estimates of the sensible heat flux derived from the SPARSE model are in close agreement with those obtained from the XLAS. These results indicate that the XLAS can be fruitfully used to validate large-scale sensible heat flux derived from remote sensing data (and residual latent heat flux), in particular for the results obtained at the satellite overpass times, providing a feasible alternative to local micrometeorological techniques for measuring the sensible heat flux and validating satellite-derived estimates (i.e., eddy correlation). Furthermore, the extrapolation from instantaneous to daily evapotranspiration is less obvious and three methods were tested based on the stress index, the evaporative fraction and the residual approach. The daily latent heat fluxes derived from the XLAS agreed rather well with those modeled using the SPARSE model, which shows the potential of the SPARSE model in water consumption monitoring over heterogeneous landscapes in semiarid conditions, and especially to locate areas most affected by water stress. However, the precision in ET prediction with the SPARSE model is restricted by several assumptions and uncertainties: for instance, the instantaneous remote sensing data, mainly T_{surf} which is paramount in stress coefficient computation, are assumed to be reliable. Moreover, there is an issue with the MODIS pixel heterogeneity and notably the distribution of components at the intersection between the square pixel and the XLAS footprint. Uncertainties are also due to half-hourly forcing (meteorological and flux data) and XLAS data as well as to the extrapolation method from instantaneous to daily results. Furthermore, the empirical estimation methods of soil heat flux G (three methods were tested) as well as the possible daily heat accumulation lead to possible errors in available energy estimation and in turn in residual LE estimation.

Even if overall results are encouraging, further work is needed to improve results by (i) being most efficient in the SPARSE model application using calibrated input data specific to our study area, especially input parameters to which the model is particularly sensitive such as the mean leaf width and the minimum stomatal resistance; (ii) taking into account the heterogeneity of the 1 km MODIS pixel by applying MODIS footprint, which is determined by the sensor's observation geometry; and (iii) using a land surface model applied at the field scale (Etchanchu et al., 2017) to analyze the scaling properties from the field to the footprint of the XLAS and the MODIS pixels similarly.

Finally, in a future work, we plan to take advantage of the complementarities between the soil water balance and surface energy balance approaches (i.e., continuous but uncertain estimates using SWB due to poor soil water content control on the one hand and sensitivity of SEB to the actual water stress on the other hand) to implement an assimilation scheme of the remotely sensed surface temperature into land surface models. In fact, in order to provide further information about distributed soil water status over the studied areas, the TIR-derived evapotranspiration products could be assimilated directly either in land surface or hydrological models.

Author contributions. SS conducted the data processing, data analysis and results interpretation. GB conducted data analysis and results interpretation. MB performed the SPARSE inputs and XLAS data processing and analysis. AB carried out XLAS data processing and analysis. ÉD conducted daily evaporative fraction computation and analysis. BM and ZLC managed the site, and PF managed site instrumentation. VS and ZLC contributed with ideas and discussions.

Competing interests. The authors declare that they have no conflict of interest.

Acknowledgements. The authors are thankful to the GDAs of Ben Salem I and Ben Salem II, which enabled the scintillometer setup and access above the two water towers. Funding from the CNES/TOSCA program for the EVA2IRT project, from the MISTRALS/SICMED program for the ReSAMEd project, from the ORFEO/CNES Program for Pleiades images (©CNES 2012, Distribution Airbus DS, all rights reserved) and from the ANR/TRANSMED program for the AMETHYST project (ANR-12-TMED-0006-01) as well as the mobility support from PHC Maghreb program (no. 32592VE/PHC 14 MAG 22) are gratefully acknowledged. We also thank the International Mixed Laboratory NAILA. This work has benefited also from the financial support of the ARTS program ("Allocations de recherche pour une thèse au Sud") of IRD (Institut de Recherche pour le Développement).

Edited by: Nunzio Romano

References

Allen, R. G., Pereira, L. S., Raes, D., and Smith, M.: Crop evapotranspiration – Guidelines for computing crop water require-

ments, FAO Irrigation and drainage paper 56, FAO, Rome, http://www.fao.org/docrep/X0490E/X0490E00.htm (last access: April 2018), 1998.

Allen, R. G., Walter, I., Elliott, R., Howell, T., Itenfisu, D., Jensen, M., and Snyder, R.: The ASCE standardized reference evapotranspiration equation, American Society of Civil Engineers, Reston, Virginia, 1–69, 2005.

Allen, R. G., Irmak, A., Trezza, R., Hendrickx, J. M., Bastiaanssen, W., and Kjaersgaard, J.: Satellite-based ET estimation in agriculture using SEBAL and METRIC, Hydrol. Process., 25, 4011–4027, 2011.

Amri, R., Zribi, M., Lili-Chabaane, Z., Szczypta, C., Calvet, J. C., and Boulet, G.: FAO-56 dual model combined with multi-sensor remote sensing for regional evapotranspiration estimations, Remote Sensing, 6, 5387–5406, 2014.

Anderson, M. C., Norman, J., Diak, G., Kustas, W., and Mecikalski, J.: A two-source time-integrated model for estimating surface fluxes using thermal infrared remote sensing, Remote Sens. Environ., 60, 195–216, 1997.

Anderson, M. C., Kustas, W. P., Norman, J. M., Hain, C. R., Mecikalski, J. R., Schultz, L., González-Dugo, M. P., Cammalleri, C., d'Urso, G., Pimstein, A., and Gao, F.: Mapping daily evapotranspiration at field to continental scales using geostationary and polar orbiting satellite imagery, Hydrol. Earth Syst. Sci., 15, 223–239, https://doi.org/10.5194/hess-15-223-2011, 2011.

Andreas, E. L.: Atmospheric stability from scintillation measurements, Appl. Optics, 27, 2241–2246, 1988.

Bai, J., Liu, S., and Mao, D.: Area-averaged evapotranspiration fluxes measured from large aperture scintillometer in the Hai River basin, in: River Basin Research And Planning Approach, Orient ACAD Forum, edited by: Zhang, H., Zhao, R., and Zhoa, H., Marrickville, Australia, 331–340, 2009.

Bai, J., Jia, L., Liu, S., Xu, Z., Hu, G., Zhu, M., and Song, L.: Characterizing the footprint of eddy covariance system and large aperture scintillometer measurements to validate satellite-based surface fluxes, IEEE Geosci. Remote Sens. Lett., 12, 943–947, 2015.

Bastiaanssen, W. G. M.: Regionalization of surface flux densities and moisture indicators in composite terrain; a remote sensing approach under clear skies in mediterranean climates, SC-DLO, Wageningen, 1995.

Bastiaanssen, W. G. M., Allen, R. G., Droogers, P., D'Urso, G., and Steduto, P.: Twenty-five years modeling irrigated and drained soils: State of the art, Agr. Water Manage., 92, 111–125, https://doi.org/10.1016/j.agwat.2007.05.013, 2007.

Boulet, G., Braud, I., and Vauclin, M.: Study of the mechanisms of evaporation under arid conditions using a detailed model of the soil–atmosphere continuum. Application to the EFEDA I experiment, J. Hydrol., 193, 114–141, https://doi.org/10.1016/S0022-1694(96)03148-4, 1997.

Boulet, G., Chehbouni, A., Gentine, P., Duchemin, B., Ezzahar, J., and Hadria, R.: Monitoring water stress using time series of observed to unstressed surface temperature difference, Agr. Forest Meteorol., 146, 159–172, https://doi.org/10.1016/j.agrformet.2007.05.012, 2007.

Boulet, G., Mougenot, B., Lhomme, J. P., Fanise, P., Lili-Chabaane, Z., Olioso, A., Bahir, M., Rivalland, V., Jarlan, L., Merlin, O., Coudert, B., Er-Raki, S., and Lagouarde, J. P.: The SPARSE model for the prediction of water stress and evapotranspiration components from thermal infra-red data and its evaluation over irrigated and rainfed wheat, Hydrol. Earth Syst. Sci., 19, 4653–4672, https://doi.org/10.5194/hess-19-4653-2015, 2015.

Bounoua, L., Zhang, P., Thome, K., Masek, J., Safia, A., Imhoff, M. L., and Wolfe, R. E.: Mapping Biophysical Parameters for Land Surface Modeling over the Continental US Using MODIS and Landsat, Dataset Pap. Sci., 2015, 564279, https://doi.org/10.1155/2015/564279, 2015.

Braud, I., Dantas-Antonino, A. C., Vauclin, M., Thony, J. L., and Ruelle, P.: A simple soil-plant-atmosphere transfer model (SiSPAT) development and field verification, J. Hydrol., 166, 213–250, https://doi.org/10.1016/0022-1694(94)05085-C, 1995.

Brunsell, N. A., Ham, J. M., and Arnold, K. A.: Validating remotely sensed land surface fluxes in heterogeneous terrain with large aperture scintillometry, Int. J. Remote Sens., 32, 6295–6314, https://doi.org/10.1080/01431161.2010.508058, 2011.

Brutsaert, W. and Sugita, M.: Application of self-preservation in the diurnal evolution of the surface energy budget to determine daily evaporation, J. Geophys. Res.-Atmos., 97, 18377–18382, 1992.

Calera, A., Campos, I., Osann, A., D'Urso, G., and Menenti, M.: Remote Sensing for Crop Water Management: From ET Modelling to Services for the End Users, Sensors, 17, 1104, https://doi.org/10.3390/s17051104, 2017.

Chahbi, A., Zribi, M., Saadi, S., Simonneaux, V., and Lili Chabaane, Z.: Classification et caractérisation de la couverture végétale dans un milieu semi aride en utilisant des images SPOT 5, Deuxième Workshop AMETHYST, 11 Février 2016, Marrakech, Maroc, 2016,

Chávez, J., Neale, C. M. U., Hipps, L. E., Prueger, J. H., and Kustas, W. P.: Comparing Aircraft-Based Remotely Sensed Energy Balance Fluxes with Eddy Covariance Tower Data Using Heat Flux Source Area Functions, J. Hydrometeorol., 6, 923–940, https://doi.org/10.1175/jhm467.1, 2005.

Chehbouni, A., Watts, C., Lagouarde, J. P., Kerr, Y. H., Rodriguez, J. C., Bonnefond, J. M., Santiago, F., Dedieu, G., Goodrich, D. C., and Unkrich, C.: Estimation of heat and momentum fluxes over complex terrain using a large aperture scintillometer, Agr. Forest Meteorol., 105, 215–226, https://doi.org/10.1016/S0168-1923(00)00187-8, 2000.

Chirouze, J., Boulet, G., Jarlan, L., Fieuzal, R., Rodriguez, J. C., Ezzahar, J., Er-Raki, S., Bigeard, G., Merlin, O., Garatuza-Payan, J., Watts, C., and Chehbouni, G.: Intercomparison of four remote-sensing-based energy balance methods to retrieve surface evapotranspiration and water stress of irrigated fields in semi-arid climate, Hydrol. Earth Syst. Sci., 18, 1165–1188, https://doi.org/10.5194/hess-18-1165-2014, 2014.

Choudhury, B. J. and Monteith, J.: A four-layer model for the heat budget of homogeneous land surfaces, Q. J. Roy. Meteorol. Soc., 114, 373–398, 1988.

Choudhury, B. J., Idso, S. B., and Reginato, R. J.: Analysis of an empirical model for soil heat flux under a growing wheat crop for estimating evaporation by an infrared-temperature based energy balance equation, Agr. Forest Meteorol., 39, 283–297, https://doi.org/10.1016/0168-1923(87)90021-9, 1987.

Courault, D., Seguin, B., and Olioso, A.: Review on estimation of evapotranspiration from remote sensing data: From empirical to numerical modeling approaches, Irrig. Drain. Syst., 19, 223–249, https://doi.org/10.1007/s10795-005-5186-0, 2005.

Clevers, J. G. P. W.: Application of a weighted infrared-red vegetation index for estimating leaf Area Index by Correct-

ing for Soil Moisture, Remote Sens. Environ., 29, 25–37, https://doi.org/10.1016/0034-4257(89)90076-X, 1989.

Clothier, B., Clawson, K., Pinter, P., Moran, M., Reginato, R. J., and Jackson, R.: Estimation of soil heat flux from net radiation during the growth of alfalfa, Agr. Forest Meteorol., 37, 319–329, 1986.

Danelichen, V. H. D. M., Biudes, M. S., Souza, M. C., Machado, N. G., d. Silva, B. B., and d. Nogueira, J. S.: Estimation of soil heat flux in a neotropical Wetland region using remote sensing techniques, Rev. Brasil. Meteorol., 29, 469–482, 2014.

De Bruin, H. and Wang, J.: Scintillometry: a review, Researchgate, https://www.researchgate.net/publication/316285424_Scintillometry_a_review, last access: 17 June 2017.

Delogu, E., Boulet, G., Olioso, A., Coudert, B., Chirouze, J., Ceschia, E., Le Dantec, V., Marloie, O., Chehbouni, G., and Lagouarde, J. P.: Reconstruction of temporal variations of evapotranspiration using instantaneous estimates at the time of satellite overpass, Hydrol. Earth Syst. Sci., 16, 2995–3010, https://doi.org/10.5194/hess-16-2995-2012, 2012.

Etchanchu, J., Rivalland, V., Gascoin, S., Cros, J., Brut, A., and Boulet, G.: Effects of multi-temporal high-resolution remote sensing products on simulated hydrometeorological variables in a cultivated area (southwestern France), Hydrol. Earth Syst. Sci., 21, 5693–5708, https://doi.org/10.5194/hess-21-5693-2017, 2017.

Feddes, R. A., Kowalik, P. J., and Zaradny, H.: Simulation of Field Water Use and Crop Yield, Wiley, USA, 1978.

Frehlich, R. G. and Ochs, G. R.: Effects of saturation on the optical scintillometer, Appl. Optics, 29, 548–553, 1990.

Gentine, P., Entekhabi, D., Chehbouni, A., Boulet, G., and Duchemin, B.: Analysis of evaporative fraction diurnal behaviour, Agr. Forest Meteorol., 143, 13–29, 2007.

Giorgi, F. and Avissar, R.: Representation of heterogeneity effects in earth system modeling: Experience from land surface modeling, Rev. Geophys., 35, 413–437, 1997.

Giorgi, F. and Lionello, P.: Climate change projections for the Mediterranean region, Global Planet. Change, 63, 90–104, 2008.

Glenn, E. P., Huete, A. R., Nagler, P. L., Hirschboeck, K. K., and Brown, P.: Integrating remote sensing and ground methods to estimate evapotranspiration, Crit. Rev. Plant Sci., 26, 139–168, 2007.

Green, A. E. and Hayashi, Y.: Use of the scintillometer technique over a rice paddy, J. Agricult. Meteorol., 54, 225–234, 1998.

Gruber, M. and Fochesatto, G. J.: A new sensitivity analysis and solution method for scintillometermeasurements of area-averaged turbulent fluxes, Bound.-Lay. Meteorol., 149, 65–83, 2013.

Gurney, R. and Hsu, A.: Relating evaporative fraction to remotely sensed data at the FIFE site, in: Symposium on FIFE – First ISLSCP Field Experiment, 7–9 February 1990, Anaheim, CA, USA, p. 5, 1990.

Hain, C. R., Mecikalski, J. R., and Anderson, M. C.: Retrieval of an Available Water-Based Soil Moisture Proxy from Thermal Infrared Remote Sensing. Part I: Methodology and Validation, J. Hydrometeorol., 10, 665–683, https://doi.org/10.1175/2008jhm1024.1, 2009.

Hartogensis, O. K., Watts, C. J., Rodriguez, J.-C., and Bruin, H. A. R. D.: Derivation of an Effective Height for Scintillometers: La Poza Experiment in Northwest Mexico,

J. Hydrometeorol., 4, 915–928, https://doi.org/10.1175/1525-7541(2003)004<0915:doaehf>2.0.co;2, 2003.

Hemakumara, H. M., Chandrapala, L., and Moene, A. F.: Evapotranspiration fluxes over mixed vegetation areas measured from large aperture scintillometer, Agr. Water Manage., 58, 109–122, https://doi.org/10.1016/S0378-3774(02)00131-2, 2003.

Hill, R., Clifford, S. F., and Lawrence, R. S.: Refractive-index and absorption fluctuations in the infrared caused by temperature, humidity, and pressure fluctuations, J. Opt. Soc. Am., 70, 1192–1205, 1980.

Hoedjes, J. C. B., Chehbouni, A., Jacob, F., Ezzahar, J., and Boulet, G.: Deriving daily evapotranspiration from remotely sensed instantaneous evaporative fraction over olive orchard in semi-arid Morocco, J. Hydrol., 354, 53–64, 2008.

Hoedjes, J., Ramier, D., Boulain, N., Boubkraoui, S., Cappelaere, B., Descroix, L., Mougenot, B., and Timouk, F.: Combining scintillometer measurements and an aggregation scheme to estimate area-averaged latent heat flux during the AMMA experiment, J. Hydrol., 375, 217–226, https://doi.org/10.1016/j.jhydrol.2009.01.010, 2009.

Horst, T. and Weil, J.: Footprint estimation for scalar flux measurements in the atmospheric surface layer, Bound.-Lay. Meteorol., 59, 279–296, 1992.

Hunink, J., Eekhout, J., Vente, J., Contreras, S., Droogers, P., and Baille, A.: Hydrological Modelling using Satellite-Based Crop Coefficients: A Comparison of Methods at the Basin Scale, Remote Sensing, 9, 174–190, 2017.

Jackson, R. D., Moran, M. S., Gay, L. W., and Raymond, L. H.: Evaluating evaporation from field crops using airborne radiometry and ground-based meteorological data, Irrigation Science, 8, 81–90, https://doi.org/10.1007/bf00259473, 1987.

Kalma, J. D., McVicar, T. R., and McCabe, M. F.: Estimating land surface evaporation: A review of methods using remotely sensed surface temperature data, Surv. Geophys., 29, 421–469, 2008.

Kipp & Zonen: LAS and X-LAS instruction manual, http://www.kippzonen.fr/Download/244/LAS-and-X-LAS-Scintillometers-Manual?ShowInfo=true, last access: 7 December 2017.

Kohsiek, W., Meijninger, W. M. L., Moene, A. F., Heusinkveld, B. G., Hartogensis, O. K., Hillen, W. C. A. M., and De Bruin, H. A. R.: An Extra Large Aperture Scintillometer For Long Range Applications, Bound.-Lay. Meteorol., 105, 119–127, https://doi.org/10.1023/a:1019600908144, 2002.

Kohsiek, W., Meijninger, W. M. L., Debruin, H. A. R., and Beyrich, F.: Saturation of the Large Aperture Scintillometer, Bound.-Lay. Meteorol., 121, 111–126, https://doi.org/10.1007/s10546-005-9031-7, 2006.

Kroes, J. G., van Dam, J. C., Bartholomeus, R. P., Groenendijk, P., Heinen, M., Hendriks, R. F. A., Mulder, H. M., Supit, I., and van Walsum, P. E. V.: SWAP version 4; Theory description and user manual, Report 2780, Wageningen Environmental Research, Wageningen, available at: http://library.wur.nl/WebQuery/wurpubs/fulltext/416321, last access: 21 March 2017.

Kustas, W. P. and Anderson, M.: Advances in thermal infrared remote sensing for land surface modeling, Agr. Forest Meteorol., 149, 2071–2081, 2009.

Kustas, W. P. and Daughtry, C. S. T.: Estimation of the soil heat flux/net radiation ratio from spectral data, Agr. Forest Meteorol., 49, 205–223, https://doi.org/10.1016/0168-1923(90)90033-3, 1990.

ustas, W. P. and Norman, J. M.: Evaluation of soil and vegetation heat flux predictions using a simple two-source model with radiometric temperatures for partial canopy cover, Agr. Forest Meteorol., 94, 13–29, https://doi.org/10.1016/S0168-1923(99)00005-2, 1999.

Kustas, W. P., Daughtry, C. S. T., and Van Oevelen, P. J.: Analytical treatment of the relationships between soil heat flux/net radiation ratio and vegetation indices, Remote Sens. Environ., 46, 319–330, https://doi.org/10.1016/0034-4257(93)90052-Y, 1993.

Lagouarde, J.-P., Jacob, F., Gu, X. F., Olioso, A., Bonnefond, J.-M., Kerr, Y., Mcaneney, K. J., and Irvine, M.: Spatialization of sensible heat flux over a heterogeneous landscape, Agronomie – Sciences des Productions Vegetales et de l'Environnement, 22, 627–634, 2002.

Lagouarde, J.-P., Bach, M., Sobrino, J. A., Boulet, G., Briottet, X., Cherchali, S., Coudert, B., Dadou, I., Dedieu, G., and Gamet, P.: The MISTIGRI thermal infrared project: scientific objectives and mission specifications, Int. J. Remote Sens., 34, 3437–3466, 2013.

Lagouarde, J.-P., Irvine, M., and Dupont, S.: Atmospheric turbulence induced errors on measurements of surface temperature from space, Remote Sens. Environ., 168, 40–53, https://doi.org/10.1016/j.rse.2015.06.018, 2015.

Leclerc, M. Y. and Thurtell, G. W.: Footprint prediction of scalar fluxes using a Markovian analysis, Bound.-Lay. Meteorol., 52, 247–258, https://doi.org/10.1007/bf00122089, 1990.

Leduc, C., Calvez, R., Beji, R., Nazoumou, Y., Lacombe, G., and Aouadi, C.: Evolution de la ressource en eau dans la vallée du Merguellil (Tunisie centrale), in: Séminaire Euro-Méditerranéen sur la Modernisation de l'Agriculture Irriguée, 19–23 Avril 2004, Rabat, Maroc, p. 10, 2004.

Lhomme, J.-P. and Elguero, E.: Examination of evaporative fraction diurnal behaviour using a soil–vegetation model coupled with a mixed-layer model, Hydrol. Earth Syst. Sci., 3, 259–270, https://doi.org/10.5194/hess-3-259-1999, 1999.

Li, Z.-L., Tang, R., Wan, Z., Bi, Y., Zhou, C., Tang, B., Yan, G., and Zhang, X.: A review of current methodologies for regional evapotranspiration estimation from remotely sensed data, Sensors, 9, 3801–3853, 2009.

Liou, Y.-A. and Kar, S.: Evapotranspiration Estimation with Remote Sensing and Various Surface Energy Balance Algorithms – A Review, Energies, 7, 2821–2849, 2014.

Marx, A., Kunstmann, H., Schüttemeyer, D., and Moene, A. F.: Uncertainty analysis for satellite derived sensible heat fluxes and scintillometer measurements over Savannah environment and comparison to mesoscale meteorological simulation results, Agr. Forest Meteorol., 148, 656–667, https://doi.org/10.1016/j.agrformet.2007.11.009, 2008.

Mauser, W. and Schädlich, S.: Modelling the spatial distribution of evapotranspiration on different scales using remote sensing data, J. Hydrol., 212, 250–267, 1998.

Meijninger, W. M. L., Hartogensis, O. K., Kohsiek, W., Hoedjes, J. C. B., Zuurbier, R. M., and De Bruin, H. A. R.: Determination of Area-Averaged Sensible Heat Fluxes with a Large Aperture Scintillometer over a Heterogeneous Surface –

Flevoland Field Experiment, Bound.-Lay. Meteorol., 105, 37–62, https://doi.org/10.1023/a:1019647732027, 2002.

Minacapilli, M., Ciraolo, G., D'Urso, G., and Cammalleri, C.: Evaluating actual evapotranspiration by means of multi-platform remote sensing data: a case study in Sicily, IAHS Publication, 316, 207–219, 2007.

Minacapilli, M., Agnese, C., Blanda, F., Cammalleri, C., Ciraolo, G., D'Urso, G., Iovino, M., Pumo, D., Provenzano, G., and Rallo, G.: Estimation of actual evapotranspiration of Mediterranean perennial crops by means of remote-sensing based surface energy balance models, Hydrol. Earth Syst. Sci., 13, 1061–1074, https://doi.org/10.5194/hess-13-1061-2009, 2009.

Moene, A. F., Meijninger, W., Kohsiek, W., Gioli, B., Miglietta, F., and Bosveld, F.: Validation of fluxes of an extra large aperture scintillometer at Cabauw using sky arrow aircraft flux measurements, in: Proceedings of 17th symposium on boundary layers and turbulence, American Meteorological Society, San Diego, 22–25, CA, 2006.

Monin, A. and Obukhov, A.: Basic laws of turbulent mixing in the surface layer of the atmosphere, Contrib. Geophys. Inst. Acad. Sci. USSR, 24, 163–187, 1954.

Mougenot, B., Touhami, N., Lili Chabaane, Z., Boulet, G., Simonneaux, V., and Zribi, M.: Trees detection for water resources management in irrigated and rainfed arid and semi-arid agricultural areas, in: Pléiades Days, 1–3 April 2014, Toulouse, 2014.

Mutziger, A. J., Burt, C. M., Howes, D. J., and Allen, R. G.: Comparison of measured and FAO-56 modeled evaporation from bare soil, J. Irrig. Drain. Eng., 131, 59–72, 2005.

Odi-Lara, M., Campos, I., Neale, C., Ortega-Farías, S., Poblete-Echeverría, C., Balbontín, C., and Calera, A.: Estimating Evapotranspiration of an Apple Orchard Using a Remote Sensing-Based Soil Water Balance, Remote Sensing, 8, 253–273, 2016.

Oki, T. and Kanae, S.: Global Hydrological Cycles and World Water Resources, Science, 313, 1068–1072, https://doi.org/10.1126/science.1128845, 2006.

Overgaard, J., Rosbjerg, D., and Butts, M. B.: Land-surface modelling in hydrological perspective – a review, Biogeosciences, 3, 229–241, https://doi.org/10.5194/bg-3-229-2006, 2006.

Pereira, L. S., Oweis, T., and Zairi, A.: Irrigation management under water scarcity, Agr. Water Manage., 57, 175–206, 2002.

Poussin, J. C., Imache, A., Beji, R., Le Grusse, P., and Benmihoub, A.: Exploring regional irrigation water demand using typologies of farms and production units: An example from Tunisia, Agr. Water Manage., 95, 973–983, https://doi.org/10.1016/j.agwat.2008.04.001, 2008.

Pradeleix, L., Roux, P., Bouarfa, S., Jaouani, B., Lili-Chabaane, Z., and Bellon-Maurel, V.: Environmental Impacts of Contrasted Groundwater Pumping Systems Assessed by Life Cycle Assessment Methodology: Contribution to the Water–Energy Nexus Study, Irrig. Drain., 64, 124–138, 2015.

Raupach, M. R.: Simplified expressions for vegetation roughness length and zero-plane displacement as functions of canopy height and area index, Bound.-Lay. Meteorol., 71, 211–216, https://doi.org/10.1007/bf00709229, 1994.

Saadi, S., Simonneaux, V., Boulet, G., Raimbault, B., Mougenot, B., Fanise, P., Ayari, H., and Lili-Chabaane, Z.: Monitoring Irrigation Consumption Using High Resolution NDVI Image Time Series: Calibration and Validation in the Kairouan Plain (Tunisia), Remote Sensing, 7, 13005–13028, https://doi.org/10.3390/rs71013005, 2015.

Samain, B., Simons, G. W. H., Voogt, M. P., Defloor, W., Bink, N.-J., and Pauwels, V. R. N.: Consistency between hydrological model, large aperture scintillometer and remote sensing based evapotranspiration estimates for a heterogeneous catchment, Hydrol. Earth Syst. Sci., 16, 2095–2107, https://doi.org/10.5194/hess-16-2095-2012, 2012.

Santanello Jr., J. A. and Friedl, M. A.: Diurnal covariation in soil heat flux and net radiation, J. Appl. Meteorol., 42, 851–862, 2003.

Schmid, H. P.: Footprint modeling for vegetation atmosphere exchange studies: a review and perspective, Agr. Forest Meteorol., 113, 159–183, https://doi.org/10.1016/S0168-1923(02)00107-7, 2002.

Shuttleworth, W. J. and Gurney, R. J.: The theoretical relationship between foliage temperature and canopy resistance in sparse crops, Q. J. Roy. Meteorol. Soc., 116, 497–519, 1990.

Simonneaux, V., Lepage, M., Helson, D., Metral, J., Thomas, S., Duchemin, B., Cherkaoui, M., Kharrou, H., Berjami, B., and Chehbouni, A.: Estimation spatialisée de l'évapotranspiration des cultures irriguées par télédétection: application à la gestion de l'irrigation dans la plaine du Haouz (Marrakech, Maroc), Science et changements planétaires/Sécheresse, 20, 123–130, 2009.

Solignac, P. A., Brut, A., Selves, J. L., Béteille, J. P., Gastellu-Etchegorry, J. P., Keravec, P., Béziat, P., and Ceschia, E.: Uncertainty analysis of computational methods for deriving sensible heat flux values from scintillometer measurements, Atmos. Meas. Tech., 2, 741–753, https://doi.org/10.5194/amt-2-741-2009, 2009.

Steduto, P., Hsiao, T. C., Raes, D., and Fereres, E.: AquaCrop – The FAO crop model to simulate yield response to water: I. Concepts and underlying principles, Agron. J., 101, 426–437, 2009.

Stöckle, C. O., Donatelli, M., and Nelson, R.: CropSyst, a cropping systems simulation model, Eur. J. Agron., 18, 289–307, 2003.

Su, Z.: The Surface Energy Balance System (SEBS) for estimation of turbulent heat fluxes, Hydrol. Earth Syst. Sci., 6, 85–100, https://doi.org/10.5194/hess-6-85-2002, 2002.

Sugita, M. and Brutsaert, W.: Regional surface fluxes from remotely sensed skin temperature and lower boundary layer measurements, Water Resour. Res., 26, 2937–2944, 1990.

Tasumi, M., Trezza, R., Allen, R. G., and Wright, J. L.: Operational aspects of satellite-based energy balance models for irrigated crops in the semi-arid US, Irrig. Drain. Syst., 19, 355–376, 2005.

Tatarskii, V. I. (Ed.): Wave propagation in turbulent medium, in: Wave Propagation in Turbulent Medium, by Valerian Ilich Tatarskii, translated by: Silverman, R. A., McGraw-Hill, UK, 285 pp., 1961.

Torres, E. A. and Calera, A.: Bare soil evaporation under high evaporation demand: a proposed modification to the FAO-56 model, Hydrological Sciences Journal – Journal des Sciences Hydrologiques, 55, 303–315, 2010.

Touhami, N.: Détection des arbres par imagerie Très Haute Résolution Spatiale sur la plaine de Kairouan Engineer, Institut National Agronomqiue de Tunisie, Tunis, 78 pp., 2013.

Tucker, C. J.: A comparison of satellite sensor bands for vegetation monitoring, Photogram. Eng. Remote Sens., 44, 1369–1380, 1978.

Twine, T. E., Kustas, W., Norman, J., Cook, D., Houser, P., Meyers, T., Prueger, J., Starks, P., and Wesely, M.: Correcting eddy-covariance flux underestimates over a grassland, Agr. Forest Meteorol., 103, 279–300, 2000.

Ventura, F., Spano, D., Duce, P., and Snyder, R.: An evaluation of common evapotranspiration equations, Irrigation Science, 18, 163–170, 1999.

Watts, C. J., Chehbouni, A., Rodriguez, J. C., Kerr, Y. H., Hartogensis, O., and de Bruin, H. A. R.: Comparison of sensible heat flux estimates using AVHRR with scintillometer measurements over semi-arid grassland in northwest Mexico, Agr. Forest Meteo-

rol., 105, 81–89, https://doi.org/10.1016/S0168-1923(00)00188-X, 2000.

Xie, Y., Sha, Z., and Yu, M.: Remote sensing imagery in vegetation mapping: a review, J. Plant Ecol., 1, 9–23, 2008.

Zribi, M., Chahbi, A., Shabou, M., Lili-Chabaane, Z., Duchemin, B., Baghdadi, N., Amri, R., and Chehbouni, A.: Soil surface moisture estimation over a semi-arid region using ENVISAT ASAR radar data for soil evaporation evaluation, Hydrol. Earth Syst. Sci., 15, 345–358, https://doi.org/10.5194/hess-15-345-2011, 2011.

PERMISSIONS

LIST OF CONTRIBUTORS

Francesco Marra and Efrat Morin
Institute of Earth Sciences, Hebrew University of Jerusalem, 91904, Jerusalem, Israel

Nadav Peleg
Institute of Environmental Engineering, Hydrology and Water Resources Management, ETH Zurich, Zurich, Switzerland

Yiwen Mei and Emmanouil N. Anagnostou
Department of Civil and Environmental Engineering, University of Connecticut, Storrs, CT, USA

Diana Lucatero and Karsten H. Jensen
Department of Geosciences and Natural Resource Management, University of Copenhagen, Copenhagen, Denmark

Henrik Madsen
DHI, Hørsholm, Denmark

Jens C. Refsgaard and Jacob Kidmose
Geological Survey of Denmark and Greenland (GEUS), Copenhagen, Denmark

Mehmet C. Demirel
Geological Survey of Denmark and Greenland, Øster Voldgade 10, 1350 Copenhagen, Denmark
Department of Civil Engineering, Istanbul Technical University, 34469 Maslak, Istanbul, Turkey

Juliane Mai
Department Computational Hydrosystems, UFZ – Helmholtz Centre for Environmental Research, Leipzig, Germany
Department of Civil and Environmental Engineering, University of Waterloo, Waterloo, Canada

Gorka Mendiguren
Geological Survey of Denmark and Greenland, Øster Voldgade 10, 1350 Copenhagen, Denmark
Department of Environmental Engineering, Technical University of Denmark, 2800 Kgs. Lyngby, Denmark

Julian Koch
Geological Survey of Denmark and Greenland, Øster Voldgade 10, 1350 Copenhagen, Denmark
Department of Geosciences and Natural Resource Management, University of Copenhagen, Copenhagen, Denmark

Luis Samaniego
Department Computational Hydrosystems, UFZ – Helmholtz Centre for Environmental Research, Leipzig, Germany

Simon Stisen
Geological Survey of Denmark and Greenland, Øster Voldgade 10, 1350 Copenhagen, Denmark

Ewan Pinnington and Tristan Quaife
Department of Meteorology, University of Reading, Reading, UK
National Centre for Earth Observation, University of Reading, Reading, UK

Emily Black
Department of Meteorology, University of Reading, Reading, UK
National Centre for Atmospheric Science, University of Reading, Reading, UK

Clement Albergel, Simon Munier and Jean-Christophe Calvet
CNRM UMR 3589, Météo-France/CNRS, Toulouse, France

Emanuel Dutra
Instituto Dom Luiz, IDL, Faculty of Sciences, University of Lisbon, Lisbon, Portugal

Joaquin Munoz-Sabater, Patricia de Rosnay and Gianpaolo Balsamo
ECMWF, Reading, UK

Guillaume Le Bihan, Olivier Payrastre and Eric Gaume
IFSTTAR, GERS, EE, F-44344 Bouguenais, France

David Moncoulon
CCR, 157 boulevard Haussmann, 75008 Paris, France

Frédéric Pons
Cerema, Direction Méditerranée, 30 rue Albert Einstein, F-13593 Aix-en-Provence, France

Alessio Pugliese, Simone Persiano, Alberto Montanari and Attilio Castellarin
Department DICAM, University of Bologna, Bologna, Italy

Stefano Bagli and Paolo Mazzoli
GECOsistema srl, Cesena, Italy

Juraj Parajka and Günter Blöschl
Institute for Hydraulic and Water Resources Engineering, TU Wien, Vienna, Austria

Berit Arheimer and René Capell
Swedish Meteorological and Hydrological Institute (SMHI), Norrköping, Sweden

Fredrik Wetterhall and Francesca Di Giuseppe
European Centre for Medium-range Weather Forecasts, Shinfield Park, Reading, UK

Felipe Hernández and Xu Liang
Department of Civil and Environmental Engineering, University of Pittsburgh, Pittsburgh, PA, 15261, USA

Cecile M. M. Kittel and Peter Bauer-Gottwein
Department of Environmental Engineering, Technical University of Denmark, Technical University of Denmark, 2800 Kgs. Lyngby, Denmark

Karina Nielsen
National Space Institute, Technical University of Denmark, 2800 Kgs. Lyngby, Denmark

Christian Tøttrup
DHI-GRAS, 2970 Hørsholm, Denmark

Fabio F. Pereira
Sustainability Science Program, Kennedy School of Government, Harvard University, Cambridge, MA 02138, USA
Department of Renewable Energy Engineering, Federal University of Alagoas, Maceió, AL, Brazil

Fabio Farinosi
Sustainability Science Program, Kennedy School of Government, Harvard University, Cambridge, MA 02138, USA
Ca' Foscari University of Venice, Venice, Italy
European Commission, DG Joint Research Centre, Ispra, Italy

Mauricio E. Arias
Sustainability Science Program, Kennedy School of Government, Harvard University, Cambridge, MA 02138, USA

Department of Civil and Environmental Engineering, University of South Florida, Tampa, FL 33620, USA

Eunjee Lee
Sustainability Science Program, Kennedy School of Government, Harvard University, Cambridge, MA 02138, USA
Goddard Earth Sciences Technology and Research, Universities Space Research Association, Columbia, MD 21046, USA
Global Modeling and Assimilation Office, NASA Goddard Space Flight Center, Greenbelt, MD 22071, USA

John Briscoe
Sustainability Science Program, Kennedy School of Government, Harvard University, Cambridge, MA 02138, USA

Paul R. Moorcroft
Sustainability Science Program, Kennedy School of Government, Harvard University, Cambridge, MA 02138, USA

Frederik Kratzert, Daniel Klotz, Claire Brenner, Karsten Schulz and Mathew Herrnegger
Institute of Water Management, Hydrology and Hydraulic Engineering, University of Natural Resources and Life Sciences, Vienna, 1190, Austria

Sameh Saadi
Centre d'Etudes Spatiales de la Biosphère, Université de Toulouse, CNRS, CNES, IRD, UPS, Toulouse, France
Université de Carthage/Institut National Agronomique de Tunisie/ LR17AGR01-GREEN-TEAM, Tunis, Tunisia

Gilles Boulet, Malik Bahir, Aurore Brut, Émilie Delogu, Pascal Fanise, Bernard Mougenot and Vincent Simonneaux
Centre d'Etudes Spatiales de la Biosphère, Université de Toulouse, CNRS, CNES, IRD, UPS, Toulouse, France

Zohra Lili Chabaane
Université de Carthage/Institut National Agronomique de Tunisie/ LR17AGR01-GREEN-TEAM, Tunis, Tunisia

Index

A

Actual Evapotranspiration, 34-36, 38-39, 43, 46-47, 50, 152, 156, 208, 220, 229

Artificial Neural Networks, 71, 191, 204, 206

B

Bandwidth Matrix, 134-135

Bed Slope, 88, 157

Budyko Framework, 150, 156, 166, 169

C

Calibration Period, 37, 40, 44, 163-164, 194-195, 197-198, 200, 202-203

Climate Change Initiative, 52, 66, 70

Climatological Grid Cell, 171-172

Community Land Model, 79, 170, 187-188

Cumulative Distribution Function, 13, 21, 120

D

Digital Terrain Models, 84, 89

Distributed Hydrology Soil Vegetation Model, 129, 136, 149

Drying-down, 55, 58, 60-61

Dynamic Scaling Function, 38, 46-47

E

Ecosystem Demography, 170-172, 184, 187-188

Ecosystem Respiration, 67

Ensemble Kalman Filters, 130

Ensemble Prediction System, 117, 148

Ensemble Streamflow Prediction, 17

F

Flash Flood Forecasting Systems, 83-84

Flow-duration Curves, 101-104, 113, 115-116, 169

Forecast-observation Pairs, 20-21

G

General Circulation Models, 17, 148, 186

Ground Clutter Filtering, 4

H

Heterogeneous Semiarid Landscape, 208

Hydraulic Parameters, 52-53

Hydrogeological Model, 101

Hydrological Forecasting, 18, 31-33, 118, 126-127

Hydrological Model, 17-18, 20, 22, 25-28, 30-32, 37, 46, 48-50, 80, 86, 100-101, 108, 113, 118-119, 123, 126, 128, 147-153, 157, 168, 186, 190, 194, 196-197, 203, 206, 229

Hydropower Dams, 108

I

Inland Water Surface Spot Heights, 155, 168

Intensity-duration-frequency, 1, 4-5, 12, 14-16

L

Land Surface Model, 48-49, 51-53, 61-62, 64-65, 67, 77, 79-80, 82, 101, 189, 225

Land Surface Temperature, 35-36, 48, 62

Latent Heat Fluxes, 35, 49, 69-70, 78, 211-213, 215, 219-220, 222, 224-225

Leaf Area Index, 37, 47, 65, 69-71, 79-80, 209, 227

Log-transformed Streamflow, 107

M

Mean Annual Flow, 105, 108-109

Mean Relative Error, 56, 120, 122-123

Mesoscale Hydrologic Model, 34-35, 37, 49

Meteorological Seasonal Forecasts, 17, 31

Mid-latitude Cyclones, 3

Multilayer Soil Diffusion Scheme, 67

N

Nash-sutcliffe Efficiency, 23, 69, 107, 110, 113, 136, 163, 170, 176, 180, 197-198

Normalized Difference Vegetation Index, 151, 208-209, 219

O

Overland Flow, 19, 136, 149, 156, 173, 175

P

Pedo-transfer Functions, 34, 37, 54-55, 58

Q

Quantitative Precipitation Estimates, 14, 83, 86

R

Rain Gauges, 1-5, 7, 15, 53, 191

Rainfall-runoff Models, 33, 49, 83, 87, 98, 102, 112-113, 152, 167

Recurrent Neural Network, 190-192, 205

Residual-duration Curve, 104-105, 113

River Flow Dynamics, 170, 184

River Flow Velocity, 157

Root-zone Soil Moisture, 52, 62, 81

S

Sacramento Soil Moisture Accounting Model, 190, 192, 195

Seasonal Meteorological Forecasts, 18, 32

Seasonal Streamflow Forecasts, 17-18, 31-32

Semiarid Climates, 1, 9, 12

Sensible Heat Flux, 75, 208, 210, 212-215, 218, 220-222, 224-225, 228-230

Soil Moisture, 17, 30, 33, 36, 38, 46, 51-73, 75-83, 99, 131, 133, 141-142, 144, 147-148, 151, 155, 165, 167-168, 171, 190-192, 195, 209-210, 217-218, 220, 227

Soil Parameter, 34

Soil Thermal, 52-53, 80

Soil Water Balance, 67, 175, 209, 226, 229

Solar Zenith Angle, 36

Spatio-temporal Scales, 2, 7

Steady-state Hydraulic Computations, 91

Stream Network, 84-87, 96, 102, 107, 113, 136

Streamflow Index, 103

Streamflow Metrics, 36, 39-40, 42, 44, 46

Surface Energy Budget, 64, 209, 216, 220, 227

T

Terrestrial Biosphere Models, 170-171, 182-184

Thermal Infrared, 53, 61, 175, 209, 217, 226-228

Thicker-tailed Distributions, 1, 6, 12

Total Runoff Integrating Pathways, 65-66, 82, 171, 188

Total Water Storage, 36, 150-152, 154, 157, 161-163, 165-166, 168

Tropical Plume, 3

Tropical Rainfall Measuring Mission, 150, 153

Turbulent Heat Fluxes, 50, 65-66, 229

V

Vapour Pressure, 192

Volumetric Soil Water Content, 51

Volumetric Time Series, 72-73, 76-77

W

Wet Radome Attenuation, 4